Sensors and their Applications XII

Sensors Series

Senior Series Editor: **B E Jones**
Series Co-Editor: **W B Spillman, Jr**

Sensors and their Applications XII

**Proceedings of the Twelfth Conference
on Sensors and their Applications,
held at the University of Limerick, Ireland, September 2003**

Edited by

S J Prosser
TRW Automotive

and

E Lewis
University of Limerick

CRC Press
Taylor & Francis Group
Boca Raton London New York

CRC Press is an imprint of the
Taylor & Francis Group, an **informa** business

CRC Press
Taylor & Francis Group
6000 Broken Sound Parkway NW, Suite 300
Boca Raton, FL 33487-2742

First issued in paperback 2019

ISBN-13: 978-0-7503-0978-3 (hbk)
ISBN-13: 978-0-367-39504-9 (pbk)

British Library Cataloguing in Publication Data

A catalogue record for this book is available from the British Library.

Library of Congress Cataloging-in-Publication Data are available

The Proceedings for S&A XII are sposored by National Instruments. ᴺᴬᵀᴵᴼᴺᴬᴸ INSTRUMENTS

Series Editor: **Professor B E Jones**, Brunel University

Visit the Taylor & Francis Web site at
http://www.taylorandfrancis.com

and the CRC Press Web site at
http://www.crcpress.com

Preface

The twelfth conference in the *Sensors and their Applications* series took place at the University of Limerick, Ireland from 2-4 September 2003. The event was organised by the Instrument Science and Technology Group of the Institute of Physics. Previous conferences in the series were held in Manchester (1983 & 1993), Southampton (1985 & 1998), Cambridge (1987), Canterbury (1989), Edinburgh (1991), Dublin (1995), Glasgow (1997), Cardiff (1999) and London (2001). The event provided a forum for academic researchers and industrial engineers in all areas of sensors to come together to review and update themselves on developments and applications in the field.

Twenty years on since the first conference, we are pleased to maintain the continuity of enthusiasm and commitment to this conference series. In total over eighty papers are included in this volume. As in previous conferences, great emphasis was placed upon exhibited poster presentations. The high quality programme, spearheaded by notable contributions from invited and keynote speakers, included optical sensing, sensor materials, system-on-chip, non-destructive testing, system and sensor networks, imaging sensors and water quality monitoring. This conference was particularly highlighted by a large number of sensor applications papers.

We take this opportunity to thank all those who have contributed to the event, including our industrial sponsors. The support from the Science Foundation of Ireland enabled a large number of students to attend. Our thanks also go to our colleagues in the Instrument Science and Technology Group for their support and encouragement, particularly in the refereeing of the abstracts. Special thanks go to Claire Pantlin of the Conferences Department of the Institute of Physics who has expertly managed the planning and organising of this Conference.

We hope that these conference papers will provide a useful snapshot of sensor developments and applications during 2003.

S J Prosser
Conference Chairman
TRW Automotive

E Lewis
Local Chairman
University of Limerick

Contents

Section 1

Optical Sensing I

Optical Fiber Sensors Based on Spectroscopic Interrogation for Process Control and Environmental Applications   (Invited)

Anna Grazia Mignani and Andrea Azelio Mencaglia

CNR-Institute of Applied Physics 'Nello Carrara', Via Panciatichi 64, 50127 Firenze, Italy

Abstract: An innovative series of optical fiber sensors based on spectroscopic interrogation is presented. The sensors are custom-designed for a wide range of applications, including gasoline colorimetry, chromium monitoring of sewage, museum lighting control, for use with a platform for interrogating an array of absorption-based chemical sensors, as well as for color and turbidity measurements. Two types of custom-design instrumentation have been developed, both making use of LED light sources and a low-cost optical fiber spectrometer to perform broadband spectral measurements in the visible spectral range. The first was designed especially to address color-based sensors, while the second assessed the combined color and turbidity of edible liquids such as olive oil. Both are potentially exploitable in other industrial and environmental applications.

Keywords: optical fiber sensors, absorption spectroscopy, colorimetry, online monitoring, process control, reflection spectroscopy

1.   Introduction

To be able to run cost effectively, plants and processes require online monitoring by components designed to optimize operating efficiency. Optical fiber sensors, capable of performing in adverse environments, tight and/or hard-to-access spaces, and/or under real-time constraints, are increasingly replacing conventional electronic-based sensors that are not always capable of doing the job. Since they can be easily interfaced with optical data communication and secure data transmission systems, optical fiber sensors are being increasing used in industrial, automotive, avionic, military, geophysical, environmental, and biomedical applications [1, 2, 3, 4, 5, 6]. In addition to steadily rising quality and increasingly competitive costs, they have other welcome advantages:

- They operate without electricity, which makes them explosion-proof and intrinsically immune to any kind of electromagnetic interference. This is a notable advantage considering the widespread use of radiating instruments.
- They are compact, lightweight, and flexible, thereby enabling them to reach what were previously inaccessible areas.
- They can withstand chemically aggressive and ionizing environments.

Spectroscopy in the visible and near infrared spectral regions is one of the most popular methods in conventional analytical chemistry [7, 8]. The intrinsic optical and mechanical characteristics of optical fibers, together with the wide availability of bright LEDs and portable spectrometers, further enhance the application areas of spectroscopy, and make it possible the implementation of compact instrumentation dedicated to the monitoring of specific parameters.

Two measurement techniques are commonly used:

- *Direct absorption spectroscopy.* In direct absorption spectroscopy, the light intensity transmitted through or reflected by the analyte is measured. This technique is especially suitable for colorimetry measurements on surfaces and in non-turbid liquids and for analyzing compounds with non-interfering spectra. In many cases, the measurement selectivity in complex media can be improved through the use of multivariate data analysis methods such as artificial neural networks, the non-negative least mean squares method, or other noise suppression algorithms.
- *Chemically mediated spectroscopy.* In chemically mediated spectroscopy, the analyte interacts with a chemical; the analyte's status is indirectly obtained by measuring the spectral properties of the chemical. Different kinds of sensors based on chemically mediated spectroscopy can be used for measuring the physical, chemical, and biochemical parameters, since the indicator chemical may be adjusted to provide selectivity to a specific parameter [9].

This paper presents examples of sensors based on optical fiber spectroscopy:

- direct absorption spectroscopy sensors for online monitoring of gasoline mixtures in petrochemical plants and for detecting chromium in industrial sewage;
- chemically mediated reflection spectroscopy sensors, activated by calibrated textiles, for providing lighting control in museums;
- an optical fiber multimeter for interrogating an array of absorption-based chemical sensors that was validated on a set of porphyrin-based materials with gas-sensor potential;
- instrumentation for simultaneously assessing color and turbidity in edible liquids which was used to fingerprint extravirgin and non-extravirgin olive oils and was capable of discriminating and clustering the oils according to type.

2.   Optical fiber sensors based on absorption or reflection spectroscopy

The fiber optic instrumentation designed for addressing absorption or reflection spectroscopy sensors complies with the following requirements:

- High degree of compactness, i.e., each optoelectronic module plus its sources, detectors, and coupling optics fit into a compact box without requiring mechanical alignments of the optical components.
- Probe using a single optical fiber configuration, rather than bundle.
- A flexible software interface friendly enough to be used by operators with little technical background.
- Low cost optoelectronic components.

The instrumentation consists of absorption and reflection spectroscopy optoelectronic modules, as shown in Figure 1, which are interfaced to a PC by an A/D board with input/output ports. The modules differ only in the optical interface for coupling the probes [10]. Both use a set of LEDs to obtain a low-cost source, and a fiber optic microspectrometer by STEAG-Microparts GmbH as detector [11]. The measuring range is from 380 to 780 nm, with 10 nm resolution.

Two- and six-LED configurations are provided, the latter affording a more uniform spectral intensity. The LEDs are temperature-controlled by Peltier cells to optimize source stability in terms of spectral shape and intensity. The six-LED source uses three LEDs with broadband white emissions and three with emission peaks at 420, 500, and 640 nm. The two-LED source uses only white and 640 nm LEDs. LED type can easily be changed to meet application requirements. The LEDs are housed in ST-compatible receptacles for easy fiber coupling; the fiber bundle from LEDs is connected to the end in a patent-pending microoptics joint allowing connection to a single optical fiber. The highest illumination efficiency is provided by a 600 μm-core optical fiber.

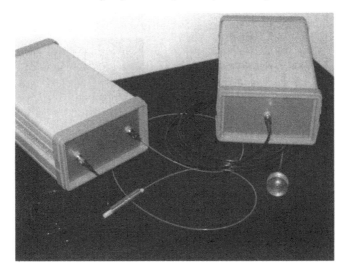

Figure 1. Optical fiber spectrophotometers for absorption (left) and reflection (right) spectroscopy.

The reflection spectroscopy module uses a microoptic joint which also incorporates the optical fiber coupled to the microspectrometer. Reflection measurements are performed by a single 600 μm-core optical fiber which is polished at one end to avoid Fresnel reflections. This module can also be used for absorption spectroscopy by means of a folded-path transmissive probe.

The absorption spectroscopy module uses separate illumination and detection fibers. The transmissive and folded-path transmissive probe designs are illustrated in Figure 2. The probes use GRIN lenses for beam collimation and focusing and are designed for path-lengths in the 10-100 mm range. Input and output connections for flow applications were also implemented.

LabView® software, programmed to perform automatic or semiautomatic measurements and data processing, is used for optoelectronic module management. Software management of the module enhances the instrumentation's flexibility, since several programs have been written to fit different measurement requirements, including the interrogation of optrode-based fiber optic sensors.

These custom instruments were the key unifying feature to address fiber optic sensor, as described in the following paragraphs [12].

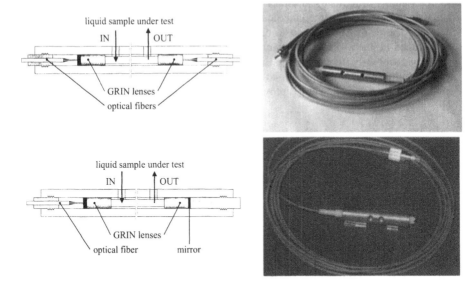

Figure 2. Optical fiber probes for absorption spectroscopy: transmissive (top) and folded-path transmissive (bottom) designs.

2.1   Direct absorption spectroscopy: online gasoline blending monitoring

Online near infrared spectrophotometry of raw and refined hydrocarbon products is useful for the optimization of refining and petrolchemical processes. The possibility of control unit installation in a non-classified area, while immersing a safe probe in inflammable and explosive environments, make optical fiber probes extremely attactive. Another important application is colorimetry in gasoline blending control, especially when different blendings are sent sequentially along the same pipe and a final electrovalve must take the decision of their addressing to the relative tanks.

The sensor we have implemented was designed to recognise the difference between lead-free (green) and normal (red) gasoline and their blendings. The absorption spectra of Agip green and red gasoline blendings are shown in Figure 3. The blendings can be measured by means of a full spectral range analysis, or by means of a dual-wavelength analysis [13].

2.2   Direct absorption spectroscopy: chromium monitoring in sewage industrial water

Environmental contamination due to chromium has become a matter of serious concern because of its frequent use in industrial processes and other waste. Chromium is considered essential as well as toxic for human beings depending upon its speciation, in particular with respect to the oxidation state, and the doses and concentration levels involved. Trivalent and hexavalent chromium (Cr(III) and Cr(VI)) are the most common oxidation states of chromium in the environment. Both Cr(III) and Cr(VI) are frequently used in industrial processes. Manufacturing of products or chemicals containing chromium releases chromium to the air, soil and water. Cr(III) and Cr(VI) enter the environment as a result of effluent discharge from a lot of industrial sources, including steel production, metal plating, leather tanning, spray painting operations and combustion sources, such as automobiles and incinerators. Most chromium in water sticks to dirt particles that fall to the bottom; only a small amount dissolves. However, residual waste waters are rarely discharged directly into the sewers, because of the high pollution level. There are several companies that deal with sewage collection and detection of Cr(III) and Cr(VI) performed during some stages of the recycling process, and an online monitoring would be welcome.

Since the absorption spectra of Cr(III) and Cr(VI) water solutions in the mg/l concentration range lies in the visible spectral range, direct absorption spectroscopy was performed by means of fiber optic spectrophotometers here presented, thus obtaining online analysis of chromium contaminated water samples [14, 15].

In addition, complex spectra like those shown in Figure 4 were measured by the fiber optic spectrophotometers. Then, these interferring spectra were processed by means of neural network data processing, thus obtaining satisfactory pollutant identification [16].

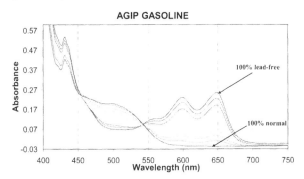

Figure 3. Absorption spectra for Agip gasoline.

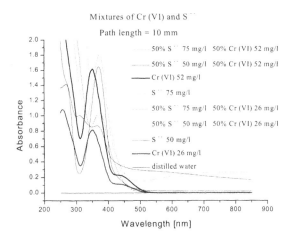

Figure 4. Absorption spectra of mixtures of hexavalent chromium and sulfur in water.

2.3   Mediated reflection spectroscopy: lighting control in museum environment

Works of art exposed to excessive lighting suffer from color changes due to photo-oxidation or photo-reduction of the pigment/dye. Chromatic damage to artworks is not only the result of exposure to overly intense light, but can also be caused by the combined action of light with other environmental factors such as thermo-hygrometric variables and pollutants, which can amplify and accelerate photo-alteration [17, 18, 19]. Therefore, the prevention of chromatic damage requires measurement of the equivalent-light dose, i.e., the dose capable by itself of producing a color variation, without the contribution of other environmental factors. Mediated reflection spectroscopy was applied to perform a continual measurement of the equivalent-light dose so as to indicate the potential for chromatic alteration of the exhibition environment.

Blue Wool Standard (BWS) dosimeters, which have been officially adopted by the International Standards Organisation to measure photo-induced color variation, consist of a set of wool samples that progressively and irreversibly fade when exposed to light. The samples are loaded with different light-sensitive blue dyes to show color fading with varying sensitivities to light exposure [20, 21]. When BWSs are exposed under normal conditions, fading occurs because of exposure to light and to the combined action of light and other environmental factors. BWSs are therefore optimal transducers for performing dosimetry of equivalent-light.

The reflection spectroscopy module, suitably adapted, was used for equivalent-light dosimetry to measure BWS color fading. The measurements were obtained from two sets of identical dosimeters housed on a disk in front of a fiber optic probe used for reflection measurements. One of the dosimeter sets was exposed to the environment, while the other was kept in the dark. Color comparisons between the exposed and unexposed dosimeters were provided by the spectrophotometer. A calibration procedure, performed by means of an accelerated ageing test, made it possible to achieve equivalent-light dosimetry from the measurement of BWS color fading [22].

The implemented optical fiber instrument is shown in Figure 5, together with the disk housing the BWSs and the fiber-optic probe for reflection measurements. Experimental results were obtained during continual measurements over 300 days, part of which at the Uffizi Gallery of Florence.

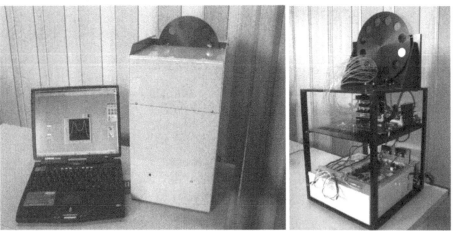

Figure 5. Optical fiber instrument for equivalent-light dosimetry.

2.4   Mediated absorption spectroscopy:
a platform for interrogating an array of absorption-based chemical sensors

A fiber optic multimeter was implemented to address an array of absorption-based chemical sensors operating in the visibile spectral range. The instrumentation is a particularly flexible device since it can be fitted with selective or non selective sensors. When selective sensors are used, quantitative measurements are achieved so that the multimeter becomes a device for multiple analysis. When sensors with limited selectivity are used, the overall response of them can be smart-processed to achieve qualitative information in order to approach artificial olfactory perception.

The multimeter, which is shown in Figure 6, consists of three independent modules: the absorption spectroscopy module for spectral interrogation of the sensor array; a mechanical module consiting of a gear that houses the sensors and provides them with exposure to the analyte being tested, and their sequential interrogation; a software module, that is an interface capable of managing the optical and mechanical modules and of achieving the spectral response of all the sensors. The gear is a revolving platform rotated by means of a step motor. The revolving platform houses 16 replaceable glass disks arranged around a ring. One disk is taken as reference for signal normalization. The other 15 disks are considered as sensors, since they can act as substrates for sensing materials such as thin films or spray- or spin-coated layers. The platform is fitted into a flow cell suitable for gas or liquid analysis. The flow cell can be opened to permit platform change or cleaning.

The multimeter was validated by means of several types of metalloporphyrins that are sensitive to volatile compounds and harmful gases, however lacking in high selectivity. The overall sensor response was processed by means of chemometric methodologies. This system was capable of distinguishing different types of vapors and gases and showed effective potential for achieving an artificial olfactory perception of these gases.

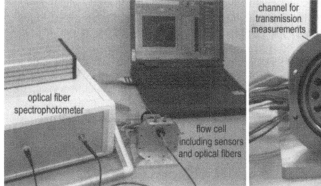

Figure 6. An overall view of the fiber optic multimeter (left) and a detailed view of the open flow cell, including the gear housing the sensors (right).

3.   Optical fiber sensor based on absorption spectroscopy and scattering measurements

The Mediterranean diet is spreading worldwide thanks to its unquestionably beneficial nutritional properties. A primary ingredient of the Mediterranean diet is extravirgin olive oil, the production of which is also one of the most important products of the Italian agricultural industry [23, 24, 25]. Extravirgin olive oil is made of a blending of oils coming from different 'cultivar', that is, from different varieties of tree species. The organoleptic properties of oil also depend on blending. Many types of extravirgin olive oils are produced by mixing oils from different origins, each type of blending being characterized not only by a different taste, but also by a different color and turbidity. In order to satisfy customer requirements, an oil from a certain producer must be easily distinguishable and identified by presenting the same taste as well as the same color and turbidity.

Spectral nephelometry is a novel optical technique that combines absorption spectroscopy and multi-angle scattering measurements performed in the visible spectral range to provide simultaneous monitoring of color and turbidity. As shown in Figure 7, the instrumentation is an optoelectronic device that measures the absorption spectra of the oil sample at different angles. Four white light sources are used, which span the 450-630 nm spectral range. A miniaturized optical fiber spectrometer serves as detector. The sources, positioned at different angles with respect to the detector, are sequentially switched on to measure, in addition to the transmitted spectrum, the scattered spectra at the given angles so as to obtain spectral nephelometry. The transmitted spectrum mainly provides information regarding color, which is also dependent on the amount of turbidity present, while the scattered spectra mainly provide information on turbidity, which is also dependent on the color present. Spectral data are processed by means of the Principal Component Analysis thus providing coordinates identifying the oil sample in a 2D or 3D map, namely the 'nephelometric map'. Consequently, when many samples of oils are analyzed, the map is populated by points, each of them representing the individual identity of an oil. Oils with similar characteristics appear in the map as clusters of points [26].

The instrumentation and data processing technique were capable of discriminating different types of oils, such as extravirgin-olive, lampante or seed oils. Clusters have been created where oils are grouped according to type. Recently, the system was used to analyze extravirgin olive oils from selected regions of Italy (Toscana, Puglia, Calabria, Sicilia). A 3D map was created, which was capable of clustering the oils according to their area of origin.

4.   Conclusions

The spectroscopic optical fiber sensors described in this work comprise PC-interfaced electrooptic units programmed to fit individual sensor interrogation schemes. Their advantages include a high degree of miniaturization and compactness, flexibility, and suitability to a broad range of applications. Thanks to the availability of several functional materials for selective molecular recognition that exhibit optical absorption in the visible spectral range modulated by interaction with the analyte, the instrumentation is potentially suitable for a wide range of sensors and for olfactory perception in process control, safety, and environmental applications. Designed for the color analysis of liquids and surfaces in industrial process control, water analysis, and food processing control, the proposed spectrophotometers can be used in other colorimetric applications by operators with little or no technical skills. The presence of colorimetric passive samplers acting as dosimeters for gaseous pollutants and acids enables the instrumentation to act as an in-situ spectrophotometric laboratory, i.e. a totally, autonomous nonstop monitor of passive samplers in a host of industrial and environmental applications.

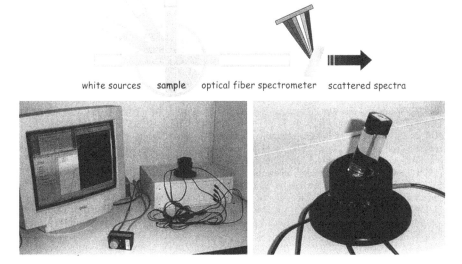

white sources   **sample**   optical fiber spectrometer   scattered spectra

Figure 7. Working principle (top) and optical instrumentation (bottom) for extravirgin olive oil traceability.

Acknowledgements

The authors are grateful to the following people and organizations for their financial support and assistance in developing the sensors:

- *Gasoline Blending*: CNR Target Project MADESS II, Cosimo Trono.
- *Hexavalent Chromium*: EU INCO-DC Programme, RAMFLAB Project (#ERBIC18CT970171), Andrea Romolini, Lorenzo Spadoni.
- *Lighting in the Museum Environment*: Regione Toscana, RIS+ Pilot Project, Mauro Bacci and Folco Senesi.
- *Spectroscopic Platform:* Ministero dell'Industria e della Ricerca (Grant #RBNE01KZZM), Patrizia Bizzarri, Arnaldo D'Amico, Corrado Di Natale, Antonella Macagnano, and Roberto Paolesse.
- *Color and Turbidity Assessment of Edible Liquids:* EU SM&T Programme, OPTIMO Project (#SMT4-CT972157) and the Regione Toscana's CARABIOTEC Project, Leonardo Ciaccheri, Antonio Cimato, Graziano Sani, and Peter R. Smith.

They also wish to thank Franco Cosi for providing the assembly of the optical systems and Daniele Tirelli for carrying out the mechanical designs.

References

[1] B. Culshaw and J.P. Dakin Eds., *Optical Fibre Sensors*, Artech House, Nordwood MA; Vol. 1: *Principles and Components*, 1988; Vol. 2: *Systems and Applications*, 1989; Vol. 3: *Components and Subsystems*, 1996; Vol. 4: *Applications, Analysis and Future Trends*, 1997.

[2] K.T.V. Grattan and B.T. Meggit Eds., *Optical Fiber Sensor Technology*, Chapman & Hall, London UK, 1995.

[3] K.T.V. Grattan and B.T. Meggit Eds., *Optical Fiber Sensor Technology*, Kluwer Academic Pbl., Dordrecht, The Netherlands, 1999; Vol.3: *Applications and Systems*, Vol. 4: *Chemical and Environmental Sensing*.

[4] D.A. Krohn, *Fibre Optic Sensors: Fundamentals and Applications*, Instrumentation Society of America Pbl., Research Triangle Park, NC, 2000.

[5] C.M. Davis, E.F. Carome, M.H. Weik, S. Ezekiel, R.E. Einzig, *Fibreoptic Sensor Technology Handbook*, Optical Technology Inc., 1986, download at http://fibreoptic.com/pubs/handbook.cfm

[6] J.M. Lòpez-Higuera Ed., *Handbook of Optical Fibre Sensing Technology*, John Wiley & Sons Ltd., Chichester UK, 2002.

[7] M.G. Mellon, *Analytical Absorption Spectroscopy*, John Wiley & Sons Inc., New York NY, 1950.

[8] R.P. Bauman, *Absorption Spectroscopy*, John Wiley & Sons Inc., New York NY, 1962.

[9] O.S. Wolfbeis, *Fiber Optic Chemical Sensors and Biosensors*, Vols. I and II, CRC Press, Boca Raton FL, 1991.

[10] A.A. Mencaglia and A.G. Mignani, 'Optical fiber instrumentation for online absorption and reflection spectroscopy', Proc. SPIE Vol. 4763 *European Workshop on Smart Structures in Engineering and Technology*, B. Culshaw Ed., 2003, pp. 248-251.

[11] STEAG-MicroParts, Germany, online: http://www.microparts.de/.

[12] A.G. Mignani and A.A. Mencaglia, 'Direct and chemically-mediated absorption spectroscopy using optical fiber instrumentation, *IEEE Sensors Journal*, Vol. 2, n. 1, 2002, pp. 52-57.

[13] A.G. Mignani, C. Trono and A.A. Mencaglia, 'Spectral analysis of hydrocarbon products using optical fiber instrumentation', Proc. SPIE Vol. 4074 *Applications of Optical Fiber Sensors*, A.J. Rogers Ed., 2000, pp. 390-394.

[14] A.G. Mignani and L. Spadoni, 'Absorption spectroscopy in the ultraviolet and visible spectral range of hexavalent chromium aqueous solutions', SPIE Proc. Vol. 3821 *In situ and Remote Measurements of Water Quality*, R. Reuter Ed., 1999, pp. 319-322.

[15] A.G. Mignani and A. Romolini, 'Absorption spectroscopy of trivalent chromium in distilled water and in buffers at different pH values in the visible spectral range', SPIE Proc. Vol. 4074 *Applications of Optical Fiber Sensors*, A.J. Rogers Ed., 2000, pp. 395-402.

[16] T. Kuzniz, M. Tur, A.G. Mignani, A. Romolini and A.A. Mencaglia, 'Neural network analysis of absorption spectra and optical fiber instrumentation for the monitoring of toxic pollutants in aqueous solutions', Proc. SPIE Vol. 4185 *14th International Conference on Optical Fiber Sensors*, A.G. Mignani and H.C. Lefèvre Eds., 2000, pp. 444-447.

[17] G. Thomson, *The Museum Environment*, Butterworth-Heinemann, London, 1986.

[18] D. Camuffo, *Microclimate for Cultural Heritage*, Elsevier Science BV, Amsterdam, 1998.

[19] T.B. Brill, *Light: its Interaction with Art and Antiquities*, Plenum Press New York, 1980.

[20] British Standard 1006: 1990 (issue 2, October 1992), 'British Standard Methods of Test for Colourfastness of Textiles and Leather', British Standard Institution, Milton Keynes, 1992, B01/1-7.

[21] The Society of Dyers and Colourists, UK, online: http://www.sdc.org.uk .

[22] A.G. Mignani, M. Bacci, A.A. Mencaglia and F. Senesi, 'Equivalent light dosimetry in museums with blue wool standards and optical fibers', *IEEE Sensors Journal*, Vol. 3, n. 1, 2003, pp. 108-114.

[23] Organizzazione Nazionale Assaggiatori Olio di Oliva, online: http://www.oliveoil.org .

[24] Consiglio Oleico Internazionale, online: http://www.poggiotoccalta.it/PT/Dep95Olio4.htm .

[25] Online: http://www.olivetree.cc/default.htm .

[26] A.G. Mignani, P.R. Smith, L. Ciaccheri, A. Cimato and G. Sani, 'Spectral nephelometry for making extravirgin olive oil fingerprints', *Sensors and Actuators B*, 2003, Vol. 90, n. 1-3, pp. 157-162.

Section 1: Optical Sensing I
Paper presented at Sensors and their Applications XII, September 2003
©2003 IOP Publishing Ltd

A Combined Optical Fibre Temperature and Colour Sensor for Automation of the Cooking Process in a Large Scale Industrial Oven

M O'Farrell, E Lewis, C Flanagan, K T V Grattan[1], T Sun[1], N Jackman[2]

Department of Electronics and Computer Engineering, University of Limerick, Ireland.
[1] City University, London.
[2] Food Design Applications Ltd. Newtown, Castletroy, Limerick, Ireland

Abstract: An Optical fibre based sensor system has been developed for the purpose of examining the core temperature and colour of food products online as they cook in a large-scale industrial oven. Core temperature is ascertained by monitoring the fluorescence decay time of a rare earth material and correlating this with temperature To determine the colour spectroscopic techniques are employed to interrogate the sensor signal and the resultant output spectral patterns are examined by an Artificial Neural Network, which is capable of classifying colours that are favourable and those that are not optimum. By combining these two technologies it is possible to control the cooking process and optimise food quality.

1. Introduction

With an ever-increasing volume in the throughput of the food industry, profit margins dictate that it is no longer acceptable for large errors to occur, as the result would be significant. This is evident in the cooking of large batches of various food types in industrial ovens of lengths up to 20 metres Fig.1. The conveyor belts of these ovens may be fully loaded with food for maximum efficiency and the cost of an error in one batch would be intolerable for the producer. Automation in the cooking of the food is critical in order to control oven temperature, conveyor belt speed, height of steamers etc. so that the food is cooked to the highest quality. There is a need for monitoring the food as it passes through the oven and determining that all the processing parameters are correct, the most important of these being the food's core temperature and the colour of the food (both internal and external).

It is no longer viable that a human can monitor these high-speed production lines. Even though a human has the intelligence, it is clear that automated control methods can ensure optimum performance over humans for the following reasons:

- They allow measurements in environments that are unsuitable for humans e.g. where there are high temperatures or in confined and/or dangerous locations.
- They are more accurate and precise in that they don't get tired or bored, resulting in flawed judgment.

- They are also capable of detecting at speeds much greater than humans.

Grading the quality of a product e.g. in terms of colour [1] or shape and size [2], ensures consistency in the product. Factors such as colour have a huge influence over a customer decision in purchasing a product, as they often have preconceived notions of how a product should look; roast chickens must be golden, flame-grilled hamburgers must have black lines and cooked sausage must be a rich brown. The fact that the food is cooked to a core temperature recommended by food safety authorities is taken for granted by the customer, but for the manufacturer it is a critical parameter indicating that all bacteria are killed and food poisoning is avoided. Therefore, it is clear that both colour and temperature, are used to determine if a product is correctly cooked.

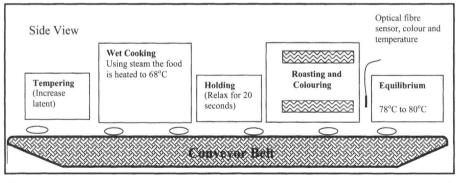

Process takes 20 minutes

Figure 1: Industrial oven used in on-site testing

The system described in this paper comprises two parts, colour measurement and temperature measurement, both using optical fibre sensors. This paper further develops results previously reported by the authors [3], [4]. Previous results have been obtained in collaboration with Food Design Applications Ltd. with products such as minced beef, high meat content patti burgers and sausages. Further investigations have been made and results were obtained during on-site testing of various food samples, i.e. high cereal content patti-burgers and puff pastry products, which are reported later in this paper.

2. Experimental Set-Up

2.1 Colour Measurements

Food producers often employ a technique to colour the food by using surface air-blowing towards the end of the belt to colour the food. It is clear from the diagram (Fig. 1) that there are five main stages involved in the cooking, represented by the five stages in the oven. The colour sensing is performed between the fourth and fifth stages, after colouring by roasting has taken place.

The colour is measured using a reflective probe and recording the light reflected in the visible spectrum, which is representative of the colour. Illumination of the food is performed with 6 fibres attached to a tungsten halogen white light source. One fibre returns the reflected light to an Ocean Optics s2000 spectrometer, controlled by a host P.C., Fig. 2(a). The data is then saved for subsequent analysis using the Stuttgart Neural Networks Simulator (SNNS) the Neural Network Software [5]. Examples of the use of artificial intelligence for decision-making in the food industry are plentiful [6], [7].

The probe for measuring the external readings is required to have a diameter that is as large as possible in order to cover a greater area. The fibres used are 1mm in diameter. For testing a single probe was used, as for initial measurements. It was important for the investigators to have full and free control over where it is placed; it is envisaged to have 5 or 6 separate probes placed around the surface of the food sample and taking an average reading. The internal readings need to be unobtrusive so it is important that the diameter is kept to a minimum. The fibres for this probe are 200/220 μm. The tip of this probe is polished at 45 Deg. This gives the tip a pointed edge for insertion and increases the area of the fibre exposed to the food, Fig. 2(b).

Both probes are encased in stainless steel. This material was chosen to allow the probe to be regularly sterilised, something that is essential in the food industry. Its robustness allows the probe to be injected at high speed into the product and puncture without causing damage to the product or itself.

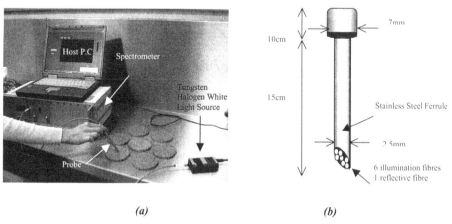

(a)   *(b)*

Figure 2. (a) Photo of the System Set-up (b) Probe for internal colour readings.

2.2 Temperature Measurements

The optical sensor scheme used in the work is based upon the results of previous research by some of the authors. It involves the monitoring of the fluorescence decay time of a rare earth material and correlating this with temperature, as discussed by Grattan and Zhang [4]. Given the temperature range under consideration in this work, it was decided to base the optical system upon the use of a thulium-doped garnet (Tm: YAG) coupled with silicon optical fibres. The system design is as shown in Figure 3(a) with the probe design shown in Figure 3(b). The fluorescent medium is excited by light from a LD light source operating at 785 nm, coupled to the active material through a silica fibre bundle. The received fluorescence emission is detected with an extended wavelength InGaAs photodiode and the lifetime data extracted using a PLD (phase locked detection) scheme reported by some of the authors elsewhere [4].

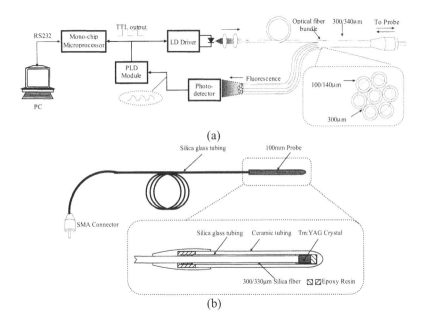

Figure 3 Optical fibre sensor (a) system design (b) optical probe design

The probe itself is designed with a stainless steel sheath to allow sterilisation between successive usages. The system has a temperature uncertainty, determined from laboratory calibration of $\pm 1°C$, when calibrated against the output of a thermocouple.

3. Results and Analysis

3.1 Colour

3.1.1 Spectroscopic Results
Results were obtained for the following products: high cereal content patti burgers and puff pastry.

1. Patti Burger
The high cereal content in this product results in less obvious colour change in the product as it cooks, due to the lack of meat, and therefore making it on of the most difficult products to classify. Fig.4 below shows the spectral variation between 390 spectra, 130 for each of the three classes i.e. raw, undercooked (65°C) and overcooked (100°C). It is clearly seen that the variations are so slight that it is quite difficult to segregate the colours 'by eye'. The raw patti poses the greatest problem as there are many dark flecks (slightly visible in Fig. 2) that result a range of spectra for the one colour (the black lines in Fig 4.), whereas the other two colours are quite well defined.

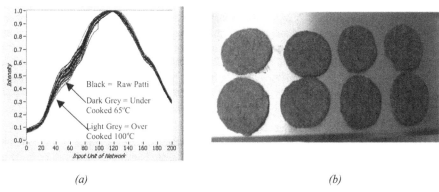

(a)   (b)

Figure 4: (a) Spectral Variation in High Cereal Content Patties. (b) From left to right; Raw patties, Undercooked (65°C), Correctly cooked (78°C) and Overcooked (100°C).

2. Puff Pastry
Tests were also conducted on puff pastry in two food samples, cornish pasties and sausage rolls. The resulting spectra are shown below. Fig 5. There are 985 spectral samples i.e.197 of each raw, too light, correct colour, too dark and burnt. The variation is much clearer in this case and it was possible to classify into 5 categories.

3. Data Conditioning
The x-axes of each of the graphs represent the amount of data points that were used to represent the spectra. Each spectral reading from the s2000 spectrometer has 2048 data samples, resulting in high resolution, in the range between 200nm and 900nm. It was possible to reduce the amount of data points whilst still preserving the original spectral shape. For the cereal burgers 200 points were taken between the wavelengths of 400nm and 850nm and for the pastry only 15 points were needed between the wavelengths of 450nm and 750 nm.

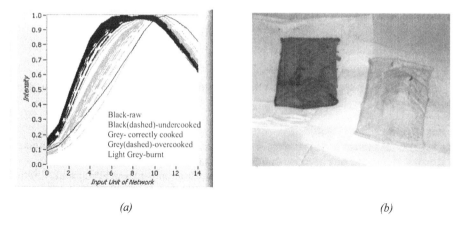

(a)

(b)

Figure 5: (a) Spectral Variations of Puff Pastry (b) Cornish Pasty; Overcooked (left) and Raw (right)

3.1.2 The Neural Network

Two Feed Forward 3 layer neural networks were configured with the number of input nodes representing the number of data points required to represent the spectra of each cooking stage, i.e. 200 for the cereal burger and 15 for the pastry. It was necessary for more data points to represent the cereal burger results since the variations in the spectra were less pronounced. At the output layer of each network one node was chosen for each condition to be sensed. This represents 3 nodes for the cereal burger neural network and 5 for the pastry neural network.

1. Neural Network for High Cereal Content Burger

Following a comparison of different configurations, a hidden layer of 10 nodes was chosen. Although there is no set method for choosing the number of nodes to be used in a hidden layer, there are limitations associated with too few hidden layer nodes, which would result in poor trainability and too many nodes resulting in a poor generalisation by the network.

Using the standard training algorithm of Backpropagation with Momentum, a total of 393 result patterns were used to train the network, 131 of each condition; raw, undercooked and overcooked. The trained network was then tested using an independent set of 69 data patterns (23 of each condition) to that which had been applied previously in the training of the network. The network correctly classified all test patterns. A sample of the results obtained from the test set is shown in Table 1(a).

2. Neural Network for Pastry

Once again various network configurations were tested and it was decided that a hidden layer of 12 units was most suitable. Using Backpropagation with Momentum the Network trained successfully, distinguishing between 5 conditions: raw, undercooked, correctly cooked, over cooked and burnt. 985 test patterns, 197 for each condition, were used to train the network. When the network was tested using a test set of 370 test patterns, 74 of each condition, it correctly classified each cooking stage. A sample of the results obtained from the test set is shown in Table 1(b).

Table 1: (a) Sample of 6 (from 69) Results from the trained ANN Test Set for the Cereal Patti Burger (b) Sample of 6 (from 74) Results from the trained ANN Test Set for the Pastry

(a)

Patti Cooking Stage	Ideal Output	Observed Activation Output
Raw (too light)	100	0.880 0.056 0.000
Raw (too light)	100	0.901 0.000 0.000
Under Cooked	010	0.004 0.845 0.064
Under Cooked	010	0.004 0.909 0.093
Over Cooked (too dark)	001	0.000 0.003 0.893
Over Cooked (too dark)	001	0.000 0.023 0.880

(b)

Pastry Cooking Stage	Ideal Output	Observed Activation Output
Raw	10000	0.923 0.085 0.041 0.002 0.000
Raw	10000	0.919 0.087 0.041 0.002 0.000
Under Cooked	01000	0.032 0.955 0.056 0.001 0.001
Under Cooked	01000	0.030 0.956 0.056 0.001 0.001
Correctly Cooked	00100	0.000 0.000 0.802 0.159 0.010
Correctly Cooked	00100	0.000 0.000 0.702 0.118 0.008
Over Cooked	00010	0.000 0.000 0.021 0.920 0.082
Over Cooked	00010	0.000 0.000 0.025 0.908 0.076
Burnt	00001	0.000 0.000 0.000 0.000 0.100
Burnt	00001	0.000 0.000 0.000 0.001 0.992

3.2 Temperature

A number of food samples (pastries) were sampled and the temperature as the sample proceeded through the final stage of cooking in the oven were recorded. The photograph of such a pastry being sampled and the resulting temperature profile during cooking are shown in figure 6(a) and (b) respectively. The results show that the temperatures achieved during this phase of cooking reach a maximum of 72°C which is an adequate for the core temperature to ensure absence of food poisoning bacteria.

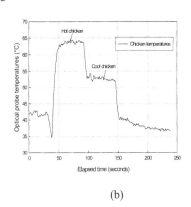

(a)   (b)

Fig 6   (a) The cooked pastry with the temperature probe inserted
(b) The resulting temperature profile measured from a chicken as it moves through the oven

4. Conclusion

A series of experimental tests have been conducted to assess the quality of food products as they cook in a large-scale industrial oven. According to food production guidelines it has been necessary to assess the colour of the food as it cooks (Internally and Externally) as well as the core temperature. The colour was assessed using an optical probe coupled to a spectrometer and PC. The resultant spectra were interrogated using Neural Networks software and these were used to successfully classify the cooked state of the food. The core temperature was measured using an optical fibre probe which achieved an accuracy of ± 1°C. These measurements are novel in the area of mass food production and are essential for guaranteeing the quality of the product. It is intended to integrate the two internal measurements (Colour and Temperature) into a single point probe for future developments of this system.

References

[1] McConnell RK Jr., Blau HH Jr. "Colour classification of non-uniform baked and roasted foods." *Proc.   Food Processing Automation IV, pp.40 –46, 1995*
[2] Aleixo N., Blasco B., Molto E., Navarron F., "Assessment of citrus fruit quality using a real-time machine vision system", *Pattern Recognition, 2000. Proc. 15th International Conference pp. 482 –485 vol.1, 2000*
[3] O'Farrell M., Lewis E., "An Intelligent Optical Fibre based System for Monitoring Food Quality" *IEEE Sensors 2002. Proc. 1st International Conference pp. 308-311 vol.1, 2002*
[4] K T V Grattan and Z Y Zhang, "Fibre Optic Fluorescent Thermometry" (Chapman & Hall, London 1995)
[5] SNNS, Stuttgart Neural Network Simulator, User Manual version 4.1 or http://www-ra.informatik.uni-tuebingen.de/SNNS/
[6] Shiranita K., Hayashi K., Otsubo A., Miyajima T., Takiyama R., "Determination of Meat Quality by image processing and Neural Network Techniques" *The ninth International Conference on Fuzzy System, pp 989 –992, vol.2, 2000*
[7] Abdesselam A., Abdullah R.C., " Pepper Berries Grading using Artificial Neural Networks" *TENCON 2000. Proceedings , pp 153 –159, vol.2, 2000*

Simultaneous and Co-located Measurement of Strain and Temperature in Optical Fibre Using a Bragg Grating and Strain-independent Erbium Fluorescence

S Trpkovski, S A Wade, G W Baxter, S F Collins

Optical Technology Research Laboratory (F119), Victoria University, PO Box 14428, Melbourne City MC, Victoria 8001, Australia.

Abstract: An optical fibre sensor for the simultaneous measurement of strain and temperature is presented. This is achieved using two sensing techniques within a single sensing element, namely a fibre Bragg grating located within a length of erbium-doped fibre, thereby achieving temperature and strain measurement at the same location.

Green fluorescence, arising from the de-excitation of erbium ions, is used to provide the temperature through the fluorescence intensity ratio method of temperature sensing. This method has the advantage of a cross-sensitivity with strain that is essentially nil. Thus shifts in the Bragg grating output due to temperature changes are readily determined so that the applied strain can be calculated. The particular sensor arrangement described offers the added attraction of requiring only one excitation source to stimulate green fluorescence and illuminate the grating.

The sensor was characterised by placing it in a tube oven and attaching weights to the end of the fibre with a pulley arrangement. Calibration of the sensor over various strain and temperature ranges resulted in temperature and strain accuracies better than 1 °C and around 15 µε, respectively.

1. Introduction

Strain measurement is important in many areas, such as in the monitoring of civil structures such as bridges, tunnels and dams. Improved safety and reliability through real-time monitoring will result in significant savings to the community. Optical fibre sensors are of much interest in these applications because of their small size and flexibility and their ability to be embedded into equipment and structures. Fibre Bragg gratings (FBG) are very attractive as sensors since an applied strain is encoded as a wavelength shift [1]. Unfortunately Bragg grating strain sensors suffer an inherent cross sensitivity with temperature. Overcoming the cross-sensitivity in Bragg grating sensors continues to be a major interest of many groups [1-2,6], most of whom obtain both parameters simultaneously from two separate sensor elements having different temperature and strain responsivities. Often two essentially identical sensors are used in close proximity, with one sensor having temperature-sensitivity only (e.g. two Bragg gratings, one not bonded to the structure [1]). However, if the sensors are not exactly at the same location, measurements may be misleading, particularly if the temperature or strain varies significantly from place to place as often occurs. Ideally these sensors should be co-located, and various co-located strain/temperature sensors involving only gratings have been reported [1]. These include a pair of Bragg gratings operating at 850 and 1300 nm, a combination of a Bragg grating with a long period grating, and the use of both the first- and second-order diffraction wavelengths of a single Bragg grating. Another approach is to combine a Bragg grating with another sensing technique; for example, by writing a grating into rare-earth-doped fibre. Recently such a sensor was reported which involved the combination of the fluorescence lifetime of erbium ions and the Bragg wavelength, both of which are strain and temperature dependent [2].

In linear systems (e.g. two Bragg gratings), if ϕ_1 and ϕ_2 are the quantities measured by the two sensors both the temperature, T, and strain, ε, can then be derived by [1]:

$$\begin{bmatrix} \Delta T \\ \Delta \varepsilon \end{bmatrix} = \frac{1}{K_{1T}K_{2\varepsilon} - K_{2T}K_{1\varepsilon}} \begin{bmatrix} K_{2\varepsilon} & -K_{1\varepsilon} \\ -K_{2T} & K_{1T} \end{bmatrix} \begin{bmatrix} \Delta \phi_1 \\ \Delta \phi_2 \end{bmatrix}. \tag{1}$$

Clearly both parameters can be obtained simultaneously provided the two separate sensor elements have different temperature and strain responsivities. However, the uncertainties in both measurands in this approach are necessarily larger than those obtainable with a single measurand sensor due to measurement errors and matrix element uncertainties [3]. Reduced uncertainties are expected if one sensor were to operate independently of one of the measurands.

The fluorescence intensity ratio (FIR) temperature sensing technique used has a strain dependence that is consistent with zero [4]. This method has been demonstrated successfully with a variety of rare-earth dopants and host materials, giving a range of sensitivities [5]. The technique exploits the thermalisation that occurs between a pair of closely-spaced excited dopant ion energy levels, denoted here as 1 and 2, when they are populated using a light source, such as an LED or laser diode, at an appropriate wavelength. Since the temperature dependent relative populations of a group of levels is described by the Boltzmann ratio, the ratio, R, of the intensities of the fluorescence emitted from the two levels, to a common final level (0), at temperature, T, is given by

$$R = \frac{\omega_{20} w_{20} g_2 \exp\left(-\dfrac{\Delta E}{kT}\right)}{\omega_{10} w_{10} g_1}, \tag{2}$$

where the degeneracies of the levels are g_1 and g_2, w_{10} and w_{20} are the temperature-independent decay rates of these levels, the angular frequencies of the fluorescence radiation from the two levels are ω_{10} and ω_{20}, and k is Boltzmann's constant. An example of these levels, for the erbium ion, is shown in Figure 1. It is therefore of interest to take advantage of the strain insensitivity of the FIR method by incorporating an FBG within a length of rare-earth-doped fibre, i.e. a co-located version of the serial arrangement reported recently [6].

Erbium was chosen as the dopant ion for this sensor since the relevant levels, ${}^2H_{11/2}$ and ${}^4S_{3/2}$, are readily excited using 800-nm radiation, via excited state absorption, as shown in Figure 1. Furthermore, in this diagram it can be seen that this pump light excites the ${}^4I_{13/2}$ state also, from which the well-known broad 1550-nm spectral fluorescence originates; this fluorescence is suitable for illuminating fibre Bragg gratings, as illustrated in Figure 2. Thus just a single pump source is required to operate both sensors.

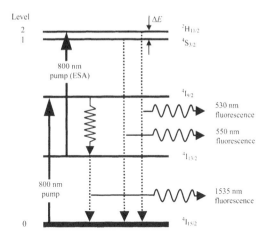

Figure 1. Energy levels of the relevant levels of Er³⁺ ions in silica fibre, showing how both the green fluorescence (used for fluorescence intensity ratio measurements) and infra-red fluorescence (for Bragg wavelength interrogation) are obtained when pumped at 800 nm

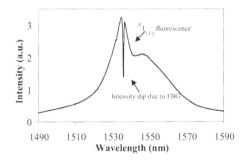

Figure 2. Measured spectrum from an OSA showing the ⁴I₁₃/₂ fluorescence from the Er³⁺-dopant ions together with the intensity dip due to the FBG at approximately 1536 nm.

2. Experimental Arrangement

Measurements of the temperature and strain dependence of the fluorescence intensity ratio and the fibre Bragg grating wavelength shift used the arrangement shown in figure 3. The sensor consisted of a 10-cm length of Er³⁺-doped fibre with a Bragg grating (0.5 cm long) inscribed 0.8 cm from one end of the doped fibre. The doped fibre had a doping concentration of approximately 1400 ppm. The centre wavelength of the FBGs post-annealing was 1536 nm, as measured at room temperature with zero applied strain. The grating was written into the Er^{3+}-doped fibre using 244-nm radiation from a frequency-doubled argon-ion laser through a phase mask.

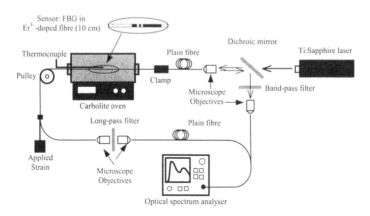

Figure 3. Schematic diagram of experimental setup for dual temperature and strain measurement using the co-located Er^{3+}-doped fibre/FBG sensor.

The relevant excited levels of the Er^{3+} ions were excited using the 800 nm output of a Ti:sapphire laser. Counter-propagating fluorescence was measured to test the dependency of the green intensity ratio on the temperature and strain. A dichroic mirror, with optimised reflection at 540 nm, was used to direct the counter-propagating fluorescence through a green-band-pass filter, which was then collected by an optical spectrum analyser (OSA) via a multimode fibre. Co-propagating fluorescence was measured to test the dependency of the FBG wavelength shift on temperature and strain. This fluorescence was directed through a long-pass filter ($\lambda_{pass} > 830$ nm) and into a single mode fibre, which was connected to an optical spectrum analyser; a representative example of a spectrum is shown in Figure 2.

The effects of temperature and axial strain, on the FIR and FBG wavelength shift were measured by placing the co-located sensor in a electronically temperature-stabilised tube oven (Carbolite, type: MTF 12/38/400). A K-type thermocouple (located alongside the sensor fibre in the tube oven) was used to monitor the temperature. The effect of strain on the FIR and Bragg wavelength was measured by using a pulley system where known weights were cycled progressively. For the fibre Bragg grating, strain and temperature were calculated by monitoring the minima of the Bragg wavelength and then averaging over 5 cycles.

3. Results and Discussion

Figure 4 shows examples of spectra in the region of the Er^{3+} green fluorescence recorded on the OSA at three representative temperatures. The intensities required to calculate the FIR were obtained by summing the intensities from two separate spectral windows on the OSA, namely 515-533 nm for the upper level ($^2H_{11/2}$) emission and 543-561 nm for the lower ($^4S_{3/2}$) radiation. The wavelength ranges were chosen to maximise signal-to-noise ratios. As anticipated from equation 2, the upper level fluorescence intensity increases with temperature while the lower level intensity decreases. The variation of the FIR with strain was also determined.

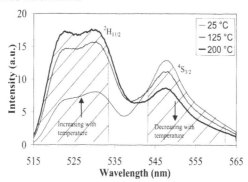

Figure 4. Measured spectra from an OSA showing the green fluorescence from the Er^{3+}-dopant ions at the three temperatures indicated.

The measured variations in the FBG wavelength and the FIR, with strain and temperature, are shown in Figure 5. Figure 5(a) exhibits the usual linear variation of the FBG wavelength with these measurands, while Figure 5(b) illustrates the non-linear dependence of the FIR on temperature and also shows that the strain-dependence is consistent with zero.

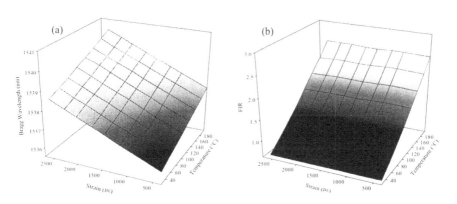

Figure 5. (a) FBG wavelength shift as a function of strain and temperature and (b) FIR measurements as a function of strain and temperature.

Three analysis methods were used to deconvolve the temperature and strain values from the experimental results. The first is the standard matrix method, described by Equation 1, which assumes linear dependencies of both measurands. The second method involved fitting a quadratic to the FIR temperature calibration data and assumed that the fluorescence intensity ratio exhibited zero strain dependence. This fit was then used as a temperature correction of the Bragg wavelength to determine the applied strain. The third method was a modified version of the second method, the difference being the inclusion of the very small strain dependence of the FIR, which was assumed linear, in addition to the quadratic. Table 1 shows the standard deviations, for the three analysis techniques used, between the applied and calculated values. It can be seen that improvement occurs when the temperature range is reduced, as would be expected as the smaller temperature range produces improved fits to the data.

Table 1. Standard deviations for temperature and strain measurement errors across two temperature ranges, for the three analyses undertaken for the co-located dual sensor

Fluorescence Intensity Ratio Dependence on:		Temperature Range (°C)			
		25 – 200		25 - 150	
Temp	Strain	ΔT (°C)	$\Delta \varepsilon$ ($\mu\varepsilon$)	ΔT (°C)	$\Delta \varepsilon$ ($\mu\varepsilon$)
Linear	Linear	1.4	18.0	1.3	14.7
Quadratic	Zero	0.8	16.9	0.6	15.2
Quadratic	Linear	0.9	18.5	0.8	16.2

When comparing the analysis methods, a significant improvement in temperature accuracy is observed for the quadratic fits in comparison to the matrix (i.e. linear-linear) method. This is expected due to the improved fit to the fluorescence intensity ratio versus temperature data using a quadratic rather than the linear fit required in the matrix method. Inclusion of a possible (very small) dependence of the FIR on strain caused the standard deviations to increase, possibly due to the large uncertainties in the parameters for this linear fit.

4. Conclusion

These results demonstrate the suitability of this sensor scheme for the measurement of temperature and strain simultaneously. These sensors have two important advantages, the first of which is the use of a strain-independent temperature measurement, which reduces computational uncertainties in determining the strain. The second advantage is that the sensors are co-located, meaning that the two measurands are determined at a single location. The accuracy of the sensor is comparable to the combined Bragg grating and fluorescence lifetime sensor [2], but with the computational complexity and associated errors of retrieving relevant measurand information reduced. Ongoing evaluation of these and other combined sensors will enable more detailed analysis of the uncertainties in the calculated temperature and strain. It is anticipated that a more compact version of this sensor scheme will find many uses in a range of structural monitoring applications.

Acknowledgement

The authors would like to acknowledge the Laboratorie de Physique de la Matière Condensée at Nice University for providing the rare-earth-doped fibre used in this work.

References

[1]   Kersey A D, Davis M A, Patrick H J, LeBlanc M, Koo K P, Askins C G, Putnam M A and Friebele E J 1997 *J. Lightwave Technol.* **15** 1442-63
[2]   Forsyth D I, Wade S A, Sun T, Chen X and Grattan K T V 2002 *Appl. Opt.* **41** 6585-92
[2]   Jin W, Michie W C, Thursby G, Konstantaki M and Culshaw B 1997 *Opt. Eng.* **36** 598-609
[4]   Wade S A, Collins S F, Grattan K T V and Baxter G W 2000 *Appl. Opt.* **39** 3050-2
[5]   Collins S F, Baxter G W, Wade S A, Sun T, Zhang Z Y, Grattan K T V and Palmer A W 1998 *J. App. Phys.* **84** 4649-54.
[6]   Trpkovski S, Wade S A, Baxter G W and Collins S F 2003 *Rev. Sci. Instrum.* **74** (May)

Section 2

System on a Chip

Lab-on-a-Chip Sensors: Polymer Microfabrication for Chemical and Biochemical Sensing   (Keynote)

Peter R Fielden and Nick J Goddard

Department of Instrumentation and Analytical Science, University of Manchester Institute of Science and Technology, PO Box 88, Manchester M60 1QD, UK.

Abstract: Polymer microfabrication for the development of Lab-on-a-chip sensors and sensor systems is presented. The potential to integrate functions such as electroseparations, electrochemical detection, optical waveguide detection and microfluidics are demonstrated through the descriptions of two all-polymer devices. A chip for isotachophoresis is described with integrated electrodes, for detection and separation, fabricated from carbon fibre loaded polymer. A polymer flow cytometer is also described that facilitates leaky waveguide illumination of cells and particles that are streamed by hydrodynamic focusing in a microchannel. The potential to fabricate ten parallel channels in a single device is reported. The utility of these systems is exemplified with simple analytical measurements.

1. Introduction

Lab-on-a-Chip devices for chemical sensing, sometimes referred to as micro total analytical systems (μ-TAS) [1,2], have traditionally been fabricated in silicon and glass by processes developed by the semiconductor industry, frequently based on photolithography [3]. Such devices have been applied to many application areas that require chemical or biochemical measurements. Microfabricated oxygen sensors have been reported for environmental monitoring applications [4]. The utilization of microfluidics for a range of bioanalysis operations has been reviewed with examples of recent trends to employ microfabrication in plastic materials [5]. Other reviews have featured the application of microfluidics to clinical chemical [6] and forensic [7] analysis, where mainly glass and some polymer devices are described. Microchip-based analytical separation systems for biomedical and pharmaceutical measurements that apply capillary electrophoresis have been reviewed, with some reported devices having been constructed from poly(dimethylsiloxane) (PDMS) [8]. Advanced use of laminated polymer sheets to fabricate microdevices, applied to drug development, have been reported as commercial disposable systems that are typically credit-card sized [9]. Chip-based microfabricated systems are beginning to play an increasingly important role in genomic and proteomic analysis [10].

The authors have concentrated on the development of lab-on-a-Chip devices based on polymer microfabrication. This has led to the fabrication of a series of devices for electroseparations, with a focus on the technique of isotachophoresis (ITP) [11]. Research into microfabricated isotachophoresis systems has included the development of integrated injection-moulded electrodes, formed from carbon-fibre loaded polystyrene [12], and an injection-moulded ITP device that incorporated integrated injection-moulded detection microelectrodes, for conductivity measurements, and integrated moulded microlenses for simultaneous optical absorbance measurements [13]. The authors have also reported the first polymer microfabricated bi-directional ITP system in which cations and anions from the same sample may be analysed simultaneously [14]. The development of polymer microfabricated optical sensing systems has led to the exploitation of leaky waveguide measurement structures. An embossed polymer device has been reported as a lab-on-a-chip detector for visible absorbance spectroscopy in chemical sensing and chemical microreactor applications [15]. Injection moulding has also been demonstrated as a rapid prototyping technique for microfabricated optical systems that incorporate integrated diffractive elements within microfluidic structures [16]. Polymer microfabrication by injection moulding, photolithography and lamination techniques has provided a means of both rapid prototyping for the development and evaluation of research lab-on-a-chip sensor designs, and for the production prototyping of replicate devices. Polymer microfabrication techniques can also facilitate the manufacture of integrated structures that combine microfluidics with both optical elements and electrochemical measurement and control elements. This is illustrated through the development of integrated disposable isotachophoresis devices and disposable flow cytometers for lab-on-a-chip applications.

2. Experimental

Injection moulding of Lab-on-a-chip devices provides planar channel structures, similar to those produced by cast moulding and lamination, but offers the additional benefit of incorporating other elements in the third dimension. The simplest example of this is to directly form the sample and reagent reservoirs within the substrate half that defines the separation channels in an ITP device. The choice of fabrication materials allows for the mass production of very low cost devices. Polymers, such as polycarbonate, polystyrene, and PMMA, are readily available as medical grade granules, and are well suited to the manufacture of miniaturized devices. Typical lab-on-a-chip devices have been injection moulded from polystyrene, onto which is laminated a sealing cover, using a compression adhesive polyester laminate film (Plastic Art, Manchester, UK). This work has used a Babyplast (Cronoplast S.A., Spain) automatic injection moulding machine, which is a precision micro-injection moulder with a shot size of up to 7g. Tool platens have been made to accommodate planar tool inserts (J.E.T. Ltd, UK), to allow rapid changeover of mould tools. This has enabled a route for rapid prototyping of injection moulded designs for Lab-on-a-chip devices, such as the ITP and flow cytometer systems described. The tool inserts are machined in-house with a CAT 3D (Datron, UK) precision milling machine, with feature sizes down to 10μm possible (the milling machine has a repeatability of 3μm). The tool inserts are made from pre-cut brass blocks and are polished by hand to remove any burring left from the milling operation. Designs are created using AutoCAD Mechanical Desktop v5.0 (Autodesk, San Jose, CA, USA) and tools-paths computed with EdgeCAM v.6.75 (Pathtrace, Reading, UK) for any specified milling tool. In our work many designs are cut with a 1mm end mill with a 2° tapered draft angle to facilitate clean ejection of the device from the mould cavity. Although 100μm end mills are available, the selection of the 1mm mill is a sensible compromise between the minimum feature radius of 500μm, the time required to cut a tool insert, and allowance for tool wear. This work employed polystyrene (Northern Industrial Plastics, Chaderton, UK) for the bulk of the lab-on-a-chip devices, and 40% carbon fibre loaded high-impact polystyrene (RTP 487, RTP Company UK Plastics, Bury, UK).

Figure 1 shows a schematic illustration of an isotachophoresis chip design. The separation channel is 44mm long and 200x200µm in cross section. The conductivity detection electrodes comprise an integrated pair of parallel-opposed rectangular electrodes (200µm height, 160µm width) pre-moulded in the carbon fibre loaded polystyrene. Figure 2 shows the detection electrode integrity with respect to the separation channel (filled with an aqueous dye solution for clarity). The liquid wells at the ends of the channel, and at either end of an injector cross channel, have embedded electrodes to drive the isotachophoretic separation process. The incorporation of the electrodes is achieved by moulding the electrode assemblies in a separate operation, followed by their integration into the main moulding process as inserts into the channel-forming mould tool. The device has been applied to a range of analytical application including the determination of heavy metals, inorganic speciation, constituents in photoprocessing solutions and amino acid profiles in urine samples. The separation process requires two electrolytes either side of the sample volume. The application of a constant current source, capable of high voltage operation (typically up to 4kV) across the main separation channel effects the separation of the ionic components on the basis of effective electrophoretic mobility. The steady state zones produced are contiguous, with sharp, self-focusing boundaries, where the zone length is proportional to the concentration of the separated ion in the sample. Careful control of the local chemical environment can be achieved to tune the separation process.

 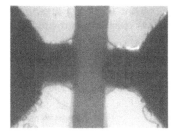

Figure 1 Schematic illustration of a fully integrated polymer isotachophoresis chip, for the analysis of ionic species in aqueous solution.

Figure 2 A photograph which shows the integration of conductivity detection electrodes either side of the isotachophoresis microchannel.

Two different fabrication methods were developed to construct the disposable flow cytometer. Firstly, dry film negative photoresist was laminated onto polymer substrates (between 0.75 and 2 mm thick) with a hot roll laminator. Multiple chip photomasks were then used to create the channel pattern by UV exposure. The exposed sheet was developed in a 1% potassium carbonate solution to remove the non-crosslinked resist. Finally, an unlaminated polycarbonate sheet with predrilled access holes was thermally bonded onto the patterned chip. This fabrication method can be used to produce multilayer 3-D channel networks for lab-on-a-chip applications. We have produced a five-layer 3-D channel networks using this method. In the second method, polystyrene injection-moulded parts were fabricated using conventionally machined brass tooling. Feature sizes of ~100 μm can be created in this way. In addition, coupling prisms and lenses can be fabricated as part of the device. In either case, analyte solution was pumped along the microfabricated channels. These channels thus formed a low refractive index region sandwiched between higher index substrates. A refractive index profile of this sort cannot support conventional optical modes. However, leaky modes can be supported by such low-index waveguides as a result of the high Fresnel reflections at glancing incidence at the sample-substrate interfaces [15]. These LW modes have the property that their optical leakage is very low, which means that they are able to propagate several centimeters without suffering significant loss.

Figure 3 shows a schematic illustration of the principle of operation of the microfabricated flow cytometer. The cell slurry is introduced into the main flow channel and is hydrodynamically focused by two sheath liquid inlets. Laminar flow enables this principle to be expanded to give parallel channel operation. The excitation of the cells, or particles, under study is achieved by leaky waveguide propagation along the length of the channel. Measurement may be made at any point along the channel, or the whole channel may be imaged with an area detector.

To test the utility of leaky wave excitation for monitoring fluorescence of cells or particles, a simple channel 500 μm wide by 30 μm deep was constructed using polycarbonate substrates and dry-film photoresist. The channel was filled with a dilute fluorescein solution and the first leaky wave mode excited using a 473 nm laser via an equilateral coupling prism. A 5000 pixel linear CCD was used to monitor the decay of fluorescence along the channel. A 1:1 imaging lens was used to form an image of the channel on the CCD. An approximately exponential decay of fluorescence along the channel was observed, with a decay length of approximately 6.3 mm.

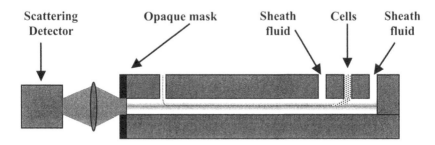

Figure 3 Schematic illustration of the microfabricated flow cytometer design.

3. Results and Discussion

Figure 4 shows an example isotachopherogram for the separation of a mixture of metal ions. The leading electrolyte is 0.02M sodium hydroxide, adjusted to pH 4.95 by the addition of propionic acid, with 0.015M 2-hydroxyisobutyric acid (HIBA) and 0.05% hydroxyethylcellulose (HEC). The terminating electrolyte is 0.01M carnitine hydrochloride. The sample ions are at a concentration of 8×10^{-4}M, except for magnesium, which is at a concentration of 1.2×10^{-3}M. The separation was achieved at a constant current of 20μA. The isotachopherogram shows that the channel length of

44mm is slightly short to achieve a steady state condition, in that the response zones are rounded, rather than square. The data obtained are still of sufficiently good quality to allow quantitation of the ionic components. Adjustment of the leading and terminating electrolyte composition would enable the separation chemistry to be modified such that fewer components could be resolved more effectively using the relatively short channel length.

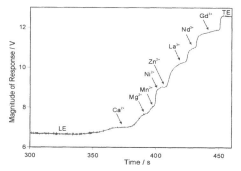

Figure 4 An example isotachopherogram obtained from the separation of a mixture of eight metal ions on the ITP chip illustrated in figure 1.

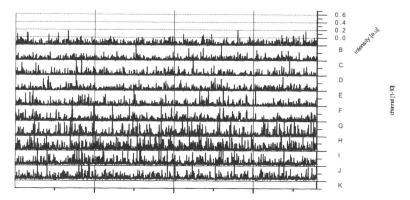

Figure 5 Data traces obtained from ten channel avalanche photodiode array using 10 channel orthogonal Leaky Wave excited laminar flow system.

The operation of the microfluidic flow cytometer requires a diluent, such as phosphate-buffered saline, and the cell suspension to be directed by air pressure or syringe pump into the chip. The cell suspension is injected behind separators through drilled access holes with a diameter of approximately 150 μm. The pressure of the diluent fluid against the cell streams aligns the particles in a *single file* and allows precise centering of these particle streams. The sample streams travel by laminar flow along the chip. The structure of the chip lends itself to optical interrogation by leaky wave modes. This means that each cell can be interrogated *individually*. Suitable detectors can then be arranged above or below the device to collect fluorescence or scattered light from the particles as they intercept the orthogonal leaky wave mode. The ease of fabrication of laminar flow devices has made it possible to produce a chip with many sample streams. We have realised a ten channel version to match currently-available avalanche photodiode arrays for the detection of the scattered and the fluorescent light of the cells. Figure 5 shows the output from each element of a ten channel avalanche photodiode array monitoring fluorescence from 5 μm labelled beads flowing in the laminar system. Each trace shows a four second segment of data obtained from this system. Such laminar flow behaviour persists along the entire length of the flow channel, some 2 cm in total.

4. Conclusions

Polymer microfabrication offers an efficient route to rapid prototyping of lab-on-a-chip systems and devices. Direct machining, photolithography and moulding techniques can support the development of research prototypes made in a variety of polymers. Our work has shown the compatibility of polymer microfabrication with the integration of sensor systems that employ electroseparations, electrochemical detection and optical waveguide detection techniques. The use of injection moulding has the further advantage of being able to fabricate replicate components to high precision, with the opportunity to explore the performance of batches of identical lab-on-a-chip devices.

Acknowledgements

The authors wish to acknowledge the input of many members of the DIAS Miniaturisation Group who have contributed to the development of polymer-based lab-on-a-chip research. Their names are recognized through the publications from the group in references [11-16]. The authors also acknowledge the research funding from EPSRC, BBSRC and the EC.

References

[1]   Reyes D R, Iossifidis D, Auroux P-A and Manz A, 2002, Anal. Chem., 74, 2623-2636.

[2]   Auroux P-A, Iossifidis D, Reyes D R and Manz A, 2002, Anal. Chem., 74, 2637-2652.

[3]   Rai-Choudhury P, Ed., "Handbook of Microlithography, Micromachining, and Microfabrication – Volume 1: Microlithography", 1997, SPIE Press Monogram PM39, Washington.

[4]   Suzuki H, 2000, Materials Science and Engineering C, 12, 55-61.

[5]   Khandurina J and Guttman A, 2002, J Chromatogr. A, 943, 159-183.

[6]   Cunningham D D, 2001, Anal. Chim. Acta, 429, 1-18.

[7]   Verpoorte E, 2002, Electrophoresis, 23, 677-712.

[8]   Gawron A J, Scott Martin R and Lunte S M, 2001, European J Pharmaceutical Sciences, 14, 1-12.

[9]   Weigl B H, Bardell R L and Cabrera C R, 2003, Advanced Drug Delivery Reviews, 55, 349-377.

[10] Sanders G H W and Manz A, 2000, Trends in analytical Chemistry, 19, No 6, 364-378.

[11] Prest J E, Baldock S J, Treves Brown B J and Fielden P R, 2001, Analyst, 126, 433-437.

[12] Baldock S J, Fielden P R, Goddard N J, Prest J E and Treves Brown B J, 2003, J. Chromatogr. A, 990, 11-22.

[13] Fielden P R, Baldock S J, Goddard N J, Morrison L, Prest J E, Treves Brown B J and Zgraggen, 2002, Proc. SPIE, 4626, 429-440.

[14] Prest J E, Baldock S J, Fielden P R, Goddard N J and Treves Brown B J, 2002, Analyst, 127, 1413-1419.

[15] Malins C, Harvey T G, Summersgill P, Fielden P R and Goddard N J, 2001, Analyst, 126, 1293-1297.

[16] Hulme J P, Mohr S, Goddard N J and Fielden P R, 2002, Lab. Chip., 2, 203-206.

Development of a fluorescence-based biochip for multianalyte immunosensing

H M McEvoy[1], P P Dillon[2], C McDonagh[1], B D MacCraith[1], R O'Kennedy[2]

[1] School of Physical Sciences, National Centre for Sensor Research, Dublin City University, Dublin 9, Ireland.
[2] School of Biotechnology, National Centre for Sensor Research, Dublin City University, Dublin 9, Ireland.

Abstract: An optical waveguide-based biochip is presented here. Using a polydimethlysiloxane flowcell, sandwich assays of capture antibodies, antigens and fluorescently-labelled antibodies were patterned onto the surface of a waveguide. Evanescent wave excitation was used to excite the labelled antibodies and the emitted fluorescence was recorded using a cooled CCD camera. The concentration of antigen present was determined from the intensity of the fluorescence recorded. Results have been obtained for a glass substrate and the next stage of work will include the progression to mouldable plastic chips and CMOS detection.

1. Introduction

In recent years, there has been considerable research in the development of immunosensors for medical diagnostics as well as for applications in environmental and food quality monitoring [1]. Immunosensors are based on the principle of specific antibody-antigen binding and offer significant advantages over conventional immunoassays [2]. Optical immunosensors offer added values of increased sensitivity and selectivity. The potential for multianalyte detection is one of the leading drivers for further development of immunosensing devices. There is increasing demand for cheap, miniature biochips that will produce rapid analysis with low sample volume.

In this work we present the development of an optical waveguide-based multianalyte immunosensor. The immunosensor is based on a sandwich assay format, which means that the antigen is sandwiched between two antibodies. The Bovine Serum Albumin (BSA)/anti-BSA system is used as a model assay for proof of principle. The development of the chip is presented under three headings: antibody immobilisation, patterning protocol and optical system.

2. Materials

Standard glass microscope slides (76x26x1mm, sand blasted) were purchased from AGB Scientific Ltd. (Dublin, Ireland). Antibody preparations were provided by the School of Biotechnology, Dublin City University. General chemical and biological reagents, including γ-glycidoxyproplytrimethoxysilane, were obtained from Sigma Aldrich Irl. (Dublin, Ireland) or Fannin Healthcare (Dublin, Ireland). All reagents were of analytical grade. Cy5 bisfunctional reactive dye was purchased from Amersham International (Bucks, England).

3. Antibody Immobilisation Protocol

Glass microscope slides were used as the sensor substrates. The slides were cleaned by immersion in concentrated NaOH overnight at room temperature, followed by rinsing in deionised water.

In order for the antibodies to attach to the glass slide it was then necessary to firstly silanise the surface. Silanes covalently bound to the slide acted as linkers between the inorganic chemistry of the glass and the organic chemistry of the compounds required for attaching the antibodies. γ-glycidoxyproplytrimethoxysilane was mixed with a 50:50 solution of methoxyethyl ether in water to give a final concentration of 10% v/v. The slides were immersed in this solution for 1 hour at room temperature, after which, they were rinsed with ethanol and water.

Protein A, an immunoglobulin-binding protein, was used to link the antibodies to the silanised surface of the glass slide [3]. A 20µg/ml solution of protein A in phosphate-buffer saline (PBS) was prepared. The silanised surface of the glass slide was exposed to the protein A solution for 1.5hrs at 37°C (or at room temperature for 2hrs). It was then exposed to 20mM NaOH for 5mins and stored in nitrogen at 4°C overnight. A major advantage of using Protein A as the antibody linker (rather than other proteins such as Biotin [4]) was that it allowed for reversible antibody attachments, so the same slide could be re-used several times for different sample solutions.

As mentioned in the introduction, the immunosensor described here is based on a sandwich assay format. Capture antibodies were passed over the slide and left for 1 hour at room temperature. The slide was then rinsed using phosphate-buffered saline containing 0.05% Tween 20 (PBST) to remove any unbound antibodies. A sample solution containing the antigen to be detected was passed over the bound antibodies and antigens present in the sample attached to the capture antibodies. After 15mins the slide was rinsed with PBST. In the final stage of the sandwich assay formation, antibodies labelled with fluorophores were passed over the slide. Fluorophores are molecules that produce a fluorescent emission when irradiated with light at a suitable excitation wavelength. The fluorescently-labelled antibodies attached to the bound antigens and after 15mins a final rinse with PBST was carried out. The overall assay configuration is illustrated in Fig. 1.

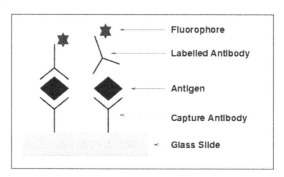

Figure 1: Illustration of antibody/antigen/antibody sandwich assay.

4. Patterning

The antibodies were patterned using a flowcell fabricated from poly(dimethylsiloxane) (PDMS). The PDMS was formed from a liquid prepolymer and a curing agent mixed together in a 10:1 ratio (weight or volume). The mixture was poured into a multi-channel mould and cured at 70°C for 1 hour. After curing, the PDMS was removed from the mould. When lightly pressed against it, the flowcell came into conformal contact with the glass slide.

In protein arrays, microlitre volumes are more common and less expensive than millilitre volumes so aluminium masks were used to create 3mm by 200µm by 600µm ridges in SU-8 photoresist. PDMS was then moulded against these ridges to create flowcells with channel thicknesses in the µm range.

The microfluidic system was completed by inserting needles into the flowcell at both ends of the channels and connecting tubing to the ends of the needles. A peristaltic pump pulled the solutions from a reservoir, through the tubing into a needle at one end of the channel and out the needle at the other end.

For multianalyte detection, multiple channels were used to pattern strips of different antibodies along the substrate. After the antibodies attached to the slide, the flowcell was rotated and the sample passed over at right angles to the strips. Any antigens present in the sample attached to their associated antibodies. Fluorescently-labelled antibodies were then passed over the sample and attached to the bound antigens. These were detected as described in Section 5. Fig. 2 illustrates the fluorescence at the crossover point of a strip of capture antibodies and a sample solution.

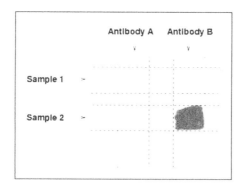

Figure 2: Fluorescence recorded from labelled antibodies attached at crossover point indicating the presence of Antibody B in Sample 2.

5. Optical System

A 635nm laser diode was used to excite the labelled antibodies immobilised on the surface of the waveguide. The excitation light was passed through a line generator so that the circular beam of light was converted to a rectangular beam. This facilitated the uniform illumination of the strips of antibodies on the substrate. A cylindrical lens was used to narrow the beam sufficiently for coupling into the end-face of the glass slide. Fig. 3 illustrates the excitation setup.

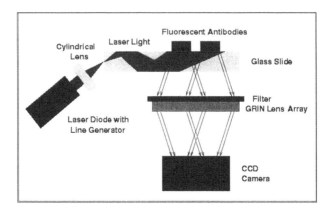

Figure 3: Excitation and detection setup

The substrate acted as a waveguide, with the light propagating by total internal reflection (TIR) at the waveguide surface. In order to satisfy the requirements for TIR, the surrounding media (substrate and cover layer) must be of lower refractive index than the waveguide. The light undergoing TIR created an evanescent wave that penetrated a short distance (100's of nanometres) outside of the waveguide [5]. Due to this short-range penetration, the evanescent wave selectively excited the fluorescently-labelled antibodies bound on the surface and discriminated against unbound antibodies in the solution.

The fluorescence emitted by the labelled antibodies was passed through a filter, which removed any excitation light and entered a graded index (GRIN) lens array. The emitted fluorescence was recorded using a Cooled Charge Coupled Device (CCD) Camera (Sensicam PCO) positioned at the effective focal length of the GRIN lens array. A cooled camera was used so as to detect the low light levels emitted by the fluorescently-labelled antibodies.

The concentration of antigen present in the sample is determined from the intensity of the fluorescence recorded.

6. Validation and Results

In order to validate the system and to demonstrate its sensing capabilities, a calibration assay was carried out. Bovine Serum Albumin (BSA) and its associated antibody were chosen as the model system. A range of BSA concentrations was prepared and a stock of antiBSA was labelled with the fluorescent dye, Cy5.

Using a silanised slide previously coated with protein A as described in Section 3, unlabelled antiBSA was flowed through a single-channel flowcell and left for one hour. The flowcell was then rotated by 90 degrees and the antigen, BSA, was passed over the slide. An image was taken at this point and used as the background image. After 15 minutes, the flowcell was rotated once more and fluorescently-labelled antiBSA was passed over the slide. The fluorescently-labelled antiBSA was left for 15 minutes before a non-fluorescing solution was passed over the slide and an image taken. Before and after the introduction of each new solution a phosphate-buffered saline was passed over the slide to wash it and remove any unbound antigens or antibodies. This was to ensure that only the bound antibodies and antigens were available for the next stage of the immunosensing [6]. As expected, an increase in antigen concentration resulted in a corresponding increase in fluorescence intensity. The dose response curve obtained is shown in Fig. 4. Currently, the working range is limited. Further work will focus on increasing the range and enhancing the limit of detection.

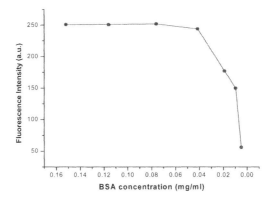

Figure 4: Dose response curve for antiBSA

7. Conclusion

A fluorescence-based biochip has been fabricated and validated. The antibody immobilisation protocol has been optimised and PDMS flowcell patterning established. Results have been obtained for a single analyte on glass substrates. Future work will concentrate on expanding the range, validating the system for multianalyte sensing and progressing to mouldable plastic chips and CMOS detection.

References

[1] Lowe C.R., "An introduction to the concepts and technology of biosensors", Biosensors, 1, pp. 3-6, (1985)
[2] P.S. Petrou, S.E. Kakabakos, I. Christofidis, P. Argitis, K. Misiakos, "Multi-analyte capillary immunosensor for the determination of hormones in human serum samples", Biosensors and Bioelectronics, 17, (2002), 261-268
[3] Anderson G.P., Jacoby A. M., King K.D., "Effectiveness of protein A for antibody immobilisation for a fiber optic biosensor", Biosensors & Bioelectronics, 12, 4, pp. 329-336, (1997)
[4] Busse S., Scheumann V., Menges B., Mittler S., "Sensitivity studies for specific binding reactions using the biotin/streptavidin system by evanescent optical methods", Biosensors and Bioelectronics, 17, 8, pp. 704-710, (2002)
[5] Plowman T.E., Durstchi J.D., Wang H.K., Christensen D.A., Herron J.N., Reichert W.M., "Multiple-analyte fluorimmunoassay using an integrated optical waveguide sensor", 71, 19, pp. 4344-4352 (1999)
[6] Rowe C.A., Tender L.M., Feldstein M.J., Golden J.P., Scruggs S.B., MacCraith B.D., Cras J.J., Ligler F.S., "Array biosensor for simultaneous identification of bacterial, viral and protein analytes", Analytical Chemistry, 71, 17, pp. 3846-3852, (1999)

A *Laboratory on a Chip* for Test Engineering Education

I A Grout, J Walsh, T O'Shea and M Canavan

Department of Electronic and Computer Engineering, University of Limerick, Ireland

Abstract: This paper will discuss the concept design and potential utilisation of a novel architecture reconfigurable integrated circuit (IC) aimed at supporting the teaching of Integrated Circuit test engineering concepts. The objective is to provide a single configurable device that allows for the downloading of specific circuit designs, along with specific target circuit faults into a digital configurable array. As such, it can provide the potential to be a highly flexible and interactive device for IC hardware test strategy development, acting as a single IC based hardware laboratory, and may complement a corresponding fault simulation study. This paper will discuss the design of a demonstrator version of the array whose functionality may be extended. The potential users will be persons investigating IC test concepts and techniques.

1. Introduction

In modern microelectronic circuit based systems, reconfigurable Integrated Circuits (ICs) form an important role within the overall design functionality, providing benefits in terms of the ability to prototype ideas and to readily modify hardware functionality. These may be discrete IC solutions, or may also be considered as macro cells within single *System on a Chip* (*SoC*)[1][2] devices for electronics system level solutions. Reconfigurable or programmable devices are today available with digital and analogue circuit functions. The ability to program/configure circuit functionality post-fabrication and in the final application, is increasingly important for the higher complexity devices, where increasing levels of circuit integration (more transistors per mm^2 of silicon "real estate") and increasing functionality is required in applications such as *Laboratory on a Chip* (LoC) scenarios.

This paper will discuss the design, development and potential utilisation of a novel architecture configurable Integrated Circuit aimed at the teaching of IC test engineering concepts[3]-[7]. The objective is to provide a single IC solution that allows for the downloading of specific circuit designs, along with specific target circuit faults into the configurable array. It is aimed to be highly interactive, allowing for different digital circuit/system designs to be generated, allowing for rapid configuration of the device (fault-free circuit), and to insert/retract specific circuit faults at any time during the device operation. As such, it may be utilised as an integral part of a CAL (Computer Aided Learning) environment. This would allow for the linking hardware and education software seamlessly via a Personal Computer control, as shown in *figure 1*. The configurable device is to be referred to as LOCTEE (Laboratory On a Chip for Test Engineering Education). It is referred to as a laboratory as it would contain the ability to create all the electronic circuit hardware that would be required to undertake the types of experiments envisaged (circuit design, build, test development, test program implementation and evaluation). Such circuits would range in size, complexity and behaviour under fault-free and fault conditions. For example, at the lower end of the scale in terms of complexity, combinational logic circuits with a small number of Primary Inputs (PIs) and Primary Outputs (POs), and a small number of internal nodes, can be designed such that they contain faults that are not detectable. This is a situation that should be avoided for device testability. These non-detectable faults would be due to circuit design features and would identify design and testability issues, along with the need to generate suitable designs that are functionally correct and with the right level of testability. At the higher end of the scale in terms of complexity, then the consideration would be for digital designs up to the complexity of small to medium size microprocessor/microcontroller cores and the modelling of memory macro cells. These would allow for consideration into the aspects of *Design for Testability* (DfT), design partitioning and test strategy development. It should be noted however that even the seemingly simplest of circuits can still highlight important aspects for circuit testing.

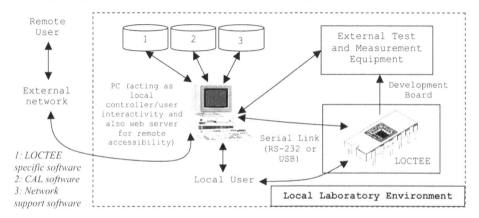

Figure 1: Basic set-up for LOCTEE (Laboratory on a Chip for Test Engineering Education)

An important aspect for the design would be the ability to temporarily introduce certain circuit/logic level faults into the circuit, either at power-up and would be present until the device is reconfigured (in order to model fabrication introduced faults), or for the faults to be introduced "on-the-fly" (in order to introduce faults occurring during operation). These could be considered as permanent (if introduced once and left in place) or as "soft-errors", which may occur under specific operating conditions. These "soft-errors" are increasingly noticeable at lower device geometries (VDSM – Very Deep Sub-Micron (towards the 90nm technology node)) and higher operating frequencies in digital processor architectures. The decision however would need to be made as to which faults are to be considered and how they are modelled and used in the final circuit. There are a number of fault models commonly used in current test program development, and the aim would be to initially utilise a sub-set of these:

- Stuck-At-Fault (SAF) – industry standard fault model but limited in modelling realism
- Bridging Fault (BF) – resistive bridges between nodes increasingly problematic
- Open Fault – increasingly problematic

- Delay Fault (both Path Delay Model (PDM) and Transition Delay Model (TDM))
- I_{DDQ} Fault – complemented by I_{DDT} testing

The Stuck-At-Fault, Bridging Fault and I_{DDQ} Fault models would be the three to be utilised since these form the basis of many test development strategies. Due to the nature of semiconductor memories (RAM: Random Access Memory and ROM: Read Only Memory), memory faults have a number of other characteristics that may also be considered. Communications between the device and PC would be established using a standard PC serial link, RS-232 or USB (Universal Serial Bus) and would be chosen for ease of use, ease of programming and allowing for two-way communications using a minimum number of device pins.

The device is to be configured on a temporary basis in volatile memory, where the configuration would be cleared whenever the power is removed. This will be for simplicity of design and use, and the device could be envisaged to be used either:

- Connected and configured via the PC.
- In a stand-alone mode where an external serial ROM (Read Only Memory) will contain the configuration data and download this on power-up, in a similar mode as to a number of standard available devices, see *figure 2*.

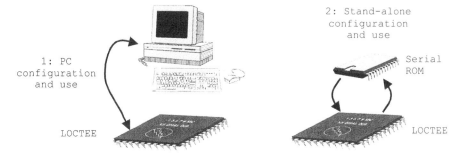

Figure 2: Configuration modes: standard approach to be taken

This paper is structured as follows. This section has introduced the concept and work-at-hand. *Section 2* will discuss programmable devices, for both digital and analogue circuits. This will be followed by a discussion into the rationale for LOCTEE in *section 3*, and *section 4* will discuss the proposed architecture. *Section 5* concludes the paper.

2. Programmable Devices

Three types of configurable device are commonly available for use. These are:

- FPGA: Field Programmable Gate Array (digital)[8][9]
- CPLD: Complex Programmable Logic Device (digital)[10]
- FPAA: Field Programmable Analogue Array (analogue and mixed-signal)[11]-[13]

In addition, the PLD   (Programmable Logic Device), PLA (Programmable Logic Array), PAL (Programmable Array of Logic) and GAL (Generic Array of Logic) are also utilised. These devices are used in an educational environment since they do offer reconfigurability and rapid prototyping benefits. These can be either "configure once" without the opportunity of modifying the design after configuration (anti-fuse based), or reconfigurable (reprogrammable memory based configuration). These will either be configured away from the final application, or In-System Programmable (ISP), typically using the on-chip JTAG Boundary Scan[14] (IEEE Standard 1149.1). The existing devices may be used to a limited extent in the purpose discussed here, **but are limited** in their ability to be highly interactive by quickly and easily inserting/retracting specific faults into/from the Circuit Under Test (CUT).

"Designing a circuit to intentionally incorporate faults is not a normal design consideration"

The ability to generate faults as the circuit is operating leads to a highly flexible device in the application of concepts for learning purposes and the ability to identify the behaviour of a design at specific points in time and under specific operating conditions. This is something that would also be limited if a simulation model of the design was to be utilised in a simulation study. The device would however also allow for comparisons of simulation model and actual circuit behaviour to be made. For the flexibility of use to be available, then a device with an architecture that allows for this has to be determined and developed. The design and utilisation of LOCTEE will aim to provide a solution for this.

3. Rationale for the Device

3.1 Introduction

The rationale for this work can be considered from three viewpoints.

- Firstly, from the teaching of test engineering[3]-[7],[14] concepts aimed at providing the skills for graduates to become effective members of a test engineering team, and to be equipped to undertake a test development role within the microelectronics industry.

- Secondly, from the viewpoint of developing new architectures and applications for configurable (programmable) and highly interactive hardware Integrated Circuits (ICs).

- Thirdly, from the user perspective of technologies in order for their effective utilisation with a cost effective means for accessibility, installation, set-up and learning.

3.2 Teaching of test engineering concepts

The education of students in the area of test in electronic and computer engineering has a number of avenues that can be taken. Firstly, considering whether it is hardware or software test that is the primary concern. With hardware test, whether the main areas of concern are device (IC) level or system level. Today, however, it should be noted that the boundary between device and system is less well defined with the advent of the *System on a Chip* (SoC)[1][2], where the primary concern is to integrate as much of an electronic system as possible onto the silicon die. At the device level, the consideration is whether the type of circuit is digital, analogue, or mixed-signal. Each choice taken opens up new avenues of investigation and requirements for the design, fabrication and ultimately the testing of the device. In the test area, then the test engineer will ultimately be required to work in a team environment, primarily with the design engineer and process engineer. The primary concern for the design engineer is to develop a product that meets the functional requirements. The primary concern for the process engineer is to ensure that the device can be fabricated with a high yield and low cost. The test engineer (i) ensures that the device can be tested to the required quality level and (ii) develops and applies the cost-effective final production test program. As such, the test engineer may be seen as a linking force with the design and fabrication communities. This requires insight into the role and requirements for design and fabrication, alongside the specific requirements for test.

3.3 Configurable Devices

Configurable (programmable) devices have made a major impact in electronic system design, with the ability to utilise a single device for multiple purposes, for prototyping design ideas and to reconfigure the electronic hardware for modification/upgrading purposes. These have been primarily in the digital domain[8][9][10], but reconfigurable analogue arrays[11]-[13] are becoming more widely adopted, but not primarily in the test area. These are primarily designed to be configured before use in the application and only to be reconfigured when the design has to be changed occasionally. These are used to develop designs without the intention of circuit faults being present and for occasional reconfiguration. Interaction with the device in the final application is limited as these devices are not aimed at a high level of user interaction, as would be ideally required in an educational environment.

3.4 Accessibility, installation, set-up and learning

The ability for an end-user to quickly access and utilise a system has to be addressed from the system design outset and this allows for a user-centred approach to the system design and development to be undertaken.

3.5 Range of circuits to be considered

There are a wide range of useful demonstrator circuits available, either derived for the problem scenario at-hand or via provided benchmark circuits such as the ISCAS-85, ISCAS-89 and ITC-99 benchmark circuits.

4. Proposed Device Architecture

In the usual approach to the design process, the design is created to be functionally correct, meeting a set design specification and is tested to ensure that no faults exist in either the design functionality or in the fabrication of the device. In design verification, this is to ensure no faults exist in the design itself and in production test, to ensure no faults exist due to fabrication defects. When considering configurable devices such as

Field Programmable Gate Arrays, Complex Programmable Logic Devices and Field Programmable Analogue Arrays, the architectures are designed to facilitate the above requirements. In this work, the proposal is to develop a device architecture which as well as implementing a functionally correct circuit, will also allow for specific faults to be inserted. This may provide a number of benefits in the education process. As such, a user may use the device as follows:

- Configuring the fault-free design and not considering any faults to be present.
- Configuring the fault-free design and specific circuit faults before device utilisation.
- Configuring the fault-free design and interactively insert/remove specific faults.

The device may be used in one of two modes:
- Locally, with the user sited in the same laboratory as the device and a great deal of interactivity available.
- Remotely, via an internet connection and accessing the device via the World Wide Web (WWW). In this mode, the user will still have a level of interactivity, but this would be limited. It does however allow for distance learning capabilities[15]-[18].

The proposed device architecture is shown in *figure 3*.

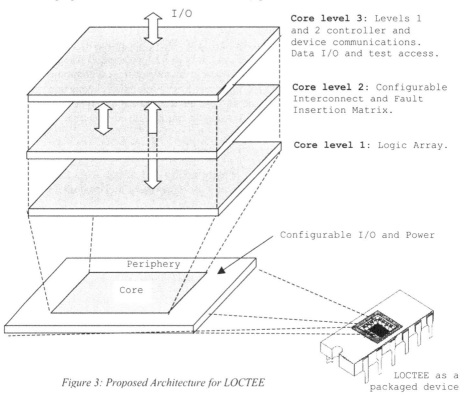

Figure 3: Proposed Architecture for LOCTEE

The architecture is based around three levels of circuit to form the overall functionality:

- **Level 1: Logic Array**. This will form the basic block of logic that will implement the digital circuit under test.
- **Level 2: Configurable Interconnect and Fault Insertion Matrix**. This will form the part of the circuit that will be configured to connect the appropriate logic in level 1 and also provide the ability for inserting/retracting faults via a fault matrix.
- **Level 3: Levels 1 and 2 controller and device communications**. This will form the part of the device that controls specific aspects of levels 1 and 2, along with providing serial communications (either RS-232 or Universal Serial Bus (USB)) with the PC.

5. Conclusions

This paper has described a novel architecture configurable device for aiding the education of IC test concepts. This configurable device is referred to as LOCTEE and is considered as a *Laboratory on a Chip* (LoC) for Test Engineering Education. This single device provides a single IC solution to a range of digital laboratory exercise scenarios that would traditionally utilise a set of discrete ICs. The concept design was introduced and a demonstrator version of the concept highlighted.

References

[1]   MacMillen D. et al., *An Industrial View of Electronic Design Automation*, IEEE Transactions on Computer Aided Design of Integrated Circuits and Systems, Vol. 19, No. 12, December 2000, pp1428-1448

[2]   Wakabayashi K. and Okamoto T., *C-Based SoC Design Flow and EDA Tools: An ASIC and System Vendor Perspective*, IEEE Transactions on Computer Aided Design of Integrated Circuits and Systems, Vol. 19, No. 12, December 2000, pp1507-1522

[3]   Hurst S., *VLSI Testing digital and mixed analogue/digital techniques*, IEE, 1998, ISBN 0-85296-901-5

[4]   Wilkins B.R., *Testing Digital Circuits An Introduction*, Van Nostrand Reinhold (UK), 1986, ISBN 0-442-31748-4

[5]   Burns M. and Roberts G., *An Introduction to Mixed-Signal IC Test and Measurement*, Oxford University Press, 2001, ISBN 0-19-514016-8

[6]   O'Connor P., *Test Engineering, A Concise Guide to Cost-effective Design, Development and Manufacture*, Wiley, 2001, ISBN 0-471-49882-3

[7]   Needham W., *Designer's Guide to Testable ASIC Devices*, Van Nostrand Reinhold, 1991, ISBN 0-442-00221-1

[8]   *Spartan FPGA*, Xilinx, Inc, USA, http://www.xilinx.com/

[9]   Venkateswaran R. and Mazumder P., *A survey of DA techniques for PLD and FPGA based systems*, Integration, the VLSI journal, No. 17, 1994, pp191-240

[10]   *CPLD Product*, Lattice Semiconductor Corporation, USA, http://www.latticesemi.com/

[11]   *FPAA product*, Anadigm, UK, http://www.anadigm.com

[12]   *Analog Array roduct*, Lattice Semiconductor Corporation, USA, http://www.latticesemi.com/

[13]   Blaszkowski A. et al., *A Comparative Analysis of Continuous and Discrete-Time Field Programmable Analogue Arrays*, Preprints of the IFAC Workshop on Programmable Devices and Systems, 11-13 February 2003, Ostrava, Czech Republic, pp190-195

[14]   JTAG Boundary Scan, IEEE standard 1149.1, http://standards.icee.org/

[15] Coleman J. et al., *Effectiveness of Computer-Aided Learning as a Direct Replacement f[9] or Lecturing in Degree-Level Electronics*, IEEE Transactions on Education, Vol. 41, No. 3, August 1998, pp177-184

[16] Enloe C. et al., *Teleoperation in the Undergraduate Physics Laboratory – Teaching an Old Dog New Tricks*, IEEE Transactions on Education, Vol. 42, No. 3, August 1999, pp174-179

[17] Shen H. et al, *Conducting Laboratory Experiments over the Internet*, IEEE Transactions on Education, Vol. 42, No. 3, August 1999, pp180-185

[18] Brofferio S., *A University Distance Lesson System: Experiments, Services, and Future Developments*, IEEE Transactions on Education, Vol. 41, No. 1, February 1998, pp17-24

Section 3

Sensing Materials

Development of Integrated MEMS devices   (Invited)

P J French

Laboratory for Electronic Instrumentation, DIMES, Delft University of Technology, Delft The Netherlands, e-mail P.J.French@its.tudelft.nl

Abstract: In the development of MEMS devices, the questions arise of whether or not to integrate and which technology should be used. Choosing the non-integrating option gives more processing freedom but may limit the possibilities of adding new functions in the system. Taking the integrated option, which combines sensor and electronics on a single chip, it is preferable to use existing processes or layers to fabricate the sensor. If not possible, the number of additional steps should be minimised. The additional process steps should be compatible with the electronics. In this paper the development of MEMS technology will be discussed taking into account the application and the integration possibilities.

1. Introduction

The first silicon mechanical sensors using isotropic etching appeared in the early 1960s [1], and later anisotropic etchants. These basic structures can be grouped as bulk micromachined devices. Surface micromachining makes use of thin films deposited on top of the wafer. The first devices also appeared in the 1960's [2], but became more prominent in the 1980's with the growth of polysilicon as a mechanical layer [3-4], and saw the first moving mechanical structures. In more recent years technologies which use the upper 10s of microns of the substrate have been developed [5]. These have lateral dimensions similar to those of surface micromachining but with increased vertical dimensions. The development of RIE has given considerably more freedom in the structures, with high aspect ratio, which can be fabricated [6-7].

When developing devices using one of the micromachining processes the question arises of whether or not to integrate with electronics. The process should be commercially viable and the application should be suitable for electronics at source. Many technological problems can arise when combining two processes (although not in all cases), and if the solution is complicated it may be impractical (e.g. the process yield drops to unacceptable levels. On the application side the designer should consider whether integration is necessary and also suitable. In some harsh environments the electronics may not operate correctly making the integrated option unsuitable. Despite these problems there are many applications where integrated devices can be an ideal option.

2. Compatibility Issues

When we talk about integration, we mean combining the mechanical structures with CMOS, bipolar or BiCMOS. The main compatibility issues are: contamination, selectivity of etching and thermal budget. The contamination issue will determine the process sequence and often requires separate process areas. The selectivity issue often requires extra masking or use of alternative etchants. The thermal budget is a critical issue, particularly with surface micromachining. A range of integrated processes have been developed where the process sequence has been determined by the above issues. The three options are; 1) pre-processing, 2) integrated processing and 3) post processing. Examples of these will be given in a later section.

3. Basic Processes

3.1 Wet bulk micromachining

The main wet bulk micromachining processes use anisotropic etchants such as KOH and TMAH. The anisotropic aspect is due to the low etch rate of the (111) plane compared to the (100) and the (110) planes (Figure 1). Both of these processes are relatively low temperature and the main consideration is cleanroom contamination in the case of KOH.

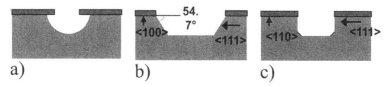

Figure 1 Three etch profiles; a) isotropic etching, b) anisotropic on (100) wafers and c) anisotropic on (110) wafers.

A wet etching technique capable of achieving high aspect ratios is macro-porous processing. Macro-porous silicon is an electrochemical etch process in HF and in n-type material forms deep straight holes. This process requires electronic holes for the etching to continue and this is provided by illumination from the backside. An interesting feature of this process is that the width of the hole is controlled by light intensity and thus true 3-D structure can be fabricated in a single etch-step [8]. This is illustrated in Figure 2.

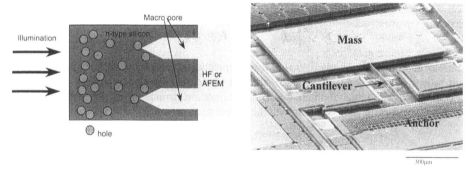

Figure 2 (right) Etch mechanism and (right) free standing structure fabricated using macro-porous processing [8].

Although mainly considered a process for n-type silicon, macro-porous silicon has also been fabricated in p-type silicon [9].

3.2 Dry bulk micromachining

Dry etching uses an ionised gas to achieve etching. This enables us to control the etch profile and thus fabricate a wide range of structures. To achieve deep etching and high aspect ratios we need to have good control of this profile and high selectivity over the mask. The cryogenic process uses temperatures round –110°C and oxygen is added to achieve continuous sidewall passivation during etching [10-11]. The Bosch process uses cycles of isotropic etching and passivation and can be characterised by the wavy sidewall [12-13] (Figure 3). Work has continued to improve, for example, profile control [14] and smoother sidewall by faster switching [15].

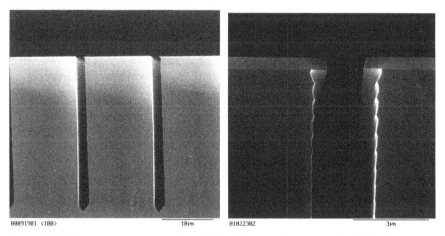

Figure 3 (left) Cryogenic process and (right) Bosch process (reproduced with kind permission, Michiel Blauw and Emile van der Drift, DIMES, Delft University of Technology.

The cryogenic process has the potential advantage of smooth walls but issues such as crystal orientation and design layer have to be considered.

3.3 Surface micromachining

Surface micromachining involves thin films deposited on the surface of the wafer and some are selectively removed to form free-standing structures. The basic process is shown in Figure 4. There are many options for sacrificial and mechanical layers and some examples of these are given in table 1, along with the appropriate etchant.

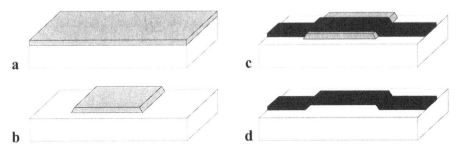

Figure 4 Basic surface micromachining process (a) deposition of sacrificial layer (SAC) (b) patterning of SAC, (c) deposition and patterning of the mechanical layer and (d) sacrificial etching.

Sacrificial layer	Mechanical layer	Sacrificial etchant
Oxide (PSG, LTO etc)	Polysilicon, silicon nitride, silicon carbide	HF
Oxide (PSG, LTO etc)	Aluminium	Pad etch
Polysilicon	Silicon nitride	KOH, TMAH
Polysilicon	Silicon dioxide	TMAH
Resist	Aluminium	Acetone/oxygen plasma

Table 1. Examples of sacrificial and mechanical layers with the appropriate etchant.

The development of surface micromachining has lead to a not only free standing structures but also a range of moving structures. An example of these structures is given in *Figure 5* [16].

Figure 5 Examples of gears and locking systems fabricated using a 5 layer polysilicon surface micromachining process. Reproduced with kind permission from Sandia labs [16].

In terms of integration, there are three options and these are listed below with examples of processing issues

- Pre-processing
  - Cleanroom compatibility, planarity
- Integrated processing
  - Cleanroom compatibility, thermal budget
- Post-processing
  - Thermal budget (usually <400°C)

An example of pre-processing is given in *Figure 6*. This uses polysilicon mechanical layers buried in wells, which are filled before the CMOS processing is started. In this case it is important to ensure that the mechanical properties of the polysilicon are optimum after the CMOS processing and no necessarily after deposition [17-18].

The integrated option is perhaps more complicated and the processing will depend on the flexibility of the IC process with which the mechanical layers are integrated. In some IC processes the micromachining can be added at the end of the thermal processing but before the aluminium [19].

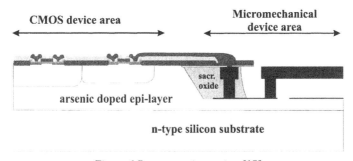

Figure 6 Pre-processing option [17]

Post processing provided two main directions. The first, and most simple, is to use existing layers. Examples include using the passivation and metal layers [20] and gate polysilicon [21] as the mechanical layer. The potential problem with this approach is that the layers in standard IC processing are not developed for their mechanical properties and the designer must compensate for this.

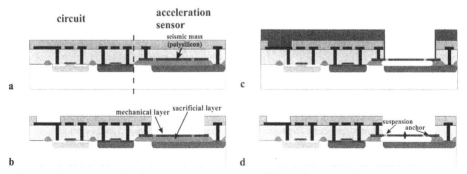

Figure 7 Integration of surface micromachining with CMOS. The gate polysilicon is used as the mechanical layer. Taken from [18].

If layers are to be added at the end of an IC process we are limited to metals, polymers and PECVD layers [22].

4. Conclusions

Silicon micromachining, using etching and deposition techniques, offers a wide range of opportunities for integration with IC devices. For each of these options there are a number of issues which have to be considered. This includes, not only the compatibility, but also complexity, yield and packaging. However, if these issues can be successfully addressed, there are many opportunities for new integrated devices with expanded features.

References

[1]   O.N. Tufte and G.D. Long, "Silicon diffused element piezoresistive diaphragm,", J. of Appl. Phys., vol. 33, (1962) p. 3322.

[2]   H.C. Nathanson and R.A. Wickstrom, "A resonant-gate silicon surface transistor with high-Q band pass properties", Appl. Phys. Lett., 7, (1965), p 84.

[3]   R.T. Howe and R.S. Muller, "Polycrystalline and amorphous silicon micromechanical beams: annealing and mechanical properties", Sensors and Actuators, 4, (1983), pp 447-454.

[4]   L-S. Fan, Y-C. Tai and R.S. Muller, "Pin joints, gears, springs, cranks and other novel micromechanical structures", Proc. Transducers 87, Tokyo, (1987), pp 849-852.

[5]   P.J. French, P.T.J. Gennissen and P.M. Sarro, "New silicon micromachining techniques for microsystems", Sensors and Actuators, A 61, (1997), pp 652-662

[6]   K.A. Shaw and N.C. MacDonald, "Integrating SCREAM micromachined devices with integrated circuits", Proceedings IEEE MEMS, San Diego, USA, 11-15 February 1996, pp 44-48.

[7]   M. de Boer, H. Jansen and M. Elwenspoek, "The black silicon method V:A study of the fabricating of moveable structures for micro electromechanical systems", Proc. Transducers 95, Stockholm, Sweden, (1995), pp 565-568.

[8]   H. Ohji, P.T.J.Gennissen, P.J.French and K.Tsutsumi, "Fabrication of beam-mass structure using single-step electrochemical etching for micro structures (SEEMS)", J. Micromech. Microeng., 10, (2000), pp 440-444.

[9]   H. Ohji, P.J. French and K. Tsutsumi, "Fabrication of mechanical structures in p-type silicon using electrochemical etching", Proc. Transducers 99, Sendai, Japan, (199), pp 1086-1089.

[10]   M.A. Blauw, T. Zijlstra, R.A. Bakker and E. van der Drift, "Kinetics and crystal orientation dependence in high aspect ratio silicon dry etching", J. Vac. Sci. Techn. B 18, 3453 (2000).

[11]   G. Craciun, M.A. Blauw E. van der Drift, P. M. Sarro and P.J. French, "Temperature influence on etching deep holes with SF6-O2 cryogenic plasma", Journal of Micromechanics and Microengineering 12, 390 (2002).

[12]   F. Lärmer and A. Schilp, "Method of anisotropically etching silicon", US Pat. Nr. 5501893.

[13]   M.A. Blauw, T. Zijlstra and E. van der Drift, "Balancing the etching and passivation in time-multiplexed deep dry etching of silicon", J. Vac. Sci. Techn. B 19, 2930 (2001).

[14]   M.A. Blauw, G. Craciun, W.G. Sloof, P.J. French and E. van der Drift, "Advanced time-multiplexed plasma etching of high aspect ratio silicon structures", J. Vac. Sci. Techn. B 20, 3106 (2002).

[15]   M. Chabloz, Y. Sakai, T. Matsuura and K. Tsutsumi, "Improvement of sidewall roughness in deep silicon etching", Microsystems Technologies, 6, (2000), pp 86-89.

[16]   www.mdl.sandia.gov/Micromachine

[17]   J.H. Smith, S. Montague, J.J. Sniegowski, J.R. Murray, R.P. Manginell and P.J. McWhorter, "Characterisation of the embedded micromachined device approach to the monolithic integration of MEMS with CMOS", Proceedings SPIE Micromachining and Microfabrication Process Technology II, Austin, Texas, USA, October 1996, vol 2879, pp 306-314.

[18]   Y.B. Gianchandani, M. Shinn and K. Najafi, "Impact of long high temperature anneals on residual stress in polysilicon", Proc. Transducers'97, Chicago, USA, June 1997, pp 623-624.

[19]   B.P. van Drieënhuizen, J.F.L. Goosen, P.J. French, Y.X. Li, D. Poenar and R.F. Wolffenbuttel, "Surface micromachined module compatible with BiFET electronic processing", Proceedings Eurosensors 94, Toulouse, France, September 1994, p 108.

[19]   G.K. Fedder, S. Santhanan, M.L. Read, S.C. Eagle, D.F. Guillou, M.S.-C. Lu and L.R. Carley, "Laminated high-aspect ratio microstructures in a conventional CMOS process ", Proceedings MEMS 96, San Diego, USA, Feb 1996, pp 13-18.

[21]   C. Hierold, A. Hilderbrandt, U. Näher, T. Scheiter, B. Mensching, M. Steger and R. Tielert, "A pure CMOS surface micromachined integrated accelerometer", Proc. MEMS 96, San Diego, USA 1996, pp 174-179.

[22]   L.S. Pakula, H. Yang, H.T.M. Pham, P.J. French and P.M. Sarro, "Fabrication of a CMOS compatible pressure sensor for harsh environments", Proc. IEEE MEMS 2003, Kyoto, Japan, January 2003, pp 502-505.

Evaluation of Glass Frit Binder Materials for use in Thick-Film Magnetostrictive Actuators

N J Grabham, S P Beeby and N M White

Department of Electronics and Computer Science, University of Southampton, Southampton, SO17 1BJ, UK

Abstract: This paper presents research carried out at the University of Southampton into the development of a magnetostrictive thick-film material suitable for use with silicon micro-machined devices. This form of magnetostrictive material has previously been deposited onto alumina substrates and this paper reports further work on migrating the technology onto silicon. The evaluation of two alternative glass frits for use as the binder within the thick-film is reported. The correct choice of the binder material is important because it is responsible for binding the active material within the film also adhering the film to the substrate. A series of tests have been applied to samples fabricated using various glass frits to assess their mechanical properties and suitability for use in micro-actuator applications.

1. Introduction

Magnetostrictive materials exhibit a dimensional change under the influence of a magnetic field. As the actuation is achieved in a non-contact manner with this material, it is not necessary to provide physical connections to the sample, as required with, say, piezoelectric materials. The potential for wire-free operation is advantageous in the field of MEMS devices, some of which have been fabricated using thin-film techniques to deposit a range of magnetostrictive materials. An example of such a device is the micropump described by Quandt and Seemann [1].

One of the disadvantages of thin-film deposition is the relatively low deposition rate, which tends to produce films typically less than 1μm in thickness. Thick-film technology is capable of producing films up to several hundred μm thick. All film technologies, however, result in the production of layers having different physiochemical properties from that of the bulk material.

Thick-films are comprised of three component parts: An active material to give the film its desired properties; A binder to bind the active material together and to the substrate; And a organic vehicle to give the paste the required viscosity for printing.

The magnetostrictive thick-film pastes used in this work contain the giant magnetostrictive material Terfenol-D, in a powdered form, as the active material. Two alternate glass frits have been evaluated for use as the binder material within the film. To improve the structure of the fabricated thick-film, an inert filler material is added to the paste to reduce voids. The active material is the same as that used in the fabrication of epoxy bonded magnetostrictive composites, such as those described by Duenas *et al* [2].

2. Fabrication

To produce the magnetostrictive pastes used in this work, the following quantities of dry materials were used: 15g of sieved Terfenol-D powder (mean particle size less than 160μm); 6g of alumina powder (inert filler); and 6.75g of glass frit binder. Sieved Terfenol-D powder is used because the material supplied by the manufacturer includes particles that exceed 300μm and are too large for compatibility with the screens used in the printing process. Use of over-size particles in the paste causes clogging of the screen, which results in incomplete prints. The preparation of the pastes, and the subsequent printing and firing of the thick-films, are achieved using the same techniques as those used for printing magnetostrictive thick-films onto alumina substrates as described by Grabham *et al* [3]. With the exception that for this current work, the alumina substrates have been replaced with 100mm diameter, 530μm thick, microelectronic grade silicon wafers. These have a 1μm thick layer of thermally-grown silicon dioxide on their surface to improve the chemical bonding between the thick-film and the silicon substrate.

To produce the individual test samples, once the firing process has been completed, the 100mm diameter wafers are sawn using a standard diamond wafer saw. This yields samples that are ready for the measuring process, having overall dimensions of 12.8mm wide by 88mm long. The thick-film region is 11mm wide by 45mm long, centred in the width of the substrate, and with the end of the thick-film situated 32mm from the unsupported end of the beam. The thick-films have a nominal thickness of 150μm after firing.

3. Choice of Binder Material

Two different glass frits were evaluated for use as the binder material in this work. The glass frits used were Corning CF7575, a lead-zinc-borosilicate devitrifying glass; and Ferro EG2760, a thick-film passivation material. The batches of magnetostrictive pastes prepared for these experiments differed only in the glass frit used, with the fabrication processes kept consistent between batches so as to ensure that any behavioural changes were only due to the properties of glass matrix.

4. Measurement Techniques

To assess the effect of the binder material on the mechanical properties of the magnetostrictive thick-films, a number of tests were carried out. These are described below.

4.1 Tape and Scratch Tests
These are standard tests that are widely used within the hybrid microcircuit industry. In the tape test, a strip of adhesive tape is placed across the surface of the fired thick-film, left for several seconds and then removed. Any material on the surface of the thick-film

that is poorly bonded will come away with the tape and is clearly evident from observation by eye. In the scratch test, a metal stylus is used to scratch the thick-film off the substrate. This tests the adhesion between the thick-film and the substrate. The amount of force required to mark the surface of the thick-film can be used as an indication of its strength.

4.2 Visual Inspection
The surfaces of the fired thick-films and, after cleaving, their cross-sections were inspected using a scanning electron microscope (SEM) to enable the structure of the fired films to be examined. This allows the distribution of the active and filler materials to be observed and the presence of any inclusions to be identified.

4.3 Young's Modulus
The Young's modulus of the thick-films were determined by loading tests and finite element analysis (FEA). The FEA approach was used so that the anisotropic nature of the silicon substrate can be taken into account. The use of computational modelling also permits samples with non-trivial geometry to be readily studied.

The loading tests require the sample to be supported in a cantilever beam configuration, which is then loaded by adding masses to the free-end. The deflection of the beam is recorded for increasing load. The results are then used to calculate a deflection for a given arbitrary loading. This is then input into the FEA model and the Young's modulus of the thick-film is adjusted until the computed deflection matches that corresponding to the chosen arbitrary loading.

4.4 Frequency Response
The frequency response of the magnetostrictive samples was recorded and this can be used, in conjunction with FEA modelling, to determine the net magnetostriction of the film under test. The frequency response is achieved by actuating a beam containing the film within a magnetic field and measuring the displacement of the free-end of a cantilever beam. An optical interferometer, based on one used to measure micromachined devices [4], is used to record the deflection. This can be used as a metric to compare the performance of different films, subject to them having similar dimensions and substrate material. The deflection can also be used with FEA modelling to enable a value for the net magnetostriction of the thick-film to be determined.

5. Experimental Results

5.1 Tape and Scratch Tests
The scratch and tape tests were performed on samples of both thick-films. In the case of both thick-film materials, a small amount of surface material was removed during the tape test and a moderate force was required to mark the thick-film using a metal stylus. In the case of the tape test, slightly less material was removed from the samples containing the EG2760 glass frit than that observed for the CF7575 based film. This suggests that the bonding of the material within the EG2760 based film is superior to that of the CF7575 based film. As the results from the scratch tests were similar, no comparisons between the two glass frits could be drawn from those tests.

5.2 Visual inspection
The surfaces and cross-sections of the thick-films were examined using a SEM and both types of film were found to exhibit similar structure. This indicates that the choice of

binder material does not affect the distribution of the active material and filler particles within the fired film. A SEM image of a typical thick-film surface is shown in figure 1. The larger, angular, Terfenol-D particles can be seen to be surrounded by the smaller alumina particles, which serve to fill in voids within the thick-film that are present because of the irregular shape of the larger Terfenol-D particles.

5.3 Young's Modulus

The Young's modulus was determined for samples of both film types using the loading technique outlined previously. The CF7575 based thick-film was found to possess a modulus of 33.5GPa, whilst the film containing EG2760 glass frit exhibited a lower modulus of 22GPa. The higher modulus of the CF7575 based film may suggest that it is more rigidly bonded as the resulting material is stiffer. The actual modulus of the glass frits, however, is not known, so the effect of this on the overall modulus of the structure cannot be quantified.

5.4 Frequency Response

The frequency response of the samples was observed for frequencies around the mechanical resonance of the sample (approximately 110Hz for an unloaded sample), when mounted in a cantilever configuration as described earlier. The excitation magnetic field was kept at the same strength for all tests and was measured to be 5.19kA/m. The frequency response measurements were performed for both types of thick-film under various pre-loading conditions, and the results then examined.

The initial pre-loadings studied were self-mass loading, where the loading is solely provided by the mass of the beam itself, in both compressive and tensile loading arrangements. Further to these experiments, the effects of increasing a compressive preload were also examined, this was achieved through the addition of mass to the free end of the sample. This additional mass was formed using small squares of alumina, which does not interfere with the magnetic field around the sample.

5.5 Effects of Tensile and Compressive Loading

The effect of using both tensile and compressive self-mass loading was investigated initially. It was found that the application of compressive loading had the effect of increasing the observed deflection from that recorded for the same sample under tensile loading. A comparison of the difference between these two loading conditions can be seen in the graph shown as figure 2, which shows the frequency responses obtained for both tensile and compressive loading of the same sample on common axis. These results lead to further investigation of compressive loading for both types of thick-film.

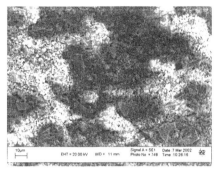

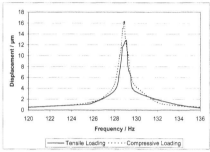

Figure 1: Typical surface of a magnetostrictive thick-film

Figure 2: Effect of changing from tensile to compressive loading on thick-film containing CF7575 glass frit

5.6 Effects of Increasing Compressive Loading

To investigate the effects of increasing the compressive loading of the thick-film, frequency responses were recorded for each of the samples with the loading varied between zero additional mass, in which case the sample's own mass produces compressive loading, and 700mg of additional mass. The loading mass was increased in 100mg steps.

For both binders it was observed that the application of additional loading caused an increase in the deflection observed at the structure's resonant frequency. This increase is present in both cases until a loading of approximately 600mg is reached. For loading in excess of this value, the observed deflection was seen to reduce for both types of thick-film under test. As these measurements were made at the resonant frequency of the test samples, it is necessary to take the Q-factor of the structure into account and compute an equivalent static deflection. The Q-factor is obtained from the individual responses for each of the loading conditions. The resulting equivalent static deflections, normalised to the deflection at zero additional loading, are plotted in figure 3 for the thick-film containing CF7575 glass frit, and figure 4 for that containing the EG2760 glass frit. The results from the thick-film containing the CF7575 glass frit show that for additional loading in the range 300-400mg the equivalent static deflection is increased by approximately 60% from that observed for zero additional loading. For loading over 400mg, the increase in static deflection starts to reduce. Such a marked improvement is not shown by the thick-film incorporating the EG2760 glass frit as the maximum increase is of the order of 30%, which is achieved with a loading of 500mg after which the deflection starts to reduce.

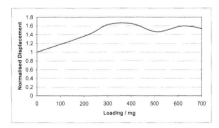

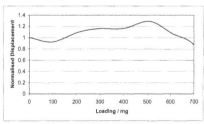

Figure 3: Normalised static deflection of thick-film containing CF7575 glass frit

Figure 4: Normalised static deflection of thick-film containing EG2760 glass frit

6. Discussion of Results

6.1 Advantage of Compressive Loading Over Tensile Loading
The effect of changing from tensile to compressive load, which was shown in figure 2, is to increase the observed deflection. It is proposed that this is attributable to the compressive loading providing a pre-load (or pre-stress) to the thick-film, thereby causing the particles within the film to be in closer contact. This in turn means that when the active Terfenol-D particles undergo a dimensional change, upon the application of the actuating magnetic field, the strains produced by the particles are better coupled into the composite as a whole. This results in an increase in the net magnetostriction of the thick-film, and hence an increase in the observed deflection.

6.2 Results of Increased Compressive Loading
The effect of increasing the compressive loading, up to a point, has been seen to further increase displacements recorded for both film types, both in the dynamic values and also in the computed equivalent static results. These increases are attributed to the structure becoming more compact and dense, thereby leading to improved coupling of the produced strains into the composite as a whole. As the loading continues to increase, a point is reached after which the benefits of pre-stressing the thick-film are outweighed by the increased loading that the material has to work against, and therefore the resulting deflection starts to reduce.

7. Conclusions

This paper has presented results of a study into the suitability of two glass frits, Corning CF7575 and Ferro EG2760, for use as the binder material in a magnetostrictive thick-film material that is suitable for deposition onto silicon substrates pre-coated with a 1μm thick interface layer of silicon dioxide. Of the two glass frits tested, the Ferro EG2760 was found to produce a thick-film, that when deposited onto silicon substrates and excited using a magnetic field of 5.19kA/m, exhibited greater actuation than comparable thick-films containing the Corning CF7575 glass frit as the binder material. It has also been seen that the application of a compressive pre-stress to the magnetostrictive thick-film can be used to increase the observed deflection, and thus the net magnetostriction of the thick-film. Pre-stress is often applied to bulk magnetostrictive materials, where a similar effect on the net magnetostriction is seen. In the case of the thick-film containing the Ferro EG2760 glass frit as the binder, a loading of 500mg was found to give the largest increase in the calculated equivalent static deflection.

Acknowledgements

The authors wish to thank the Engineering and Physical Sciences Research Council (EPSRC) for their financial support under grant number GR/R43327. We also gratefully acknowledge the support and assistance given to us by Newlands Technology Ltd.

References

[1] E. Quandt, K. Seemann, Magnetostrictive Thin Film Microflow Devices, Proceedings of Micro Systems Technologies 96, (1996), 451-456.
[2] T.A. Duenas, L. Hsu, G.P. Carman: Magnetostrictive Composite Material Systems Analytical/Experimental, Materials Research Society Symposium Proceedings, Vol. 459 (1997), 527-543.
[3] N.J. Grabham, S.P. Beeby, N.M. White: The formulation and processing of a thick-film magnetostrictive material, Measurement Science and Technology, Vol. 13 (2002), 59-64.
[4] M.V. Andres, K.H.W. Foulds, M.J. Tudor: Analysis of an interferometric optical fibre detection technique applied to silicon vibrating sensors, Electronics Letters, Vol. 23 (1987), 774-775.

Section 3: Sensing Materials
Paper presented at Sensors and their Applications XII, September 2003
©2003 IOP Publishing Ltd

A Sensor System for Oil-Water Separators: Materials Considerations

J M Hale[1], T Dyakowski[2], A J Jaworski[3], N M White and N R Harris[4]

[1] School of Mechanical & Systems Engineering, University of Newcastle upon Tyne, UK
[2] Department of Chemical Engineering, UMIST, Manchester, UK
[3] School of Engineering, The University of Manchester, UK
[4] Department of Electronics and Computer Science,
The University of Southampton, UK

Abstract: The development of a "dipstick probe" for measuring the position of phase interfaces in high temperature and pressure oil/water separators is described. This uses dual modality (capacitance and ultrasonics) measurement for this difficult application.

The issues of material selection for use in the aggressive high temperature environment are presented. In particular the problem of sensor materials was overcome by the use of thick-film and ceramic materials for all sensor components. A novel combined ultrasonic and capacitance thick-film sensor is described.

Finally, the construction of the probe is described, together with the test programme to be undertaken.

1. Introduction

When crude oil is extracted from the ground it is not in a state to be refined into the desired products of heavy oils, gasoline, etc. It is usually mixed with water, contains gas and holds entrained particulates. These phases have to be separated out so that the water can be cleaned and disposed or reinjected, and the oil and gas can be taken off for further processing.

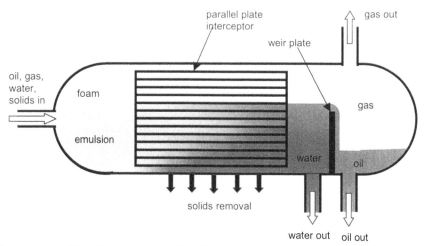

Figure 1.   Schematic representation of an oil separator vessel

The gas and water are progressively separated from the oil as the mixture flows from left to right across the plate assembly.

This separation is not a trivial matter, not least because the process fluids may be at high temperature and pressure. It is done in a "separator vessel", a pressure vessel in which the oil/water/gas mixture is allowed to settle into its various phases, encouraged by chemical additives. The vessel contains a series of plates and weirs to separate the phases as they flow from left to right. The arrangement is shown schematically in Figure 1.

The process engineer faces a serious problem in operating these vessels, in that there is currently no instrument to tell him where the phase interfaces lie. This is partly because of the difficulty of making measurements in the aggressive environment, but also because the interfaces are not clear; they are a graded emulsion of oil and water, and a foam of oil and gas, respectively.

This is the problem that has been addressed in the research described in this paper.

2. Background

The UMIST/Manchester members of the team have considerable experience of process instrumentation, including capacitance tomography for imaging phase distributions in chemical equipment [1]. They have worked previously with Newcastle to incorporate advanced composite materials into their measurement systems and have produced a "dipstick" incorporating a column of capacitance sensors.

The dipstick was tested in a separator at a low temperature oil production facility and found to give good results until the sensor electrodes became fouled with oil residues [2]. However, it was shown that capacitance measurement alone is not sufficient to characterise the "fuzzy" interfaces (emulsions and foams) under all conditions and preliminary tests indicated that ultrasound would provide the necessary added measurement modality.

It is common for oil wells to operate at high temperature and pressure. It was thus decided to design for a working environment of hydrocarbons and water at 150°C and 150 bar (15MPa). Previous work at Newcastle [3] has shown that polymer composites (glass reinforced plastics or GRP) are not stable in these conditions and so the use of multi-layer printed circuit board to form the electrodes (the technique used previously) would not be viable. The alternative proposed was to use thick-films deposited onto ceramic tiles. The University of Southampton has considerable experience of the design and fabrication of thick-film sensors, although the demanding specification of high temperatures within a liquid environment was an area that had not previously been comprehensively researched. In view of their expertise in sensor fabrication, they were invited to complete the team.

3. Sensor Configuration

It was decided to develop a dual modality advanced dipstick probe incorporating both capacitance and ultrasonic sensors, and using high power ultrasonic cleaning of the sensor surfaces. One sensor unit is shown schematically in Figure 2.

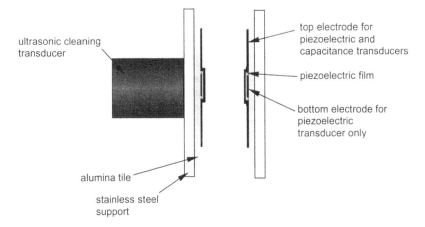

*Figure 2. **Advanced dipstick probe: schematic representation of one sensor location***

Thick-film materials used for combined ultrasonic and capacitance sensors are shown (the tile and film thicknesses are exaggerated for clarity). A high power ultrasonic actuator is used to clean the sensor surfaces.

The capacitance and ultrasonic transducers are wholly realised in thick-film form. The bottom electrode of the ultrasonic transducers defines their effective size; small devices in the region of 10×10mm ($10^{-4}m^2$) have been found to be most effective in this application. Larger electrodes of 35×35mm ($1.2×10^{-3}m^2$) were required for the capacitance sensors. This large area has no effect on the size of the ultrasonic transducers.

These combined transducers are deposited onto alumina tiles, which are mounted on stainless steel plates to give the assembly stiffness. The transducer pairs face one another across a 20mm gap through which the process fluid flows.

The high power ultrasonic transducer used for cleaning the sensor faces is mounted on the back face of one of the stainless steel support plates. The ultrasound excites the adjacent sensor tile and couples into the process fluid, through which it propagates to the other sensor tile. In this way one high power transducer cleans both tiles.

4. Sensor Materials Testing

The environment, water and hydrocarbons at high temperature and pressure, is very aggressive. There was reason to believe that thick-films would be able to withstand it, but they had not been used in this way before and it was necessary to test them at an early stage in the work. To this end, thick-film specimens of increasing complexity were made at Southampton, characterised at Manchester, exposed at Newcastle and finally retested at Manchester. Various materials were tried and finally the combination described below was established as a robust configuration.

The breakthrough made in this work was the development of thick-film piezoelectric transducers that can be used as ultrasonic actuators as well as sensors. The conventional view is that thick-film piezoelectric materials of this type, known as 0-3 composites [4], make adequate ultrasonic sensors but poor actuators. We had expected to have to find a way to use conventional ceramic piezoelectric actuators, possibly mounted on the back of the support plate. The arrangement described here, which has been found to work well in this application, has simplified the sensor design considerably.

5. Probe Design

Apart from materials considerations, the main instrumentation concern was the wiring. The high temperature precludes any possibility of locating electronics inside the probe close to the sensors. This means that all sensor signals must be taken several metres in an unamplified state to the signal conditioning equipment. This is particularly critical for the capacitance sensors which will be required to detect changes of a few pico farads, and the problem is exacerbated by the ultrasonic signals which will have to be carried in parallel cables. This makes it vitally important to screen all leads properly.

Fortunately the specification of 150°C is comfortably within the working range of PTFE *(Teflon)*, and PTFE insulated signal wire, both screened and unscreened, is readily available. However, PTFE is notoriously porous and would absorb the process fluid if exposed to it. It was thus decided to design the probe in the form of an inverted pressure vessel with the interior, containing all wiring, dry and at atmospheric pressure, and the outside exposed to the process environment.

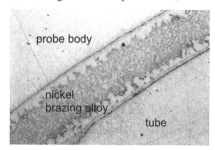

Figure 3. Micrograph of a typical nickel brazed bond between a stainless steel tube and probe body

The whole probe had to be designed to be inserted through a "6 inch flange" in the circular shell of a 3m diameter separator vessel. It will hang vertically from the flange across the diameter of the vessel. The penetration is actually smaller than the name appears to imply, and an envelope only 100mm diameter is available for the probe.

It was decided to carry the wires in stainless steel tubes, which, being relatively small diameter, are easily capable of withstanding the pressure of 150 bar (15MPa). These tubes also provide the strength and stiffness for the probe. The main difficulty was terminating these tubes because conventional compression fittings are too bulky for the limited space available. The alternative of brazing the tubes into place was investigated and found to be problematical as normal brazing materials would be attacked by the high temperature water. A special purpose nickel based brazing compound was identified, but this has to be furnace brazed in a vacuum. Fortunately a large vacuum furnace facility, normally used for fabricating steam turbine blades, was found in Newcastle at the Siemens Parsons Works. Pressure tests and a metallurgical examination were conducted and the process found to give an excellent bond with no evidence of environmental degradation. Figure 3 shows a typical micrograph of a section through a tube bond, taken at the Corrosion Research Centre at Newcastle University.

By this means, tubes can be packed very close together so that it would be possible to carry each signal wire in its own small bore tube if this became necessary in the future to obtain ever greater noise immunity. However, the current design uses 9mm diameter tubes, each carrying up to eight PTFE insulated screened wires.

The length of the probe is 3m in order to monitor the whole diameter of the vessel. The prospect of transporting such a long slender instrument to a remote oil production facility was not appealing, so it was decided to produce a modular design. The module length was set at 750mm, dictated by the size of the Siemens vacuum furnace.

Modules are connected by spigot joints, sealed with high temperature *Viton* O rings and retained by screw collars. In fact, the screw collars are redundant in operation as the high pressure in the vessel would hold the joint together. They are only required to ensure integrity during loading of the probe into the vessel.

A prototype probe section, made to test the brazing and sealing techniques, is shown in Figure 4. This has a single 9mm diameter tube and eight tubes of 3mm diameter, six of which are packed in a bundle at 4mm centres. This test showed both the difficulty of handling a large number of small tubes, but also the feasibility of doing so should it prove necessary.

Electrical penetration to connect the sensor electrodes to the signal wires inside the probe body presented a problem. Eventually a company specialising in this type of work, Wesley Coe Ltd, was found. They have supplied small stainless steel units with nickel plated copper through pins bonded in place with glass. The glass provides the insulation and the body will provide the required electrical screening.

Figure 4. Probe component showing close packing of furnace brazed tubes

These units are intended for mounting by electron beam welding. However, it has been found that for this application, where the pressure is always in one direction, it is possible to use ordinary epoxy adhesive to seal the units into a carefully designed seat. The seal was tested in water at 150°C and 150 bar, and although the epoxy must have been greatly softened at the operating temperature, particularly in view of the reduction in T_G (glass transition temperature) that will have occurred due to water absorption, it was found that a good seal was maintained.

6. Sensor Testing

Specimen sensors have been tested by exposure to water at 150°C and 150 bar (15MPa), which is the more hostile phase of the proposed environment. These were tested as ultrasonic sensors before and after exposure to determine its effect. By this means the combination of conductor and piezoelectric thick-film materials given above has been identified and shown to be viable in this application.

At the time of writing a proof of concept model probe is being built to test all the components in combination. When this is ready it will be tested in an autoclave at Newcastle, first for pressure integrity and then, using the capacitance and ultrasonic instrumentation from Manchester, for viability of the sensors in simulated working conditions. Results of these tests should be ready for presentation at the conference.

7. Materials

The final materials selected were:

All parts of the probe body in contact with the process environment	316 stainless steel
brazing compound	Johnson - Matthey HTN2 (BS EN1044)
substrate for thick-film transducers	alumina
conductive thick-film for sensor electrodes	ESL 8836 gold paste
piezoelectric thick-film for ultrasonic transducers	PZT-5A/glass thick-film paste made at Southampton University (PZT powder supplied by Morgan Electroceramics Ltd)
O rings	*viton*
wiring insulation	PTFE

8. Field Trials

Finally, the full scale probe, comprising four 750mm sections and all necessary instrumentation, will be transported to a live production facility in Canada or Norway (the site will depend on availability). There it will be assembled and inserted into a separator vessel and tested over several weeks.

References

[1] Dyakowski T, Jeanmeure L F C & Jaworski A J 2000 Powder Technology 112 174-192

[2] Jaworski A J, Dyakowski T & Davies G A 1999 Measurement Sci and Tech 10 L15-L20

[3] Hale J M, Gibson A G & Speake S D 2002 J of Composite Materials 36 no3 257-270

[4] Newnham R E, Skinner D P & Cross L E 1978 Materials Research Bull 13 525-536

Metallic Triple-beam Resonant Force Sensor with Thick-film Printed Piezoelectric Vibration Excitation and Detection Mechanisms

T Yan, B E Jones⁺, R T Rakowski, M J Tudor[1], S P Beeby[1], N M White[1]

The Brunel Centre for Manufacturing Metrology, Brunel University, Uxbridge, Middlesex UB8 3PH, UK. ⁺E-mail: barry.jones@brunel.ac.uk
[1] Department of Electronics and Computer Science, University of Southampton, Highfield, Southampton, Hampshire, SO17 1BJ, UK.

Abstract: A metallic resonant force sensor consisting of a triple beam tuning fork structure with thick film piezoelectric elements to excite and detect the vibrations is presented. The resonating element of the sensor was fabricated in 430S17 stainless steel with a length of 15.5 mm and width of 7 mm. The resonator has an operational resonance frequency at 6.9 kHz under 15 N pretension. The pretensioned sensor was further loaded up to about 50 N, exhibiting a sensitivity of 13.0 Hz/N and a Q-factor greater than 1400 in air. The fabrication of the device involves combination of a photochemical etching technique and a standard screen-printing process, which allows mass production of the device at low costs.

1. Introduction

Resonant sensors have been used in a wide range of sensing applications, such as load, pressure, torque and fluid flow characteristics. The key element of these sensors is the resonator, an oscillating structure, which is designed such that its resonance frequency is a function of the measurand. The most common sensing mechanism is for the resonator to be stressed as a force sensor. The applied stress effectively increases the stiffness of the resonator structure, which results in an increase in the resonator's natural frequency. The resonator provides a virtual digital frequency output, which is less susceptible to electrical noise and independent of the level and degradation of transmitted signals, offering good long-term stability. The frequency output is compatible with digital interfacing, requiring no analogue-to-digital conversion and therefore maintaining inherent high accuracy and low cost. Resonator sensors often have a high mechanical quality factor (Q-factor), which leads to a high resolution of frequency and hence high sensitivity. A high Q-factor also implies low energy losses from the resonator and therefore low power requirements to maintain the resonance, and better noise rejection outside the resonance frequency bandwidth, which simplifies the operating electronics. Resonant sensors have been made in a wide range of types, sizes and materials [1]. This paper describes a metallic triple-beam tuning fork resonant force sensor with screen-printed thick film lead zirconate titanate (PZT) drive and sense elements [2] and presents initial results from the device.

2. Design and Fabrication

Figure 1 shows a photograph of the metallic resonant force sensor with bonding pads for applying forces. The sensor consists of a suspended, well-balanced, triple beam vibrating structure maintained at resonance by a closed-loop feedback-control electronic circuit. The sensor is designed to oscillate in a differential mode where the central beam vibrates in anti-phase with the outer beams to minimise mechanical energy losses from the resonator. Figure 2 shows the operational modal behaviour of the sensor modelled by finite element analysis (FEA).

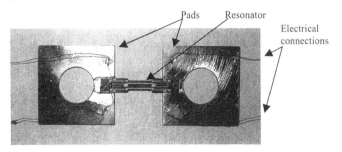

Figure 1. Photograph of the metallic triple-beam resonant force sensor.

Figure 2. Operational modal behaviour of the sensor modelled by FEA.

The resonator triple-beam tuning fork was photochemically etched from a 0.5 mm thick 430S17 stainless steel thin wafer, with a top pattern to define the layout of the resonator and a bottom pattern to etch in a standoff distance in the section of resonating element. The resonating element has a length of 15.5 mm, a thickness of 0.25 mm and beam widths of 2mm for central tine and 1 mm for outer tines. The distance between the beams is 0.5 mm. The tuning fork was excited into resonance by a thick-film printed piezoelectric element at one end of the resonator and the oscillation was sensed by a second thick-film printed piezoelectric element at the other end of the resonator both being on the central beam. The frequency of the detected vibration forms the sensor output and this signal is fed back to the drive mechanism via amplifiers and phase shift circuits so as to maintain the structure at resonance in the required mode.

For preliminary tests the resonator was bonded to a 430S17 stainless steel pad at each end using cyanoacrylate adhesive so that loads can be applied. The resonator presented here also has thick film piezoelectric elements printed on the outer beams for purposes of evaluation and comparison of different drive and pickup configurations.

3. Experimental Results

The resonant force sensor has been tested both in open-loop and closed-loop configurations. The open-loop test identifies the natural frequencies of the tuning forks and their mechanical quality factors and confirms successful operation of the device and the drive and sense mechanisms. The resonator has been designed to have a maximum load capacity of 100 N with a safety factor greater than 2. During tests, the resonator in air was pre-tensioned by 15 N from a hanger structure and further loaded and unloaded between 0 N and about 50 N. Figure 3 shows a typical open-loop frequency response of the TBTF over a frequency range of 2-10 kHz under a load of 25 N. A dominant resonance at frequency 7.2 kHz can be seen with two other resonances just visible at frequencies 4.3 kHz and 5.5 kHz. According to FEA predictions, these resonances correspond to the third, the first and the second vibration modes of the resonator respectively. In mode one, the three beams vibrate in phase. In mode two, the central beam does not vibrate while the outer beams oscillate at a phase of 180 degree with respect to each other. In mode three, i.e. the differential mode, the central beam oscillates in anti-phase with the outer beams. The differential mode is far more dominant than the others as this is due to the favourable dynamic structure balance associated with the mode. The Q factor of the experimental resonator at the load of 25 N for the differential mode was measured to be 2180, which is excellent when compared to a Q-factor of 140 of a metallic double beam resonant force sensor vibrating in air [3] or a Q-factor of 400 of a silicon triple beam resonant force sensor operating in air [4] or the Q-factors of other resonators in air [5,6]. The Q-factor of the resonator with no load has been measured to be as high as 3100 in air.

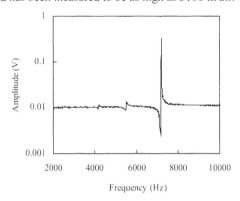

Figure 3. Amplitude-frequency response of the resonant sensor.

Figure 4 shows the response of the sensor to variable load and Table 1 summarises the typical characteristics of the sensor. The sensor has a sensitivity of 13.0 Hz/N. It should be noted that the performances of the sensor, in terms of repeatability, linearity, stability and hysteresis, are very much limited by the use of an inadequate weight-hanging test rig, which takes time to stop the swinging. Suitable structural mechanisms with appropriate clamping methods will be needed to incorporate the resonator to sensor structures in order to obtain the full potential of the resonator performances [7-9]. The use of better spring materials such as 17-4H stainless steel and beryllium copper to etch the tuning fork would also improve the general performance of the sensor. Printing of thick film piezoelectric elements onto these materials to fabricate the resonators and employing of suitable structural mechanisms with appropriate clamping methods to embed the resonators for load cells, weighing machines and torque sensors are currently under investigation.

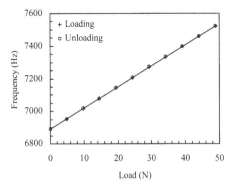

Figure 4. Load response of the force sensor.

TABLE 1. Characteristics of the force sensor.

Parameter	Value	As % of span (0 - 49 N)
Natural frequency at zero load	6890 Hz	
Mechanical Q	>1460*	
Frequency shift	635 Hz	
Sensitivity	13.0 Hz/N	
Max Hysteresis	2 Hz	0.3
Repeatability	5 Hz	0.8
Max non-linearity	2.5 Hz	0.4
Stability (over 30 minutes)		
zero load	0.4 Hz	0.07
half load	0.4 Hz	0.07
full load	0.5 Hz	0.07

*lowest Q-factor over the applied load range.

A feedback-control electronic circuit has also been designed to operate the sensor in a closed-loop configuration. The system comprises a PZT sensing element, a charge amplifier circuit, followed by a digital 90-degree phase shift circuit and a second stage amplification circuit all fabricated on a single circuit board. The output from the second stage amplification was fed back to the other PZT element for driving the vibrations. In such a way, the resonator was maintained at resonance in the required differential mode of vibration. Figure 5 shows the digital frequency output from the sensor under a load of 25 N in such a closed-loop configuration, which is easy to be interfaced to sensing instrumentation.

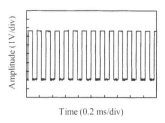

Time (0.2 ms/div)

Figure 5. Digital frequency output from the sensor in closed-loop.

4. Conclusions

The device presented here is the first prototype of its kind, a metallic resonant force sensor using a triple beam tuning fork structure with thick film PZT elements to drive and pickup the resonant vibrations. The prototype sensor has been preliminarily tested both in open-loop and closed-loop configurations and the results are encouraging. The pretensioned sensor exhibited a sensitivity of 13.0 Hz/N over the applied load range up to about 50 N, having a Q-factor greater than 1400 which compares favourably with other reported resonant sensors of similar structures operating in air. The resonator substrate of the sensor was fabricated by a double-sided photochemical etching technique and the thick film PZT elements were deposited by a standard screen-printing process. The combination of these two batch-fabrication processes presents benefits for low-cost mass production of the device. Further work is under way to print the thick film PZT elements onto better spring materials for resonator fabrication and to employ suitable structural mechanisms with appropriate clamping methods to incorporate the resonators for load cell, weighing machine and torque sensor applications.

Acknowledgements

The authors wish to acknowledge the support of EPSRC (Grant GR/R51773) and the industrial collaborators within the Intersect Intelligent Sensing Faraday Partnership

Flagship Project (2002-2005) entitled "Resonant Microsensor Modules for Measurement of Physical Quantities (REMISE)"
(Project website: www.brunel.ac.uk/research/bcmm/remise/).

References

[1] Langdon R. M. "Resonator sensors – a review", J. Phys. E: Sci. Instrum. 18 pp103-115, (1985)

[2] Jones B. E. and White N. M. Metallic resonators. Patent Application GB0302585·5 (2003)

[3] Barthod C., Teisseyre Y., Gehin C. and Gautier G. "Resonant force sensor using a PLL electronics", Sensors and Actuators A 104 pp143-150 (2003)

[4] Th. Fabula, H.-J. Wagner, B. Schmidt, Triple-beam resonant force sensor based on piezoelectric thin films, Sensors and Actuators A 41-42 pp375-380 (1994)

[5] Randall D. S., Rudkin M. J., Cheshmehdoost A. and Jones B. E. "A pressure transducer using a metallic triple-beam tuning fork", Sensors and Actuators A 60 pp160-162 (1997).

[6] Beeby S. P. and White N. M. "Thick-film PZT – silicon micromachined resonator", Electron. Lett. 36 (19) pp1661-1662 (2000)

[7] Nishiguchi Y., Uchiyama S. and Kobayashi M. "Piezoelectric mechanism for converting weight into frequency", U.S. Patent # 4544858, 1 October 1985.

[8] Kirman R. G. and Spencer S. A. "Vibrating force sensor", European Patent #0333377B1, 13 November 1991.

[9] Ford M. W. "A load cell", UK Patent #02265198B, 17 May 1995.

Section 4

Sensor Applications I

Pressure sensitive paint (PSP) measurements for aerodynamic applications

J Hradil, C Davis, K Mongey, D Gray, T Dalton[1], C McDonagh, B D MacCraith

Optical Sensors Laboratory, National Centre for Sensor Research, Dublin City University, Glasnevin, Dublin 9, Ireland.
[1] Stoke's Research Institute, University of Limerick, Ireland

Abstract: Pressure-sensitive paint (PSP) technology is an important new technique, which allows pressure mapping of surfaces under aerodynamic conditions. The pressure profile is obtained by measuring the surface oxygen concentration. Temperature correction is an established problem in PSP measurements. In this work, a temperature-corrected pressure profile is achieved by co-doping an oxygen permeable sol-gel coating with both a temperature (manganese doped magnesium fluorogermanate(MFG) and an oxygen $(Ru(dpp)_3^{2+})$ sensitive luminophore. A luminescence lifetime approach was used and the luminophores were selected to facilitate the use of a single excitation source and single gated, image-intensified camera. The system thus provides a temperature-corrected surface pressure profile.

1. Introduction

Pressure measurements in aerodynamical testing are carried out by applying a series of individual pressure taps or transducers across the surface of a model in a wind tunnel. The number of taps which can be applied is limited. This gives discrete pressure readings resulting in an incomplete pressure map. As well as being an expensive process, there are practical limitations in using pressure taps around thin edges and sharp corners, which tend to be the areas of most interest.

More recently, pressure sensitive paint (PSP) has been used in wind tunnel testing[1]. The PSP technique has many advantages over pressure tap technology. The test surface is coated with a paint containing an oxygen-sensitive dye. The luminescence of the dye is quenched by the oxygen in the airflow around the surface. By imaging the illuminated surface with a CCD camera, the oxygen profile, and hence the pressure profile of the surface, is mapped. PSP technology is cheaper, non-invasive and relatively easy to apply compared to pressure tap technology, as well as providing a continuous pressure distribution compared with measurements at discrete points[2]. Luminescence quenching affects both the intensity and the lifetime of the luminophore and both approaches can be used for PSP applications. Current PSP techniques employ mainly an intensity based approach which is susceptible to drift and aging of both light source and detector as well as effects due to paint inhomogeneities. In this work, a lifetime approach is used which eliminates many of the problems encountered in intensity imaging. Since the lifetime is an intrinsic property of the luminophore, unlike intensity, it is virtually independent of external perturbations.

Both luminescence intensity and lifetime are dependent on temperature, which can fluctuate considerably across a model in a wind tunnel as well as during prolonged testing. Many techniques have been developed to correct for this temperature dependence. Generally, reported temperature correction systems have used separate paints[3], separate light sources and in some cases separate cameras[4]. In this paper, a new approach is reported, which employs a dual-luminophore paint, where both luminophores, an oxygen sensitive complex and a temperature-sensitive phosphor, have overlapping absorption bands, thus enabling the use of a single excitation source. The luminophores have been selected to have temporally separated lifetimes which allows lifetime-based detection with a single CCD camera. A porous sol-gel film is used as the binder for the paint. The key feature of this paper is the demonstration of a single camera, temperature-corrected, lifetime-based PSP system utilising a sol-gel paint.

2. Background

Generally, PSP techniques involve the use of oxygen-sensitiveluminescent dyes. In this work, the widely-used luminescence complex, [Ru II-tris(4,7-diphenyl-1,10-phenanthroline)]$^{2+}$, abbreviated to [Ru(dpp)$_3$]$^{2+}$, was chosen as the oxygen-sensitive luminophore. The luminescence intensity and lifetime of the complex are quenched in the presence of oxygen. Hence these quantities are a measure of the partial pressure of oxygen at a surface. Lifetime sensing minimises effects such as light source and detector drift, changes in optical path due to paint inhomogeneities and drift due to degradation of the dye. This measurement approach, which is independent of excitation intensity, positioning of the test piece and fluctuations due to uneven paint thickness, is used here.

Currently, temperature effects are a major limiting factor in the accuracy of PSP measurements. The most widely-used temperature correction strategy is to use paint containing an oxygen insensitive temperature dependent phosphor. The temperature response of the phosphor is used to correct the oxygen pressure profile for temperature fluctuations over the test surface. Many PSP systems, require two excitation sources and two cameras depending on the spectral properties of the two luminophores. Ideally, the temperature-correction luminophore should be (i) excitable at the same wavelength as the pressure sensitive luminophore, (ii) be insensitive to oxygen (iii) have a decay time which differs by several orders of magnitude from the pressure sensitive luminophore enabling temporal separation of the signals. All of the above criteria are satisfied in the system reported here.

We report a lifetime-based, dual-luminophore, single paint PSP system where the spectral properties of the luminophores have been selected to allow LED excitation with a single source and detection by one camera. To facilitate the recording of both lifetimes using a single gated CCD camera, the luminophores were selected to have lifetimes which differ by several orders of magnitude. The temperature phosphor chosen here is magnesium fluorogermanate (MFG) which has a lifetime of ~ 3ms compared to τ of ~ 5µs for the oxygen sensitive Ruthenium complex.

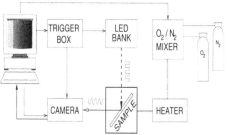

Figure 1 Schematic diagram of PSP measurement system

3. Experimental Details

Figure 1 shows a schematic diagram of the dual lifetime calibration system. The test piece was placed in a flow chamber containing an optically transparent window, into which a controlled mixture of nitrogen and oxygen gas was delivered. The relative ratio and flow rate of gases was computer-controlled via a pair of mass flow controllers C7300 (Unit Instruments, Ireland). Experiments were carried out at constant flow rate of 2 litres per minute. The temperature inside the flow cell was controlled by a gas heater which heated the mixture before it entered the chamber.

The excitation source used was an externally triggered multi-LED bank PRS100B (Photonic Research Syst., UK), which consisted of 100 LEDs emitting at 460nm. The test piece was directly illuminated and the emitted luminescence was collected through a standard objective into a 12 bit image-intensified gated DiCAM Pro CCD camera (PCO, Germany). The camera is capable of exposure times down to 3 ns and the delay time between the triggering pulse and exposure being variable in steps down to 20 ns. All camera parameters were computer-controlled.

The ruthenium complex tris-(4,7-diphenyl-1,10-phenanthroline) ($[Ru(dpp)_3]^{2+}$) was synthesized and purified as described elsewhere[5] and manganese-activated magnesium fluorogermanate 3.5MgO 0.5MgF$_2$ GeO$_2$: Mn (or MFG) was purchased from Meldform Metals Ltd. The sol-gel precursor methyltriethoxysilane (MTEOS) was purchased from Aldrich.

4. Results and Discussion

4.1. Selection of luminophores

The highly emissive ruthenium complex, $[Ru(dpp)_3]^{2+}$, has been widely used in oxygen sensor applications. This complex has strong absorption in the blue-green part of the spectrum ensuring good compatibility with high-brightness blue LED's. The complex is optically stable and is easily incorporated into a variety of binder matrices. Magnesium fluorogermanate (3.5MgO 0.5MgF$_2$ GeO$_2$: Mn) or MFG, is a thermographic phosphor which is commonly used for colour correction of high pressure mercury-vapour lamps[6]. The lifetime of this complex is ~3ms and, as with the ruthenium complex, the absorption band (420nm) has considerable overlap with the blue LED output, enabling the use of a single excitation source.. Emission peaks at 655 nm and 630 nm are spectrally close to the 610 nm emission of the ruthenium complex. The temporally separated lifetimes, 5 µs for the ruthenium complex and 3ms for MFG, allow the use of a single gated camera for the acquisition of both luminescent decays. The MFG lifetime was found to have no oxygen dependence.

4.2. Preparation of the sol-gel paint

Most PSP formulations use polymer-based paints. The most widely used have been silicone polymers due to the high oxygen diffusivity of this matrix[7]. In this work, a single porous sol-gel paint has been used, which is codoped with the oxygen sensitive ruthenium complex and the oxygen insensitive MFG phosphor. Advantages of sol-gel paint over polymers include better control of film composition and low cost fabrication of large area films[8].

The sol-gel paint was prepared by the hydrolysis and condensation of the precursor methyl-triethoxysilane (MTEOS) in an ethanol solution using a water:precursor ratio of 4. The MFG, in powder form, was added to the sol containing the ruthenium complex 90 mins after initial mixing. This allowed hydrolysis and condensation to proceed prior to the addition of the MFG powder and resulted in a relatively homogeneous paint. The final concentrations of $[Ru(dpp)_3]^{2+}$ and MFG in the sol-gel were 0.2 and 5 mass percent, respectively. The relative concentrations of the luminophores were chosen to optimise the signal-to-noise ratio while avoiding agglomeration effects which would affect the measured lifetimes. The paint was spray-coated onto the test pieces with a thickness again chosen to optimise the signal-to-noise ratio. Films were cured overnight at 70°C to accelerate the drying process. The paint can also be cured at room temperature.

4.3 Experimental measurements

The calibration system shown in figure 1 was used to generate oxygen and temperature calibration data for the test pieces. As discussed previously, the spectroscopic properties of the luminophores were selected to allow all measurements to be made in the time domain with one camera. The timing protocol for the camera and LED bank allowed for the collection of luminescence in sequence from both luminophores. For each measurement sequence, the image is integrated usually over 256 triggering cycles and several images are averaged. Typical dual lifetime data are shown in figure 2. The data in this figure were acquired at ambient pressure, and lifetimes for three different temperatures are shown. It is clear from the figure that the lifetimes of the two luminophores are sufficiently separated in time to enable the single camera measurement. Note that the left and right sides of the graph have different scales. The variation of the ruthenium complex lifetime with temperature(left side of figure), seen in figure 2, illustrates the need for temperature correction of the pressure profile. The decay times are calculated by a least squares fit to a single exponential decay, with variable baseline.

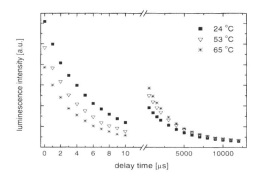

Figure 2 Decay curves for $[Ru(dpp)^3]^{2+}$ and MFG (right side) at ambient pressure

Pressure calibration data were generated by measuring $[Ru(dpp)_3]^{2+}$ lifetimes over the range 0-100% oxygen partial pressure at ambient pressure and over the temperature range 20°C to 65°C. The data in figure 3 represent an 8x8 pixel region of the surface at three representative temperatures. The entire data were fitted to generate numerically a quasi-continuous set of lifetime data over a continuous range of temperature and oxygen pressure. A typical PSP measurement using this single camera system involves acquiring a 2-D $[Ru(dpp)_3]^{2+}$ lifetime image (pressure profile) and an MFG lifetime image (temperature profile) of the test surface. Using the pressure calibration data, the measured MFG temperature dependence and the temperature profile, the pressure profile of the test surface can be temperature-corrected. While the protocol for producing temperature-corrected PSP images using this single-camera system

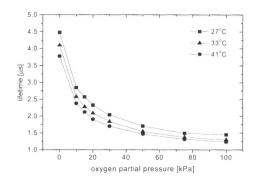

Figure 3 Calibration curve for $[Ru(dpp)_3]^{2+}$ as a function of pressure at three different temperatures

has been established and while satisfactory uncorrected pressure profiles can be generated, currently there are limitations, particularly with respect to temperature measurement, as a result of insufficient signal-to-noise ratio (SNR). The error in temperature, based on the standard deviation in MFG lifetime, is currently ±3°C. Our objective is to achieve an error better than ±1°C. While aspects of the system such as concentration of luminophores and paint thickness have been optimised, there remain a number of strategies which could lead to further improvements in SNR. These include further optimisation of camera parameters such as exposure times and binning protocols as well as image post-processing strategies.

5. Conclusion

A novel technique for the production of temperature-corrected oxygen pressure profiles for PSP applications has been demonstrated. The principal feature of this technique is the use of a single camera and a single excitation source. A new sol-gel based paint formulation has been developed which incorporates both a pressure-sensitive and a temperature-sensitive phosphor. As well as having overlapping absorption bands, requiring a single excitation source, the luminophores have been selected such that their lifetimes can be temporally separated in order to facilitate the use of a single CCD camera. Preliminary calibration data have been generated over the range of temperatures and pressures that would typically occur in wind tunnel testing. While satisfactory pressure profiles can be generated, noise limitations in the system prohibit the production of temperature-corrected profiles to the level of accuracy required. Future work will address this problem.

6. Acknowledgements

This work was supported by Enterprise Ireland.

References

[1]   M. M. Ardasheva, L. B. Nevskii and G. E. Pervushin, *J. Appl. Mech. Tech. Phys.*, **26**, 469-474, 1885.

[2]   B. G. McLachlan and F. H. Bell, *Experimental Thermal and Fluid Science*, **10**, 470-485, 1995.

[3]   L. A. Creswell and M. N. Cripps, *Proc. 8th Annual Pressure Sensitive Paint Workshop*, NASA Langley, Hampton VA (Unpublished)

[4]   G. M. Buck, *AIAA*, 91-104, 1991

[5]   R. J. Watts and G. A. Crosby, *J. Am. Chem. Soc*, **93**, 3184, 1971.

[6]   K. K. Gopinathan, R. Lakshminarayanan, N. Rajaram, M. I. A. Siddiqd and C. C. Suryanarayana, *Indian J. Phys.* **51B**, 423-440, 1977.

[7]   S. Burns and J. Sullivan, *Proc. 16th Int Congress Instrumentation in Aerispace Simulation Facilities (ICIASF), IEEE,*Wright-Patterson AFB, Dyton, OH, **32.1- 32.14**, 1995

[8]   C. McDonagh, B. D. MacCraith and A. K. McEvoy, *Anal. Chem.* **70**, 45-50, 1998.

On-Line Fuel Identification Using Optical Sensors and Neural Network Techniques

Lijun Xu[1], Yong Yan[1], Steve Cornwell[2], Gerry Riley[2]

[1] Centre For Advanced Instrumentation and Control, School of Engineering, University of Greenwich at Medway, Chatham Maritime, Kent ME4 4TB, UK.

[2] Innogy plc, Windmill Hill Business Park, Whitehill Way, Swindon SN5 6PB, UK.

Abstract: This paper presents a novel approach to on-line fuel identification using optical sensors and neural network techniques. A special flame detector containing three photodiodes is used to derive multiple signals covering a wide spectrum of the flame from the infrared to ultraviolet regions through the visible band. Advanced digital signal processing and neural network techniques are deployed to identify the dynamic "finger-prints" of the flame both in the time and frequency domains and ultimately the type of fuel being burnt. A series of experiments was carried out using a $0.5MW_{th}$ combustion test facility operated by Innogy plc, UK. The results obtained demonstrate that this approach can be used to identify the type of fuel being burnt under steady combustion conditions.

1. Introduction

Power plants are increasingly firing a more diverse resource of coals under tighter economic and environmental constraints. Experience has shown that boiler optimization packages can help plant operators to optimize the combustion process and hence improve its efficiency for a given type of fuel [1]. Power stations that burn consistent coal diets have demonstrated that the coal combustion process can be optimized in terms of reduced NO_x emissions and carbon-in-ash levels. However, a power station can have a wide range of coals in its stock but what type of coal is being fired at any moment is often unknown and even unpredictable. The application of the optimization packages is thus seriously limited by the wide variation of the coal diet. Therefore, on-line fuel identification at a power station where a wide range of coals is used would improve the performance of the optimization packages leading to increased combustion efficiency and reduced pollutant emissions.

On-line coal analyzers operating on radiometric, microwave and infrared methods are available on the market [2-3]. Passive tagging techniques have also been adopted for on-line fuel tracking [4]. However, these systems are very expensive and require complex installation as their operations involve either taking samples from the fuel feeding system or detecting tracer particles that have been added into the fuels.

Flame radiation covers a wide range of wavelengths from the infrared (IR) to ultraviolet (UV) through the visible band [5, 6]. As different fuels are of different physical and chemical properties, flames generated by firing such fuels are expected to have different features. Such features may be extracted from the flame signals as unique signatures to identify the types of fuel. As the relationship between the features and fuel type is complex and nonlinear, it is impossible to obtain an explicit relationship between the both. Back-propagation (BP) neural networks with multiple neuron layers and nonlinear transfer functions for neurons, are suited for approximating any functional relationship between its input and output. Therefore, a BP neural network has been used to map the features extracted from the flame signals into individual type of fuel.

In this paper an approach is reported that modifies the existing flame detectors on all power stations and utilizes advanced signal processing and neural network techniques to identify the coal or "family" of similar coals that is being fired.

2. Methodology

2.1. System description

The monitoring system consists of a flame detector, a signal conditioning circuit and a PC-based signal processing unit. Fig.1 shows the structure and main constituent elements of the system. The flame detector is designed in order to extract as more information about the combustion flame as possible. The new flame detector has the same installation specifications as the conventional one so that it can be easily fitted to the existing sight tube (Fig.1) that is normally mounted to prevent the detector from excessive thermal radiation from the flame and provide mechanical support for the detector [7]. Three photodiodes covering the IR, visible and UV spectral bands are utilized to obtain flame signals containing the dynamic "finger-prints" of the flame and hence the type of fuel.

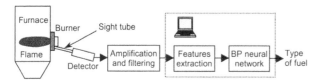

Fig.1. Block diagram of the system

2.2. Feature extraction

The flame features are extracted from both the time and frequency domains. In the time domain the DC and AC levels indicate the intensities of DC and AC components of the flame signal respectively. The number of zero crossings of the AC component is a measure of the dynamic characteristics and time structure of the signal [8]. In addition, the 2nd to 4th moments of the signal can be used to describe the variance, skewness and kurtosis of the signal respectively.

In the frequency domain, the quantitative flicker [9] describes the average flickering frequency of the flame. To characterise the PSD distribution of the signal, the entropy and the shape factor are often determined.

In general, all the features described above can be used to identify the "finger-prints" of the signal. However, not all the features are equally sensitive to the fuel type. Through experimental tests the features that are most sensitive to the fuel type are selected.

2.3. BP neural network

A BP neural network is used to map the representative features extracted from the flame signals into the corresponding type of fuel. This type of neural network has multi-layer feed-forward structure and its training is carried out by means of back-propagation. Fig.2 shows the topological structure of the BP network that has been used in this study. There are two layers in the network: a hidden layer and an output layer and Log-sigmoid transfer

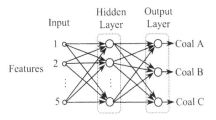

Fig.2. Topological structure of the BP network

functions [10] are used in each layer. In the output layer the number of neurons is the same as the number of fuel types, while in the hidden layer the number of neurons is determined by the complexity of the problem and in this case it is set as the number of the input features. In this particular study, the neural network has five input features and three output nodes (See section 3). Once fully trained, the complex, nonlinear mapping relation from the flame features onto the fuel types will be represented by and stored in the weights and the neural network can then be applied to infer fuel type from flame features.

3. Results and discussion

A series of experimental tests was conducted under steady combustion conditions on a $0.5MW_{th}$ combustion test facility operated by Innogy plc, UK. The fuels under test were Coal A, Coal B and Coal C. The coal feeding rate was 67kg/h whilst the total air flow rate (sum of the primary, secondary and tertiary air flows) was 700-720 kg/h. The flame signals were sampled at a frequency of 2 kHz with each data sequence having 4096 data points.

The DC and AC levels of the signals in the IR, visible and UV bands for each fuel are shown in Fig.3(a) and Fig.4(a) respectively. Their average values across the whole spectrum are plotted in Fig.3(b) and Fig.4(b). The "error" bars indicate the uncertainty of each feature. As expected, different fuels have indeed produced different spectral radiation distributions. The difference between the signals both in the average DC level and in the average AC level is evident irrespective of their fluctuations.

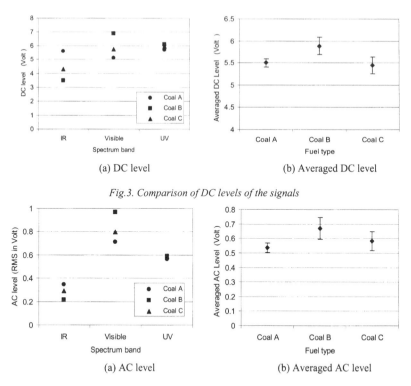

(a) DC level

(b) Averaged DC level

Fig.3. Comparison of DC levels of the signals

(a) AC level

(b) Averaged AC level

Fig.4. Comparison of AC levels of the signals

The variance, skewness, kurtosis, number of zero crossings in the time domain and the PSD distribution, flickering frequency, entropy and the shape factor of the PSD were also obtained for each fuel. It was found that these features did not contribute equally to fuel identification. In addition to the averaged DC and AC levels, the number of zero-crossings and the kurtosis in the visible band and the flickering frequency of the IR region were found to be most sensitive to the fuel type and hence chosen to be the feature parameters for fuel identification. Fig.5 illustrates the comparisons between the above three parameters for the three types of fuel. It is clear that the flame signals have different average DC and AC levels. Although no individual feature is indicative of the type of fuel alone due to its uncertainty in value, an effective combination of all the features through the use of the BP neural network allows unambiguous identification of the type of fuel.

Ninety-six randomly selected data vectors, each of which consists of the features and the corresponding fuel type, were used to train the neural network. The tendency line of the residual error, i.e. the deviation of the network output from the expected target, is depicted in Fig.6. It can be seen that the residual error decreases rapidly with the training epoch. After 26 epochs the residual error has decreased to 5×10^{-15}, which means the network has been fully trained. Twenty-four other data vectors, eight for each coal, were used to evaluate the trained neural network. The evaluation results are illustrated in Fig.7, where the first eight sample points correspond to Coal A, the middle eight to Coal B and the remaining to Coal C. The results show that the network performs very well. It can also be seen that the output of the network at the fourth test point for Coal B deviates relatively significantly from the expectation implying that the test data unseen by the network have resulted in a slight uncertainty in the fuel identification. Nevertheless, the type of fuel has still been correctly identified.

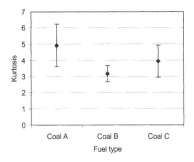

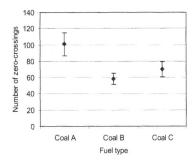

(a) Kurtosis in the visible band   (b) Number of zero-crossings in the visible band

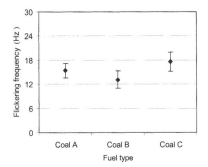

(c) Flickering frequency in the IR band

Fig.5. Feature parameters for the three different coals

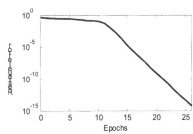

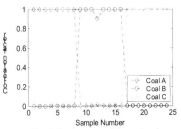

Fig.6 Residual error of the network output during training   *Fig.7 Evaluation of the trained neural network*

4. Conclusion

It can be concluded that the features of the flame signals derived from an improved flame detector vary with the type of fuel. A combination of five such feature parameters, including averaged DC and AC levels, number of zero-crossings, kurtosis in the visible band and flickering frequency of the IR band, has been found most effective for fuel identification. A back-propagation neural network has been established, which accepts the features extracted to infer the type of fuel being fired. Experimental results obtained on a $0.5MW_{th}$ combustion test facility have demonstrated that the prototype system is capable of identifying the three types of fuel in question under steady combustion conditions.

Acknowledgement

Acknowledgment is made to the UK Department of Trade and Industry for a grant in aid of this research, but the views expressed are those of the authors, and not necessarily those of the Department of Trade and Industry.

References

[1] Khesin M and Girvan R 1997 Demonstration tests of new burner diagnostic system on a 650 MW coal-fired utility boiler *Proceedings of American Power Conference*, Chicago, **59** 325-330.

[2] Wilpo Ltd, On-line coal analyser. From http://www.wilpo.com.pl/english/analysers.html.

[3] Peetz-Schou J, Economics of on-line ash, coal and unburned carbon monitors in coal-fired power plants. (M&W Asketeknik ApS, Denmark). From http://www.netl.doe.gov/publications/proceedings/98/98flyash/PEETZ.PDF.

[4] Lauf D, RFID (Radio frequency identification) coal tracking technology. (Transponder News). From http://rapidttp.com/transponder/tcoalsrt.html.

[5] Matsuzaki A, Yamanaka T, Hashimoto A et al. 1991 New products for boiler safety operation *International Power Conference*, San Diego, CA, pp.1-7.

[6] Willson P M and Chappell T E 1985 Pulverised fuel flame monitoring in utility boilers *Measurement and Control* **18** 66-72.

[7] Xu L, Yan Y, Cornwell S and Riley G S 2003 On-line fuel identification using digital signal processing and soft-computing techniques *IMTC 2003-Instrumentation and Measurement Technology Conference*, Vail, CO, USA, 20-22 May.

[8] Tan C K 2002 *The monitoring of near burner slag formation*. PhD Thesis, the University of Glamorgan, UK.

[9] Huang Y, Yan Y, Lu G and Reed A 1999 On-line flicker measurement of gaseous flames by image processing and spectral analysis *Measurement Science and Technology* **10** 726-733.

[10] Hagan M, Demuth H and Beale M 1996 *Neural Network Design* New York: Brooks/Cole Publishing Company.

Use of Flex Sensors in a Hand Function Biofeedback System

J Dempsey, G Lyons, Annette Shanahan[1]

Department of Electronics and Computer Engineering, University of Limerick, Ireland.
1 Abbey Physiotherapy Clinic, Charlott's Quay, Limerick Ireland.

Abstract: This paper describes a desktop system which analyses a patient's finger movements using flex sensors. The system displays the movements onscreen as well as recording them to a database. It also allows the prerecording and simultaneous display of exercises to be performed and provides biofeedback to the patient.

This system is designed to encourage the patients, by providing biofeedback, which shows them when they are meeting or exceeding the goals set by the physiotherapist. It can also help physiotherapists to monitor how well a patient is recovering over a period of time.

1. Introduction

A reduction in finger range-of-motion (ROM), decreases the overall functionality of the hand, which can severely affect a person's ability to perform routine tasks. As part of a hand function rehabilitation program, sets of exercises are usually assigned by a physiotherapist, to the patient, to increase the range of movement in his/her fingers.

The purpose of the system, described in this paper, is to accurately record and analyse finger movement to provide biofeedback to the patient on how well they are performing a set of prescribed hand exercises. It has been shown that biofeedback can help a person undergoing physiotherapy to recover more quickly, by keeping the subject more interested in the exercise, by providing with simple indications that they are making good progress and informing/rewarding them with a sight or sound when they complete an exercise correctly[1-3].The system also provides the user with animations demonstrating how particular exercises should be performed. Finally the system, uses a patient database, to keeps track of patient progress over the rehabilitation period by logging finger joint angle data and deviations from the desired joint angle trajectories over each exercise session.

Flex sensors attached to a glove are used to measure the angles of the finger joints on the affected hand. Using these angles, the full range of movement for each joint can be seen. Exercises given to the patient to perform can be recorded and shown onscreen as the patient attempts them, to show them the correct form of the exercise and to indicate to them their deviation from the desired result. The physiotherapist can set ROM goal angles that the patient tries to reach. The program continuously shows the current position of the patient's fingers as well as the pre-recorded exercise (if enabled), so the patient can see how close they are to the correct movements.

2. System Description

A block diagram of the developed system is shown in Figure 1.

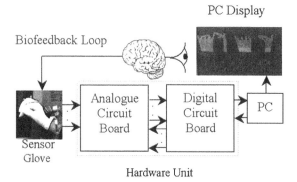

Figure 1: Biofeedback System Block Diagram

A hardware unit interfaces to the sensor glove and transmits the sensory data in digital form to the PC using the RS232 port. Software resident on the PC, reads the sensor data and controls animation of the computerised version of the subject's hand under sensor control. The target hand is animated under software control using pre-stored exercise routines. The exercise routines can be generated automatically if the therapist wears the sensor glove and carries out the desired hand function when the system is operating in training mode.

3. Hardware description

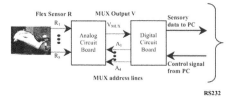

Figure 2: System Hardware Block Diagram

The hardware consists of three major components.
- The Sensor Glove with the flex sensors attached.
- The Analogue Circuit Board, containing an analogue multiplexer, three 5V regulated supplies, one ±15V supply, RS-232 level conversion and opto-isolation between the PC and subject.

- Digital Circuit Board, incorporating the ADuC812S evaluation board, based on the ADuC812S microcontroller, which features an eight channel multiplexed 12-bit ADC and serial output.

3.1 The Sensor Glove

The glove is based around a tight fitting yet flexible leather glove to allow a full range of finger motion while also allowing the sensors to accurately measure finger joint angles. Eight sensors are fitted, two on each of the four fingers (thumb motion is not monitored). The sensors used are Flex Sensors™ (Jameco Inc.), whose resistance increases with bending (Figure 3). The resistance ranges from approximately 10 kΩ when straight to approximately 35 kΩ when bent to ninety degrees. The sensors were positioned along the glove fingers and secured using plastic ties. The wires used to connect the flex sensors to the circuit board were wire-wrap wires, as they were light and would not impede movement and also could be easily soldered to the sensor terminals.

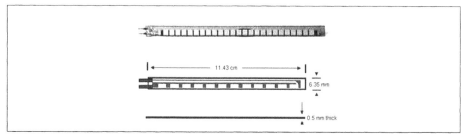

Figure 3: Flex sensor, showing the dimensions of the device

The sensors are configured in a voltage divider arrangement shown in Figure 4, with the sensor in the pull-up position. One end of each sensor is connected to 5V (supplied from the analogue circuit board) and the other end is input to the analogue circuit board, where the pull-down resistors (33kΩ) are located. With this arrangement, the voltage divider output ranges from 2.8V to 4V when the sensor is being flexed from 0° to 90°.

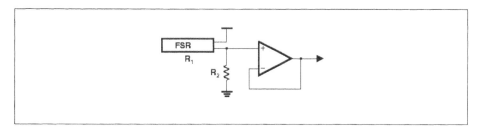

Figure 4: Flex sensor configuration with voltage divider and unity gain voltage follower

3.2 The Analogue Circuit Board

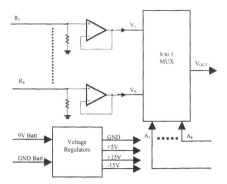

Figure 5: Analogue Circuit Board Block Diagram

The analogue circuit is powered by one 9V rechargeable PP3 battery and voltage regulation is used to provide regulated 5V and ±15V supplies.

The outputs of the sensor voltage follower circuits are fed to an 8 input analogue multiplexer (DG408) and the three address lines of the multiplexer are connected to one of the digital ports of the AduC812S microcontroller on the Digital Circuit Board. The output of the multiplexer is input to the Digital Circuit Board for analog-to-digital conversion.

3.3 The Digital Circuit Board

The digital Circuit Board is implemented using a ADuC812S micro-controller evaluation board from Analog Devices Inc. This device features a built-in 12-bit ADC and DAC.

The Digital Circuit Board performs three main functions:
1. Convert the sensory data from analog to digital form
2. Transmit the digital representation of the sensory data to the computer and receive control signals from the computer.
3. Drive the address lines of the analog multiplexer on the Analog Crcuit Board.

The ADuC812S evaluation board is under software control, resident in the on-board memory. Micro-controller code was written in assembly language, which was precompiled on the development computer and then downloaded onto the micro-controller via the serial port. The analogue port of the ADuC812S receives the output from the analogue multiplexer, from the Analogue Circuit Board. The data is output from the ADuC812S evaluation board is in RS-232 format (±10V).

An important feature of the system hardware is the incorporation of optical isolation. To ensure the absolute electrical safety of subjects using the equipment, there must be galvanic isolation of greater than 5,000 V between the subject (where the sensors are fitted) and the mains powered PC. This is achieved using two stages of optical isolation achieved using opto-couplers (CNY74-2H, CNY74-4H). Each opto-isolation layer is powered by its own isolated 5V regulated power supply. Each stage of isolation provides 5,000V of isolation, giving 10,000V of isolation in total.

4. Software Description

The system software has three different elements:

- The OpenGL Application Programming Interface (API) is used with C++ to display the hand graphics on the computer screen.
- The Win32 API is used with C++ to setup windows and to read to and from both the serial port and data files.
- Assembly language is used to program the ADuC812S microcontroller.

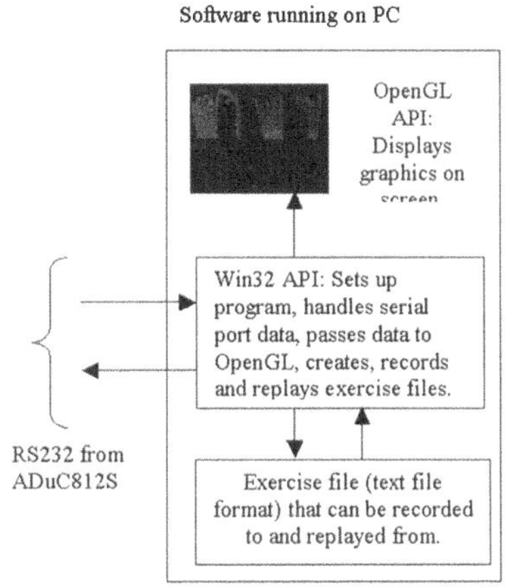

Software running on PC

OpenGL API: Displays graphics on screen

Win32 API: Sets up program, handles serial port data, passes data to OpenGL, creates, records and replays exercise files.

RS232 from ADuC812S

Exercise file (text file format) that can be recorded to and replayed from.

Figure 6: Software Block Diagram

4.1 OpenGL API Using C++

OpenGL stands for Open Graphics Library, this is a platform independent API, so it can be used over a wide range of computers [4]. The graphics software displays four hands onscreen, two coloured red and two coloured blue. The red hands display movement of the fingers by the subject wearing the glove, the blue hands display pre-recorded movements that have been saved to a file. For both the actual movements and the pre-recorded movements, a side view and a back view of the hand are shown on-screen. Beneath the display of the hands, a real-time graph, also coloured red and blue, is displayed.

4.2 Win32 API Using C++

Programming with the Win32 API provides a set of standard functions and commands that can be used by programmers to ensure that programs created will perform in the same manner on all versions of Windows, from Windows 95 upwards, as all these operating systems support the Win32 API.

In this project, the Win32 API was used to setup the window that the program runs in and to handle any events that occur while it is running. It was also used to create, read and write to and from files, for recording data, and to setup, send and receive data through the serial port to the ADuC812S micro-controller board.

4.3 Assembly Language

Assembly code was used instead of C code to program the ADuC812S as it is more efficient. When the micro-controller receives a signal from the computer it samples all the sensors, converts the data from analog to digital and then sends the data, in the RS-232 format, to the serial port of the computer.

5. Results

5.1 Sensor Calibration

The sensors were calibrated by taking ten readings from each sensor un-flexed ($\theta = 0°$) and flexed to 90° ($\theta = 90°$). These readings provided offset and sensitivity values for each sensor, as variation in these parameters from device to device is to be expected due to manufacturing tolerance. These measured sensor parameters were stored in a look-up table on the system software.

5.2 Bench Testing

The results from the bench testing were promising with he onscreen graphics accurately followed the movement of the finger joints being monitored.   More comprehensive characterisation of sensor glove is currently underway.

5.3 Full System Test

The full system test was performed using a healthy subject. The system recorded and replayed exercises to specification, with the pre-recorded exercise played exactly as performed by the subject. The system also displayed the actual movement of the subject alongside the pre-recorded exercise and real-time graph accurately and without any delay.

6. Conclusion

A working hand-function bio-feedback system has been developed. The system has several novel features, namely the ability to record an exercise programme and to replay this exercise as a guide for the subject during rehabilitation. Two forms of biofeedback are provided to the users, a graphical representation of the hand is presented in two views and angle information is provided in graph format. Biofeedback on finger motion was successfully delivered to the subjects when performing exercise programmes. A compact desktop hardware unit has been developed with opto-isolation incorporated into the design. Future work includes detailed characterisation of the sensor glove and a comprehensive clinical evaluation of the biofeedback system in assisting a hand function rehabilitation programme.

References

[1]   B.H. Cho, J.M. Lee, J.H. Ku, D.P. Jang, J.S.Kim, I.Y. Kim, J.H. Lee, S.I. Kim. "Attention Enhancement System using Virtual Reality and EEG Bio feeback", *Proceeding of the IEEE VirtualReality 2002 (VR'02)*, 1087-8270/02

[2]   Brown DM, Nahai F. "Biofeedback strategies of the occupational therapist in total hand rehabilitation", *In Biofeedback: Principles and Practice for Clinicians, 2nd Edition,* 90-106, 1975.

[3]   Louis G. Durand, Gelu D. Ionescu, Michel Blanchard, Jocelyn Durand, Sylvie Tremblay, Jocelyne Caya, Robert Guardo. "Design and preliminary evaluation of a portable instrument for assisting physiotherapists and occupational therapists in the rehabilitation of the hand", *Journal of Rehabilitation Research and Development* Vol. 26 No.2: 47-54.

[4]   Jeff Molofee. "Nehe Productions", *NeHe.gamedev.net,* present.

[5]   Analog Devices, Inc. 2002. "ADuC812
        MicroConverter®, Multichannel 12-Bit ADC with Embedded Flash MCU" data-sheet (web address: http://www.analog.com/UploadedFiles/Datasheets/160742467ADuC812_d.pdf).

EM Wave Monitoring Sensor for the Oil Industry

A Shaw[1], A I Al-Shamma'a[1], R Tanner[2], J Lucas[1]

[1] Dept. of Electrical Engineering & Electronics, University of Liverpool, Liverpool, UK.
[2] The Industry Technology Facilitator (ITF), Aberdeen, Scotland.

Abstract: Reliable sand monitoring is an important component of offshore sand management. Existing designs of sand monitor are widely used and often provide useful results, but they can require regular calibration and may struggle to provide reliable quantitative measurements. At the University of Liverpool the authors investigating the propagation of EM waves through various mediums within the oil pipe including oil, gas, water and sand in order to quantify the percentage of the media mixtures. This paper will describe the microwave system experimental set-up including the transmitter and receiving antennae. Experimental results of various mixtures of oil and sand will be presented.

1. Introduction

The presence of sand in offshore oil and gas production is an increasing problem in the North Sea. It is particularly associated with damage to poorly consolidated oil bearing rock. Sand production is most likely to be a problem in later field life due to changing differential pressures across the oil bearing rock as reservoir pressure reduces and due to entrainment of sand in formation water. Sand production also arises in the case of failure of downhole sand control measures. There are a number of serious problems resulting from the presence of sand in the produced fluids:

- Sand-in-liquid slurries are very erosive especially at higher velocities, for instance with high gas-to-oil ratios. The wellhead choke, valves, pipe work bends and other restrictions are particularly vulnerable. This is a serious potential safety issue. If deposited sand promotes corrosion, combined erosion/corrosion can be a particularly severe problem.
- If sand is present in the fluids it is liable to settle out, particularly where the fluid velocity is reduced. The most common location is the primary separator where sand can reduce separation efficiency due to reduced residence time and can provide active sites for corrosion. Plugging by sand can cause valves to jam and seize. Level instrumentation may also become inoperable. Deposition in pipe work can reduce flow capacity and cause erosion through increased velocities.

There are several potential disposal routes for the sand but each has its own associated problems. If some remains in the oil product, it may result in the Base Sediment and Water (BS&W) specification being exceeded and cause problems at the onshore processing terminal. If sand leaves with the produced water, erosion of the produced water system is a concern and there may be environmental implications if the sand is discharged overboard untreated. If re-injected with produced water, it can plug the formation. If deposited sand has to be removed from topsides equipment manually, it creates a safety hazard due to the presence of entrained oil and of radioactive solids known as low specific activity (LSA) scale.

Mature fields produce petroleum fluids that are prone to have a high water cut (high percentage of water in the pipe) and unwanted sand content. The EM wave sand sensor (EM SAND) aids the optimisation of the production rates in oil, gas and multiphase wells. The system provides real time quantitative monitoring of sand in any production flow, thus helping the operator to control erosion in values, inline process equipment and flow levels. Figure 1a shows the various components of the piping section that could be subject to erosion and an estimation of the system degradation could be estimated from the accumulated sand monitoring results over its lifetime. This would contribute to prediction of safety and would be included in a general monitoring of the field.

Figure 1a: Corrosion by sand [1]

In addition to the safety and production, technical aspects of unwanted/ uncontrolled sand production, the active use of on-line sand monitoring equipment will also have a cost reducing impact on future installations since piping tolerances with respect to erosion may be reduced by strengthening strategic sections.

Process equipment could also be filled up due to the settling of sand. This introduces a sand disposal problem at the platform. As this sand is usually contaminated with oil particles, it forms an additional environmental problem.

It is common knowledge that water breakthrough can lead to increased sand content in the well fluid. This is illustrated in Figure 1b, which shows results obtained from the Tordis well over six month period. The sand production corresponds quite well with the increase in water cut over the same period.

Figure 1c shows how the sand production and hence erosion varies with time and choke settings through a six month period on the Heidrun platform in the North Sea. All choke settings are shown on the figure, indicating how the well flow is gradually reduced in order to reach the optimum sand free production rate at about 17% choke opening.

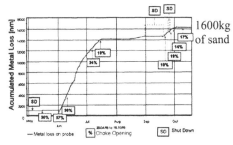

Figure 1b: Sand Production

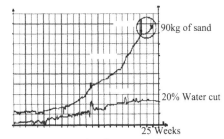

Figure 1c: Sand and water production

An important component of the sand handling strategy is sand monitoring and it is common to have a sand monitor installed on topsides flow lines after the production choke on potentially sand producing wells. Reliable sand monitoring allows timely actions to be taken to handle sand production. These actions include:

- Increased inspection to detect erosion and sand build-up;
- Reduction of flow rates from particularly sandy wells and in extreme cases taking wells off line;
- Installation of new or implementation of existing sand handling systems such as wellhead desanders and water jetting systems.

2. Status of Sand Monitoring

There are 2 generic types of sand monitor in common use:

- Ultrasonic monitors analyse the sound made by sand particles when they collide with the pipe wall. The acoustic sensor is located after a bend where the particles are forced out of the flow and hit the pipe wall. They are calibrated to provide direct indication of the sand concentration;
- Erosion monitors consist of a probe installed within the flow stream that is eroded if sand is present. The electrical resistance across the probe is monitored. The change in resistance is a measure of metal loss and thus can be related to the sand concentration in the hydrocarbon fluids.

These monitors are valuable tool in sand management especially if they are regularly calibrated to increase reliability. However offshore oil and gas operators have indicated that there is room for improvements in sand monitoring. This view has been made in the literature [1] and was highlighted at a seminar on offshore sand handling organised by IGL Engineering [2].

Ultrasonic monitors are non-intrusive and therefore do not interfere with the flow and can be relatively easily relocated. However they can struggle to provide repeatable quantitative measurements and are affected by process noise such as changing gas-to-liquid ratios. Erosion monitors, whilst intrusive, can provide reliable quantitative measurements.

3. EM Wave Sand Monitor

The concentrations of oil, gas, water and solids within a multi-phase fluid flow pipeline can be obtained by measuring the resonant frequency for the propagation of electromagnetic waves within the fluid [4,5]. Ongoing work into the use of electromagnetic waves for multiphase metering in the Department of Electrical and Electronic Engineering at Liverpool University has demonstrated that this technique is appropriate for sand concentration monitoring.

The real time, non intrusive sensor system will be constructed (as a resonant cavity), as illustrated in figure 2, in an oversized concentric section of pipeline of length (L) and diameter (D) attached to the outside of the pipeline. The cavity section is periodically isolated from the pipeline multi-phase fluid flow using a novel robust hydraulic valve. The cavity has a transmitter (T) and a receiver (R) antenna for the propagation of low power (mW) electromagnetic (EM) waves through the fluid. A series of EM wave resonances occur whose frequency is dependent on the multi-phase fluid dielectric constant (ε) and the node number (p = 1, 2, 3, etc). The resonant frequency (f) for one of the nodes will be measured. When the cavity is isolated from the pipeline there will be a frequency shift Δf caused by the gravity separation of the sand. The mathematical relationship is given by,

$$\frac{\Delta f}{f} = \frac{1}{2}\left(\frac{4.2}{\varepsilon_r} - 1\right)C \tag{1}$$

where ε_r is the dielectric constant of a medium (e.g. air =1 and sand = 4.2) and shows that the sand concentration (C) is directly related to the frequency shift. The sensor undertakes a series of sand concentration readings at regular intervals. The sensor system can be fitted to any diameter of oil pipelines and in any orientation and fluid flow regimes.

Figure 2: Laboratory trials of electromagnetic sand monitor

The resonant frequency is directly related to the average value of the dielectric constant for all the constituents. Preliminary experimental demonstrator results in figure 3 show that sand at the percentage level can be easily detected.

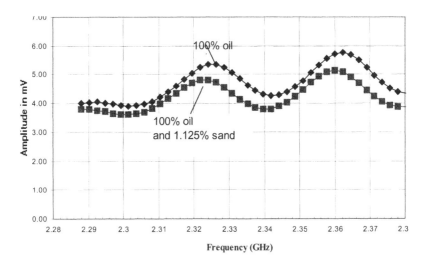

Figure 3: Sample results for electromagnetic sand monitor

At this concentration there is a frequency shift of several thousand kHz. As the resolution of the frequency-measuring instrument is 1 kHz, measurements of sand concentrations at the part per million (ppm) level should be possible as shown in figure 4. This figure shows the experimental frequency shift obtained when sand was added to oil in various concentrations. This is presented by one of the EM wave nodes and gives a frequency shift of 1860kHz when 1 in 10^3 of the mixture by volume is sand in seawater.

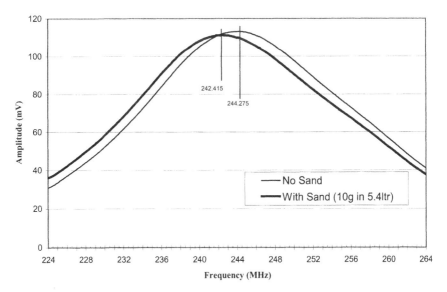

Figure 4: A resonance peak occurring at a low frequency in an 8" diameter cavity in seawater/sand mixture.

The monitor, which operates on a few watts of power, measures the volumetric sand concentration and can offer a completely non-intrusive system. The monitor is installed around a non-metallic spool piece or one with quartz windows as illustrated in figure 5.

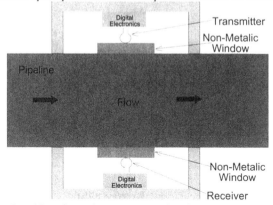

Figure 5: Possible industrial arrangements for electromagnetic sand monitor

4. Conclusions

The proposed system is based upon using electromagnetic wave resonance of a cavity. It is non intrusive and continuously samples the fluid. Its resolution capability is 1ppm and is applicable to all fluid flow regimes especially if the ISA mixer plate [6] is used. The technique is to measure the resonance frequency (f) for the propagation of EM waves within a section of the pipeline of radius R. The operating principle is that the resonance frequency (f) is related to the net dielectric constant ε_{net} of the fluid and the velocity (c) of EM waves. This is expressed by

$$f = \frac{C}{\sqrt{\varepsilon_{net}}\,(2\pi R)} \quad , \text{with } \varepsilon_{net} = C_{oil}\varepsilon_{oil} + C_{gas}\varepsilon_{gas} + C_{water}\varepsilon_{water} + C_{sand}\varepsilon_{sand}$$

Where C is the volumetric fractions and ε is the dielectric constant for each constituent. Preliminary results for sand in both oil and water have clearly shown this frequency shift over the full frequency spectrum as shown in figure 4. The sensitivity of the system depends upon the accuracy of measuring the resonant frequency and the frequency shift. Further for the water/sand mixture f=242,415 kHz and Δf =-1,860 kHz for a sand volumetric concentration of 1000ppm. The resolution of the frequency measuring instrumentation is 1 kHz hence in theory a resolution of 1ppm is possible. For these low ppm measurements it is necessary to undertake signal averaging over a short time period of about 1 second.

References

[1]   http://www.flowprogramme.co.uk.
[2]   Salama M.M, "Performance of Sand Monitors", Conoco Inc., NACE INT Corrosion Conf (Corrosion 2000), Paper No. 85 2000, Orlando, Florida, 2000.
[3]   "Sand Handling Seminar", organised by IGL Engineering Ltd., Aberdeen, September 1999.
[4]   Lucas J. and Al-Shamma'a A.I., "Apparatus for Determining Dielectric Properties of an Electrically Conductive Fluid", British Patent PO 66650GB 21 January 1999.

[5]   Hogan B, Lucas J. and Al-Shamma'a A. I., "Real-Time Non-Intrusive Multiphase Metering using Microwave Sensors", 2nd North American Conference on Multiphase Technology, page 281-295, ISBN 1 86058 252 4, Banff, Canada, 21-23 June, 2000.

Novel Gas Sensing for Early Fire Detection

C G Day, Emmauel Eweka[1]

Fire Research Group, QinetiQ, Haslar, Gosport, UK
[1]Power Sources Group, , QinetiQ, Haslar, Gosport, UK

Abstract: This paper aims to evaluate the effectiveness of novel gas sensors and determine whether they could form part of a fire detection system that employs gas sensing, rather than particle detection methods, as a means of early fire detection. It has been shown in previous work [1] that certain gases are released from electrical wires and electronic components before smoke particles. It is hoped that the detection of these gases will give a very early indication of an immanent fire within electronics cabinets and reduce false alarms.

1. Introduction

Over the last decade fire detection concepts have undergone radical changes with a large percentage of novel research projects concentrating on specialist detection with gas sensors. Traditionally, fire detection has been carried out by particle sensing (optical and ionisation), heat detection or flame detection methods. However, detection of low energy fires and smouldering combustion is difficult using these technologies, as traditional ceiling mounted detectors rely on high plume strengths to bring combustion products to the detector. In the case of low energy fires smoke can stratify, particularly in cooled/well ventilated areas and may not reach the detector. One way in which smaller fires can be detected is with an increased number of sensors. This leads us to the idea of detection within the equipment.

The availability, selectivity, quality and robustness of commercial gas sensors have improved dramatically, facilitating current research and developments in fire detection. Gas sensing methods have many advantages to offer, especially in specific and relatively contained environments. A suite of gas detectors can be selected and tailored to detect specific vapours of interest such as chemicals emitted from overheating electrics.

The environmental changes that occur during the development and occurrence of a flaming fire are called fire signatures. The main fire signatures are the emission of heat, light, smoke particles and gas. The early detection of fire relates to the detection of pre-fire conditions such as over-heating or smouldering during which vapour, combustion products and smoke are released, but a full-scale fire has yet to develop.

The aim of recent work at QinetiQ has been to further evaluate the effectiveness of novel gas sensors and determine whether they could form part of a fire detection system that employs gas sensing, rather than particle detection methods. The finite detection time could then, in theory, be quicker than particle detectors as fire precursors are produced prior to smoke particles.

To date QinetiQ have tested and evaluated a range of novel metal oxide sensors and a specially developed SAW (Surface Acoustic Wave) sensor unit for early fire detection applications. Most recently QinetiQ have been evaluating conducting polymer sensors developed in house. The final aim of the on going project is to produce an array of sensors suitable for the application of very early fire detection within electronic cabinets.

2. Targets for Novel gas sensor detectors and arrays

Ideally the novel sensor array would have the ability to detect gaseous vapours from an electrical system that is overheating before the smouldering, combustion stages of a fire has commenced. The array's response to the fire signatures from overheating electrical systems should be repeatable and not show interference effects that would bring about false alarms. The array would have to perform better than current laser point optical smoke detectors in terms of detection times, showing a clear advantage for using novel gas sensor technology. Importantly, the array should have repeatable response times and correct alarm levels, when exposed to a range of environmental condition such as fluctuating temperature and humidity, so that it can demonstrate long term effectiveness. The sensors themselves should have a degree of physical robustness, have inherent low maintenance and long shelf life.

From an earlier characterisation study, it was determined that the sensors would need to be sensitive to many types of gaseous hydrocarbons. It was necessary, therefore, to pre-select those species deemed specific to overheating electrical components and not to other organic gases that may be present in the background atmosphere or emitted from other equipment such as a photocopier etc. Therefore an objective was set to work out what gases were required and what sensors are needed to detect those gases.

3. Technology evaluated

A total of thirteen novel mixed metal oxide sensors were tested, five purchased from a commercial supplier, four provided by Dr Thomas Starke from the university of Swansea and four were purchased from Dr Ratcliffe at the university of West England, Bristol. These sensors were tested alongside a SAW sensor unit and newly developed conducting polymer sensors.

The metal oxide sensors tested were all of a conventional design, however instead of using tin oxide as the gas/solid interface, a variety of mixed metal oxides have been utilised. Metal oxide semiconductor sensors have been used as gas sensors for many years and have been considered by QinetiQ for fire detection applications on many occassions. Previous work has shown that tin oxide sensors (the forerunner to mixed metal oxides) had a humidity dependency [2] and that the sensors were unspecific [3]. However, it is claimed by manufacturers that the move from tin oxide to mixed metal oxides has potentially resolved these problems and thus made the investigation of this technology necessary.

The SAW unit tested was a second generation unit provided by Marconi Applied Technologies. The unit comprises of an array of six SAW sensors, two of which act as reference sensors. The remaining four sensors each have a different polymer coating to make it selective to different classes of organic vapour.

The conducting polymer sensors produced in-house by QinetiQ Haslar are fabricated from substrates which are supplied by Windsor Scientific Ltd. The substrates consist of four individually addressable gold microband electrodes, which are formed by patterning high-grade gold films on an insulated silicon substrate, see figure 1. The silicon substrate is supplied mounted on a printed circuit board header to allow easy testing of the sensor, see figure 2. The substrates are set up in a three electrode system using the gold microbands as the working electrode, a large area platinum gauze as the counter electrode and a calomel reference electrode. The three electrode system is set up in an electrochemical cell, using a 0.1 molar solution of 1-butanesulfonic acid as the base electrolyte, the monomer (pyrrole) is then added and the air is purged from the cell by bubbling through with argon. The potential of the electrode is then stepped to a suitable oxidising value, using a Solartron 1286 electrochemical interface, for a predetermined period of time in order to grow the polymer layers. Once the polymer has been coated onto the microband electrodes, it requires activation before use as a chemical sensor. The conducting polymer layers are prepared by an electrochemical technique called potentiostatic deposition.

Figure 1 Silicon substrate

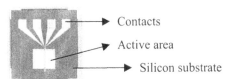

Contacts

Active area

Silicon substrate

Figure 2 mounted microelectrodes

4. Experimental

Initial tests were carried out with toluene, methanol and acetone vapours, and CO and toluene gases, to check for sensitivity, selectivity and repeatability. Crude temperature and humidity dependency tests were also carried out. The final test conducted on all gas sensing methods was the hot wire test, which is used to determine the sensors response to realistic stimuli that would be produced from an electrically overheated wire.

During evaluation the sensors were fitted within a plastic enclosure with internal dimensions of 345 x 310 x 310mm. The enclosure has been aged over several years; hence it was unlikely that it was still out-gassing.

4.1   Crude Laboratory Tests for Vapours

All sensor types were exposed to vapours emitted from small amounts of acetone, methanol and toluene. These crude tests allowed the sensors to be quickly evaluated and hence determine how they might respond in future tests, with more complex vapours.

The sensors were allowed to log data for 120 seconds before 20ml of methanol, acetone or toluene was placed in the enclosure. The liquid was removed after 300 seconds, at which point the enclosure's door was opened and the fume cupboard extraction was turned on. Data logging was stopped approximately 300 seconds after the liquid had been removed from the enclosure.

4.2   Laboratory Tests for gas analysis

In the second set of tests the sensors were exposed to 50ppm of toluene and then 500ppm of carbon monoxide (CO). Toluene was selected as it had appeared in the original list of chemicals produced from the overheated military specification wire [4]. Although it has been seen that CO will not be produced in an electrical fires [5], it was deemed useful to carry out this test since two of the commercially available sensors measured CO. Fires in more general areas, especially where soft furnishings are present, will produce CO and so evaluation of the sensors to this gas gives information on future application potential.

The sensors data was logged for 120 seconds before the gas was allowed to flow into the cube. The gas supply was turned off after 60 seconds, at which point the cube's door was opened and the fume cupboard extraction turned on. Data logging was stopped approximately 250 to 300 seconds after the gas supply was turned off.

4.3   Laboratory Tests to Overheated Equipment Wire

The tests carried out in this section ascertained sensor response to realistic stimuli that would be produced from an electrically overheated wire. A Rayfast military specification wire, rated at 4 amps was used, as it would be a likely primary fuel source within RN electronics cabinets. The wire was wound into a repeatable sized coil to create a "hotspot" ensuring smoke was produced from the same location every time. All the wire bundles were stored in a conditioning room, which was maintained at 23°C and 50% relative humidity. This measure ensured that at the onset of a test, every wire was in approximately the same condition. The wire was overheated within two perforated tubes.

The data from the sensors was logged for 120 seconds before power was supplied to the wire. Power was supplied at 20 Amps and 2 Volts for 390 seconds. Data logging was stopped approximately 600 seconds after the power was turned off. Figure 3 shows the rising scale of damage from the original unheated wire (0) to the destroyed wire (5). Each test wire has been given a grade based on these examples and corresponds to the description given in table 1 below.

Wire Damage Gradings	Description of Damage
0	Original, unheated test wire
1	Very little visible damage
2	Slight discoloration and glazing, individual coils still clearly visible
3	Initial merging of individual coils
4	Extreme degradation and disintegration of sheath, early signs of inner wire being exposed
5	Sheathing almost entirely destroyed, inner core of wire exposed and charred

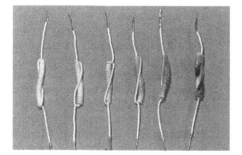

Table 1: Wire damage gradings   *Figure 3: Sample wires*

4.4   Temperature Dependencies

The SAW sensor unit was placed in a gas-chromatogram (GC) oven where it was subjected to varying temperatures. The oven temperature was initially maintained at 30°C and was raised in 2°C steps up to 40°C and then lowered in 2°C steps back to 30°C. The second test raised the temperature to 40°C in one step where it was maintained for a fixed period of time, and then lowered back to 30°C. Unfortunately, due to restraints on both time and the availability of the GC oven, it was not possible to test the remaining sensors in the same manner as the SAW sensors, and a cruder test had to be used. A hotplate was placed in the cube and set to its maximum setting with power being supplied immediately. The temperature was recorded every 20 seconds from the temperature/humidity meter, to allow any correlation between temperature and sensor response to be seen. The hotplate was removed when a temperature of 40°C was reached, at which time the fume cupboard extraction was turned on.

4.5 Humidity Dependencies

The sensors were subjected to changes in humidity to establish if a relationship exists between sensor responses and humidity changes. A 75ml beaker of water was placed on the hot plate, 350mm below the sensors, with the humidity meter probe being located next to the sensors. The hotplate and water were removed when 100% humidity was reached, at which time the fume cupboard extraction was turned on.

5. Conclusions

The results have shown that the majority of sensors tested showed some form of temperature and humidity dependency. The commercially available metal oxide sensors are specific to their target gases and would be suitable for inclusion into a novel gas sensing array.

The conducting polymer sensors have shown promising results. Two of the three sensors produced showed good responses to the vapour tests, see figure 4 below, and the third sensor responded well to the hot wire tests. The most promising results however were from the temperature and humidity tests, showing that conducting polymer sensors are not greatly affected by temperature and humidity changes. By further tailoring the monomer selection, along with the film thickness and doping of the polymer, it will be possible to further improve the sensors performance for fire detection.

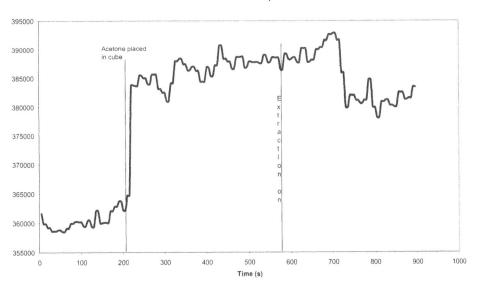

Figure 4: Example response of polypyrrole sensor to acetone vapour

The work has demonstrated that modern gas sensing technologies can successfully detect fire precursors. A place exists in further research for both conducting polymer and metal oxide semiconductor gas sensing techniques. It is evident that a combination of sensing methods will produce better results than relying on one type of gas detector. However it is accepted that further research and development of the technologies studied will be required, before selection of the gas sensors suitable for an array can be made.

References

[1]   J Riches, D Beardmore, The Analysis of Gaseous Emissions from some Electronic Components at Smouldering Temperatures DERA/SSES/C21.13.3
[2]   Fister G, Detection of smoke gases by solid state sensors – A focus on research activities. Fire safety journal, Vol.6 165-174, 1983
[3]   Riches J, Chemical sensor systems for the early detection of fire in electronic equipment DERA/CBD/CR980231/1.0. August 1998
[4]   Riches J, Chemical sensor systems for the early detection of fire in electronic equipment DERA/CBD/CR980231/1.0. August 1998
[5]   Willat B M Chemical sensors for carbon monoxide detection in fires, EUSAS workshop on the detection of combustible gases in connection with fire detection. July 1998

Section 5

Non-Destructive Monitoring

A microcontroller-based sensor for measuring accumulated fatigue damage in welded steel structures

B Fernandes[1], Michael Burdekin[2] Yan Hui Zhang[3] and Patrick Gaydecki[1]

[1] Department of Instrumentation and Analytical Science, UMIST, Manchester, UK
[2] Department of Civil and Construction Engineering, UMIST, Manchester, UK
[3] TWI Granta Park, Great Abington, Cambridge, CB1 6AL, UK
Email: patrick.gaydecki@umist.ac.uk

Abstract: A sensor system is described that performs long-term monitoring of fatigue damage in load-bearing welded components on structures such as earth-moving vehicles. The sensor is a steel coupon containing a fatigue pre-crack, and is permanently attached to the structure. Cyclic stresses induced in the coupon by in-service loading on the component cause the crack to lengthen. Monitoring the crack length gives a measure of the fatigue accumulated by the structure, allowing prediction of its remaining life. A battery-powered microcontroller system embedded within the sensor achieves this. The data stored on-board can be downloaded to a PC for analysis.

1. Introduction

Despite the attention given during the design stages of structural component development, failure in service continues to occur. This is an expensive problem with estimates of the consequential costs for the USA alone being put at about 4% of their GDP in 1983[1]. Fatigue cracking under cyclic loading is the most common mode of failure in metallic structures, welded structures being particularly susceptible because of the relatively low fatigue strength of the welded joints by comparison with that of the parent material. By monitoring a sensor attached adjacent to a critical joint, the rate at which the design life of the structure is being expended can be determined and, hence, the useful remaining life evaluated. In this project, the development of a sensor capable of registering cumulative fatigue damage in welded steel components is being undertaken.

The sensor concept is based on a theoretical design developed in the department of Civil and Structural Engineering at UMIST and realised by the Fatigue Monitoring Bureau (FMB). The device consists of a steel coupon containing a fatigue crack, which is attached to the structure adjacent to a critical joint. Cyclic stresses induced in the coupon cause the fatigue crack to extend. The physical size of the coupon, its shape and method of attachment result in direct proportionality between crack growth in the coupon and expended fatigue life of the target joint. Sensitivity of the device is selected to suit the target joint. The device is entirely passive with no external connecting wires or power requirement.

The sensor will have wide application in many industrial sectors including structures such as bridges, cranes, offshore installations, and in a range of transport industries. In the present project its use will be validated by focusing on earthmoving machinery and axle/suspension systems for heavy goods vehicles (HGVs), but the results will be relevant to other interested industries. The most important benefit is the safety implications of sensor use, resulting from its ability to predict and prevent structural failure.

This publication focuses on the development of a system for measuring the crack length in the sensor. The system is being designed to be autonomous and capable of operating for over a decade without connection to an external power source. It uses a microcontroller to interrogate the sensor and determine the crack length. It will 'wake up' whenever the crack length increases by a finite, predetermined amount, and will record the date and time that event occurs. Data will continue to be stored in non volatile memory until the crack has propagated to the peripheral extents of the sensor coupon. On demand, an inspector will be able to download the stored data to a PC ready for analysis. Monitoring crack length in this manner will give a measure of the fatigue damage accumulated by the structure, and will allow predictions of remaining life under similar loading conditions to be made. The total sensor system will be of a size similar to that of a credit card.

2. Material fatigue analysis and sensor design

Any engineering structure will be subjected to external forces arising from service conditions or the environment in which it works. If it is in equilibrium, the resultant of the external forces will be zero, but the structure will nevertheless experience a load that will tend to deform it. Internal forces set up within the material will react upon this load.

Fatigue is the failure of a material under fluctuating stresses each of which is believed to produce minute amounts of plastic, irreversible, strain [2]. The behaviour of materials under fatigue is usually described by a *fatigue life* or *S-N* curve in which the number of cycles, N to produce a failure with a stress peak of S is plotted against N. This curve has the profile similar to an exponential decay that becomes asymptotic to a minimum stress level S_n, which is the stress value at which the component under test is assumed to have an infinite life. This means that stresses below this value applied to the member will never cause it to fatigue. Structures are usually designed to keep the maximum stress to a level below S_n. This usually implies over-design in terms of physical size and material usage in consideration of cases where the member occasionally experiences loading that exceeds S_n. However, in some cases like in the case of aircraft component design, weight is a critical issue constraining the design to the point that only allows it to have a finite life. It is therefore clear that the number of times S_n is exceeded and by how much will have an important bearing on the prescribed life of the component. Such designs inevitably require frequent testing to establish whether the design-prescribed life of the component has been shortened.

Almost all the preliminary work in designing the fatigue sensor has been carried out by the Fatigue Monitoring Bureau (FMB) at UMIST. Here, a brief description of the sensor is given.

The rate of growth of the crack in the gauge can be determined from fracture mechanics principles using the Paris law as follows [3]:

$$\frac{da}{dN} = C(\Delta K)^m \tag{1}$$

where C and m are material constants, and ΔK is the stress intensity factor range in the gauge. A typical value for m for many metallic materials is $m = 3$ and values of C can be chosen to allow for different materials, mean stress and environmental conditions.

Where the gauge design is such that the stress intensity factor is independent of the crack length, the crack growth equation can be integrated directly. This gives the relationship between change in crack length in the gauge, Δa, gauge factor, q, stress range, S, and number of cycles, N, in the underlying structure as

$$\frac{da}{dN} = C(qS)^3 N \qquad (2)$$

showing that the crack growth in the gauge is directly proportional to stress in the structure and the number of cycles, and can be expressed as $S^3 N$. By choice of the appropriate gauge factor q from Finite Element analysis, the gauge geometry can be chosen to give an appropriate change in crack length to match weld design performance. In the case of variable amplitude fatigue loading, the fatigue damage can be represented as $\Sigma S_i^3 n_i$ where n_i is the number of cycles at stress range S_i.

Since the principles rely on change in crack length in the gauge to indicate fatigue damage, it is necessary to have an initial crack present of sufficient length for constant gauge factor performance ($a/W>0.3$). The sensor operates by simply following the flexions of the component under test. Since it already has a predefined crack in it, cyclic loading will cause the crack to propagate along the width of the coupon. The rate at which this occurs will be dependant on the magnitude of the inflexions and the material properties of the sensor and hence the component under investigation.

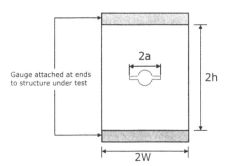

Figure 1. The fatigue sensor coupon.

3. Crack length measurement

In order for the crack length to be determined, several electrically conductive tracks are laid on the coupon in a direction perpendicular to that of the crack. These tracks are similar to those on electronic circuit boards and their spatial resolution can be adjusted at design time depending on the number required for the specific coupon used. In practice, a minimum of 30 tracks will be used, 15 being on either side of the central circle on the coupon shown in Figure 1. Initially, each track will have a potential of 3 Volts across it. In terms of Boolean logic, all tracks can be considered to be initially at logic 'high'. Therefore, the 15 lines on either side of the new sensor will output logic 1 or decimal equivalent 32,768. In the event of a track being broken due to crack propagation, the decimal number will decrease, since the logic level on that particular port pin will be zero. Crack propagation is determined by noting the decimal value represented by the 15 lines and comparing it to the original value of 32768. Then, based on the spatial resolution of the tracks, the crack length can be calculated.

The microcontroller used for sensor interrogation and data logging the Motorola™ 68HC908. In addition to having on-board volatile memory (RAM), flash memory and a serial communications interface, it has four 8-bit bi-directional parallel ports and can operate in power down mode, during which it consumes 1 μA of current. The complete system incorporates a power management system to enable it to operate without any maintenance, using a long life battery.

4. Method and Results

In the development of the interrogation system, the microcontroller has been used to read the logic status of the lines connected across the sensor. In the experimental set-up, a DIP switch was used to simulate the fatigue sensor. The arrangement is as shown in Figure 2, where the 68HC908 evaluation board is shown. In addition to the microcontroller and associated peripheral microchips, it has a breadboard section where circuits can be constructed. In this set-up, eight lines were used to connect port B to the DIP switch. The other end of the switch was connected to a power line from port C via a bank of resistors. Broken tracks were simulated by turning off the relevant line on the DIP switch. In the case of a line being on, the relevant pin on port B received 5 Volts from the supply. In the event of an open switch on that line, simulating crack propagation, the pin on port B would be at 0 Volts.

Figure 2. Experimental set-up for testing the microcontroller programs using the 68HC908 evaluation board. The dip switch is used to simulate broken or unbroken circuit tracks.

The prompt to read port B was provided by an external interrupt to the microcontroller. The interrupt itself was provided by a pulse generator in the form of a 50 ns pulse every 10 seconds. This triggered the microcontroller every time it sensed a low going edge. Having received the interrupt, the microcontroller executed a software routine which read the status of port B and passed the value read to the PC via the RS232 interface.

Although the microcontroller system was tested using a DIP switch, work is underway to optimise the method of laying of circuit tracks on the sensor coupon, so that they separate completely upon being broken by the crack. Figure 3 shows the result of crack propagation on one of the samples tested.

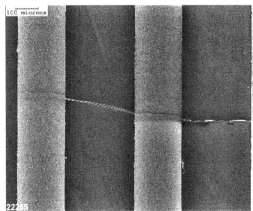

Figure 3. Photograph of a sensor showing crack propagation past two circuit tracks.

The prototype system replaces the pulse generator with a real time clock chip that will be used to wake up the microcontroller at a predetermined interval. This will enable the system to operate in standby mode, thus saving energy. The timer chip will be connected in alarm mode and the entire system, including the timer itself and the microcontroller, will be in standby mode, consuming approximately $2\mu A$. The timer alarm register is configured to transmit an interrupt to the microcontroller at regular intervals, typically once per day or once per month. This will wake up the HC908, to read the crack data. Since the alarm output is normally logic high, falling low in response to an alarm, current will flow continually through a pull-up resistor. The resistor will be in the order of 500 kΩ to minimize energy losses

5. Energy budget

A single lithium battery rated at 160mA hours (576 A seconds) will be used to power the circuit. This will theoretically allow the HC908 to operate for 18 years in sleep mode at 1 μA. The timer system will also consume approximately 1 μA. Hence over a ten year period, the timer and microcontroller would each require 315 A seconds, giving a total of 630 A seconds. This therefore reduces the operational life, to 9.1 years. However, the timer chip can be powered by another dedicated battery which, although increasing the chip count, will allow the sensor to operate for more than 10 years.
In summary therefore, the final system will be able to read 30 lines from the sensor and store the data in memory. The system will operate on minimum power and will be awoken whenever the crack has propagated past a track or whenever a request for data arrives via the serial communications interface. When awake, it will read the input port and store the data in the next available space in memory. If the request is to read the contents of memory, it will dump the data sequentially to the communications port. Initially, the software and hardware is being developed to a fully working state without

using the actual fatigue gauge.   It will then simply be attached to the actual sensor for testing.

6. Conclusion

The initial stages of this development project have generated results which demonstrate that it is feasible to realise a full scale working gauge. The initial sensor development work has been completed and the electronic data recording methodology has been established.   What essentially remains therefore is software refinement and encapsulation into an integrated hardware/software unit ready for on-site testing.

Acknowledgement

The authors wish to express their thanks to the Department of Trade and Industry (DTI) for financially supporting this work.

References

[1]   Reed R P et al, The economic effects of fracture in the United States, NBS Report 647, Washington 1983.
[2]   Hearn. E.J., Mechanics of Materials, Vol. 2, Pergamon, 1985
[3]   P. Lukáš, M. Klesnil, J. Polák, High cycle fatigue life of materials, *Mater. Sci. and Eng.,* **15**, 1974, 239

Online Acoustical Abnormality Monitoring System for a Total Power Plant

R Ohba, I Goto[1], Y Tamanoi[2] and S Tsujimoto[3]

Applied Physics Division, Graduate School of Eng., Hokkaido Univ., Sapporo, Japan.
[1] IPP Division, Iron and Steel Department, Kobe Steel Co. Ltd., Kobe, Japan.
[2] AAM Division, Yamatake Corporation, Yokohama, Japan.
[3] Kansai Branch, Yamatake Corporation, Osaka, Japan.

Abstract: A 700 MW coal-fired power plant located nearby a dense residential area of the city of Kobe has adopted an acoustical abnormality monitoring system to detect any abnormality of the total plant machinery round the clock to guarantee a stable operation by implementing a preventive maintenance system without any experienced engineer. To ensure reliable, 24/7 monitoring of plant equipments so as to detect predictive signs of malfunction prior to any failure, a distributed sound monitoring system is installed. Any sign of suspecting abnormal operation is detected, the system immediately notifies staff so that timely preventative maintenance can be performed, minimizing downtime and maximizing efficiency.

1. Introduction

We have been developing techniques so as to allow reduced manpower monitoring of possible abnormal conditions of plants. An inverse filter based abnormal signal segregation and extraction technique (*IF-ASSET*[*]) has been developed based on inverse filtered acoustic/vibration signals and it has been proved to be effective to monitor abnormal conditions in units such as rotary machine, compressor, piping and high pressure vessel [1-3]. Combining IF-ASSET and pattern recognition techniques, a principle of an intelligent sensor system has been proposed for detecting possible leakage of high pressure gases and its feasibility has been illustrated in our previous papers [4, 5]. This technique has been applied to an intelligent robot system for monitoring leakage of high pressure gases in oil refineries [6-9]. It is also possible to develop an online abnormality monitoring system for a group of machines such as compressors in a compressor station which frequently change their operational conditions applying a learning function along with IF-ASSET [10, 11].

Shinko Kobe Power Inc., a subsidiary of Kobe Steel Co. Ltd. (Kobelco) and Japan's largest IPP (Independent Power Producer), operates a coal-fired power plant to supply the city of Kobe with electricity. Kobe is one of Japan's major ports for international trade and the nearby Hanshin industrial zone, which flanks Osaka Bay, is equally important for the production of ship, machinery, and iron and steel. Kobe is a large city with a population of over 1.4 million and in the summer season its electricity

---

[*] IF-ASSET is a registered trademark of Yamatake Corporation.

consumption peaks at 1.7-1.8 GW. To meet this large demand, the city should have a stable power supply, preferably in the immediate area. Kobelco, one of leading steel manufacturers in Japan, has identified IPP as a unique business opportunity and realizes that it can employ its own resources and experience to serve the community by generating electricity using the latest, environment-friendly technologies. Kobelco has possessed already in Kobe ground, quays and other infrastructures needed for such a venture. So construction began in March 1999, and commercial operation of Shinko Kobe Power Station has started in April 2002. At present, output is 700 MW, and a second unit scheduled to come on line in 2004 doubles output and effectively serves as much as 80% of the city's demand.

Shinko Kobe Power Inc. that runs the plant puts priority on its responsibilities to the community, so every effort has been made to ensure safety, minimize environmental load, and optimise energy utilization. For example, some of the steam produced by the plant is supplied to a neighbouring brewery. Other important responsibilities include making available information about the plant's operations and employing the best available preventative maintenance procedures and *PlantWatcher-type A*[*], which is an online abnormality monitoring system designed on the basis of IF-ASSET, is adopted to ensure a predictive maintenance. In this paper the online acoustic abnormality monitoring system for the power plant is described.

2. Online Monitoring System

2.1 Power Plant

The power plant to be monitored consists of a main turbine/generator of 700 MW, a sub generator for the armature current of the main generator, a coal-fired boiler, coal mills, coal crashers, super heaters, fans and a group of pumps to mainly serve operations of the boiler. Most machineries of the plant are housed in an independent four-storey building of 80 meter high to reduce divergence of operational noises in the environments in order to keep noise power level well less than that restricted by its location at nearby a dense residential area of the city of Kobe.

Ambient noises by operating machineries change considerably due to their operating conditions. For examples, the operational noise of a pump changes owing to its output and an on/off of a soot blower of the boiler drastically changes ambient noise conditions near it. Therefore, a learning scheme is to be equipped in an IF-ASSET based system for preparing a suitable inverse filter at each monitoring site in order to cope with noise changes relating to changes in operating conditions in the power plant such as

    i)   Output/load change of the objective machinery, and
    ii)   Combination of on/off states of nearby machineries and/or valves.

It is extremely important to implement thorough preventative maintenance measures so as to guarantee a stable power supply for the city of Kobe that relies on this new plant. Operator of a power plant knows from his experience that the most common problems it would confront are caused by the corrosion of boiler pipes. A burst pipe leads to the loss of steam and water, and results in shutting down the plant

---

[*]PlantWatcher-type A is a registered trademark of Yamatake Corporation.

for a long period, which might have a fatal consequence. Conventional means to diagnose such a problem relies on a combined technique as vibration and temperature sensors, and periodic inspections by maintenance engineer. However, it is almost impossible for the means to detect a problem *in advance*. While it is true that very limited engineer with a special talent can sometimes detect unusual mechanical sounds which might indicate abnormal operation and the possibility of equipment failure, it is unpractical to have such an especially talented staff to patrol the plant round the clock. The new plant has been equipped with both the latest machinery and control systems, and naturally Shinko Kobe Power Inc. wanted to be implemented also with the most advanced monitoring system to detect potential problems beforehand.

2.2 Predictive Maintenance – PlantWatcher-type A

Fortunately, Shinko Kobe Power Inc. discovered a solution of this problem as a monitoring/diagnostic system, PlantWatcher-type A supplied by Yamatake Corporation. It is an innovative distributed online monitoring system based on IF-ASSET relying on analyses of signals collected by sensors on the spot. It is easy for the monitoring system to accomplish fine-tuning the system, updating the software, and adding/deleting sensors as the objective equipment is changed. Its clear display of up-to-date information means that even inexperienced personnel can easily understand the diagnosis and displays. Moreover, the incoming data is archived for later reference. Reliability and flexibility add to the appeal of this monitoring system, which provides the staff with a comprehensive and objective measures of operational status 24 hours a day and 7 days a week. This type of remote monitoring system is ideal for covering a large plant, especially for potentially dangerous area to personnel.

Real plants are full of powerful operational noises and one of solutions for the problem to detect a possible abnormality of the plants is to distinguish the faint noises generated by the abnormal machines from powerful ones generated by the others in normal operation. Noises from machines in normal operation are quite powerful and are both of continuous and intermittent. In signal processing, it is well known that a stationary noise can be whitened by the inverse filtering. Principle of IF-ASSET is to detect noises in a certain frequency band, after reducing sonic components of stationary ambient noise from normal operation of machines by the inverse filter designed on the basis of the noise signal collected under the normal operation. The spectrum of any stationary noise signal from normal operation can be whitened with the inverse filter, however, in any abnormal condition, the signal which has not been included in the ambient noise signal used to design the filter exists, and the signal can not be whitened by the inverse filter and gives some amount of coloured residual signal. PlantWatcher-type A is possible to monitor any abnormal conditions of a group of machines by the coloured signal gained by IF-ASSET.

TABLE 1: Site, number and objective of sensors.

Site	Nr. of sensors	Objective
Large fans	7	Bearing of motors/fans
		Leak of fluids
Burners	4	Blow out of coal powder
Coal mills/crashers	5	Malfunction of mills/crashers
		Large crashing sound
		Slipping of belts
Boiler feed water pumps	3	Bearing of turbines/motors
		Leak of water
Boiler tubes	4	Burst of boiler tubes
Super heaters of supply water	4	Leak of steam/liquids

PlantWatcher-type A for the new power plant has been equipped with 27 sensors (microphones), 24 of which are fitted with a parabolic sound collecting hood to improve the SNR. They literally listen into boilers, water pumps, fans and other important machineries. Table 1 shows sites, numbers and objectives of the sensors. Figure 1 shows an example of sensor at a water charger pump site. The collected signal — which includes proximity information for pinpointing sound sources — is sent for analysis to the monitoring computer installed in the central control room via Ethernet. In the computer, the incoming sound data is compared every two minutes with the data generated by the model to produce the residual signal, and power of which is displayed on the monitor screen.

Fig. 1 One of microphones with a sound collecting parabolic hood at a pump site.

Figure 2 shows examples of the computer monitor screen for (a) normal and (b) simulated abnormal operation-noises, respectively. The sound signal for the simulated abnormal operation is generated by adding a 2-8 kHz band noise with the SNR of 0 dB to the normal operation noise. Each screen shows incoming sound data at a point (upper right), data generated by the model (upper left) and power trend of the residual signal (bottom). The residual signal is the difference of the incoming data (upper right) and data generated by the model (upper left), in the present case. It can be clearly known by comparing the both figures that the power level of the residual signal (bottom) is as low as 0 dB for the normal operation noise (a), whereas more than 12 dB higher for the simulated abnormal case (b). Then, the level can provide a good indicator for the operational status of the objective machinery.

A monitoring system based on PlantWatcher-type A offers many advantages. The most important of these is IF-ASSET, that is, a mathematical model is constructed for an operational noise associated with normal operation. The computer checks incoming sound data using this model to detect any abnormality. IF-ASSET effectively simulates how the talented engineer would listen for signs of malfunction. However, unlike a human being, it is able to filter out background noise and detect very faint sounds. And in addition to greater accuracy, it performs its duties round the   clock,   without   rest.

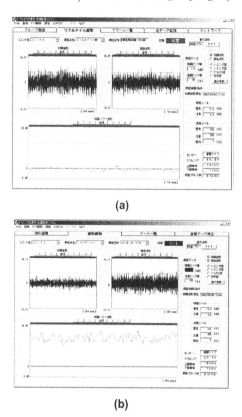

(a)

(b)

Fig.2 Examples of the computer monitor screen (a) for a normal and (b) for a simulated abnormal operations. Each screen shows incoming sound data at a point (upper right), data generated by the model (upper left) and power trend of the residual signal (bottom).

Should the system detect any sign of impending failure, it automatically notifies staff so they can take any necessary preventative measures. The clear display of up-to-date information means that even inexperienced personnel can easily understand the diagnosis and displays. Moreover, the incoming data is archived for later reference. Fine-tuning the system, updating the software, and adding more microphones/sensors as a new equipment is installed — all of these are easily accomplished by adopting the PlantWatcher-type A. Reliability and flexibility add to the appeal of this sophisticated monitoring system, which provides the staff with a comprehensive and objective measures of operational status 24 hours a day and 7 days a week.

2.3 Performance test

After the test run and tuning period, a preparatory performance test of the monitoring system was carried out on commercial running power plant from April to September, 2002. It was confirmed possible to detect any change in the operational noises by the present system though there were several erroneous alarm signals during the performance test. Some of the possible reasons for the erroneous alarm are as follows:

i)   Error for large-fans/boiler feed water pumps is due to power changes by the temperature change,

ii)   Error for boiler-tubes is due to nearby intermittent sound.

The first case has been resolved by adopting seasonal inverse filters, and for the second case, it is also resolved by changing inverse filters if the timing of an intermittent sound can know. It may, however, require some learning schemes to cope with if the timing cannot know.

3. Conclusions

An acoustical abnormality monitoring system, which is introduced to detect any abnormality of the total plant machinery of a 700 MW coal-fired power plant, is described. PlantWatcher-type A is adopted to guarantee a stable operation of the plant round the clock by implementing a preventive maintenance system without any experienced engineer. The monitoring system installs a distributed sound monitoring system and it enables 24/7 monitoring of plant equipments so as to detect predictive signs of malfunction prior to any failure. Any sign of suspecting abnormal operation is detected, the system immediately notifies staff so that timely preventative maintenance can be performed, minimizing downtime and maximizing efficiency.

The monitoring system has been at work for only a half year and the power plant itself is new as well, and it has experienced no opportunity yet to prove its full power. As plant equipments age, however, the importance of reliable 24/7 monitoring will grow steadily. And with the construction of the second unit, the Shinko Kobe Power Inc. is looking forward to further adopting of another PlantWatcher-type A based monitoring system to ensure that the city of Kobe can enjoy a reliable, uninterrupted supply of electricity for many years ahead.

References

[1] Ohba R., Tamanoi Y., Ohtsuka T., Komatsu M. and Maeda T. "Machine diagnosis by acoustic signal processing", *Sensors VI - Technology, Systems and Applications -* (K. T. V. Grattan and A. T. Augousti eds.) IOP Publishing (Bristol), pp. 193-199 (1993).

[2] Tamanoi Y., Ohtsuka T. and Ohba R. "Machine diagnosis using acoustic signal processing techniques and special sound collecting hood", *Trans. IEICE Jpn.* **E78-A**, pp.1627-1633 (1995).

[3] Ohba R. and Tamanoi Y. "Method and apparatus for machine diagnosis", *Jpn. Patent No. 3,020,349* (1999). (in Japanese)

[4] Ohba R., Tanaka T., Tamanoi Y., Matsumoto M., Shinagawa T., Ohtsuka T. and Noguchi Y. "Leak detection under overwhelming ambient noise conditions", *Sensors and their Applications VII* (A. T. Augousti ed.) IOP Publishing (Bristol), pp.119-124 (1995).

[5] Ohba R. and Tamanoi Y. "Subject identification method, apparatus and system", *U. S. Patent Nr. 5,798,459* (1998).

[6] Ohba R., Tamanoi Y., Tokui I. and Takeshima M. "Intelligent robot watchdog for leak monitoring in oil refinery", *Sensors and their Applications X* (N. M. White and A. T. Augousti eds.) IOP Publishing (Bristol), pp.239-244 (1999).

[7] Ohba R., Tamanoi Y., Kuwahara T. and Edanami K. "Intelligent robot watchdog for gas-leak monitoring", *Proc. of ISA* **4** Safety and Environmental (2000).

[8] Ohba R., Tamanoi Y., Hotta H. and Uchinuma S. "Development of an intelligent robot for monitoring gas leakage", *Proc. of Joint technical conference of ISA, JEMIMA and SICE 2001*, SE-1 (2001).

[9] Ohba R., Tamanoi Y., Kuwahara T. and Edanami K. "Gas-leak monitoring using an intelligent robot watchdog system", *Valve World*, **6**(3), pp. 53-57 (2001).

[10] Ohba R., Uchinuma S., Tamanoi Y., Kuwahara T. and Edanami K. "Robot for Abnormality Monitoring of Group Compressors", *Sensors and their Applications XI* (K. T. V. Grattan and S. H. Khan eds.) IOP Publishing (Bristol), pp.155-161 (2001).

[11] Uchinuma S., Tamanoi Y., Kuwahara T., Edanami K. and Ohba R. "Development of a robot for monitoring abnormalities of multiple compressors", *Proc. of Joint technical conference of ISA, JEMIMA and SICE 2001*, RE-1 (2001).

Section 5: Non-Destructive Monitoring
Paper presented at Sensors and their Applications XII, September 2003
©2003 IOP Publishing Ltd

Multi-frequency current injection model inversion for determining coating thickness and electromagnetic material properties of cylindrical metal bar or wire

S J Dickinson and A J Peyton

Engineering department, Lancaster University, Lancaster, Lancs LA1 4YR, UK

Abstract: This paper discusses the development of a non-distructive multi-frequency current injection method for determining conductive coating thickness and characteristic electromagnetic material properties on metal bar or wire. The development of an analytically derived forward model and the inversion of this model is described, and inversion results for a range of characteristic material properties are discussed.

1. Introduction

Non-destructive measurement techniques for the determination of conductive coating properties and thickness have many uses including production process control, quality assurance and inspection. This paper considers the theoretical development of a method for determining the thickness of a conductive coating on a bar or wire together with material properties using four wire impedance measurement data. The method requires complex impedance data taken over several decades of frequency as input, and provides five parameters: coating thickness; coating electrical conductivity; coating magnetic permeability; core conductivity; and core permeability; as output.

The method relies on the well known phenomena of eddy currents which push the bulk current flowing in a conductor towards surface as frequency is increased. This is known as the skin effect has been widely used for non-destructive testing (NDT) employing inductive coils and magnetic/eddy current techniques for sub-surface inspection [1] and crack detection [2,3]. More recently the use of spectral methods has enabled the accurate measurement of coating thickness with coatings of similar electrical conductivity [4], indicating the power and potential of employing spectroscopic techniques. The skin effect means that at higher frequencies the electromagnetic properties of the material near the centre of the bar or wire will have less effect on the overall impedance than at lower frequencies. Scanning the frequency in this way thereby allows enough independent data to be obtained to determine the material electromagnetic properties over the cross-section of the bar or wire. In this paper, the scope of the problem has been limited to a solid bar consisting of a uniform material and a single coating of another uniform material having a uniform thickness. In this way a high degree of information redundancy can be guaranteed which is necessary to minimise the effects of random signal noise.

Inductive methods have the significant advantage of requiring no electrical contact with the object under test, however, the measurement coils are subject to parasitic impedances, which limits their physical size and frequency range while requiring expert users to perform the measurement and calibration procedures. Direct current injection and resistance measurements can be used where access to a conducting surface is permitted. Unfortunately, possibly as a consequence of the perceived difficulties in obtaining reliable electrical contacts, this approach has possibly received less attention. Nevertheless, single frequency methods are successfully used to measure crack depth in industry [5]. This paper now discusses the development of a forward model, an inversion algorithm and presents inversion results for a wide range of material scenarios.

2. Forward Model

The bar or wire is modelled as two homogeneous layers, where layer one the coating and layer two is the core, each of which with a known electrical conductivity and magnetic permeability σ_1, σ_2, μ_1 and μ_2. Each layer has an outer radius R_1 and R_2 and the outer layer has a thickness $t_1=R_1-R_2$. The purpose of the forward model is to facilitate the calculation of theoretical electrical impedance values from assumed parameter values R_1, t_1, σ_1, σ_2, μ_1 and μ_2 for direct comparison with impedance values obtained by physical measurement of the bar or wire. This model is subsequently inverted in order to determine parameter values t_1, σ_1, σ_2, μ_1 and μ_2 from measured impedance values, which can be obtained using the standard four wire measurement method and outer radius R_1.

The system has circular symmetry over the bar or wires cross section, and as long as the bar or wire is long with respect to its radius then translational symmetry can be assumed around a region central to its length. This means that there will be field solutions that are independent of polar angle θ and z and consequently the system can be modelled as a one-dimensional problem where fields are only a function of the radius r. There is an appropriate solution of Maxwell's equations where the electric field and current are normal to the cross section and that the magnetic flux is circumferential.

The impedance per meter of the bar or wire is calculated for a particular frequency by first determining the total current flowing through the bar for a given applied electric field strength $\mathbf{E_1}$ along its surface (in the z direction) and then applying ohms law. An expression for the total current flowing through the bar may be obtained by integrating the current density $\mathbf{J}$ over the area of the bar's cross section, i.e.

$$I = \int_0^{R_1} 2\pi r \mathbf{J}_r\, dr = \int_0^{R_1} 2\pi r \sigma_r \mathbf{E}_r\, dr \tag{1}$$

where R_1 is the outside radius of the bar and $\mathbf{J_r}$ is the current density at radius r. In order to perform this integration, a function that relates $\mathbf{E_r}$ at all radii ($<R_1$) to $\mathbf{E_1}$ is needed and this is obtained by solving Maxwell's equations. According to Maxwell's equations from Faraday's law (for non-magnetic materials) and Maxwell's curl equation from Ampère's circuitial law (in the case of a good conductor where $\sigma >> \omega\varepsilon$)

$$\nabla\times\mathbf{E} = -j\omega\mu\mathbf{H} \text{ and } \nabla\times\mathbf{H} = \sigma\mathbf{E}. \tag{2}$$

Curling both sides of equation 2a and substituting the vector identity $\nabla(\nabla\bullet\mathbf{E}) - \nabla^2\mathbf{E} = \nabla\times\nabla\times\mathbf{E}$ and then combining equations 2a and 2b gives

$$\nabla^2\mathbf{E} - j\omega\mu\sigma\mathbf{E} = 0, \tag{3}$$

as $\nabla\bullet\mathbf{E}$ is zero in a conductor.

Expanding equation 3 using the standard Laplacian expansion for cylindrical co-ordinates with zero terms removed gives the diffusion equation for a cylindrical conducting media

$$\nabla^2\mathbf{E} - j\omega\mu\sigma\mathbf{E} = \frac{\partial^2 E_Z}{\partial r^2} + \frac{1}{r}\frac{\partial E_Z}{\partial r} + \gamma^2 E_Z = 0 \tag{4}$$

having the applied E field oriented in the z direction, where $\gamma^2 = -j\omega\mu\sigma$. The standard solution to equation 4 is

$$E_Z = A\,\mathrm{J}_0(\gamma\, r) + B\,\mathrm{Y}_0(\gamma\, r) \tag{5}$$

where $J_0()$ and $Y_0()$ are the Bessel functions of order zero of the first and second kind respectively, A and B are coefficients determined by the boundary conditions and $\gamma = (1-j)\delta^{-1}$, where $\delta = \left(\sqrt{0.5\omega\mu\sigma}\right)^{-1}$ which is generally known as the skin depth equation.

At the centre of the bar, i.e. when $r = 0$, the Bessel function $Y_0(\gamma\, r) = \infty$. However E_Z is finite at the centre, therefore the B coefficient for the central layer must be zero, i.e. $B_2 = 0$, hence for the inner and for the outer layers respectively

$$E_Z = A_2 J_0(\gamma_2 r) \quad \text{for } r \le R_2, \text{ and} \tag{6}$$

$$E_Z = A_1 J_0(\gamma_1 r) + B_1 Y_0(\gamma_1 r) \quad \text{for } R_2 \le r \le R_1 \tag{7}$$

The unknown coefficients A_1, B_1 and A_2 can be determined for a particular set of model parameter values and boundary conditions. In this case we have three unknowns therefore we require three independent equations to solve. Three suitable equations are

$$E_1 = A_1 J_0(\gamma_1 R_1) + B_1 Y_0(\gamma_1 R_1) \tag{8}$$

which specifies the relationship between the electric field on the surface and r,

$$A_1 J_0(\gamma_1 R_2) + B_1 Y_0(\gamma_1 R_2) = A_2 J_0(\gamma_2 R_2) \tag{9}$$

which specifies that the electric field on the outside of the inner layer is the same as the electric field on the inside surface of the outer layer, and

$$A_1\gamma_1 \frac{\partial J_0}{\partial r}(\gamma_1 R_2) + B_1\gamma_1 \frac{\partial Y_0}{\partial r}(\gamma_1 R_2) = A_2\gamma_2 \frac{\partial J_0}{\partial r}(\gamma_2 R_2) \tag{10}$$

which specifies that the tangential magnetic field on either side of the material interface is continuous, i.e. the gradient of E with respect to radius on the outside of the inner layer is the same as the gradient of E on the inside surface of the outer layer.
Solving equations 8, 9 and 10 gives

$$A_1 = E_1 \frac{-J_0(\gamma_1 R_2)\gamma_2 J_1(\gamma_2 R_2) + J_0(\gamma_2 R_2)\gamma_1 Y_1(\gamma_1 R_2)}{den} \tag{11}$$

$$B_1 = E_1 \frac{J_0(\gamma_1 R_2)\gamma_2 J_1(\gamma_2 R_2) - J_0(\gamma_2 R_2)\gamma_1 J_1(\gamma_1 R_2)}{den} \tag{12}$$

$$A_2 = E_1 \frac{J_0(\gamma_1 R_2)\gamma_1 Y_1(\gamma_1 R_2) - J_0(\gamma_1 R_2)\gamma_1 J_1(\gamma_1 R_2)}{den} \tag{13}$$

where
$$den = -J_0(\gamma_1 R_1)J_0(\gamma_1 R_2)\gamma_2 J_1(\gamma_2 R_2) + J_0(\gamma_1 R_1)J_0(\gamma_2 R_2)\gamma_1 Y_1(\gamma_1 R_2) + $$
$$J_0(\gamma_1 R_2)Y_0(\gamma_1 R_1)\gamma_2 J_1(\gamma_2 R_2) - \gamma_1 J_1(\gamma_1 R_2)Y_0(\gamma_1 R_1)J_0(\gamma_2 R_2)$$

The total current is now determined by expanding equation 1 using equations 6 and 7, thus

$$I = \int_0^{R_2} 2\pi r A_2\sigma_2 J_0(\gamma_2 r)dr + \int_{R_2}^{R_1} 2\pi r\sigma_1 [A_1 J_0(\gamma_1 r) + B_1 Y_0(\gamma_1 r)]dr \tag{14}$$

and performing integration gives the final equation for total current

$$I = 2\pi \left[\frac{A_2 R_2\sigma_2}{\gamma_2} J_1(\gamma_2 R_2) + \frac{A_1\sigma_1}{\gamma_1}\{R_1 J_1(\gamma_1 R_1) - R_2 J_1(\gamma_1 R_2)\} + \frac{B_1\sigma_1}{\gamma_1}\{R_1 Y_1(\gamma_1 R_1) - R_2 Y_1(\gamma_1 R_2)\} \right] \tag{15}$$

3. Model inversion

The inverse problem is highly non-linear and potentially ill-conditioned, therefore an iterative algorithm based on the Newton-Raphson method and employing the Tikhonov regularisation method [6] has been developed to invert the forward model previously described. The inversion algorithm takes n complex impedances at different frequencies as input, and provides values for the five unknown parameters as output.

The inversion method is outlined in figure 1 and is an iterative process consisting of the three main operations: 1. Find the inverse partial derivatives at the current operating point by inversion of the Jacobian matrix; 2. Find the error at the current operating point; and 3. Take a small step in the direction to reduce the error, clipping to predefined limits. This process is then repeated until the algorithm converges.

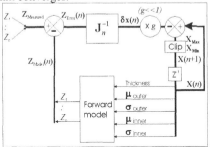

Figure 1 : The inversion loop

The forward model is highly non-linear, therefore the Jacobian matrix is generated on each model inversion iteration. The inverse model partial derivatives are found by inversion of the Jacobian matrix, however as this is not square it cannot be inverted without some reshaping, and as it is also potentially ill-conditioned, a degree of regularisation is also required. Pre-scaling is also required as the Jacobean matrix's columns differ by many orders of magnitude. The initial pre-scaling is performed by dividing each column by the mean of each column

$$\mathbf{J}_S = \mathbf{J}./\mathbf{M} \tag{16}$$

where $\mathbf{M}$ is a row vector containing the mean values of each column and the "./" operator means perform a divide on an element by element basis (in this case, column by element). The regularised equation for determining $\delta\mathbf{x}$ from a given error $\mathbf{Z}_{Error}$ (i.e. $\delta\mathbf{Z}$) is therefore

$$\left(\left[\mathbf{J}_S^T\mathbf{J}_S + \mu\mathbf{I}\right]^{-1}\mathbf{J}_S^T\delta\mathbf{Z}\right)/\mathbf{M}^T = \delta\mathbf{x} \tag{17}$$

where $\mathbf{I}$ is an identity matrix, μ is a regularisation parameter. A typical inversion run with sixteen octave spaced frequencies ranging from 200 Hz to 6.5 MHz is shown in figure 2. The first two plots show the phase and magnitude of the 'hidden' and 'found' impedances superimposed. The other plots show the paths of the five parameters, the condition of the inverted Jacobian matrix and the maximum differential, which is used to terminate the process.

4. Results and Conclusions

The model inversion algorithm outlined here with appears to work well for all parameter combinations tried so far. Figure 3 shows results obtained for a wide range of core and outer coating material conductivities, all with a relative permeability of one. All of these inversions converged on the correct values in a fashion similar to that shown in figure 2 and the variation in accuracy may be attributed to the termination criteria employed, which in this case was a threshold for the maximum normalised partial derivative of the inverted Jacobian matrix.

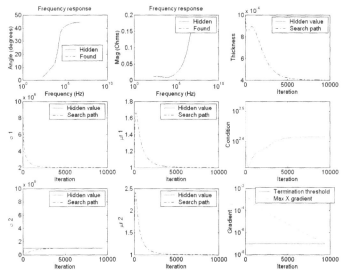

Figure 2: Inversion algorithm progress in operation (layer 1 is outer, layer 2 is inner)

Figure 4 shows the number of iterations taken to obtain the previously mentioned results. One possible explanation for the consistently sudden dramatic change in required iterations for lower coating conductivities is that a critical ridge may occur in the parameter space causing an initial decent on the 'wrong' side of the 'hill', therefore causing a long and tedious search path. This will need further investigation.

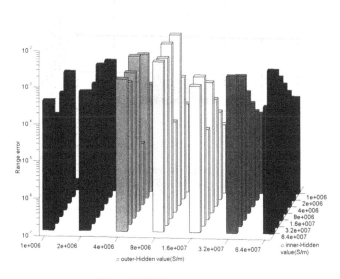

Figure 3: Maximum range error

Initial trials with fewer and lower frequencies failed for certain combinations of non-ferromagnetic materials due to the associated skin depths being significantly larger than the minimum coating thickness limit, i.e. effect of the coating was not seen by the algorithm. This problem has been solved by setting a finite minimum limit on the coating thickness (10 μm was used for the results shown), and increasing the frequency range upward to 6.5 MHz. Likewise the lowest frequencies (from 200 Hz) are needed for the successful inversion of ferromagnetic materials with associated smaller skin depths together with thicker coatings.

It is intended that future investigations will address issues such as the optimal choice of frequency range and frequency spacing for particular ranges of material parameters, the effects of signal noise on accuracy and performance, and the improvement of inversion algorithm speed.

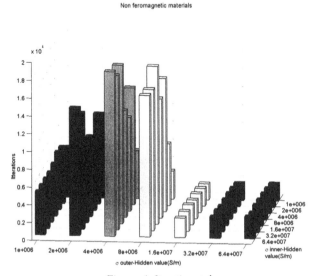

Figure 4: Iterations taken

5. Acknowledgements

The authors would like to thank the UK Engineering and Physical Sciences Research Council for their financial support of this work and Amos Dexter for his invaluable advice.

References

[1]   UZAL E. AND ROSE J.H., (1993), The impedance of eddy current probes above layered metals whose conductivity and permeability vary continuously, In IEE trans' on magnetics 29, pp1869-73.

[2]   BOWLER J.R., (1997), Review of eddy current inversion with application to non-destructive evaluation, In Int.J.Appl. Electromag. & Mech., 8, pp.3-16.

[3]   STANLEY R.K., (1996) Magnetic methods for wall thickness measurement and flaw detection in ferromagnetic tubing and plate, Insight, 38, pp 51-5.

[4]   Antonelli G. Ruzzier M. and Necci F., (1998) Thickness measurement of MCrAlY high temperature coatings by frequency scanning eddy current eddy current technique, Trans. ASME, J.Eng. Gas Turbines & Power, 120 pp.537-42.

[5]   Deutsch V. Ettel P. Platte M. and Cost H., (1996) *Crack depth measurement. State of the art measurement technology for an established method* (in German) Materprufung, 38, pp.306-10.

[6]   TIKHONOV A.N., LENOV S and YAGOLA A., (1998), Non-linear ill-posed problems, Chapman & Hall publications vol.1 & 2.

Section 6

Imaging Sensors

Section 6: Imaging Sensors
Paper presented at Sensors and their Applications XII, September 2003
©*2003 IOP Publishing Ltd*

Measuring pedestrian trajectories using a pyroelectric differential infrared detector

A Armitage, T D Binnie, J Kerridge, L Lei

School of Computing, Napier University, 10 Colinton Road, Edinburgh, EH10 5DT
Email: a.armitage@napier.ac.uk

Abstract: Low-cost, low-resolution infrared detectors have been used for measuring the trajectories of pedestrians. The detectors have been designed for counting the number of pedestrians crossing a line. The use of these detectors has been extended to provide complete trajectories across measurement areas approximately 3 metres square. This provides an effective way of rapidly measuring large numbers of pedestrian movements. Current work involves extending the effective area by combining trajectories from multiple detectors. Matching across detectors has been made difficult by the presence of edge effects. Progress is being made on the algorithms needed to track across fields of view.

1. Introduction

Accurate measurement of pedestrian trajectories is useful in a variety of situations. In retail environments, information about customer behaviour and preferences can be derived from knowledge of customer's movements. In crowded locations (such as sports venues, railway station concourses and so on) pedestrian movement influences safety concerns. Our primary concern is in the general planning of pedestrian movements within the urban environment (shopping malls, shopping streets, pedestrian concourses and the like). A related project at Napier University (Pedflow[1]) has been concerned with predicting pedestrian flows in such settings. Validation and verification of such models requires data of real pedestrian movements, both in the initial formulation of behavioural rules, and in the testing of the resultant models. Traditional methods of gathering such data are laborious, and difficult to automate. The overall aim of this project is to automate this process, allowing for the gathering of large amounts of data for use in model derivation and validation.

2. Detecting Pedestrians

Pedestrian movement can be monitored by a variety of means. Observers with clipboards can be used, though it is difficult to get detailed trajectory information in real time. If the flow can be restricted, break-beam technology can be used to count pedestrians passing by. This method becomes very inaccurate as the width of area covered increases. This is because several people may pass through side-by-side, with inevitable masking of the beam.

Video recording is a very useful technique, as it allows scenes to be replayed, or slowed down, while an observer notes the trajectory of individuals. Unfortunately, there is currently no reliable method for automatically extracting behaviour using software. Urban scenes at visible wavelengths are often very confused and cluttered. Changes in lighting and shadows (as the sun goes behind a cloud, for instance) all go to reducing the accuracy of software methods. Software is available that can help with the extraction of detailed information, but it must be used by an operator who can guide the process. Although time-consuming, this can be a reliable method for extracting detailed measurements[2].

3. The Infrared Detector

Long-wavelength infrared detectors are difficult to fabricate in arrays. Semiconductor arrays of Mercury Cadmium Telluride, Indium Antimonide and the like have been used for some time in astronomical and other applications, but at a high cost. This is partly due to the costs of the exotic semiconductor, but also to the cooling system that is needed for noise reduction. Recently, uncooled arrays of microbolometers have become available, leading to lower system costs. Despite this, the cost of an infrared detector is still of the order of 100 times that of a detector working at visible wavelengths.

The detectors used in this study have been developed by Irisys[3] as an attempt to provide a relatively low cost infrared array detection device (the detectors described on the company web site are actually a later model than the ones used here). The detector is based on an array of pyroelectric ceramic detection elements[4,5]. The array is a square format, with 16 rows and 16 columns of pixels. This is a relatively small number of elements, resulting in a low-resolution image (figure 1). However, the cost of the detector is of the order of 1/20th of that of a traditional IR array.

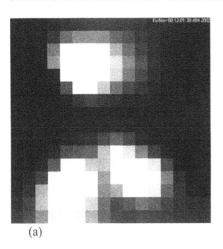 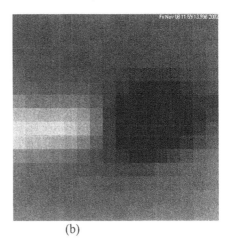

(a)   (b)

Figure 1. In image (a), three pedestrians are moving from right to left. This is the normal situation, where pedestrians are hotter than the background. In (b), a pedestrian has just entered an indoors environment and is moving left to right. Taken in Edinburgh in November, the outer layers of the pedestrian are colder than ambient, and they show as a dark mass. Note the faint warm 'wake'. This is caused by the detector being sensitive to changes in temperature. In this case the change from cold pedestrian to warm ambient produces a temperature increase, i.e. an apparent warm wake.

Unusually for a pyroelectric detector, there is no optical chopper. Pyroelectric devices respond to change in temperature. When imaging an infrared scene, pyroelectric detectors view the radiation after it has been chopped, usually by a rotating mechanical device. The detector then effectively measures the difference between the scene temperature and the temperature of the chopper blades. To keep the system costs down, the Irisys detector contains no chopper (a more expensive version is available that includes a mechanical chopper). This means that the detector responds only to changes (positive or negative) in the scene temperature. For pedestrians walking at normal rates, this is not a problem. The time constant of the detector is several seconds, so a pedestrian that stands motionless disappears from the scene after a few seconds, and this does cause difficulties when monitoring queues.

In normal use, the detector is suspended at a height of about 3 metres above the area to be measured. At ground level, a square approximately 3 metres by 3 metres is imaged. The images can be relayed via a standard serial connection to a computer for display and recording. The detector is intelligent, with an on-board DSP chip that processes the image. Pedestrians are monitored by fitting an ellipse to candidate 'blobs' in the image area. The DSP then sends, via the serial interface, details of candidate targets. These details include position, change in position, status information, and the characteristics of the ellipse that has been fitted to the target. The status information shows whether the target is hotter or colder than ambient, whether it is a new target in the scene, and whether the candidate target has fragmented (this happens, for instance, if pedestrians hold an arm out). Figure 2 shows a typical scene with a number of pedestrians that have been identified.

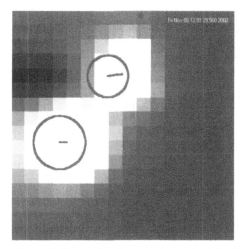

Figure 2. Two pedestrians moving to the right. Note the visual indication by a short bar of the current velocity. There is a slight delay between the time at which the image is taken, and the time at which the ellipses are measured. This results in an apparent mis-registration between the two on the image. Although this occurs on the displayed information, the recorded trajectory information is actually accurate.

4. Matching between detectors

The pedestrian counter is normally intended for use counting targets moving across a straight line (horizontal or vertical in the field of view). As such, it functions well counting people entering supermarkets, moving through corridors, and passing across railway and airport concourses. In our application area, we are interested in pedestrian movements within an urban environment (e.g. shopping streets). We would like to be able to derive trajectories for pedestrians as they move in arbitrary directions through a scene. In principle, tracking a pedestrian in the field of view of a single camera is not difficult. The real difficulty arises when the scene is so large that it has to be covered by multiple cameras. Tracking between adjacent cameras is not easy, as the low resolution of the image means that it is not possible to identify individuals. Matching has to be done on the basis of trajectories, and matching across boundaries. It would be possible to overlap fields of view, so that pedestrians appear simultaneously in the fields of view of two detectors. However we wished to avoid this, as it substantially reduces the area covered by the detectors.

5. Limitations of the detector

The obvious limitation of the detector is the low resolution. As far as our application is concerned, this is not a great problem. The use of an algorithm to fit ellipses round targets means that the average centroid of the target is located with sub-pixel accuracy. With a field of view of 3 metres, each pixel spans approximately 19 cms, and the location of the target is known to an accuracy of the order of 10 cms.

The detector has a wide field of view in order to cover a reasonable area on the ground, without having to be mounted too high up. This wide angle inevitably results in distortion, as can be seen in figure 3. This distortion is not a problem, as it can be allowed for.

Figure 3. The effect of the wide-angle lens can be seen in this cumulative plot of trajectories. Although this has been taken in a straight corridor, with the bulk of pedestrians moving from side to side, the apparent width of the corridor is less at the left and right hand sides than in the centre.

Another problem is the low frame rate of the detector. Internally, the DSP runs at a frame rate of 33Hz. Limitations in the communications mean that only every tenth frame can be transmitted to the PC where the images are displayed, recorded and processed. For pedestrians walking briskly through the field of view, this means that typically only four or five points are recorded as they pass through. This problem is compounded by the fact that the DSP can not always be confident of identifying a target as it enters a frame (the pedestrian may be half in and half out of the frame). Consequently, some of the frames at the edges are marked as invalid. This effect can be seen in figure 4. Irisys have recently introduced a new version of the detector, the IRC 1004. This has an improved communications link, and should not suffer from this limitation.

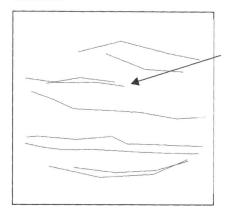

Figure 4. Trajectories are constructed from a relatively small number of points. Some trajectories start a long way in from the edge. This is normally due to the DSP in the detector rejecting the first or last point.

The area covered by a single detector is approximately 3 metres square. Ideally, we would like to be able to cover larger areas by tracking pedestrians between detectors. To do this, we have been working on matching targets across the boundaries. There is not enough detail in the images to be able to recognise individuals, so decisions must be based purely on the matching of trajectories between fields of view. The edge effects mentioned above have made this harder than we had originally anticipated, as we have to search quite far in for a candidate target. Initially, we follow a target out of one field of view, then look for candidate targets appearing within three to four pixels of the boundary of the next field of view. Targets that appeared within the correct area within one second were considered for matches. However, this missed matches because first and last readings were nor valid, resulting in a larger than expected gap between fields of view. Figure 5 illustrates the improvement that results when increasing the permissible time gap between a target disappearing from one field of view, and appearing in a new field. The number of matches increases up to about 1.5 seconds, and then does not increases further, as all possible candidate targets have been found within this time. Although adjusting the matching algorithm has improved the number of matches found, we still lose track of too many targets between frames (success rates are of the order of 60%). Current work is therefore focussing on improving the algorithms further. Our initial plan was to be able track across four detectors. As this means some pedestrians will cross three boundaries, we will need to improve the success rate for the technique to be successful. However, the most recent versions of the detector produce more trajectory points per second, and this should improve the readings at the edge of the fields of view.

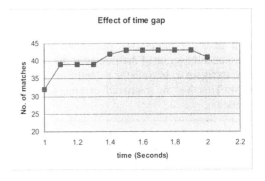

Figure 5. Effect of increasing the permissible time gap between a target moving from one field of view to the next.

6. Results

The detector works well at gathering large numbers of pedestrian trajectories. Figure 6 illustrates measurements taken over a 10 minute period in a crowded corridor. An obstruction had been placed in the middle of the corridor. Approximately 85 pedestrians passed under the detector in the time. On occasions, seven or eight pedestrians may be in the field of view at the same time; the detector manages to track them all successfully. Although coordinate data is available for the individual trajectories, a simple plot such as that of Figure 6 provides a good impression of the overall flow. Ultimately we want to be able to derive accurate measurements that can be used as the basis of pedestrian behaviour in the agent-based models we are currently developing[1].

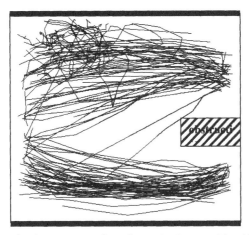

Figure 6. Flow of pedestrian around an obstruction in a corridor. Note the appearance of a 'loitering' pedestrian in the upper left quadrant.

7. Conclusion

The Irisys detector efficiently measures pedestrian movements over an area of approximately 3 to 3.5 metres square. The data that is produced is in a readily analysed format. Slight distortions in the coordinates due to the optical arrangement are not important. The main limitation is the errors found in some measurements at the edges of the field of view. This leads to difficulties when trying to correlate the movements of pedestrians across multiple fields of view. New versions of the detector should not have these problems

Acknowledgements

Funding for this project has been provided under the LINK Future Integrated Transport Project: FIT002. The cooperation of Irisys in providing equipment and information is also gratefully acknowledged.

References

[1]   Kukla, R. & Kerridge, J., "PEDFLOW: Development of an Autonomous Agent Model of Pedestrian Flow", 80th Annual Meeting TRB, Washington, DC, USA, 2001,
[2]   Kukla, R. & Kerridge, J., "Laying the foundations: the use of video footage to explore pedestrian dynamics in PEDFLOW", Pedestrian Evacuation and Dynamics, 2001 Duisburg, Germany, 2001,
[3]   www.irisys.co.uk
[4]   Hollock, S,. Galloway, J. L., " Smart Sensors Using Array Based Infrared Detectors", Sensors and their applications XI/ISMCR 2001 - Institute of Physics, City University London, September 2001
[5]   Mansi, M. V., Porter, S. G. Galloway, J. L. &. Sumpter N., "Very low cost infrared array based detection and imaging systems" (SPIE) Aerosense 2001, Orlando, Florida USA, 17-19 April 2001

Concurrent Measurement of Mass Flow Rate and Size Distribution of Pneumatically Conveyed Particles Using Combined Electrostatic and Digital Imaging Sensors

R M Carter and Y Yan*

Centre for Advanced Instrumentation and Control, School of Engineering
University of Greenwich at Medway, Chatham Maritime, Kent. ME4 4TB.
UK.
E-mail: Y.Yan@gre.ac.uk

Abstract: On-line, concurrent measurement of mass flow rate and size distribution of particles in a pneumatic suspension is desirable in many industries. This paper presents the basic principle of and initial results from a novel instrumentation system that uses a combination of electrostatic and digital imaging based sensors in order to achieve these goals. An inferential approach is adopted for the mass flow measurement of particles where velocity and volumetric concentration of particles are measured independently. The velocity of particles is determined by cross correlating two signals derived from a pair of electrostatic sensors, whilst the volumetric concentration of particles is obtained from a digital imaging sensor. The imaging sensor also provides particle size distribution data. Results obtained from a pneumatic conveyor are presented which show good performance of the system for both mass flow metering and particle sizing. Variation of instantaneous result data is discussed and a possible link with imaging frame rate and volumetric concentration of particles introduced. In general, the results obtained are encouraging and the system shows great promise.

Keywords: pneumatic conveying, particulate flow, particle size, digital imaging, mass flow rate, electrostatic sensors

1. Introduction

Pneumatic conveyance of particulate materials is an important technique widely used in many industries. In order for pneumatic conveying systems to deliver all of their potential benefits it is essential to know the parameters of the flow. Much research work has been carried out in the field of particulate flow metering [1] and many techniques have been developed, ranging from gamma rays to optical attenuation [2]. Instruments based on these techniques are designed to measure various flow parameters but few of them can measure *absolute* mass flow rate of particles and particle size distribution on-line [2, 3].

It is clear that an instrumentation system must be developed that can measure mass flow rate and particle size distribution on an on-line continuous basis and that can be installed in a non-intrusive manner (i.e. the sensor does not project into the duct or modify the flow in any way). It is recognised that such an undertaking would be 'technically challenging' [2]. A novel digital imaging based system capable of measuring particle size distribution and volumetric concentration of particles on an on-line and non-intrusive basis has already been developed [4]. This in itself represents a major advance over current sizing techniques but, when combined with well established electrostatic velocity metering technology [5], it is possible to derive the *absolute* mass flow rate of particles. The volumetric concentration of particles is measured by the imaging sensor and the velocity of particles derived from the electrostatic instrument. This paper presents the key design aspects of both sensing techniques along with experimental results demonstrating the effectiveness of the methodology.

The present research attempts to combine the benefits of each sensing type or strategy whilst avoiding possible shortcomings, thus producing a system that is worth more than the sum of its parts. Electrostatic sensors, whilst highly reliable for velocity measurement, are unsuitable for *absolute* measurement of volumetric concentration of particles due to the highly unpredictable nature of the magnitude of electrostatic charge found on the particles. An imaging based approach, on the other hand, is well suited to performing particle sizing and, therefore, concentration based measurements. Whilst the measurement of particle velocity is possible with an imaging system a different sensing arrangement is required to a concentration measurement set-up and so two systems would still be required. The work presented, in its present form, uses a novel, complimentary, combination of proven and cost-effective sensor types in order to achieve its goals.

2. Sensors and Measurement Principle

2.1 Digital Imaging Sensor

When particles are flowing through a pipeline it is possible to illuminate a 'slice' of the flow using a laser sheet and to acquire images of that slice using a CCD camera. The basic concept and sensing arrangement are illustrated in figure 1. Once the images have been obtained, digital image processing techniques can be applied to extract information such as particle size distribution and solids concentration. Such a system has been found capable, in its present form, of achieving a basic accuracy of ±1.5% when used with particles in the 150µm to 25mm range [4].

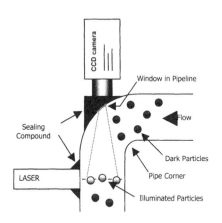

Figure 1. Imaging sensor arrangement

2.2 Electrostatic Sensor

The movement of particles in a pneumatic pipeline generates a certain level of net electrostatic charge on the particles due to particle-particle interactions, particle-wall collisions and particle-air friction [5]. Although the *amount* of charge carried on particles is usually unpredictable, the *dynamic variation* of charge can be detected by an insulated electrode in conjunction with a suitable signal processing circuit. As the charged particles pass through the electrode, a constantly changing electrical signal is generated. If two signals are acquired from a pair of axially spaced circular electrodes that are a known distance apart, then the velocity of particles can be derived through cross correlation between the two signals [1]. The general idea is illustrated in figure 2. Extensive evaluation of such a system has established its high repeatability, good linearity and fast response time [5].

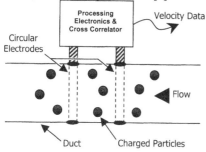

Figure 2. General electrostatic sensor arrangement

Since the velocity and volumetric concentration of particles are now known, these quantities may now be combined, thus allowing the mass flow rate to be deduced, using the following equation:

$$q_m(t) = A \rho_t V(t) \beta(t) \tag{1}$$

Where, q_m is the mass flow rate of particle (kg/s), ρ_t is the true density of particles (kg/m^3), V is the velocity of particles (m/s), β is the volumetric concentration of particles (%), A is the cross-sectional area of the duct (m^2), and t is time.

3. Experimental Set-Up

In order to provide controllable conditions for assessing the new instrumentation system a particulate flow test loop was set up. Figure 3 shows the layout of the test loop along with the locations at which the two sensors were installed. The total flow length is roughly six meters.

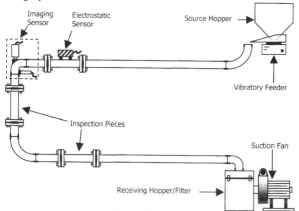

Figure 3. Layout of the particulate flow test loop

A vibratory feeder was used to provide accurate and consistent control of the introduction of particulate material to the duct. The internal duct diameter used in this case was 40mm and a variable speed suction fan provided control over material velocity within the loop. The electrostatic and imaging sensors were installed as close to one another as possible in order to prevent time delay errors.

4. Results and Discussion

In order to test the system a small quantity of domestic table salt was fed through the system. A pure salt, with no additives, was used since this material has a known absolute density of $2163kg/m^3$ [6] and also sports a fairly consistent mean particle diameter which, in this case, is about $196\mu m$ (see size distribution later in this section). A 50g sample was fed through the flow loop – this took 20 seconds at the selected feed setting giving an average mass flow rate of 0.0025kg/s.

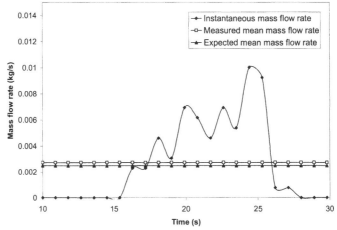

Figure 4. Measured instantaneous and mean mass flow rate compared to expected mean value

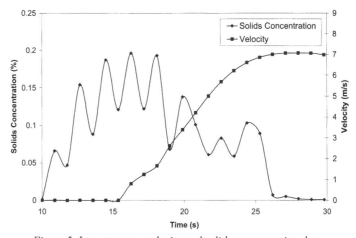

Figure 5. Instantaneous velocity and solids concentration data

Figure 4 shows the variation of the measured instantaneous mass flow rate of particles with time during the run. Also shown are the time-averaged mass flow rate (i.e. mean of the instantaneous data) and the expected mean mass flow rate (0.0025kg/s – see above). The time axis does not start at zero since the material was introduced into the rig some time after data logging commenced. The time during which significant material was flowing is clearly visible. Although the full 20 seconds of the run is shown here it took a few moments for the flow to build up and afterward, to subside. The variable nature of the instantaneous curve is thought to be due to a combination of low volumetric concentration of particles (see figure 5) and low image processing frame rate (about 1f/s) leading to a fairly low number of particle image samples. However, the time-averaged mass flow rate is very close to the expected mean deviating by only 0.0002kg/s absolute which gives, in this case, 8% relative error. This may seem high but is very promising in view of the dilute nature of the flow. It is quite possible that human error whilst timing the 20s material run may be contributing to the error – also the original 50g sample was weighed out using a balance accurate only to the nearest gram.

Figure 5 shows the raw data from which figure 4 was derived. By comparing figures 4 and 5 it is clear to see the relationship between concentration, velocity and mass flow rate. It should be pointed out that, during the initial five seconds, the velocity is so low that, despite the concentration values, only an insignificant portion of the sample was used. A longer run would avoid this problem since the velocity would reach a steady state. It is possible that this minor issue could also have contributed to the inaccuracy in the mass flow rate (see above) in this case – this source of error would be eliminated in a real, continuous flow, situation.

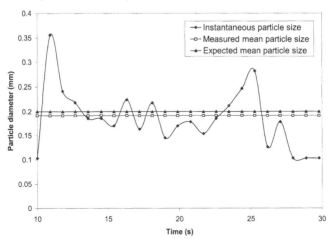

Figure 6. Measured instantaneous and mean particle diameter compared to expected mean particle diameter

Figure 6 shows the variation of the measured particle diameter with time during the run and also the time-averaged value and the expected mean value. The measured particle diameter varies during the run but only within believable boundaries. The mean value is very close to the expected value, deviating by 5μm giving 2.5% relative error.

Figure 7 shows details of the particulate size distribution for this material. It can be clearly seen that the mean size lies within the range 180-212μm. The centre of this range is 196μm which was used as the expected mean value in this case. The measured mean is well within the correct range and the relative error of 2.5% is given only as a guide to accuracy. All results and errors are summed up in table 1.

Table 1. Results comparison

	Expected	Measured	Relative Error
Mean mass flow rate	0.0025kg/s	0.0027kg/s	8%
Mean particle diameter	196µm	191µm	2.5%

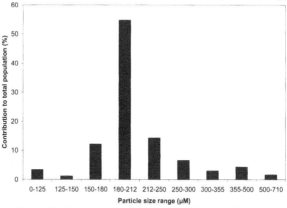

Figure 7. Measured particle size distribution of sample

5. Conclusions

This paper has presented the basic concept, operational principle and initial evaluation of a novel measuring system based upon the combination of digital imaging and electrostatic sensors. The system is intended to be used to measure size distribution and mass flow rate of particles in a dilute pneumatic suspension. Results obtained have shown that the mean particle size measurement is accurate to within 2.5% and that mass flow rate has been measured to within 8%. It has been shown why the particle size error should be considered as a guide only and it has been pointed out that the mass flow rate error may have been increased by human error and inaccuracies in the reference data. The on-line tests performed demonstrated that the system is capable of real world operation without impeding the flow in any way.

The potential benefits of this system are clear. Both imaging and electrostatic sensors were originally designed to be cost-effective and so a system based upon this work could, in its final form, provide a relatively low-cost solution to many long-standing particulate flow monitoring problems.

Acknowledgements

The authors would like to thank PCME Ltd who provided the electrostatic sensing head.

References

[1] Yan Y and Stewart D 2001 Guide to the flow measurement of particulate solids in pipelines, *The Institute of Measurement and Control,* London.
[2] Yan Y 1996 Mass flow measurement of bulk solids in pneumatic pipelines *Meas. Sci. Technol.* Vol 7 pp1687-1706.
[3] Pohl M 2001 Particle sizing moves from the lab to the process *PBE International.* May 2001 pp25-31.
[4] Carter R M and Yan Y 2003 On-line particle sizing of pulverised and granular fuels using digital imaging techniques, *Meas. Sci. Technol,* accepted for publication.
[5] Ma J and Yan Y 2000 Design and evaluation of electrostatic sensors for the mass flow measurement of pneumatically conveyed solids, *Flow Measurement and Instrumentation*, Vol 11, No 3, pp195-204.
[6] The salt institute, web-site: - http://www.saltinstitute.org/

©2003 IOP Publishing Ltd

Imaging techniques for structures using X- and γ-photons

J Sun[1], M Campbell[1], D Bailly[1], N Poffa[1], G H Galbraith[1], R C McLean[2],
C Sanders[3] and G G Nielsen[4]

[1] School of Engineering, Science and Design, Glasgow Caledonian University,
Glasgow, UK
[2] Department of Mechanical Engineering, University of Strathclyde, Glasgow, UK
[3] Building Research Establishment, Scottish Laboratory, East Kilbride, Glasgow, UK
[4] GNI Ltd, Denmark

Abstract. This paper describes methods of imaging objects using a low energy γ-tomograph and a newly acquired X-ray system.. The original objective in this work was to develop the system to diagnose faults in dense flow regimes within pneumatic conveying lines. The methodology requires the system to be initially optimised using phantoms rather than moving objects. However, in the case of the γ-tomograph, a technical fault in one of the thirty detectors produced some interesting data and corresponding images. The X-ray system has been used to generate images within wooden slices, although the time required to accumulate the data for these images is exceedingly long compared with the tomograph.

1. Introduction. –Part 1 γ-tomograph.

Gas-solids flow characteristics in a conveying pipeline depend on (i) the nature of the material (which may range from coarse granular materials to very fine powders), (ii) the gas flow-rate and (iii) the pressure drop in the line [1]. Flow regimes range from dilute phase systems (where the solids are generally fully suspended in the carrier gas and conveyed at high gas velocities) to dense phase systems where the solids concentrations are high, gas velocities are low, and non-suspended flows of solids. Pictures (figures 1(a) and 1(b).) taken by others [2] using a high speed NAC 500 camera and a transparent pipeline, illustrate some of the effects observed in dense phase flow.

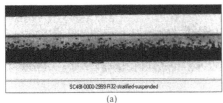

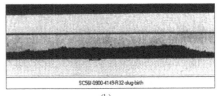

(a)

(b)

Fig. 1 NAC 500 camera images [2] (a) dense phase flow   *(b) slug birth*

1.1 Dense phase flow.

In this regime, lower gas velocities cause less material degradation of the conveyed product and, advantageously, result in reduced power consumption. The exact nature of non-suspension flow is however specific to both material and pipeline but may be divided into moving bed type flows and slug type flows [3]. The traditional approach to investigating dense phase flow involves (i) measuring the mass flow rate of the solids when the material is conveyed using a wide range of gas flow-rates and pressures and (ii) using this data to construct a set of conveying characteristics for the material. The approach adopted here [4] and by others [5,6] was to develop a tomograph system to obtain a time-varying image of the material density in the pipeline cross-section i.e. detailed qualitative and quantitative information on the flow regimes in dense phase flows.

1.2 The need for modelling.

Apart from the observations of events such as slug birth, other unusual conditions have been observed in pneumatic pipelines such as granular jump (figure 2a) which may have grave consequences for conveying high value products such as enteric coated drugs. Simulation programmes (figure 2b) have been developed at Glasgow Caledonian University [7], which depict dense phase flow but do not explain effects such as granular jump.

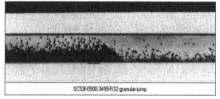   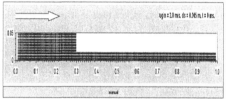

Figure 2(a) Granular jump [2]   *2(b) dense phase flow simulation*

2. Image Reconstruction.

The approach adopted here is based on an established [8.9].

Now the object $f(r,\phi)$ and the fan-beam projection $P_\beta(\gamma)$ can be shown to related by

$$f(r,\phi) = \int_0^{2\pi} \frac{1}{L^2} \int_{-\gamma_m}^{\gamma_m} D\cos\gamma\, P_\beta(\gamma) \cdot \frac{1}{2}\left(\frac{\gamma'-\gamma}{\sin(\gamma'-\gamma)}\right)^2 h(\gamma'-\gamma)d\gamma\, d\beta \qquad (1)$$

The inner integral represents a filtering operation, where the function $h(\gamma)$ is the frequency response of the filter. In this case:

$$h(t) = \int_{-\infty}^{\infty} |w| e^{j2\pi\omega t}\, d\omega \qquad (2)$$

In practice, the filtering operation can be performed by FFT method or directly convolution algorithm (see e.g. [10]. The projection data sequences and filter sequences need to be padded with a sufficient number of zeros before using Fourier transform method in order to avoid inter-period interference artifacts. In the discrete case, if an N element projection is zero padded to make it 2N–1 elements long, it can be shown that the interperiod interference will not occur. Usually, further padding may become necessary to make the sequence lengths 2^n in order to use the standard power-of-2 FFT algorithm. The filtering operation implemented in the frequency domain can now expressed as:

$$Q_\beta(n\alpha) = \alpha \times IFFT\left\{ FFT\left[P_\beta^{'}(n\alpha)\,withZP \right] \times FFT\left[h(n\alpha)withZP \right] \right\}$$

(3)

where FFT and IFFT denote the fast Fourier transform and inverse fast Fourier transform, respectively, and ZP stand for zero padding. Superior reconstructions are usually obtained when some smoothing operation is also incorporated within the filter process.

$$Q_\beta(n\alpha) = \alpha \times IFFT\left\{ FFT\left[P_\beta^{'}(n\alpha)\,withZP \right] \times FFT\left[h(n\alpha)withZP \right] \times \left[smoothing\,window \right] \right\}$$

(4)

When implemented within the frequency domain this smoothing filter may be expressed as a simple cosine function or a Hamming window: In the present case, a Hamming window function have been chosen for the smoothing process:

$$\omega(n) = 0.54 + 0.46\cos(2\pi n / N) \qquad n = -N/2,...,0,...N/2$$

(5)

The outer integral in equation 9 presents a weighted back-projection operation. Each filtered projection is smeared back to the image plane along fan beam. The sum of all the projections is used to produce the image of the object since,

$$f(r,\phi) = \int_0^{2\pi} \frac{1}{L^2} Q_\beta(\gamma')d\theta$$

(6)

In practice, if the value of γ' calculated with (8) does not correspond to one of angles for which projection data is known, then interpolation is needed, although in the present case linear interpolation is sufficient.

3. Experimental arrangement for the γ- tomograph

This consists of an axial array of six fans, each of which contains:a 3.7 GBq ^{241}Am enclosed source and five NaI(Tl) detectors with good geometry collimators as illustrated in figures 3{a}. The six fan layout is illustrated in figure 3(b).

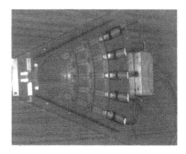

Figure 3(a) Interior of a fan showing detectors fans

3(b) Schematic layout of the six

4. Results and discussion for the γ-tomograph

Several types of phantom were tested using this system. The object in Figure 4(a) is a thin walled aluminium tube which is filled with fine sand. The image of this object has good contrast as shown in figure 4(d). The next object (figure 4b) is a wooden square and its image is shown in figure 4(e). A fault appears in the system at this time due to one detector generating spurious data. This manifests itself in the form of a distortion to the lower right of the image. The final object was chosen to illustrate a further limitation to the imaging capability of the system. It consists of both of the previous objects placed side by side. The differential density difference is such that the wooden square is barely recognisable in the presence of the higher density sand.

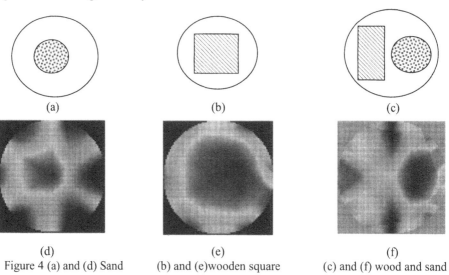

(a) (b) (c)

(d) (e) (f)

Figure 4 (a) and (d) Sand (b) and (e)wooden square (c) and (f) wood and sand

5. Part 2. X-Ray facility. Abstract.

The new experimental X-ray system shown in figure 5, built by GNI of Denmark, now makes it possible to study the transmitted X-ray spectrum for different substances at any given photon energy.

Figure 5 . The X-ray absorption system

This is because the different natural elements have a distinct absorption of photons at low energy levels. In this X-ray absorption system this is used to differentiate between the different natural elements. The system is the same as the one delivered to NIST with a few additions/modifications. Most importantly, the X-ray chamber has 50 mm insulation and there will be a climate system for control of the internal climate. Along with this an additional climate chamber of volume 0.3 m^3 is available to extend the range of temperature and relative humidity been developed which is capable of investigating materials of dimensions up to 420 mm thick under temperature and humidity regimes. The system is the third such system in the world and represents an improvement on its precursors which were installed at the Danish Technical University (DTU) in 1998 and the National Institute of Standards and Technology (NIST) in Maryland, USA in 2000.

6. Imaging details

The first object to be tested was a wooden square with a smaller aluminium square embedded within it (figure 6(a). The object was rotated at 5 degree intervals over a 180 degree range so that an appropriate amount of data could be used for the image reconstruction. While the system recognises that there is an area of differential absorption, , it has difficulty in getting the shape right (figure 6(b). This may be due to the very sharp change in density between the aluminium and the wood

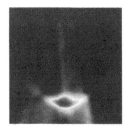

Figure 6(a) aluminium in wood   6(b) image

One possible application of this technique is the location of knots in living trees. As an initial test, part of a cross-section of trunk taken from a 200 mm diameter tree was placed on a rotary table and placed in the chamber. Ffigure 7(a) is a digital camera photograph of the top surcafe of the piece. A series of X-ray measurements were made at a distance of 10 mm below the surface and it certainly compares well with the detail on the top surface.

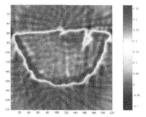

Figure 7(a) Digital photograph of wood   7(b) X-ray image at 10 mm below surface

7. Conclusions

The imaging capabilities of both γ-tomography and X-ray absorption have been investigated. The time taken to obtain data was much longer in the case of X-rays but this problem is being addressed. The images generated using X-rays is seen as having great potential for applications such as knot detection.

References

[1]   Woodcock, C.R., Mason, (1987)., J.S., *Bulk Solids Handling*, Leonard Hill.
[2]   Jaworski, D and Dyakowski, T.A., 2001, 2nd World Congress on Process Tomography, Hannover, Germany.
[3]   Mason, D.J. (1991), A study of the modes of gas-solids flow in pipelines. PhD thesis, Thames Polytechnic.
[4]   Bell, T. A. *et al.* (1993), *Powder and Bulk Solids*, Rosemont, Chicago, p.177.

[5]   Crowther, J. M. *et al.* (1995). *Process Tomography –95:Implementation for Industrial Processes*, M. S. Beck, B. S. Hoyle, M. A. Morris, R. C. Waterfall, R. A. Williams (eds.), pp. 166-175, UMIST, Manchester,

[6]   Johansen, G.A. *et al.* (1995a) , *Chem. Eng. J.*, pp. 56 -75.

[7]   Li, Jintang, 2001, PhD Thesis, Glasgow Caledonian University

[8]   Kak, A. C., (1979), *Proc. IEEE*, 67, 9, pp. 1245-1270.

[9]   Jain, A.K., (1989), *Fundamentals of Digital Image Processing*, Prentice-Hall International Editions.

[10]   Nussbaumer, H.J., (1982), *Fast Fourier Transform and Convolution Algorithms*, Springer-Verlag..

©2003 IOP Publishing Ltd

Three-Dimensional Quantitative Characterisation of Luminous Properties of Gaseous Flames

H C Bheemul[1], P M Brisley[1], G Lu, Y Yan[1*], S Cornwell[2]

[1] Centre for Advanced Instrumentation and Control, School of Engineering, University of Greenwich at Medway, Chatham Maritime, Kent ME4 4TB, UK
[2] Innogy plc, Windmill Hill Business Park, Whitehill Way, Swindon SN5 6PB, UK.

Abstract: This paper presents the application of a digital imaging system to the three-dimensional (3D) quantitative characterisation of luminous properties of gaseous flames. The system comprises three monochromatic CCD cameras, three short-wave-pass optical filters, a frame grabber and a computer with dedicated software. The three cameras, placed equidistantly and equiangular from each other around the flame, capture 2D images of the flame simultaneously from three different directions. Various image processing techniques including posterization, contour extraction and mesh generation are applied to reconstruct the 3D models of a flame from the 2D images. 3D luminous characteristics of a flame such as luminosity distribution, brightness and non-uniformity are quantified from the models generated. A series of experiments was conducted on a gas-fired combustion rig to evaluate the performance of the system. The results obtained demonstrate that the system is capable of measuring 3D luminous parameters of a flame over a range of combustion conditions.

1. Introduction

A flame is the central reaction zone of a combustion process and its geometrical, luminous and thermodynamic characteristics provide instantaneous information on the quality and performance of the combustion process. Monitoring and characterisation of combustion flames have, therefore, become increasingly important to combustion engineers for an improved understanding and subsequent on-line optimisation of combustion conditions. A number of instrumentation systems based on digital imaging and image processing techniques have recently been developed for the measurement of a range of parameters of fossil fuel fired flames [1]. Such systems have been tested under real plant conditions [2]. However, the systems use a single CCD camera, the information obtained is therefore limited to two-dimensions - the third dimension has not been taken into account. Since a flame is generally an asymmetrical 3D flow field, a two-dimensional (2D) imaging system does not provide overall information on the structure and dynamic properties of the flame [3]. This entails that the flame should be characterised three-dimensionally.

Very limited work has previously been reported on 3D visualisation and characterisation of flames using digital imaging techniques. Some 3D vision systems are indeed being developed and applied to various situations [4], but deriving meaningful 'measurements' in terms of quantitative parameters from flames remains challenging. The design, implementation and experimental evaluation of a digital imaging system that is capable of reconstructing 3D models of a flame and subsequently quantifying its geometric characteristics have recently been reported [5]. In the system three identical CCD cameras are placed around the flame. As a natural extension of this work, the present research focuses on the establishment of a methodology for the quantification of 3D luminous parameters of a flame. Each 2D flame image derived from an individual camera is posterized into 8 different grey-levels ranges. A contour extraction method has been devised to extract its inner and outer contours from the 2D images. A maximum of 8 models are reconstructed from the three 2D flame images using the same principle described in [5]. A set of 3D luminous parameters is defined and computed from the 3D flame models using various image processing techniques. Experimental results obtained on a combustion test rig under a range of operating conditions are also presented.

2. System Description

2.1 System set-up

The imaging system, as shown in figure 1, consists of three identical monochromatic CCD cameras, three short-wave-pass optical filters, a synchronisation circuit, a frame grabber, and a microcomputer with associated software. The cameras are placed at fixed locations A, B and C, separated by equal angles of 120°. Each camera has a $^2/_3$-inch CCD sensor with a resolution of 816×606 pixels. Each optical filter has a cut-off wavelength of 550nm. The synchronisation circuit ensures that the three cameras capture the images simultaneously from the three different locations. The frame grabber converts the analogue image signals from the cameras into digital images with 8-bit digitisation at up to 45MHz. Combined with the high performance computer system, the frame grabber supports a transfer rate of 132 MB/s, providing real-time transfer of the images to the host computer memory. The entire imaging system provides a frame rate of 60 frames per second.

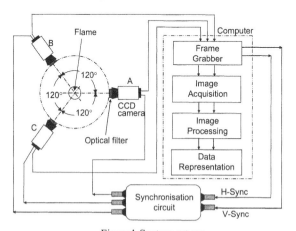

Figure1 System set-up

2.2   Brightness Calibration

An instrumentation system for luminosity measurement of an object is usually calibrated using a standard light source. In this case, however, an absolute calibration of the imaging system is impractical because of the unavailability of a standard flame source. Also, the luminosity of a flame image captured depends upon a number of factors such as iris setting, optical lens and shutter speed. It is believed that, in this particular application, a relative calibration of the imaging system, i.e. tuning the cameras to ensure they produce identical luminosity response, is more important than an absolute calibration.

The relative calibration is achieved by the use of a purpose-designed template with strips of eight distinct grey scales[1] (g_1-g_8), as illustrated in figure 2(a). Each strip represents 12.5% of the area of the template. The template is placed where the flame is usually fired. The objective length between each camera and the template is 280mm. A light source, suspended from the roof of the combustion rig, provides enough light for the camera to capture images of the template. An optical filter is mounted on the camera lens to cut out unwanted background light. The iris of the camera is manually tuned until there is no difference in grey-level g_1 between the template and its image captured by the camera. At that instant, the position of the iris is kept fixed. A total of 25 images of the template are then captured. The average area of the template representing each grey-level as well as their standard deviations (figure 2(b)) is computed. For instance, the normalised standard deviation for g_1 observed by camera A is 1.2%. The maximum deviation of the measured grey-level from its reference value is 5.5% and occurs at g_8. It is also observed that percentage of the measured area is not equal to that on the template. This is caused partly by the non-uniform light generated by the light source, which also projects a shadow on the template. The above procedure is subsequently repeated for cameras B and C. It is pragmatic that, although the grey-levels observed by the cameras are not perfectly consistent with those of the template (except in the case of g_1), the cameras have generated identical grey-levels.

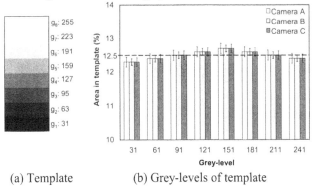

| (a) Template | (b) Grey-levels of template |

Figure 2 Brightness calibration of the system

---

[1]   The choice of the number of grey scales is a compromise between a concise presentation and a clear indication of luminous parameters. It is found that 8 grey-scales are suitable for both laboratory-scale gaseous flames and industrial gas- and coal-fired flames [6].

3.   Definition and Quantification of Luminous Parameters

Three-dimensional characterisation of the flame luminosity is based upon 3D reconstruction of flame models using inner contours of the flame. Firstly, the grey-level variations in a flame image are condensed into eight layers using posterization techniques, with the brightest pixels condensed into the inner layer of the flame image and darkest pixels into the outside layer. Since all pixel values are mapped into eight grey-level ranges following this operation, a 'contoured' appearance is observed in the image. Using contour extraction method proposed in [5], inner contours of the flame are extracted from its image. Each contour corresponds to a grey-level range, $g(i)$. A 3D model of the flame is reconstructed from each contour using the same principle described in [5]. Therefore, a total of eight models can be generated from a set of 2D flame images. Figure 3 illustrates six of the flame models reconstructed at different grey-levels. The volume of each model is computed to determine the luminous parameters of a flame.

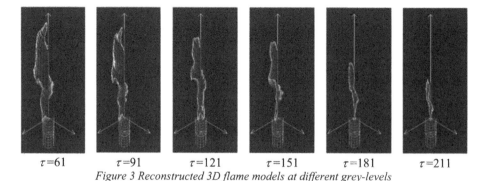

τ=61 τ=91 τ=121 τ=151 τ=181 τ=211

Figure 3 Reconstructed 3D flame models at different grey-levels

(1) Luminosity distribution $\mu(g)$: Luminosity distribution of a flame is the representation of the total number of pixels in each grey-level range. It is calculated from the ratios of the volume of a flame representating a grey-level range, $Q_f(g(i))$, to the volume of the entire flame Q_f, i.e.

$$\mu(g(i)) = \frac{Q_f(g(i))}{Q_f} \quad (i=1, 2,..., 8)$$

(1)

where $\sum_{i=1}^{8} \mu(g(i)) = 1$.

(2) Brightness (B_f): Brightness is defined as the average grey-level of the flame, normalised to the brightest grey-level (255) of the imaging system. The grey-level of each pixel of a flame image represents the individual luminosity of the flame at that pixel. To determine the brightness of a flame, the grey-levels of all pixels should be taken into account. The brightness of each grey-level range, $B_f(g(i))$, can be obtained as follows,

$$B_f(g(i)) = \frac{1}{255} \mu(g(i)) \left(\sum_{\tau=\eta_1}^{\eta_2} \tau d(\tau) \right) \times 100\%$$

(2)

where τ is the luminosity of individual pixels in grey-level range $g(i)$, η_1 and η_2 are the lower and upper limits of grey-level range $g(i)$ respectively, $d(\tau)$ is the ratio of the number of pixels with grey-level τ to the total number of pixels in $g(i)$. The brightness of an entire flame is, therefore, expressed as,

$$B_f = \sum_{i=1}^{8} B_f(g(i)) \tag{3}$$

(3) Non-Uniformity (NU$_f$) is defined as the averaged deviation of the grey-levels of individual pixels over the luminous-region from the brightness of a flame, i.e.

$$NU_f = \frac{\sum_{\tau=0}^{255} n(\tau) |\tau - B_f|}{MB_f} \tag{4}$$

$n(\tau)$ is the number of pixels with grey-level τ (0~255), and M is the number of pixels in the flame.

4.   Results and Discussion

To evaluate the performance of the system, a series of experiments was conducted on a gas-fired combustion rig. Commercial butane, stored in a cylinder, was used as fuel for the tests. Different combustion conditions were achieved by varying the air-fuel ratio (**r**). Under each particular condition a total of 25 sample images were captured from the three different locations. Following the reconstruction of the 3D flame models, the average luminous parameters as well as their uncertainties [5] were calculated.

Figure 4 illustrates the variation of luminous parameters with air flow rate. At **r** = 0 the luminosity distribution in g_8 is large, but gradually decreases with air flow rate until **r**=30, whereas the reverse is observed for g_1. This is because, with an increase in air flow rate, the flame changes from diffusive to premixed. However, it is perceived that g_8 increases slightly after **r**>30. The basis of this sudden augmentation is that the vigorous chemical reaction that occurs between the gas and air causes the flame root to become more luminous, resulting in a slight increase in luminosity distribution of g_1. Similar trend is also observed for the flame brightness. Also at **r**=0, the non-uniformity of the flame is high, indicating that the grey-level ranges in the flame have high luminosity distribution at that condition.

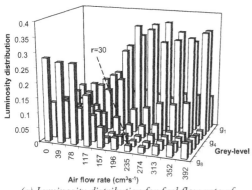

(a) Luminosity distribution for fuel flow rate of 7.8cm³s⁻¹

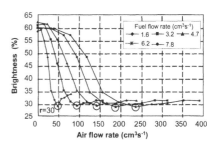

(b) Brightness

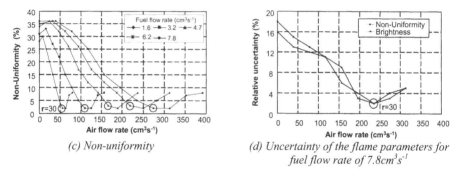

(c) Non-uniformity

(d) Uncertainty of the flame parameters for
fuel flow rate of $7.8cm^3s^{-1}$

Figure 4 Variation of flame parameters with air flow rate

However, as **r** increases, a dramatic decrease in the non-uniformity is observed. This is because the flame exhibits only two grey-level ranges of very high luminosity distribution, whereas the other grey-level ranges have significantly smaller. As **r**>30, non-uniformity also slightly increases due to the increase in luminosity distribution in g_1. As seen in figure 4(d), the uncertainty of the brightness and non-uniformity decreases with air flow rate until **r**=30, and increases slightly as **r**>30. Therefore, it can be accounted that the luminous parameters of a premixed flame is more stable than a diffusion flame, with **r**=30 being the most prominent condition.

5.   Conclusion

The method of characterising luminous properties of a flame using a multi-camera imaging system has been described. Each flame image is posterized into a maximum of eight different layers so that the inner contours can be extracted from the image using contour extraction methods. Each inner contour, corresponding to a particular grey-level range, is used to reconstruct 3D models of the flame, which allow 3D visualisation of the inner circulation zone of the flame. Based on the computed volumes of these models, the measurement principles for the luminous parameters of a flame are derived. The results obtained from a series of experiments have demonstrated that the system is capable of measuring 3D luminous parameters of the flame. As expected, the quantitative characteristics of the flame are dependent upon the operational conditions of the combustion rig. It is envisioned that reconstructed 3D flame models in conjunction with the measured flame parameters will enable combustion engineers to have a better understanding of the dynamic behaviour of flames under different operating conditions.

Acknowledgments

Acknowledgment is made to the British Coal Utilisation Research Association and the UK Department of Trade and Industry for a grant in aid of this research, but the views expressed are those of the authors, and not necessarily those of BCURA or the Department of Trade and Industry.

References

[1]   J. S. Marques, P. M. Jorge, "Visual inspection of a combustion process in a thermoelectric plant," *Signal Processing*, Vol. 80, pp. 1577-1589, 2000.

[2]   Y. Huang and Y. Yan, 'Transient two-dimensional temperature measurement of open flames by dual-spectral image analysis', *Transactions of the Institute of Measurement and Control*, Vol. 22, No. 5, pp. 371-384, 2000.

[3]   H. C. Bheemul, G. Lu and Y. Yan, '3D Monitoring and Characterisation of Gaseous Flame Using Contour Extraction Methods', *Proceedings of 11th International Conference on Sensors and their Applications XI/ISMCR 2001*, London, Sept. 2001.

[4]   F. Moratti, M. Annunzotia and S. Giammartini, 'An artificial vision system for the 3D reconstruction and the dynamical characterisation of industrial flames', *ENEA-C.R.E. Casaccia*, 1997.

[5]   H. C. Bheemul, G. Lu and Y. Yan, 'Three-dimensional visualization and quantitative characterization of gaseous flames ', *Measurement Science and Technology*, Vol.13, pp. 1643-1650, 2002.

[6]   G. Lu, 'Advanced Monitoring and Characterisation of Fossil Fuel Flames', PhD Thesis, the University of Greenwich, 2000.

Section 7

Optical Sensing II

An Overview of Automated Sensor Systems for Marine Applications (Keynote)

J. Lucas

The University of Liverpool, Department of Electrical Engineering and Electronics, Liverpool L69 3GJ, UK.

Abstract: Sensor systems for automated marine applications have been developed from a variety of optical, acoustic and rf arrangements. The optical sensors rely on relatively clear water and hence are short range and are used for positioning and process monitoring. Acoustic systems can operate over long distances varying from kilometers to meters. They are primarily used for communications, vehicle positioning and navigation. Rf sensors provide a new range of techniques whose mode of operation depends on how the rf waves are launched into the seawater. They have the potential to complement the role of acoustic sensors. Finally, there is a large number of process sensors used by the scientific community, in particular for measuring water clarity and pollution levels.

1. Introduction

A lot of marine observations are obtained using remotely operated vehicles (ROVs), autonomous underwater vehicles (AUVs) or fixed buoys. There is a vast range of sensor systems used, which vary from vehicle positioning, navigation and IMR (Inspection Maintenance and Repair) to scientific observations of seawater properties. For this review the sensor systems will be broken down into four areas namely: Optical sensing, Acoustic sensing, RF sensing and other sensors.

2. Optical Sensing

These use CCD and SIT camera systems and laser illumination for positioning or work piece inspection. These systems produce quality 2D images and are also able to give range information for parts of the image. The main problem when using the sensors is the attenuation or scattering of the illuminating light by the turbidity of the water. Also both the camera and the light source have to be housed in one atmosphere enclosed boxes, hence system size becomes an issue.

2.1 CCD Camera

The Hitachi KP-D8 CCD camera [1] is very suitable due to its compact size and the wide range of electronic shutter settings (1/50s to 1/180000s), which could be selected electronically. Additionally, there is an extensive range of flexible optics that can be used to provide a motorised iris, focus and zoom. The system is shown in figure 1.

The camera system is mounted inside a compact pressure housing of length 200 mm and cross section 66 mm x 73 mm, which operates at a pressure of 250 Bar. The housing has a Sapphire viewing window 40 mm diameter by 10 mm thick. All the functions of the camera are remotely controlled, by using an RS422 communications bus system.

Figure 1. The Hitachi Camera System

2.2 Laser Types

The preferred system   for sub-sea experiments is the diode pumped Nd:Yag laser producing multi-mode Q switched outputs up to hundreds of multi-joules at wall plug efficiencies of up to 6%. The infrared emission (1064nm) can be used directly or frequency doubled to emit within the blue/green spectral area (532nm) for which attenuation by seawater is a minimum. The Adlas DPV 315 Diode Pumped Laser has the following characteristics: Wavelength 532nm, Power >100mW, Beam Divergence 2.2m rad, Stability 2%, Size 100mm x 40mm x 32.5mm, Power Consumption 20W.

2.3 Laser Triangulation System for Object Recognition

Whilst it is possible to estimate the shape of the joint from the 2D image in certain situations it can be very difficult to determine the exact profile or the shape of the sub-sea object. A common technique used for these measurements is laser triangulation, where a laser line is projected across the object thus revealing its shape [2]. The laser line is obtained by scanning a laser point source over the object to create a line as shown in figure 2. By altering the laser position the results from multiple scans can be fused to indicate the presence of an object as illustrated in figure 3.

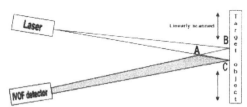

Figure 2. Synchronous scan imaging method, showing a greatly reduced back scatter volume within the region ABC

No Illumination   Laser Line Seam Illuminated

Figure 3. Optical Images in Murky water

The overall range of the system depends critically on the turbidity of the water and is usually kept below a distance of 10m.

2.4 The Welding Arc under Hyperbaric Conditions

Hyperbaric welding is an important sub-sea activity within the oil and gas industry requiring control by a remote operator. Several welding techniques are currently being used namely MIG, Plasma and TIG. In each case the image had an intense emission from the arc and a much weaker reflection from the surrounding workpiece. The objectives are to vastly reduce the contrast ratio between the arc region and the workpiece region so as to allow simultaneous observation of both regions by an operator [3].

2.4.1 Scene Observation Using YAG Laser Illumination (1064 nm and 523 nm)

The principle of the system is to totally attenuate the arc light by using a combination of a neutral density filter and a narrow band filter at the chosen YAG laser wavelength of 1064 nm or 532 nm when used in frequency doubling mode. The Microlite device by Photonic Systems operating at 25 Hz frequency with 18 J and 6 ns pulses is the preferred system. The timing circuit developed which linked the CCD camera electronics and the laser electronics is shown in Figure 4 The shutter time and location could be adjusted relative to the Q-Switch of the laser. By synchronising the camera shutter and laser on times clear definition pictures can be obtained. Three such images are shown in figure 5 for TIG and MIG welding.

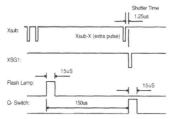

Figure 4. Timing diagram of the laser and shutter control circuit.

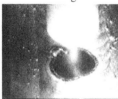

Laser Illuminated TIG   MIG Welding (Mild Steel)   MIG Welding (Aluminium).

Figure 5. The Weld Pool Sensor

3. Sonar Applications

These are mainly for determining the position of a vehicle, the position of a seabed object underwater and communications. The speed of sound in seawater is approximately 1500m/s and the attenuation linearly increases with frequency. This means that the sensor systems operate at frequency appropriate to the range being measured. These values are given in table 1.

Frequency Band	Frequency Range	Maximum Range	Range accuracy
Low Frequency (LF)	8kHz to 16kHz	>10km	2m to 5m
Medium Frequency (MF)	18kHz to 66kHz	2km to 3km	0.25m to 1m
High Frequency (HF)	30kHz to 60kHz	1500m	0.15m to 0.25m
Extra High Frequency (EHF)	50kHz to 110kHz	<1000m	<0.05m
Very High Frequency (VHF)	200kHz to 300kHz	<100m	<0.01m

Table 1. Comparison of Operational Frequency vs Operational Range

3.1 Vehicle Positioning

• LBL (Long BaseLine)

An LBL system has two parts or segments [4]. The first segment comprises a number of acoustic Transponder Beacons moored in fixed locations on the seabed. The positions of the Beacons are described in a co-ordinate frame fixed to the seabed. The distances between them form the "baselines" used by the system. The second segment comprises an acoustic transducer on a Transceiver, which is normally temporarily installed on the vessel - or on a Tow Fish. The distance from the transducer to a Transponder Beacon can be measured by causing the transducer to transmit a short acoustic signal which the Transponder detects and causes it to transmit an acoustic signal in response. The time from the transmission of the first signal to the reception of the second is measured. As sound travels through the water at a known speed, the distance between the transducer and the Beacon can be estimated. The process is repeated for the remaining Beacons and the position of the vessel relative to the array of Beacons is then calculated or estimated.

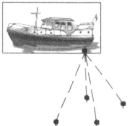

Figure 6. Long BaseLine Sonar

Figure 7. Short BaseLine Sonar

- SBL(Short BaseLine)

An SBL system is normally fitted to a vessel such as a barge, semi-submersible or a large drilling vessel. A number of (at least three but typically four) acoustic transducers are fitted in a triangle or a rectangle on the lower part of the vessel. The distances between the transducers (the "baselines") are made as large as is practical, typically they are at least 10 metres long. The position of each transducer within a co-ordinate frame fixed to the vessel is determined by conventional survey techniques or from the "as built" survey of the vessel. If the distances from the transducers to an acoustic Beacon are measured as described for LBL, then the position of the Beacon, within the vessel co-ordinate frame, can be computed. Moreover, if redundant measurements are made, a best estimate can be determined which is, statistically, more accurate than the basic position calculation.

3.2 Object Scanning

Although it is possible to achieve highly accurate positional references with the Baseline methods, it is not practical to position a vehicle close to the seafloor where large metal structures can cause some "dead spots" (caused by the spurious reflections of the acoustic signals thus preventing the wanted signal being received). Therefore, to be capable of positioning a vehicle relative to a metal structure on the seabed, alternatively positioning methods have to be employed. A typical sensor is the Tritech dual frequency profiling sonar head (DFP), illustrated in figure 8[5]. This has two mechanically scanned profiling sonars at frequencies of 0.58MHz (for use in waters containing suspended particles or when a large profiling range is required) and of 1.2MHz (for shorter range work or use in clear water). The DFP has a scan rate up to 180° per second with a resolution 0.45° to 1.8° steps. The pulse length is 20 to $300\mu s$ with a beam width of 1.4° to 3°. The device is compact 105mm (L) and 259mm(D) and weighs 1.4kg in water and can operate at a depth of 4000m.

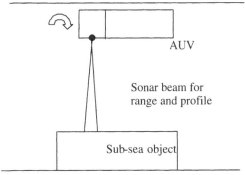

AUV

Sonar beam for
range and profile

Sub-sea object

Figure 8. Sonar Object Scanning

4. Electromagnetic Waves

4.1 Introduction

This is a relatively new area of investigation. Maxwell's equations predict the propagation of electromagnetic waves travelling in seawater. A linearly polarised plane electromagnetic wave travelling in the z direction may be described in terms of the electric field strength E_x and the magnetic field strength H_y with,

$E_x = E_o \exp. (jwt - \gamma z),$ $\qquad\qquad$ $H_y = H_o \exp. (jwt - \gamma z)$ $\qquad\qquad$ (1)

The propagation constant (γ) is expressed in terms of the permittivity (ε), permeability (μ) and conductivity (σ) by;

$$\gamma = jw\sqrt{\varepsilon\mu - j\frac{\sigma\mu}{w}} = \alpha + j\beta \qquad\qquad (2)$$

where α is the attenuation factor, β is the phase factor and $w = 2\pi f$ is the angular frequency.

The term $\varepsilon\mu$ arises from the displacement current and the term $\sigma\mu/w$ from conduction current. It is convenient to consider the solutions for the conduction band ($\sigma/w > \varepsilon$), and the dielectric band ($\varepsilon > \sigma/w$). For sea water the condition $\varepsilon = \sigma/w$ occurs for f = 1 GHz. Investigations of the parameters σ and $\varepsilon(=\varepsilon^1 - j\varepsilon^{11})$ over the full EM frequency spectrum have been obtained in electrolytic solutions by using a wide variety of experimental techniques.

Figure 9. Antenna Configurations

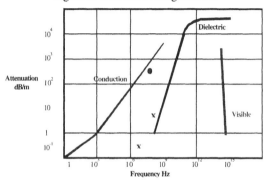

Figure 10. The EM Wave Propagation in Sea Water

4.2 Sensor Systems

Sensor systems have been developed as type A, B or C as illustrated in figure 9. For type A, the transmitter and receiver antenna were placed in direct contact with the seawater. For type B, the seawater was contained in an insulating container and the antennae placed in air but close to the container walls. For type C, both the antenna were coated with an insulator and placed within the sea for horizontal water transmissions. The most popular techniques for type A experiments involve the use of coaxial transmission lines and waveguide cavities. If the seawater is in direct contact with the walls of the line or cavity then the conduction band losses apply. For example [6], a coaxial transmission line filled with seawater gave an attenuation coefficient α = 300dB/m for f = 67MHz which is in good agreement with the conduction band solution as shown in figure 10. (This result is shown by the symbol o). In type B experiments where the seawater is placed in an insulating container within the waveguide, the conduction losses are extremely small and the dielectric solution applies. (These results are shown by the symbol x). Hence combining these two experimental techniques will allow the simultaneous evaluation of a range parameters for sea water such as

σ, ε^1 and ε^{11} and the wave velocities v_G and v_e. Such an investigation will allow a detailed documentation of the properties of coastal seawater (clear, murky and polluted) to be investigated for a wide range of temperatures, depths and frequencies. Because of the longer wavelength of the EM waves, to be used in this study, it is anticipated that the effect of murky water or pollutants will be less severe than for the optical systems. Type B sensors are used to measure the concentrations of oil, gas and water in oil production pipelines [7] as the resonance frequency within the pipeline cavity is critically dependent on the mixture content. The resonant frequency (f) as shown in figure 11 depends on the dielectric constant (ε) $\left(f = f_o / \sqrt{\varepsilon} \right)$ and the constituents have a wide range of values from water (72), oil (2.4) and gas (1.0). Hence the resonant frequency is a sensitive indicator of mixture composition.

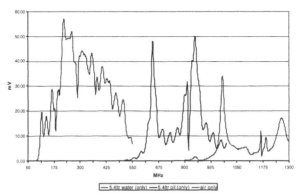

Figure 11. The spectrum of the resonance frequencies for water, oil and air.

4.3 Propagation Sensing

Type C sensors are currently being undertaken to produce communication between an AUV and a platform up to distances of 50m. The optimum theoretical and experimental system operates between 1 and 5MHz [8]. The results have severe attenuation in the near field but there is sufficient residual EM wave field strength to allow FM communications to be undertaken. Figure 12 illustrates this effect.

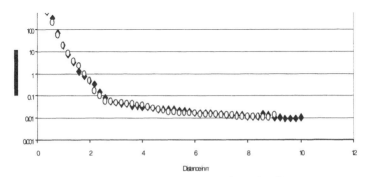

Figure 12. Received signal attenuation with distance

5. Additional Marine Sensors

There is a wide range of sensors for scientific observations. These include:

Ambient light	Hydrocarbons	Sound velocity
Conductivity	Oil on water	Temperature
Current	Other chemicals	Tide/wave motion
Dissolved gases	OH	Turbidity
Dissolved oxygen	Pressure	Volatile organics
Heading	Redox	Waterline

5.1 CTD Measurements

The standard device is the CTD instrument measuring conductivity (range 25mS/m to 1000mS/m, ±1mS) [9], temperature (±5 10^{-3} °C with response time 85ms) and pressure (±0.005FSD to response time 10ms). The pressure sensor is used to measure depth since 1 bar pressure approximately corresponds to 10m depth so that at 2500m the pressure is 250 bar. The temperature range varies from just above freezing to 100°C at deep sea vent sites.

5.2 Anodic Stripping Voltammetry

Anodic Stripping voltammetry (ASV), particularly with the use of pulsed potential waveforms, is one of the most sensitive, convenient, and cost effective analytical method for detection and quantitation of metal contaminants in rivers, lakes, process streams and drinking water [10]. Another advantage is that several metals such as Cu, Pb and Cd can be analysed simultaneously. ASV has been used for several years to analyse heavy metals such as Cu, Pb, Cd and Zn. ASV can be thought of as a small scale electroplating experiment. The metals, as ions in solution, are plated onto an electrode by applying a negative potential (deposition potential) for a specific period of time. The deposition serves to concentrate the metal ions from the solution onto the electrode in the metallic form. If the electrode is Hg, the metals often form an amalgam. After deposition, the potential is scanned toward positive potentials. Current peaks appear at potentials corresponding to the oxidation of metals as they are oxidized (stripped) from the electrode back into the solution.

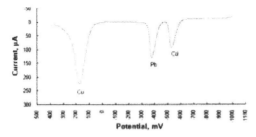

Figure 13. Ion Identification and Concentration Measurement

As shown in figure 13, the peak height or area can be correlated with the concentration of the metal ions in the solution. It is necessary to calibrate the procedure with standard solutions containing known quantities of the metal ions. The technique was originally developed with a hanging Hg drop electrode. However, to limit the quantity of Hg needed, thin Hg films can be pre-deposited onto an electrode such as glassy carbon or co-deposited with the analyte metal ions. Use of films circumvents the need for much Hg, which is toxic. With such films, sensitivities in the low part-per-billion (ppb) level can be achieved.

5.3 Environmental Optics

Good sensors already exist for most physical variables (conductivity, temperature, flow velocity). Chemical sensors (for nutrients, dissolved oxygen, trace metals) are undergoing rapid development. Populations of fish and zooplankton can be measured, once they reach sufficient biomass density, by acoustic techniques. Obvious gaps in this list include phytoplankton, low densities of zooplankton including eggs and larvae, and aggregations of organic material derived from faeces or senescent phytoplankton blooms. In theory, these 'missing variables' can all be measured by optical instruments [11] including fluorometers, scattering and absorption meters, particle analysers and imaging systems.

References

[1] Hitachi Colour Camera KP-D8, mini 3.1 Zoom Lens, Lens Innovation Ltd. http://www.photonics.com/spectra/newprods

[2] Tetlow. S, "Use of Laser Light Stripes to Reduce Backscatter in an Underwater Viewing System", PhD Thesis, Cranfield, 1993.

[3] Lucas W. and Smith J.S. "Keeping an electronic eye on automated arc welding" Welding and Metal Fabrication, Vol. 68 No. 4, pp 6-13, 2000.

[4] Sonardyne, "Acoustic Theory", http://www.sonardyne.co.uk

[5] Tritech Seaking Dual Frequency Sonar, http://www.tritech.co.uk

[6] Bogie S., "Conduction and Magnetic Signalling in the Sea", The Radio and Electrical Engineer, 42, 447-452, 1972

[7] Al-Shamma'a A.I., Shaw A., Lucas J., *"EM Wave Sensor for Wet Gas Metering"*, Multiphase Technology, 3^{d} North American Conference, ISBN 1 85598 041 X, Banff, Canada, 6th-7th June, page 391-398, 2002.

[8] Lucas J., "Transmission of Electromagnetic Waves through seawater for imaging, parameter measuring and communications", Eur OCEAN 2000, Hamburg, 29 August – 2 September, Vol II, pp 524-529, 2000.

[9] CTD Instruments (Conductivity, Temperature, Pressure), http://www.appliedmicrosystems.com

[10] Graziottin F., "Development and Validation for Remote Automatic Analysis of Heavy Metals (Zn, Cu, Pb, Cd) in Estuary and Coastal Water", EC Marine Sciences and Technologies, Vol 2, pp 916-925, 1995.

[11] Schmidt H. and Kronfeldt H.D., "Fibre-Optic Sensor systems for Coastal Monitoring", Sea Technology, Nov, pp 51-55, 1999.

Spread Spectrum Techniques in Optical Sensors and Sensor Networks

S Abbenseth[1], S Lochmann[2], H Beikirch[1]

[1] Dept. of Electrical Eng. and Information Technology, University of Rostock, Rostock, Germany

[2] Dept. of Electrical Eng. and Computer Science, Hochschule Wismar, University of Technology, Business and Design, Wismar, Germany

Abstract: In recent years, there has been an increasing interest in optical spread spectrum techniques in fibre-optic multiple access networks. The advantages regarding the simplicity of implementation, utilisation of resources as well as the disturbance insensitivity in relation to interfering users and multi-path propagation can also be utilised in the optical sensor area.

Up until now this area hasn't been much explored yet.

This paper provides an overview of optical code-division multiple access (OCDMA) principles and focuses on the major benefits of merging CDMA with established sensor technologies, particularly fibre Bragg grating networks, optical time-domain reflectometry and interferometry.

1. Introduction

Optical and fibre-optical sensors are introduced into more and more application areas, which had been dominated by electrical sensors so far. The advantages are well-known: higher dynamic range and larger bandwidth, high isolation against electromagnetic interference, security in highly explosive environment, a galvanic separation of the components by the medium of quartz glass, polymer or air and also the corrosion resistance of the transmitting medium. Space requirement and weight of the most sensors and fibres are very small in comparison to the electric counterparts - an important condition e.g. in aircraft construction /1/, /2/, /19/.

To save space and cost, in some cases a network of sensors is desirable, whereby the information of the individual sensor may be affected as less as possible. So far time division multiple access (TDMA) or wavelength division multiple access (WDMA) have mostly been the means of choice /3/, /4/. However, in communication networks optical code division multiple access (CDMA) techniques become interesting due to their very good characteristics primarily in networks with a high number of users /5/. Even if sensor networks are in many cases not comparable with the communications networks, the advantages of the CDMA can also be used in the sensor technology.

In this paper the potentials of merging optical CDMA with optical and fibre-optic sensors are pointed out. In section two an overview of the present techniques of optical CDMA will be given. Section three will illustrate the combination of sensor technology and optical CDMA with examples in OTDR, grating sensors, interferometry and transmission/reflection measurements.

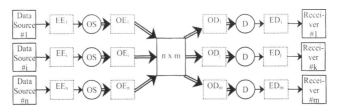

Fig. 1. A universal schematic diagram of optical CDMA.
EE: electrical encoder; OE: optical encoder; OD: optical decoder; ED: electrical decoder;
OS: optical source; D: Detector; →: electrical connection; ⇒: optical connection

2. Optical CDMA

The research in optical CDMA started nearly 20 years ago and becomes now important to enhance the capacity of existing WDMA and TDMA systems /6/. The reasons are particularly the simplicity of implementation, utilisation of resources as well as the disturbance insensitivity in relation to interfering users and multi-path propagation. Another term for CDMA is spread spectrum technique, because of the widened original data spectrum in the frequency domain. This spreading can be realized with different methods and usually contains the address information. In that way the given bandwidth can be used simply and efficiently.

Fig. 1 illustrates a general optical CDMA network. The address coding takes place either in the electrical (EE) or in the optical domain (OE). To avoid confusion the bits of the code sequence are called chips. All channels use the same optical band at the same time (synchronous or asynchronous). At the receiver side the decoding happens by means of correlation with the desired code, again, either in the optical (OD) or electrical

Table 1
optical CDMA techniques

Data-signal	optical wave properties	correlation process		practical implementation
		en-/decoder	signal/code polarity	
pulse	incoherent	electrical/ electro-optical coding	unipolar-unipolar	On-Off-keying
			unipolar-bipolar	optical correlation (SIK)
		optical coding ladder network parallel/serial	unipolar-unipolar	identical/inverse optical network on receiver side (parallel/serial)
			unipolar-bipolar	separate paths for polarities at the receiver (SIK, Balanced diode)
			bipolar-bipolar	two phys. paths (e.g. polarization)
	coherent	elektro-optical coding	unipolar-unipolar	Lithium niobate Mach-Zehnder phase modulator
		ladder network parallel/serial	unipolar-unipolar	identical/inverse optical network on receiver side (parallel/serial)
			bipolar-bipolar	
		spectral slicing	bipolar-bipolar	subpicosecond pulses en-/decoded by phase masks
conti-nuous	wavelength	fixed wavelength		prism/diffraction grating and mask; n lasers/filters
		frequency-hopping		n lasers or opt. filters with switch; tuneable laser
		chirp		manipulation of frequency chirp by laser current
	phase			PSK (e.g. injection locked laser) and homodyne detection
	polarization			switching of polarization
	holographic			encoding: with holographic filters, decoding: processing of the hologram

(ED) path. In the autocorrelation case the data signal is rebuild ideally, whereas in the cross-correlation case the unwanted data signals remains in the spread condition. Table 1 shows the most important coding and correlation techniques /7/, /8/.

Most common and described are the pulsed CDMA techniques. It can be distinguished into coherent and incoherent techniques and furthermore according to the kind of coding and decoding, namely uni- or bipolar. The systems that can be realized most simply are the incoherent applications in which the code is impressed as simple unipolar OOK sequence in the electrical domain or by using electro-optical modulators in the optical domain. On the receiver side the correlation (multiplication) can take place purely electrically or electro-optically, hence unipolar too /9/. By splitting the optical path into two and applying an optical multiplication with the desired code and/or its complement (e.g. sequence inversed keying, SIK) together with a subsequent detection using balanced diodes the unipolar chip sequence can be correlated with a bipolar code. This unipolar-bipolar correlation process maintains the properties of a bipolar-bipolar one but is much easier to implement /10/.

Since positive and negative values extinguish themselves in the bipolar correlation process the signal to interference ratio (SIR) and the signal to sidelobe ratio is much better than with unipolar correlation where the multi-user interferences (MUI) always add. Thus, the bipolar correlation is to be preferred due to the smaller bit error probability.

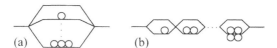

Fig. 2. ladder networks (a) parallel (b) serial

Ladder networks consist of different delay lines that follow eqn. 1

$$\partial_n > \sum_n \partial_x \tag{1}$$

where δ_n means the delay of the n-th stage and δ_x the delay of the x-th stage /11/. The architectures shown in Fig. 2. are same in coherent and incoherent cases. Only the consequences are different: There are n pulses in the first case and n phase changes or n pulses with different phases in second, respectively, which result from one pulse entering the network. Although serial networks need less space e.g. in integrated optics for realising a similar number of chips the parallel networks are more common because of their higher degree of freedom in choosing codes. Serial ladder networks are restricted to codes with 2^n pulses.

Since in the incoherent scheme the powers are added and in the coherent one the wave amplitudes are superimposed, the SIR of both techniques are related to each other by a squaring operation. Thus, more users can send at the same time with same code length and bit error probability in the coherent case compared to the incoherent. Furthermore, coherent networks can be built by means of integrated optical phase modulators. Incoherent networks make always use of delay fibres with e.g. lengths of approx. 0.2m for 1ns pulses. However, unfavourable high efforts regarding structure and stability as well as the associated high costs work against the wide spread implementation of coherent systems.

A special coherent pulse technique is spectral slicing. Here a subpicosecond pulse is coded using a phase mask which also leads to a temporal spread. The decoder contains a complex conjugated phase mask that dispreads the impulse only in the matched case /12/.

Particularly for sensor applications the use of continuous waves can be of interest. Using a fixed or dynamically allocated wavelength is one coding scheme for that propose. The fixed case can be realized simply with FBGs or diffraction gratings in connection with a mask /13/. Dynamic coding is possible by means of frequencies hopping or by utilising the frequency chirp of a laser diode /14/. Another kind of coding a continuous wave is the direct change of phase conditions using phase modulators or current variations of an injection locked laser as a more cheaper variant, respectively /15/. All these coherent techniques can apply homodyne or heterodyne decoding schemes.

In recent time also some attempts were undertaken to code the wave fronts by means of holographic filters and an appropriate electrical evaluation /16/.

In case of changing the requirements regarding the address range or the BER a CDMA network can be adapted more flexibly over the code and the code length. Predominantly orthogonal or quasi-orthogonal pseudorandom (PR) codes are used in order to reduce MAI and improve BER /7/. However, the overall address space is also reduced tremendously. Thus, other schemes allowing standard balanced code sequences should be used for large code set requirements. To enhance the capacity it is also possible to encode the data into more dimension (so called hybrid techniques), e.g. using different wavelengths in an incoherent ladder network /17/.

Furthermore, if electrical or optical hard limiters were used the influence of MUI can be reduced. The received uncorrelated optical power or its electrical pendant is limited to that of one chip. Thus the height of the ACF peak is identical to that without interfering users. On the other hand the MUI is limited to the minimum of the code polarity configuration /18/.

3. Merging optical CDMA and Sensors

The general advantages of optical sensors were already mentioned above. In this section the optical sensor technology and optical CDMA are to be linked. For better characterization the optical sensor technology has been divided into four classes: Fibre Grating sensors, interferometry, optical time or frequencies domain reflectometry (OTDR/OFDR) and transmission/reflection measurement procedures. Each group is treated individually in the following.

Fibre Grating Sensors: Part of this group are all applications where a structure shaped into a fibre serves as sensor element, particularly these are fibre Bragg gratings (FBG), sampled FBG and long period gratings (LPG). Usual sensor applications are temperature -, stress/strain- and bending curvature measurements, but also optical microphones /19/. It is a quite new discipline and at present in the centre of the research.

Within the group of grating sensors the multiplex has to be distinguished between serial sensor locations (within one fibre) and parallel locations (different fibres). Until now WDMA and TDMA are the most common methods of serial multiplexing due to the frequency-selective and reflection characteristics of the Gratings. However, TDMA has the need of an exact time management and in an the case of WDMA the spectral width of the source limits the maximum number of addressable sensors in connection with the dynamic range of each sensor. An alternative realisation by means of CDMA is given in /20/. Here, the simplest CDMA configuration (incoherent, electro optical, unipolar-unipolar) is used which results in a small sensitivity in relation to multi-path propagation. The pulse sequence is reflected by all sensor elements and appears as a time-shifted code at the receiver. The correlation using the original code with a variable time-shift results in the answer of the desired sensor with a simultaneous suppression of other sensors.

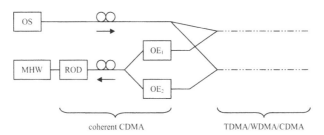

Fig.3 A coherent pulse coded FBG network
OS: optical source; OE: optical encoder; ROD: reconfigurable optical decoder; MHW: measurement hardware

The idea of a more complex parallel network is represented in Fig. 3. One pulse or a pulse train is fed to all sensor fibres, reflected from the individual sensors and in the following coherently coded. The temporal distance of two reflected impulses has at least to correspond to the running time in the longest delay fibre of the Encoder. All branches could then be connected to one transmitting fibre which transmits the coded optical signals to the measuring hardware. Now the correlation happens in a reconfigurable optical network where all codes can be made available for example by phase modulators. In this way the expenditure of hardware as well as the losses due to splitting the path into different decoder branches can be kept low. Also the network can be extended easily.

Interferometer: Into this group all sensors belong in which the interferometer is the sensor element itself. Mach-Zender, Michelson, Sagnac or so-called „White light" Interferometers are used for temperature, stress/strain, vibration, rotation and also for high electric current measurements /25/. The interferometry is a basic and one of the oldest measuring methods of optics.

Like in fibre grating sensor applications WDMA and TDMA prevail in interferometer networks. Regarding CDMA the simple structure mentioned above has also been described /21/. But there are more attractive possibilities like applying a coherent phase coded continuous signal to that scheme. Another favoured scheme consists of a coherent serial or parallel ladder network where the sensors affect the phase in each branch. Then, the decorrelation provides status information of all sensors at once.

OTDR: The development of the OTDR/OFDR has been in the centre of the scientific interest since 1980. Originally developed for the characterisation and error tracing of optical transmission circuits today e.g. liquid or distributed temperature sensors use OTDR procedures /22/.

Already at the beginning of OTDR the advantages of coherent and incoherent CDMA were used /23/. Nevertheless an enormous application potential remains. Particularly in the insitu measurement for example amid the frequency band of WDMA networks optical CDMA could be used since the spectrally spread optical power can be adapted easily to the required minimum SNR. Furthermore, nonlinearities can impair the result when entering high-power pulses. By means of CDMA theoretically the necessary energy could be lowered with identical or higher power at the measurement side.

Reflection/Transmission measurement: The last group consists of most techniques that deal with distance measurement (e.g. laser triangulation), speed measurement (laser Doppler method) as well as colour analyses or measurements of changes of the optical power within optical paths due to outside circumstances (e.g. light barriers) /24/.

In this group optical CDMA techniques are mostly easy to be integrated into the electrical path. Thus the component density and fail-safe characteristics can be enormously increased by coding transmitters and receivers. In an other case it is possible to make statements about the surface textures if several coded signals are transmitted in a spatially distributed manner. Therefore, the reflected and also spatially measured light can be easily assigned to one optical source by correlation. Thus, the way of light including multiple reflections could be calculated.

4. Conclusion

Spread spectrum techniques are very powerful for providing multiple access in networks. In recent time they show an increasing importance in the optical domain, too.

The different available realizations have been categorized and explained in an overview. Most properties of light propagation and optical waves can be included in coding and decoding techniques, e.g. time, phase, wavelength, polarization or wave fronts. In most cases the techniques can be easily integrated and adapted to individual requirements and possibilities. Thus, a very good scalability is inherent to this schemes.

Optical sensors are connected ever more frequently to networks because of space, weight and cost reasons. Due to the multiplicity of spread spectrum techniques the possibility of implementing CDMA exists also in sensor networks which mainly work with WDMA and TDMA at present. In addition measuring speed, dynamics and security of individual sensors can be improved. Examples are given for fibre gratings, interferometry, OTDR/OFDR and reflection/transmission measurements.

Optical CDMA in regard to optical sensor technology has not been studied on a scientific level until now. Its potentials are much higher as could be shown here, particularly when costs and efforts are taken into account.

References

[1]   N. Barbour, G. Schmidt, "Inertial Sensor Technology Trends," *IEEE Sensors Journal*, Vol. 1, No. 4, pp. 332-339, 2001

[2]   D. Betz, L. Stautigel, M. Trutzel, M. Schmuecker, E. Huelsmann, U. Czernay, "Test of a Fiber Bragg Grating Sensor Network for Comercial Aircraft Structures," *15th Optical Fiber Sensors Conf.*, pp. 55-58, 2002

[3]   L.C.G. Valente, A.M.B. Braga, A.S. Ribeiro, R.D: Regazzi, W. Ecke, Ch. Chojetzki, R.Willsch, "Time and Wavelength Multiplexing of Fiber Bragg Grating Sensors Using a Commercial OTDR", *15th Optical Fiber Sensors Conf.*, pp. 325-328, 2002

[4]   B.J. Vakoc, M.J.F. Digonnet, G.S. Kino, "Demonstration of a 16-Sensor Time-Division-Multiplexed Sagnac-Interferometer-Based Acoustic Sensor Array with an Amplified Telemetry and a Polarization-Based Biasing Scheme", *15th Optical Fiber Sensors Conf.*, pp. 325-328, pp. 151-154, 2002

[5]   N. Wada, H. Sotobayashi, K. Kitayama, "Error-free 100km transmission at 10Gbit/s in optical code division multiplexing system using BPSK picosecond-pulse code sequence with novel time-gating detection", *Electronics Letters*, Vol. 35, No. 10, pp.833-834, 1999

[6]   W. Huang, M.H.M. Nizam, I. Andonovic, M. Tur, "Coherent Optical CDMA (OCDMA) Systems Used for High-Capacity Optical Fiber Networks-System Description, OTDMA Comparison, and OCDMA/WDMA Networking", *Journal of Lightwave Technology*, Vol. 18, No. 6, pp. 765-778, 2000

[7]   K. Iversen, J. Mückenheim, D. Hampicke, "A Basic Theory of Fiber-Optic CDMA", *IEEE 4th Int. Symposium on Spread Spectrum Techniques and Applications*, pp. 431-437, 1996

[8]   Jawad A. Salehi, "Code Division Multiple-Access Techniques in Fiber Networks - Part I: Fundamental Principles ", *IEEE Trans. on Comm.*, Vol. 37, No. 8, pp. 824-833, Aug. 1989

[9]   G. Vannucci, S. Yang, "Experimental Spreading and Despreading of the Optical Spectrum, *IEEE Trans. On Comm.*, Vol. 37, No. 7, pp. 777-780, 1989

[10]   T. O'Farrell, S. Lochmann, " Performance analysis of an optical correlator receiver for SIK DS-CDMA communication systems", *Electronics Letters*, Vol. 30, No.1, pp.63-65, 1994

[11]   Y.L. Chang, M.E. Marhic, "Fiber-Optic Ladder Networks for Inverse Decoding Coherent CDMA", Journal of Lightwave Technology, Vol. 10, No.12, pp. 1952-1962, 1992

[12]   J.A. Salehi, A.M. Weiner, J.P. Heritage, "Coherent Ultrashort Light Pulse Code-Division Multiple Access Communication Systems", *Journal of Lightwave Technology*, Vol. 8, No. 3, pp. 478-491, 1990

[13]   M. Kavehrad, D. Zaccarin, "Optical Code-Division-Multiplexed Systems Based on Spectral Encoding of Noncoherent Sources", *Journal of Lightwave Technology*, Vol. 13, No. 3, pp. 534-545, 1995

[14]   E. Inaty, H.M.H. Shalaby, Paul Fortier, Leslie A. Rusch, "Multirate Optical Fast Frequency Hopping CDMA System Using Power Control", *Journal of Lightwave Technology*, Vol. 20, No. 2, pp. 166-177, 2002

[15]   S. Benedetto, G. Olmo, " Analysis of an optical code division multiple access scheme employing Gold sequences", *IEE Proceedings-1*, Vol. 140, No. 3, pp. 211-219, 1993

[16]   M. Abtahi, J.A. Salehi, "Spread-Space Holographic CDMA Technique: Basic Analysis and Applications", *IEEE Trans. on Wirelss Comm.*, Vol. 1, No. 2, pp. 311-321, 2002

[17]   L. Tancevski, I. Andonovic, "Hybrid Wavelength Hopping/Time Spreading Schemes for Use in Massive Optical Networks with Increased Security", *Journal of Lightwave Technology*, Vol. 14, No. 12, pp. 2636-2647, 1996

[18]   Jun-Jie Chen, Guu-Chang Yang, "CDMA Fiber-Optic Systems with Optical Hard Limiters", *Journal of Lightwave Technology*, Vol. 19, No. 7, pp. 950-958, 2001

[19]   R. Willsch, W. Ecke, H. Bartelt, "Optical Fiber Grating Sensor Networks and their Application in Electric Power Facilities, Aerospace and Geotechnical Engeneering", *15th Optical Fiber Sensors Conf.*, pp. 49-54, 2002

[20]   K.P. Koo, A.B. Tveten, S.T. Vohra, "Dense wavelength division multiplexing of fibre Bragg grating sensors using CDMA", *Electronics Letters*, Vol. 35, No. 2, 1999

[21] B.J. Vakov, M.J.F. Digonnet, G.S. Kino, "A Novel Fiber-Optic Sensor Array Based on the Sagnac Interferometer", *Journal of Lightwave Tech.*, Vol. 17, No. 11, pp. 2316-2326, 1999

[22] W.B. Lyons, D. King, C. Flanagan, E. Lewis, H. Ewald, S. Lochmann, "A 3 Sensor Multipoint Optical Fibre Water Sensor Utilising Artificial Neural Network Pattern Recognition", *15th Optical Fiber Sensors Conf.*, pp. 463-466, 2002

[23] Aasmund SV. Sudbo, "An Optical Time-Domain Reflectometer with Low-Power InGaAsP Diode Lasers", *Journal of Lightwave Technology*, Vol. LT-1, No. 4, pp. 616-618, 1983

[24] C. Fitzpatrick, E. Lewis, A. Al-Shamma'a, J. Lucas, "An Optical Fibre Sensor for Germicidal Microwave Plasma Powered UV Lamps Output With Potential For On-Line Temperature Control", *15th Optical Fiber Sensors Conf.*, pp. 455-458, 2002

[25] F.Farahi, D.J. Webb, J.D.C. Jones, D.A. Jackson, "Simultaneous measurement of temperature and strain using fibre optic interferometric sensors", *SPIE Vol. 1314 Fibre Optics '90*, pp. 270-277, 1990

An approach towards measuring distributed strain in fibres

A Hresha, M Campbell, S Canning and A S Holmes-Smith

School of Engineering, Science and Design, Glasgow Caledonian University, Glasgow, UK

Abstract. While measurements of distributed strain are now possible using multiplexed fibre Bragg gratings, systems which use single fibres are proving to be a much more elusive prospect. One approach to this goal, based on frequency modulated continuous wave technology and the optical Kerr effect, is currently being investigated. The methodology requires remote coupling between $HE_{11}{}^x$ and $HE_{11}{}^y$ eigenmodes in a bowtie fibre which arises due to changes in the refractive index of the material. Some difficulties have arisen in this development, although some progress has also been made towards the final goal of measuring distributed strain with single fibres.

1. Introduction

The measurement of distributed strain with single fibres in large remote structures has proved to be an elusive, yet highly desirable, objective for industry and academe. This view seems to be supported by the potentially large number of high value applications in structural and oil-related engineering which need to be addressed. It has been shown in the past that frequency modulated continuous wave (FMCW) technology, used in conjunction with the mechanically induced elasto-optic effect (EOE) in high birefringence (HiBi) fibres, creates a very versatile method of constructing fibre optic distributed stress sensors and non-distributed strain sensors [1-3]. Such sensors produce a beat signal which contains information regarding the magnitude and position of applied stress or the magnitude of the strain. Independently, the optical Kerr effect (OKE) has also been shown to be a useful mechanism for eigenmode coupling of continuous wave laser beam probes within such fibres [4-5]. The main advantage of OKE in the current study is its ability to couple eigenmodes in applications where the fibre is inaccessible for most of its length, including the distal end. As previously mentioned, the prospect of long range distributed strain measurement for (i) large structures such as buildings, bridges and oil platforms and (ii) laying long lengths of pipelines is an intriguing and inviting one. However, to date, attempts to measure distributed strain have been confined to relatively short lengths using multiplexed in-line Bragg grating fibres [6,7]. The present authors believe that there is now a need to explore the use of OKE in Hi-Bi fibres, in conjunction FMCW technology, to formulate a distributed strain sensor based on a single fibre. It is acknowledged that a second unbound fibre would be required for temperature compensation since both strain and temperature have the same effect on the output signal.

1.1 Previous systems. Initially two-mode fibre interferometers were developed as optical fibre strain sensors because they have (i) higher resolution compared with intensity modulation-based strain sensors and (ii) a relatively simple configuration since their reference and signal beams propagate within a single fibre without the need for a separate reference arm. Non-distributed fibre optic strain sensors were developed using equally intense LP_{01} and LP_{11} modes within e-core HiBi fibres, however, these systems tended to suffer from interfering environmental factors such as temperature. Ratiometric, Bragg grating-based Er-doped laser sensors have gone some way towards addressing this problem.

1.2 FMCW approach. Although the FMCW approach (figure 1), with conventional EOE couplers, can measure strain and distributed stress, it cannot measure distributed strain easily due to difficulty in applying the EOE coupler (force) along the fibre length.

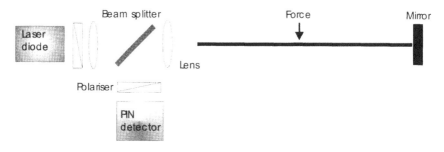

Figure 1(a). Conventional EOE coupled system for stress.

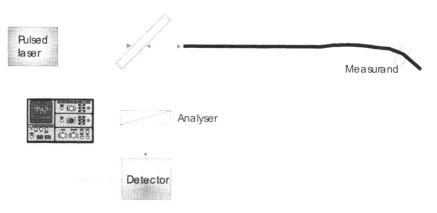

Figure 1(b). Conventional system for strain

As such, that technology would prove to be no better than the most advanced strain gauge quasi-distributed system. A mode coupling method for linearly polarised (LP) light is required which allows remote coupling access to the whole length, bearing in mind that many of the applications will have the fibre embedded within the structure which is to be monitored.

1.3 OKE. The OKE is a non-linear effect involving rotation of the eigenaxes which produces a subsequent phase shift between the polarised light components propagating in the fibre. OKE can be induced in bulk optical materials but requires very high power densities which are close to the damage threshold of these materials. The same effect can, however, be produced in polarisation maintaining optical fibres at much lower power levels. The obvious choice of medium is therefore HiBi fibres which have two orthogonal eigenaxes with non-significant crosstalk (<30dB km^{-1}). LP light launched into one eigenaxis of such a fibre should emerge as LP light as indicated in figure 2.

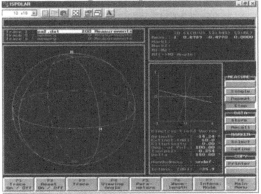

Figure 2. LP light shown as a point on the equator of the Poincare sphere at angle Ψ

Significant eigenaxes crosstalk results in elliptically polarised light as shown in figure 3.

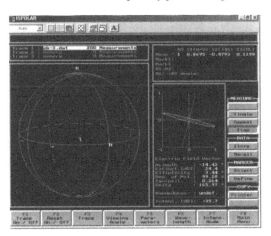

Figure 3. Elliptically polarised light shown in the upper hemisphere

Others [5] have shown that OKE can produce coupling between $HE_{11}{}^x$ and $HE_{11}{}^y$ beams using the configuration shown in figure 4. In these experiments equal intensities of LP light were launched from a probe (CW) laser, typically HeNe, into the two orthogonal eigenaxes of polarisation maintaining fibres which then acted as a high order waveplate. By launching LP light from a second (pumping) laser source consisting of high power, short duration pulses into one axis of the fibre, a resulting phase change ($\Delta\phi$) can be produced which is given by $\Delta\phi = \dfrac{2\pi L(\Delta\delta n)}{\lambda}$ where $\Delta\delta n$ is the difference in the refractive indices of the two eigenaxes induced by the OKE, L is the effective fibre length and λ is the probe vacuum wavelength.

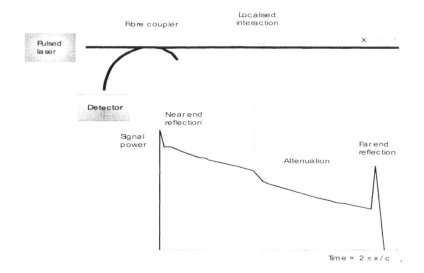

Figure 4. OKE induced eigenaxes coupling (after [5])

2. The Fibre optic system

This consists of a 100m length of bowtie fibre with two 3dB polarisation-maintaining couplers attached, in tandem, to proximal end and a mirror attached to the distal end. The advantage of this arrangement is that it enables all the instrumentation to be attached to the proximal end of the fibre. In the course of this development, the probe beam, which will ultimately be an FMCW driven diode laser, was substituted, for simplicity, by a 10W CW HeNe. The pump beam (which produces the OKE) is supplied by a 2μJ per pulse, frequency doubled Nd:YAG laser operating at 20kHz. Alignment of the HeNe beam to one of the fibre eigenaxes was achieved using a projection arc and laser pointer as shown in figure 5.

The effect of the Nd:YAG pumping beam is to produce in the probe beam a time dependent set of beat signals in a time scale of 10µs per pump pulse. Any strain applied to the fibre would cause the beat frequency patterns to change in a manner which enables the section of fibre being strained, to be determined. This means that the position of the strain would be uniquely located each time a pump pulse is sent down the fibre. The signal analysis is, of course, very complex, but such technology would find immediate application in several high value off-shore projects. Ultimately, the system could be used for intelligent actuation during installation of oil rigs and for monitoring dry risers strains.

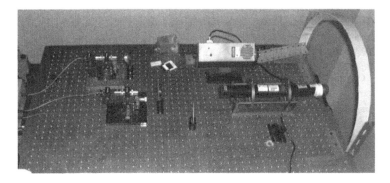

Figure 5. The experimental arrangement with laser pointer alignment system

Initially, measurements with a RPA 2000 polarisation analyser suggested that the Ψ variations occurred with a range of (7-10)°. Further measurements, using a 1 ns risetime photodiode in conjunction with a linear polariser, revealed that the intensity variations of the emergent light were comparable to that predicted by the Law of Malus (~4%) and that the Ψ variations were in the reduced range of (0.5-2)°.

3. Discussion

The authors believe that the proposed system could represent an elegant solution to the problem of measuring distributed strain over large distances. The unexpected production of Ψ variations in emerging LP and elliptically polarised light is puzzling in itself and also potentially damaging to the proposed methodology.

If LP light is launched into an eigenaxis of a polarisation maintaining fibre then it should emerge as LP light. If it partially fills the other eigenaxis then light with a small ellipticity should emerge. The authors do not know of any reason as to why LP light launched into an eigenaxis should emerge as LP light with a 2 Hz variation in its azimuthal angle. Even when elliptically polarised light is produced it seems to have the same Ψ variation at 2 Hz. The effect has been seen in e-core and panda fibre but is very much reduced.

In the proposed system, measurement of distributed strain relies on the unambiguous analysis of complex beat signal patterns. It may well be that these Ψ variations render that process inoperable.

References

[1]   G. Zheng, M. Campbell and P.A. Wallace, 1996, *Appl. Opt.* Vol. 35, No. 28, 5722-5726.

[2]   M. Campbell and G. Zheng, 1998, *Int. J. Electr.* Vol. 85, No. 4, 545-552.

[3]   M. Campbell, G. Zheng, A.S. Holmes-Smith and P.A Wallace, 1999, *Meas. Sci Technol.*, 10, 218-224

[4]   J. Dziedzic, R.H. stolen and A. Ashkin, Optical Kerr effect in long fibers, 1981, *Appl. Opt.* Vol. 20, No. 8, 1403-1406.

[5]   I. Cokgor, V.A. Handerek and A.J. Rogers, 1993, *Opt. Lett.* Vol. 18, No. 9, 705-707.

[6]   F.M Araujo, L.A Ferreira, J.L.Santos and F. Farahi, 2001, *Meas. Sci Technol.*, 12,829-833.

[7]   B.A.L. Gwandu, L.Zhang, K. Chisholm, Y. Liu, X. Shu and I. Bennion, 2001, *Meas. Sci Technol.*, 12, 918-921

Vector Bending Sensors Using Long-Period Gratings Written in Special Shape Fibres

D Zhao, X Shu[1], L Zhang, I Bennion, G M H Flockhart[2], W N MacPherson[2], J S Barton[2], J D C Jones[2]

Photonics Research Group, Electronic Engineering, Aston University, UK
[1] Indigo Photonics Ltd, Faraday Wharf, Holt Street, Birmingham, UK
[2] Department of Physical Sciences, Heriot-Watt University, Riccarton, Edinburgh, UK

Abstract: Vector bending sensors using long-period gratings (LPGs) UV-written in two types of special shape fibres: flat-clad and 4-core fibres were demonstrated. Our experiments reveal a strong fibre-orientation dependence of the spectral response when such LPGs are subjected to dynamic bending. The bending sensitivity with directional recognition suggests that the devices have applications as vector bend sensors for many smart structure applications.

1. Introduction

In-fibre grating-based sensors have many advantages over conventional electric and alternative fibre optic sensor configurations [1]. They are relatively straightforward and inexpensive to produce, immune to electro-magnetic interference and interruption, lightweight and small in size, and self-referencing with a linear response. Moreover, their wavelength-encoded multiplexing capability allows for arrays of gratings to be embedded in structural materials for smart structure applications. Strain measurements within such structures are often used to determine structural deformation, where bending is inferred from the measured strain [2]. In such applications, the bending induced strain is direction-dependent. Thus, it is important to implement optical bend sensors capable of curvature measurement with direction recognition. To date, most demonstrated fibre grating-based (including Bragg and long-period structures) optical bend sensors are limited to the measurement of curvature amplitude and few have provided the necessary directional information [3-6].

In this paper, we report the implementation of vector bending sensors using UV-inscribed long period grating (LPG) structures . We have studied the bending sensitivity characteristics of the LPGs in two special fibre types: flat-clad fibre (FCF) and 4-core fibre (4CF). The LPGs in both fibres show strong spectral dependence not only on the degree of curvature but also on the direction of the applied bending.

2. Fibre structures and LPG spectra

Fig.1(a) and (b) show the image and dimensions of the cross section of FCF, respectively. This fibre was provided by the Prim Optical Fiber Corporation. The cladding of the fibre has a near-rectangular shape with two flat and two curved sides. The dimensions of the cladding cross-section are 145μm ×96μm and the core is located at centre of the fibre with a diameter of 10.2μm.

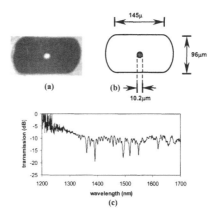

Fig. 1 (a) Image, and (b) dimensions of cross-section of FCF and (c) FCF-LPG transmission spectrum.

The FCF was photosensitised by H2 loading before the UV-inscription. The LPG structure, with a period of 296μm, was written into the FCF using a 244nm UV laser and point-by-point exposure. Fig.1(c) gives the typical transmission spectrum of the LPG in FCF, showing several resonant loss peaks originating from coupling between the core and cladding modes. The irregular distribution in the spectral spacing of the loss peaks is due to the non-circular shape of the cladding. Coupling strengths of 5 – 12dB were achieved for the loss peaks generated in the wavelength range shown in Fig. 1(c).

The 4CF was developed by France Telecom, originally for applications in telecommunications. As shown in Fig.2, the fibre consists of four cores arranged in a square with an adjacent core separation of 44μm. The diameter of each core is ~10μm and outer diameter of the fibre is ~145μm. All four cores were single-mode at 1550nm [7].

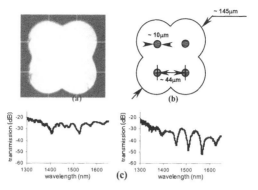

Fig. 2. (a) Image and (b) dimensions of the cross-section of 4CF. (c) LPG transmission spectra for two cores.

The 4CF was also photosensitised by a standard H2-loading process before the LPG inscription. The fibre was mounted on a rotation stage during inscription to facilitate selection of the individual cores. Several 4CF samples were fabricated with LPG structures in one, two and three cores, respectively, or in all four cores. A fan-out coupling device connecting 4CF to a single mode fibre was specially fabricated and used to interrogate all the 4CF-LPG devices. Fig.2 (c) shows LPG responses in two cores generated from two grating structures with periods of 450µm and 400µm. Again, we note that the loss peaks are not regularly distributed, but the resonances have exhibited strengths up to ~15dB.

3. Bending Experiment

A four-point bend system was produced for carrying out curvature measurements: Fig.3 illustrates the geometric configuration of this system. The LPG-containing fibre was attached with no twist to a 0.5mm thick, 20mm wide and 150mm long metal plate. In order to eliminate the axial strain, only the two ends of a 10cm central section of the fibre were fixed loosely on to the metal plate. The fibre was bent by depressing the centre of the metal plate with a micrometer drive. Experimental characterisation has indicated that the curvature varies linearly with the bending depth, h, using this arrangement [4].

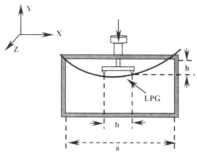

Fig. 3 The geometric configuration of the four-point bending system used in the investigation (a = 120mm, b = 20mm, and bending depth h = 0~8mm)

Bend measurement experiments were conducted to examine the spectral responses of FCF- and 4CF-LPGs to variations of the curvature at different fibre orientations. The grating transmission spectrum during bending was monitored using a broadband light source and an optical spectrum analyser. For FCF-LPGs, the spectral response under dynamic bending was measured for the bending directions along the flat and curved sides, respectively. In the case of 4CF-LPGs, the fibre was rotated through a series of angles from $0°$ to $315°$ and the bending-induced loss peak shift measured at each angular position.

4. Experimental Results and Discussion

We have observed strong directional bending sensitivity characteristics in both the FCF- and 4CF-LPG structures using the arrangement described above.

4.1 FCF-LPG

Fig. 4. plots the bending sensitivities of a measured FCF-LPG when it was bent along the flat- and curved-sides, respectively. It can be seen clearly from this figure that the respective bending sensitivity characteristics exhibit opposite slopes, indicating that the LPG resonance shifts towards longer wavelengths when it is bent along the flat side whereas the resonance moves towards shorter wavelengths when it is bend along the curved side. The figure also shows that the bending sensitivity is significantly larger for the flat side ($\sim$1.5nm/m-1) than for the curved side bending direction ($\sim$0.5nm/m-1). These distinct bending responses, clearly related to the geometric shape of the fibre, suggest that the FCF-LPG device can be usefully employed in smart structures for monitoring deformation along two axes.

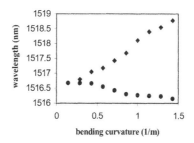

Fig. 4 Wavelength shift of FCF-LPG against curvature
• Bending applied to the curved side
♦ Bending applied to the flat side

4.2 4CF-LPG

With the 4CF-LPG device, we have carried out initial investigations with the LPG structure written in just one of the four cores. Since the core containing the LPG is not located on the fibre axis, we made the bending measurements at six different fibre orientations to characterise the relationship between this orientation and the bending sensitivity. Fig. 5 shows the six fibre positions corresponding to the rotation angles used in the experiment. For each angle, the LPG response was examined for a series of curvatures from 0 to 1.71m-1.

We have found that the bending responses of the 4CF-LPG were strongly dependent on the fibre orientation. The LPG loss peak (1) remained insensitive to bending and its attenuation decreased slightly only when the fibre was set at 0° and 180°, and (2) shifted in wavelength in opposite directions when the fibre was rotated from 0° to 90° and from 180° to 315°, respectively.

Fig. 6 shows the wavelength shift against curvature for the six fibre orientations used. It is clear that all of the responses show good linearity. The maximum bending sensitivities occurred at 90° for the increasing wavelength response and at 270° for the decreasing wavelength response. Shifts of up to 23nm and 26nm in wavelength were observed with an applied curvature of 1.71m-1 for the 90° and 270° fibre orientations, corresponding to bend sensitivities of 13.2nm/m-1 and -14.9nm/m-1, respectively.

The above results were obtained in the 4CF with the LPG written in just one core. Experiments investigating devices with LPGs inscribed in two/three cores, and with FBG and LPG inscribed in different cores are in progress. It is anticipated that devices using gratings in multiple cores will lead to practical vector sensor devices with temperature sensitivity decoupling.

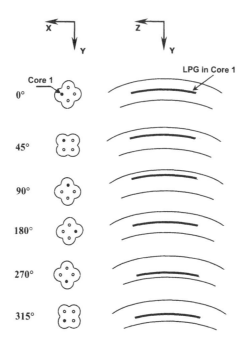

Fig. 5 Six fibre orientations used in the bending sensitivity investigation using the 4CF-LPG device.

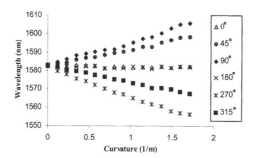

Fig. 6 Bending-induced wavelength shifts against curvature for the six fibre orientations.

5. Conclusion

We have studied the bending sensitivity characteristics of LPGs UV-written in two special fibre types - FCF and 4CF. The experimental results clearly indicate that the spectral responses of the LPGs depend strongly not only on the curvature amplitude but also on the fibre orientation. The bending sensitivity with directional recognition suggests that the devices have applications as vector bend sensors for many smart structure applications.

6. Acknowledgements

This work was supported by the UK EPSRC and the Ministry of Defence.

References

[1] L. Zhang, W. Zhang, I. Bennion, "In-fiber grating optic sensors", Chapter 4 in Fiber optic sensors, Marcel Dekker Inc., (2002)

[2] R. T. Jones, D. G. Bellemore, T. A. Berkoff, J. S. Sirkis, M. A. Davis, M. A. Putnam, E. J. Friebele, A. D. Kersey, Smart Mater. Struct., 7, p.178 (1998)

[3] M. G. Xu, J. L. Archambault, L. Reekie, J. P. Dakin, Int. J. Optoelectron, 9, p.281 (1994)

[4] Y. Liu, L. Zhang, J. A. R. Williams, and I. Bennion, "Optical bend sensor based on measurement of resonance mode splitting of long-period fiber grating", IEEE Photon. Tech. Lett., 12(5), pp. 531-533 (2000)

[5] H. J. Patrick, C. C. Chang, and S. T. Vohra, "Long period fiber gratings for structural bend sensing ", Electron. Lett., 34(18) pp. 1773-1774 (1998)

[6] J. J. Gander, W. N. MacPherson, R. McBride, J. D. C. Jones, L. Zhang, I. Bennion, P. M. Blanchard, J. G. Burnett, A. H. Greenaway, "Bend measurement using Bragg gratings in multicore fibre", Electron. Lett., 36(2), pp. 120-121 (2000)

[7] B. Rosinski, J. W. D. Chi, P. Grosso, and J. Le Bihan, "Multichannel transmission of a multicore fiber coupled with vertical-cavity surface-emitting lasers", J. Lightwave Technol., 17(5), pp. 807-810 (1999)

Surface roughness measurement using fibre grating assisted fibre interferometer

F Xie, X Q Jiang, W Zhang[1], L Zhang[1], and I Bennion[1]

School of Engineering, University of Huddersfield, Huddersfield, HD1 3DH, UK
[1] Photonics Research Group, Aston University, Birmingham, B4 7ET, UK

Abstract: A fibre-optic interferometer using fibre Bragg gratings are used for measuring the profile of rough surface. Wavelength-multiplexed fibre Bragg gratings are employed in the system. A synthesised phase information is calculated based on a two-wavelength technique that allows unambiguous measurement range much larger than the optical source wavelength. The wavelength-tuning feature of fibre Bragg grating is used in this work to set the wavelength difference in the two-wavelength technique to achieve different measurement range. Due to the distinct phase detection technique both large measurement range and high sensitivity are available in the same measurement system.

1. Introduction

In today's increasingly competitive environment, the ability to control product quality in manufacturing processing is critical. Ensuring customer satisfaction through proper manufacturing metrology is widely recognised to be a key business strategy and fundamental to a successful manufacturing operation. With the overwhelming requirement from ultra precision industries (including the micro-mechanic system and nano technology), it is necessary to develop measurement system with smaller-size, high-tolerance and ultrahigh surface accuracy. More importantly it has to be addressed for industrial on-line measurement.

Generally speaking non-contact measurement is preferred in such application. Optical interferometers have been widely used for non-contact surface roughness measurement [1-3]. Meanwhile fibre optic interferometers have been of interests of many researchers [4, 5] since they have shown a proven track on measuring a number of parameters such as acoustic, magnetic, current, strain and temperature perturbation with high sensitivity. Also the optical fibre down lead that connects the sensor to the source gives more flexibility that is important in workshop environment.

A well-known problem with optical interferometric technique is the interference phase ambiguity. Since the speckle pattern generated by a monochromatic light source on a rough surface has an essentially random phase content with a standard deviation larger than 2π, conventional interferometry will not yield any useful information about the profile of a rough surface that is larger than the source wavelength. One way to extend the range of applications for interferometry is to measure the interferometric phase at two distinct wavelengths [e. g., 3]. There have been a number of proposed systems based on this idea. In this work we present a fibre-optic interferometer based measurement system using wavelength-multiplexed fibre Bragg gratings. This technique can greatly meet the requirement to the light source used in two-wavelength technique. Since fibre Bragg grating is flexible in wavelength tuning it is easy to set a wavelength difference to accommodate required measurement range.

2. Measurement Principle

The detected interference phase from a monochromatic light source illuminated interferometer can be described as,

$$\theta = Mod(\frac{2\pi n}{\lambda} L) \qquad (1)$$

where n is the effective refractive index of the fiber, λ the wavelength of the source, L the optical path difference of the interferometer where the surface roughness information is included. The function *Mod* returns the remainder modulo 2π. Obviously the unambiguous phase change is limited to 2π, indicating a surface roughness measurement range only ~1.55µm while operating wavelength around 1.55µm. When the interferometer is illuminated with two wavelengths $\lambda 1$ and $\lambda 2$, a synthetic phase can be obtained with an expression similar to (1),

$$\Theta = Mod(\frac{2\pi n}{\Lambda} L) \qquad (2)$$

where $\Lambda = \lambda 1 \cdot \lambda 2 /(\lambda 1 - \lambda 2)$, is the synthetic wavelength. There is no ambiguity in the measurement as far as nL is less than Λ. For a wavelength difference of 1nm now the measurement range extends to 2.4mm.

3. Experiments and results

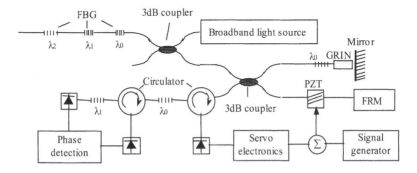

Figure 1 Experimental arrangement

This technique is demonstrated with the arrangement shown in Fig.1. A Broadband light source (EDFA) is used in the experiment. The fibre Michelson interferometer is used and illuminated by the reflection of three fibre Bragg gratings with the wavelengths of λ_0, λ_1 and λ_2. In one arm (reference arm) of the fibre interferometer a Faraday Rotating Mirror (FRM) is used to reflect all the wavelength components back. In the other arm a GRIN lens collimates the lights (λ_1, λ_2) onto a mirror that is mounted on a micrometer to simulate a surface roughness variation. The light with wavelength λ_0 is reflected by a corresponding FBG located just before the GRIN lens.

At the receiving end three optical signals with different wavelengths are de-multiplexed and photo-detected respectively by combining two fibre gratings with two circulators. The optical signal with λ_0 is used to stabilise the fibre interferometer therefore the optical signal at this wavelength has no phase information caused by varying the mirror but the phase fluctuation caused by environmental perturbation to the fibre interferometer. This signal is processed using the active phase tracking homodyne (ATPH) technique in which a PZT fibre phase modulator is incorporated into the reference arm of the interferometer and act as part of a servo feedback loop to maintain the fibre interferometer locked at its quadrature status [6]. On other hand the optical signals of λ_1, λ_2 contain both phase fluctuation from the fibre and the phase change from the mirror. Once the fibre interferometer is locked there are no phase fluctuation in λ_1, λ_2 but useful phase information of surface roughness.

The reflection spectra of the fibre Bragg gratings used in this work are as shown in Fig.2. All the gratings have a 3dB bandwidth of ~0.2nm. λ_1 is well away from λ_1, λ_2. The initial wavelength difference between λ_1 and λ_2 is around 0.5nm. The phase information of λ_1, λ_2 were detected separately using the technique in [6] and then processed to give the phase information at the synthesised wavelength Λ.

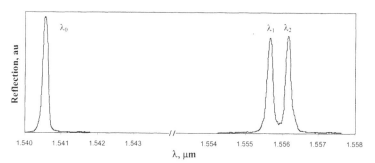

Figure 2 Reflection spectra of the fibre gratings used in this work.

A measured result over 200μm range is given in Fig.3. As the wavelength tuning of fibre Bragg grating is easy to realise, simply straining the gratings can increase the wavelength difference between λ_1, λ_2, consequently achieving different measurement range and sensitivity. A strain was applied to the grating of λ_2 so that the wavelength difference was increased to 2nm. The obtained synthesised phase is depicted in Fig. 3 as well.

Obviously the measurement range is much larger the light source wavelength. The initial optical path difference between the two arms of the interferometer is around 1.1mm, which was assessed by measuring time response of the fibre interferometer using a Lightwave Components Analyzer.

The measurement resolution is usually limited by the phase detection. Here the synthesised phase actually is a differential phase of optical signals at two wavelengths λ_1, λ_2. Since the phase information of individual wavelength is measured based on ATPH technique, it could be very sensitive (e.g., 10^{-6} rad) if the initial OPD is in quadrature and the phase change from the mirror position variation is small. However, the actual phase change for each wavelength is much larger than 2π. As a result the synthesised phase measurement produces a measurement resolution with the order of source wavelength.

Nevertheless it implies that it is likely to achieve large measurement range and high sensitivity in the same system. As the phase information of individual wavelength is measured, high sensitivity measurement can be achieved by setting one of the two wavelengths to a proper value so that the corresponding OPD is in quadrature. Further results on both large measurement range and high sensitivity will be presented on the conference.

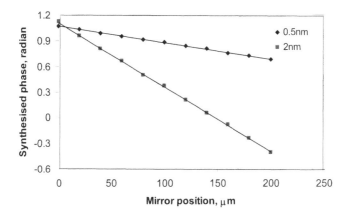

Figure 3 Experimental results of synthesised phase over 200µm range

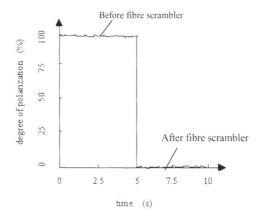

Figure 4 Polarisation degree reduction for eliminating polarisation fading effect

It should be pointed out that the stability of fibre interferometer plays a very crucial role in the measurement. The ATPH technique can compensate most random phase fluctuation caused by environmental perturbation. However, the polarisation fading always exists in the fibre interferometer, which causes serious error in the measurement. This problem was addressed by inserting a fibre polarisation scrambler between light source and fibre interferometer. The polarisation degree of the light launched into the fibre interferometer was measured before and after using the polarisation scrambler. As shown in Fig. 4. There is a dramatic reduction of polarisation degree. This effectively eliminated the polarisation fading effect during the measurement, however, at the cost of reduced interference visibility that could have some impact for high sensitivity measurement.

4. Conclusion

We have demonstrated a fibre interferometer for surface roughness measurement in which wavelength-multiplexed fibre gratings have been used for locking the fibre interferometer and achieving large measurement range without ambiguity. The outcome of this work predicts the measurement capability with both large range and high sensitivity.

References

[1] N. Bobroff, "Recent advances in displacement measuring interferometry", *Meas. Sci. Technol.*, v.4, 1993, pp. 907-926
[2] M. Adachi, H. Miki, Y. Nakai, and I. Kawaguchi, "Optical precision profilemeter using the differential method", *Opt. Lett.*, 12(10), 1987, pp. 792-794
[3] P. de Groot, "Interferometric laser profilemeter for rough surface", *Opt. Lett.*, 16(6), 1991, pp. 357-359
[4] D. P. Hand, T. A. Carolan, J. S. Barton, and J. D. C. Jones, "Profile measurement of optically rough surfaces by fibre-optic interferometry", *Opt. Lett.*, 18(16), 1993, pp. 1361-1363
[5] G. P. Brady, K. Kalli, D. J. Webb, D. A. Jackson, L. Zhang, and I. Bennion, "Extended-range, low coherence dual wavelength interferometry using a superfluorescent fibre source and chirped fibre Bragg gratings", *Opt. Comm.*, v.134, 1997, pp.341-348
[6] D. A. Jackson, A. Dandridge, and S. K. Sheem, "Measuremens of small phase shifts using a single mode optical fiber interferometr", *Opt. Lett.* 5, 1980, p.139

Interferometric interrogation of in-fiber Bragg grating sensors without mechanical path length scanning

D F Murphy[1], D A Flavin[1], R McBride[2], and J D C Jones[2]

[1] Department of Physical and Quantitative Sciences, Waterford Institute of Technology, Waterford. Ireland.
[2] School of Engineering and Physical Sciences, Heriot-Watt University, Edinburgh EH14 4AS. U.K.

Abstract: Hilbert transform processing has proved particularly effective in the demodulation of fibre Bragg grating sensors, producing high-resolution measurement for a short scan of optical path difference. Here we demonstrate the application of the technique based on a stationary interferometric configuration, in which the optical path difference is a function of position on a diode array detector. Absolute measurements of the Bragg wavelength are derived from the phase of the analytic signal of the detected interferogram. The method is demonstrated in application to temperature sensing, and is shown to provide inherent correction for resolution-limiting effects normally associated with stationary interferometers.

1. Introduction

Fibre Bragg grating sensors are one of the most interesting recent developments in optical fibre sensing. A broad range of fibre Bragg grating sensor demodulation schemes have now been reported [1,2]. The sensors have been particularly effective and measurement of dynamic strain, and are now widely explorer in the context of smart structures and structural health monitoring [3]. Absolute measurements of mean Bragg wavelength, employing an all-fiber Fourier transform spectrometer (FTS) were made using a Michelson interferometer configuration with a Nd-YAG laser reference for optical path scan stabilisation [4]. In this case, optical path delay (OPD) scans of 300 mm resulted in a 15 pm resolution of the grating wavelength. The long OPD scans required for the FTS approach introduced polarisation-fading problems that were overcome by incorporating Faraday reflection mirrors. Hilbert transform techniques have also been applied to absolute measurements of mean resonant grating wavelength [5-7]. Here measurements are based on the analytic signal phase of the Bragg grating interferogram recovered from a temporally scanned Michelson that incorporates a HeNe laser reference. OPD scans of only 1.2 mm have resulted in 5 pm resolution for single grating interrogation [5]. Similar resolutions were achieved for the interrogation of serial grating arrays [6,7]; resolutions of approximately 1 pm were reported for a 20.7 mm OPD scan.

The OPD scanning in the above schemes is performed using PZT fiber phase modulators or motorised mechanical translation stages, introducing significant complexity to the experimental designs and limiting measurement bandwidths.

A stationary or 'no moving parts' diffraction grating spectrometer was reported for the wavelength demodulation of an FBG array [8]. The diffracted Bragg grating spectra were imaged onto a CCD pixel array and curve-fitting algorithms were applied to the imaged spectra to measure the mean Bragg wavelengths. Resolutions of 1 pm were reported for a spectrometer with a nominal resolution of 100 pm, as defined by the Rayleigh criterion.

While the measurement bandwidth is high in this case, the performance of this technique depends on optical power levels and spectral widths of the grating reflections.

In this paper we propose and demonstrate a spatially-scanned interferometric, high-resolution Bragg grating demodulation technique based on a Twyman-Green tilted mirror interferometer. Absolute measurements of the mean Bragg wavelength are made based on a short OPD scan of ~ 200 μm. Wavelength measurement is derived from the phase of the analytic signal of the short spatial interferogram, obtained by the Hilbert transform processing of the interferogram. In the spatially-scanned configuration both the experimental design complexity and size are minimised. The interferograms are captured on a CCD array facilitating rapid capture and high measurement bandwidth. We use a colinearly propagating HeNe laser beam both as a wavelength reference and for correction of errors arising from non-uniformities in the beamsplitter and mirror surfaces, and in the spatial distribution of pixels in the sensing array.

2. Theory

Fig. 1 illustrates the experimental configuration of the proposed demodulation system. The resolution of the technique depends critically on associating a highly precise sequence of delay imbalance values with the individual pixels on the detector array. In the bulk optic tilted Twyman-Green interferometer section the incident beam is amplitude divided by the hybrid beamsplitter BS. Mirror M1 is tilted at a small horizontal angle α to the incident beam, and M2 is orthogonal to its incident beam. The interferogram is formed on the detector plane Σ, in the overlap between the recombining beams. When the interferometer is illuminated by a spatially coherent beam from a narrow linewidth reflective Bragg grating, at a mean optical angular frequency ω_B, the intensity distribution along the y-direction on the detector plane Σ is

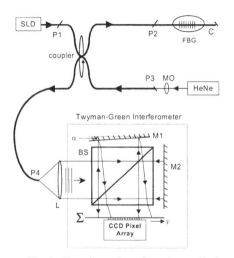

Fig. 1: Experimental configuration, a tilted Twyman-Green interferometric interrogation of fiber Bragg grating (FBG), with interferograms imaged on a CCD array. SLD, superluminescent diode; P1, P2, P3, P4, fiber coupler ports; BS, beamsplitter; M1. M2. mirrors.

$$I_B(y) = I_0\left[1 + V(y)\cos\{k_B(y)\phi_B(\tau) + \phi_{CB}\}\right] \qquad (1)$$

Here the function $V(y)$ varies slowly with y, and phase $\phi_B(y)$ is given by $\{k_B(y)\phi_B(\tau) + \phi_{CB}\}$, where $\phi_B(\tau) = \omega_B\tau$ is the phase for delay imbalance τ. The phase correction term, ϕ_{BC}, covers the case where the origin of y may not coincide with the origin of delay τ; $k_B(y)$ would be constant under the ideal conditions of perfect optical surfaces, negligible dispersive interferometric imbalance, uniformly distributed array pixels, and negligible tilt angle α. Some surface non-uniformity, dispersive delay imbalance residual, and pixel distribution non-uniformity will exist, and α is non-zero to produce the OPD required; hence we explicitly include the y-dependence of $k_B(y)$. If the interferometer is further illuminated by a colinearly propagating laser beam at frequency ω_R, then a superimposed interferogram will be formed on the detector plane; by analogy with Eq. 1 the phase of this interferogram will be $\phi_R(y) = \{k_R(y)\phi_R(\tau) + \phi_{CR}\}$. Assuming that the values of $k(y)$ differ negligibly at the frequencies ω_R and ω_B, i.e. $k_B(y) = k_R(y)$, then combining the equations for the two phases $\phi_B(y)$ and $\phi_R(y)$ yields the relationship

$$\phi_B(y) = \left\{ \left(\frac{\omega_B}{\omega_R} \right) \phi_R(y) \right\} + \left\{ \phi_C - \left(\frac{\omega_B}{\omega_R} \right) \phi_{CR} \right\} \qquad (2)$$

The second term on the right-hand side of Eq. 2 is constant for a given optical frequency ω_B, thus allowing us to write the Bragg wavelength as

$$\lambda_B = \frac{2\pi c}{\omega_B} = 2\pi c \left/ \left\{ \left(\frac{d\phi_B(y)}{d\phi_R(y)} \right) \omega_R \right\} \right. \qquad (3)$$

Thus the Bragg wavelength can be measured as the gradient of a linearly square is fit of the phase sequences $\phi_B(y)$ and $\phi_R(y)$. The individual interferograms can be extracted from the composite spatial interferogram in the frequency domain. The two phase sequences can then be determined from the unwrapped phases of the complex analytic signals of the extracted spatial interferograms using the Hilbert transform processing approach [5]. The analytic signals of the laser interferogram yield high-resolution delay mapping of the individual pixels in the detector array.

3. Experimental and Signal Processing

The broadband source is a pigtailed superluminescent diode (SLD) source with 0.16 mW output and ~ 62 nm FWHM (763 nm to 825 nm); the fibre Bragg grating (FBG) has a nominal resonant wavelength of 810 nm. HeNe light is launched into the coupler Port P3. The HeNe and grating reflection beams exit the directional coupler DC at P4, are collimated by lens L, and propagate through the interferometer. The composite spatial interferogram is detected by a 700 x 20 pixel CCD (8-bit resolution). The central array pixel corresponded approximately to zero OPD. The tilt angle α was adjusted to ensure sampling within the Nyquist criterion at ~ 2.25 pixels per HeNe fringe and ~ 2.86 pixels per Bragg fringe. This corresponded to an OPD scan of ~ 200 μm in the interferometer over the array sensing width.

We measured temperature-induced wavelength shifts over the range $30\,^\circ$C to $80\,^\circ$C. For any single measurement, the composite HeNe/FBG interferogram was captured and summed over the 20 rows of the CCD array to yield the measured sampled composite interferogram vector.   The modulus of the fast Fourier transform (FFT) of this interferogram was then obtained (Fig. 2 inset). The composite spectrum is filtered in the frequency domain to give the individual spectra. An inverse FFT of the individual spectra yields the individual complex analytic signals, the real and imaginary components of which are the individual interferogram and its Hilbert transform respectively. We recover the respective FBG and HeNe analytic signal phases from the arctan of the imaginary part of the signal divided by the real part [2]. These phases are unwrapped to yield sampled vectors for $\phi(y)$ and $\phi_R(y)$; Fig 2 illustrates the 700 values of each phase vector for one temperature.

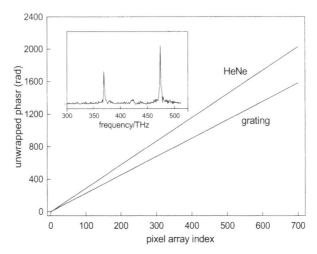

Fig. 2: Inset: magnitude of the spectrum obtained after the first Fourier transform of a composite interferogram.
Main graph: phases of the complex analytic signals for the 700 array locations, extracted and unwrapped from the composite HeNe and Bragg grating interferogram.

4. Bragg wavelength measurement

We performed a linear least squares fit to the measured values of $\phi_B(y)$ against $\phi_R(y)$ for each temperature and determined the respective Bragg wavelength from the fit slope, via Eq. 3. The phase fits were highly linear with correlation coefficient values of 0.9999998, thus supporting the assumptions underlying Eq. 2. Standard error values for each slope were calculated to be < 25 pm for λ over the full temperature range. Fig. 3 shows Bragg wavelength versus temperature between 29 °C and 81 °C. The error bars, calculated from the standard errors on the phase fit slopes, are consistent with the measurement variations in the Bragg wavelengths. The measured temperature dependence of the grating wavelength is ~ 6.35 pm/°C.

As we indicated earlier, inconsistencies in the beamsplitter or mirror surfaces, or in the spatial distribution of pixels in the sensing array give rise to a nonlinear relationship between pixel array position and group delay imbalance in the interferometer. However, by our processing of the HeNe interferogram to yield the phase of its analytic signal at each pixel position, we can immediately determine the group delay imbalance at the position. Fig.4 indicates the phase residual at each pixel based on the phase measurement there, derived from the analytic signal of a HeNe interferogram.

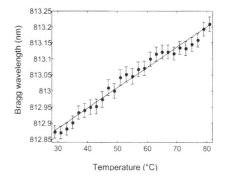

Fig. 3: Measured values of mean resonant reflected Bragg wavelength versus temperature. The error bars represent standard errors for the linear fits to the measured values of Bragg phase against HeNe phase at each temperature.

Performance degradation in temporal scanning interrogation schemes due to non-colinear illumination of the interferometer by the Bragg reflection and reference beams has been highlighted [3]. We launch the HeNe light through the fiber so that it exits the fiber at the same point as the Bragg beam gives accurate colinear illumination of the interferometer.

5. Resolution and Bandwidth Issues:

For the very limited OPD scan of ~ 200 μm scan available with the 700 pixel array, we achieved a resolution, defined by the standard error, of ~ 25 pm. This represents an improvement in resolution by more than two orders of magnitude relative to standard Fourier transform spectroscopy (FTS) and compares very favourably with the reported 15 pm resolution for a 300 mm OPD scan using the FTS approach and with previously reported temporally-scanned Hilbert transform approaches. Larger-scale spatially-scanned interferometers, using longer high-speed CCD arrays, can be readily constructed. Such schemes maintain the advantages of the stationary configuration and enable high-resolution and high-frequency demodulation of multiplexed FBG sensor arrays. CCD linear arrays of 8192 and 14404 pixels are now available commercially [6]. An 8192 pixel array would increase the OPD to ≈2.3 mm for the interferogram spatial sampling frequency that we used. Simulations, applying the noise levels met in our experiments, but based on a 2.3 mm OPD give a reduction in standard error figures from 25 pm to 2 pm. There is evidence of a periodic effect in the resulting measurements shown on Fig.3.

Using simulations of an ideal "noiseless" interferogram case, we found this effect to be caused by the combination of horror "close to Nyquist" sampling and the short OPD. For the case of simulated wavelength measurement in the range 800 to 802 nm for a 700 pixel array at a sampling density of 2.25 points per HeNe fringe; an error amplitude of as high as 12 pm is predicted. However this error decreases strongly with increasing OPD (pixel array size); for an 8192 pixel array, also at a sampling density of 2.25; the periodic effect is < 0.1 pm.

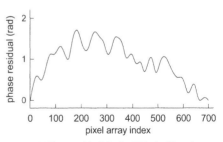

Fig. 4: Phase residual for the 700 pixel locations, derived from the HeNe analytic signal

The technique bears comparison with reported stationary interrogation schemes using CCD diffraction based grating spectrometers [8]. Our scheme, using a large number of pixels for any single reading, is less susceptible to ageing of individual pixels than the grating spectrometer scheme, which relies on a small number of pixels to detect spectral shifts. In spectrometer schemes, a range of optical powers and spectral widths from a grating sensor array may also degrade the resolution. Our configuration does not suffer from such degradation, as the spectral peaks are only observed when the interferogram is Fourier transformed. Diffraction gratings are sensitive to polarisation drifts and, for high-resolution measurements, it is usual to employ a depolariser. Hybrid beamsplitters are less susceptible to such drifts, so our scheme exhibits higher immunity to polarisation changes in the detected signal.

6. Conclusion

We have demonstrated the interferometric demodulation of Bragg grating sensors based on a stationary interferometer.   Our method overcomes the 'moving parts' and measurement bandwidth limitations associated with temporally scanned interferometric schemes. The wavelength measurement is derived from the phase of the complex analytic signal of spatial interferograms, in contrast to spectral measurement by FTS.   The technique incorporates precise delay mapping based on the analytic signal of a laser interferogram, thus overcoming limitations of conventional spatial interferometers caused by nonuniformies in optical surfaces and in the imaging array. We achieved a resolution of 25 pm for a limited OPD scan of 200 mm. This result compares favourably to reported temporally-scanned schemes and indicates potential for far higher resolutions and measurement bandwidth in larger spatial interferometer designs using long high-speed imaging arrays. We have further reported our investigation of the fundamental limits of the accuracy of the approach when using sampling densities close to the Nyquist limit.

References

[1]   K. T. V. Grattan and T. Sun, "Fiber optic sensor technology: an overview", *Sensors and Actuators*, **82**, pp 40-61, 2000

[2]   A. D. Kersey *et al.*, "Fiber grating sensors", *J. Lightwave Technol.*, **15**, pp. 1442-1463, 1997.

[3]   R. M. Measures, *Structural Monitoring with Fiber Optic Technology*, Academic Press, San Diego, 2001

[4]   M. A. Davis and A. D. Kersey, "Application of a fiber Fourier transform spectrometer to the detection of wavelength-encoded signals from Bragg grating sensors", *J. Lightwave Technol.*, **13**, pp. 1289-1295, 1995.

[5]   D. A. Flavin, R. McBride and J. D. C. Jones, "Short optical path scan interferometric interrogation of a fibre Bragg grating embedded in a composite", *Electron. Lett.*, **33**, pp. 319-321, 1997.

[6]   K. B. Rochford and S. D. Dyer, "Demultiplexing of interferometrically interrogated fiber Bragg grating sensors using Hilbert transform processing", *J. Lightwave Technol.*, **17**, pp. 831-836, 1999.

[7]   D. A. Flavin, R. McBride and J. D. C. Jones, "Short-scan interferometric interrogation and multiplexing of fibre Bragg grating sensors", *Opt. Comm.*, **170**, pp. 347-353, 1999.

[8]   A. Ezbiri, S. E. Kanellopoulos and V. A. Handerek, "High resolution instrumentation system for fibre-Bragg grating aerospace sensors", *Opt. Comm.*, **150**, pp. 43-48, 1998.

[9]   DALSA Corporation, Advance Product Information sheet on product Piranha2 quad-output Line Scan Cameras, 2001, and Kodak Data Sheet, 14,440 Pixel Trilinear Image Sensors, 1999.

Section 8

System & Sensor Networks

Measurement Techniques to improve the Accuracy of Smart Sensor Systems   (Invited)

Gerard C M Meijer and Xiujun Li

Electronic Instrumentation Laboratory, Faculty of Electrical Engineering, Delft University of Technology, Mekelweg 4, 2628CD Delft, The Netherlands. E-mail: G.C.M.Meijer@its.tudelft.nl; X.Li@its.tudelft.nl

Abstract: The paper reviews how the application of appropriate measurement techniques can improve the accuracy of smart sensor systems. The smart sensor systems under consideration consist of a number of multiplexed sensor elements, sensor-specific front-ends, modifiers and a microcontroller or a digital signal processor (DSP). It is shown that in an overall design approach the availability of memory and calculation functions enables the application of powerful measurement methods, such as chopping, autocalibration, dynamic element matching and dynamic division. Moreover, it has been shown how switched-capacitor front ends can extend the linear range for strong sensor signals. It is shown that the proposed techniques can be implemented with simple integrated circuits as well as with a special family of universal sensor interface circuits.

1. Introduction

In smart sensor systems the functions of sensors and their interfaces are combined. These functions include sensing, signal conditioning, analogue-to-digital conversion, and bus interfacing. Also, functions at a higher hierarchical level, such as self-testing, auto-calibration, data evaluation and identification can be included.

With respect to the accuracy of systems, two types of errors ha to be encountered: systematic errors caused by, for instance, the inaccuracy of the system parameters and random errors as caused by, for the traditional way of calibration, the sensor-under-test is compared to another one of superior instance, interference, noise and instability. The random errors can be eliminated by filtering, separating the common-mode and differential-mode signals, autocalibration (see section 3) and properly designing the system.

Systematic errors can be eliminated by calibration or trimming. In the traditional way of calibration, a sensor-under-test is compared to another one of superior quality. To make accurate measurements possible, the data resulting from this test can be stored in a computer memory, a controller or simply written down and used for the remainder of the sensor's life. With trimming, the sensor behavior is altered permanently to make its characteristics match the nominal ones as closely as possible, thus eliminating the need for recording individual sensor behavior.

Although calibration and trimming are useful ways to improve accuracy, there are still a number of shortcomings. For example, calibration and trimming have to be performed under certain conditions with respect to the temperature, supply voltage and time, and these conditions can differ from the conditions during actual sensor operation. Systematic errors may change due to changes in the environmental conditions and to long-term drift. The best way to deal with the basic problem is to eliminate the influences of the conversion parameters, except for those required for the measurements and those that are sufficiently reliable, stable and accurate. In smart sensor systems elimination of the influences of the conversion parameters can easily be achieved by applying advanced measurement techniques, such as chopping, autocalibration, dynamic element matching and autocalibration. In this paper a review of these techniques and related hardware is presented.

2. Advanced chopping techniques

Chopping in combination with synchronous detection is a good way to reduce the effects of low-frequency interference and noise, including $1/f$ noise, offset and offset drift and cross talk of the mains. A chopper can be implemented with a simple quad of switches (the commutator), which interchange the wires of a signal source at a frequency higher than that of the disturbing signals. In a common chopper the wires and thus the signal is switched in a +, -, +, -, +, -,....sequence. Afterwards the chopped signal passes a filter that removes the low-frequency components. Finally, after amplification, the signal is de-chopped to get the amplified signal back in the original frequency band. The chopper can be improved by changing the sequence in a +, -, -, +, +, -, -, +, ...order [1, 2]. This chopping sequence results in a filter operation with a filter transfer function, which in the z-domain, is equal to:

$$1 - z^{-1} - z^{-2} + z^{-3}. \tag{1}$$

With $z = \exp(j\omega T_{mod})$ and T_{mod} being the modulator period, Eq.(1) can be transformed into an expression in the frequency domain. We obtain a second-order low-pass filtering for low interference frequencies.

However, in [3] Bakker and Huijsing have shown, that feedthrough of the clock signals causes a new problem: Even after filtering and demodulation a residual low-frequency effect is induced by the control signals of the input chopper. To solve this problem they propose to apply a novel technique, called nested chopping. In a nested chopper a pair of inner commutators is operated at a frequency higher than the $1/f$ corner frequency, while another pair of outer commutators is operated at a rather low frequency. The latter pair of commutators has been applied to remove the dc component in the spikes generated by clock feedthrough. The technique requires an additional low-pass filter, that can easily be realized as a digital filter in the software of the applied microcontroller. In line with this idea it is possible to extend the nested-chopper idea and to apply three or more commutator pairs (Fig.1). The additional commutator pair can be operated at a medium frequency to reduce interference from a source with an intermediate frequency, for instance of the mains.

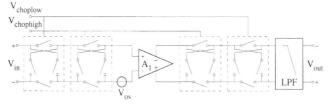

Fig.1. Nested-chopper amplifier principle.

Chopping techniques are not suited to reduce the effects of systematic errors caused by inaccuracy and drift in the transfer parameters. However, other techniques, such as autocalibration, are suited to eliminate such errors.

3. Autocalibration

Applying an autocalibration can eliminate the undesired effects of changes in the transfer parameters. Autocalibration can be performed in various ways, for instance by processing two or three signals [4, Chapt. 6].

a) The two-signal approach, in which the system is used to measure a reference signal S_1 in exactly the same way as the signal (or parameter) S_x which is being measured. The measurement result is the ratio $M_2=S_x/S_1$ or the difference $M_2=S_x-S_1$. This eliminates the influence of the multiplicative or additive parameters and errors, respectively.

b) The more accurate three-signal approach, in which two reference signals S_1 and S_2 and the unknown signal S_x are measured in an identical way, where S_1 is allowed to be zero or to equal $-S_2$. The final measurement result is the ratio $M_3=(S_x-S_1)/(S_2-S_1)$, which is insensitive to both multiplicative and additive parameters and errors. These principles have been applied in many of the sensor systems described in this paper. Unfortunately, it is not always possible to apply them to complete systems. However, even partial application can considerably reduce the effect of many drift and offset parameters.

Example: When measuring a temperature T_x, it is not practical to measure two reference temperatures T_1 and T_2. Therefore, the temperature is accurately converted to, for example, a resistor value. Then two reference resistors R_1 and R_2 can be applied for further processing (Fig. 2), thus continuing with the three-signal method described above as early as possible.

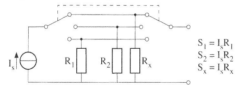

$$S_1 = I_s R_1$$
$$S_2 = I_s R_2$$
$$S_x = I_s R_x$$

Fig. 2. Measurement of a resistor R_x (four-wire method). In the three-signal approach two reference resistors R_1 and R_2 are also measured.

4. Dynamic amplification and division

During autocalibration the three or more signals are processed in an identical way. The system should be linear over the full signal range. This poses a problem when the signals are not in the same range of magnitude. In many practical situations the ratio between a small and a large signal has to be established. For instance, in a resistive-bridge sensor the ratio between the large bridge-supply voltage and a small output voltage has to be determined. In that case, to achieve a high accuracy the signal processor should have a very high dynamic range. Often such a requirement cannot be met. In that case the small signals have to be amplified or large signals to be divided with a scaling factor A, so that the resulting signals are in the same order of magnitude and the full dynamic range of the rest of the circuit can be exploited to obtain the best signal-to-noise ratio for the full range of the measurand.

In this approach the scaling divider or amplifier is not included in the auto-calibrated part of the circuit, so that the next problem is to realize a scaling factor A without loosing precision. The use of a dynamic-feedback (DEM) instrumentation amplifier [5, 6] can solve this problem.

The principle of this amplifier is shown in Fig. 3. The resistive feedback circuit consists of a chain of K matched resistors. The chain will be made rotating by addressing of the appropriate switches. The feedback is realized by u, v, and w resistors, respectively. Since the resistors are connected as a chain, a resistive load will be present which consists of z resistors. Therefore, it holds that:

$$K \equiv u + v + w + z. \tag{2}$$

By applying force and sense wires, the effect of the ON resistances of the switches S_1-S_6 is completely eliminated. The dynamic feedback is made by rotation of the resistor chain. In each of the K position of the chain a measurement is performed.

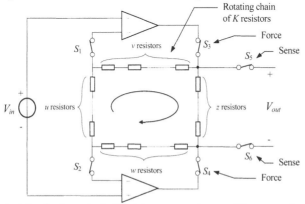

Fig. 3. The principle of a dynamic-feedback instrumentation amplifier, according to [13].

So, a resistor that is part of the load will become part of the feedback later. For this reason, this load resistor is of vital importance for the functionality of the dynamic feedback. The measurement results are processed in another part of the circuit and converted to a signal that can be read out by a microcontroller. In the microcontroller the averaged result of K successive measurements is calculated. The average gain $\overline{G}$ of this amplifier over K successive states is equal to:

$$\overline{G} = 1 + \frac{v + w}{u}. \tag{3}$$

Mismatches between the resistors hardly affect the average gain because there is a compensating effect when the resistors move along the chain. The resistor chain is controlled by a digital-state machine, which addresses the appropriate switches. Every successive state, the chain rotates one position, with a frequency of about 50 kHz. The control of the resistor chain requires that there are 6 switches connected to a single point between every two resistors, which results in a total of 6×K switches.

The output voltages are converted into the time domain, where for instance, a microcontroller takes care for the digitization and algorithmic processing.

In [7] and [8] a similar DEM amplifier has been presented, implemented with switched-capacitors (SC) in the feedback loop. A special feature of this amplifier is that the input signals can be handled over the full rail-to-rail common-mode input voltage range.

As an alternative, instead of an amplifier for the smallest signal, also amplification also a divider for the strongest signal can be applied. In [9] a dynamic voltage divider is proposed (Fig. 4).

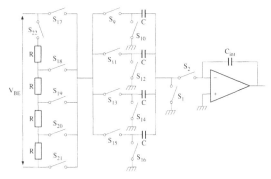

Fig 4. The dynamic voltage divider according to [16].

This simple circuit consists of a resistive voltage divider combined with a capacitive voltage divider. The divider is realized with N_R resistors and N_C capacitors, resulting in an accurate division ratio of $\alpha_d=N_C N_R$.

Implementation of the DEM techniques requires implementation of signal and data-processing circuits, such as switch controllers, memory and calculation circuits. Applying a microcontroller is a convenient way to enable implementations of these functions. Once a microcontroller is used, besides switch control it is also easy to perform data processing, in the frame of an overall system design. The data processing steps can include, for instance, averaging, non-linearity compensation, auto-calibration, self-testing and filtering.

5. Switched-capacitor front-ends with a wide dynamic range

An interesting recent development to increase the dynamic range for signal processors is to increase the dynamic range of the front-end amplifier, using switched-capacitor techniques. In [10] Meijer and Iordanov show that, for instance, in the front-end amplifiers for capacitive sensors and small-voltage sources the maximum voltage output swing of the front-end amplifier limits the dynamic range. This is due to the fact that a large amount of charge is transferred to the integrator. Also in σ-Δ converters such phenomenon can occur.

Figure 5(a) shows the basic circuit diagram of an improved input integrator circuit.

In this circuit clipping of the input amplifier output is avoided by using a negative-feedback circuit, which controls the discharge of the sampling (or sensor) capacitor C_S, which guarantees that the integrator opamp is always operated in its linear region. With this input circuit for large signals the dynamic range can be increased with a few orders of magnitude. With a proper switch control a linear relation between the period time t_p and the measurand V_X or C_S is obtained. The period time can be measured with for instance a microcontroller. The microcontroller also takes care for other important data-processing functions.

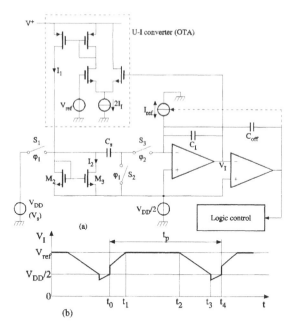

Fig.5. (a) Basic circuit diagram of a novel SC front-end with wide dynamic range; (b) output signal V_I of the integrator.

6. Universal Sensor Interfaces

The above-mentioned techniques can be implemented with simple electronic components assembled on a printed circuit board. These techniques have been also been implemented in a family of universal sensor interfaces developed at TUDelft [1, 2, 11]. These interfaces make it easy for a user to develop in a short time advanced sensor systems for a wide range of applications. One of these interfaces, the so-called UTI, is available at the market [12], including evaluation boards and microcontroller and PC software. At present research is performed on complementary products, such as the so-called universal sensor interface USI [11]. The output circuit of these interfaces is based on a period-modulated oscillator [2, 12]. The signals of the connected sensing elements are converted into period-modulated signals, similar to that shown in Fig. 6. Many of these signals can be transferred over a single output wire. For identification purposes the offset period time T_o has been made much shorter than any of the other signals, and acts as a marker for a long series of signals. The period-modulated signals are microcontroller and DSP compatible. No additional analog-to-digital converter is required. Because the internal oscillators are asynchronous, no clock line is required, thus eliminating the need for high-frequency cables and reducing EMI.

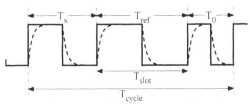

Fig. 6. Output signal of the universal interfaces.

The USI will provide interfacing for:

o   Capacitive sensors in a variable range up to 300 pF,
o   Single or multiple platinum resistors, for instance of the type Pt100, Pt1000,
o   Thermistors with a 0 °C value in the range of 1 kΩ ~ 25 kΩ,
o   Resistive bridges in a rage of 250 Ω ~ 10 kΩ with maximum imbalance +/- 4% or +/- 0.25%,
o   Conductance in the rage of 0.1 µS ~ 100 mS,
o   Single or multiple thermocouples with a voltage range of 0 mV ~ 200 mV, ±200 mV, ±25 mV and ±2 mV,
o   Voltage in the range of 0 V ~ 1 V, 0 mV ~ 200 mV, ±200 mV, ±25 mV, ±2 mV,
o   pH sensors,
o   Additional resistive temperature sensors for temperature compensation,
o   Platinum resistors with a long cable wire up to 20 m.

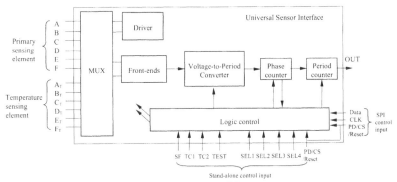

Figure 7. Diagram of the USI.

Figure 7 shows a diagram of the USI. In this figure, pins A – F, represent the connecting terminals for the primary sensing elements and pins A_T – F_T, the connecting terminals for the RTD temperature-compensation sensing element. The USI has a number of input terminals for control functions

The control functions concern PD/CS/Reset for power down/chip selection/reset, SF for slow/fast mode selection, TC1 and TC2 for temperature compensation selection, TEST for linearity test, SEL1 – SEL4 for mode selection. Optionally, also the SPI control bus can be used, so that all control functions can be selected using only three wires PD/CS/Reset, Data and CLK.

The signal processing in the USI is as follows: At first, the input sensor signal is selected by the multiplexer (MUX). Next, this sensor signal is converted into a chopped voltage signal by the corresponding front-end circuit, using the advanced chopping technique. Finally, the voltage signal is converted into a period length. The phase counter and the period counter perform the phase control and period multiplication (frequency division), respectively. In the USI modes where a large dynamic signal range is required, dynamic dividers and amplifiers are used. The logic circuit will control the mode of the USI

7. Conclusions

In the smart sensor systems presented in this paper, measurement techniques are implemented using a limited number of low-cost, low-power integrated circuits only. The applied techniques enable selective detection of the measurand with a high immunity for parasitic effects of the sensing elements, the connecting wires. By applying synchronous detection and autocalibration and advanced chopping also high immunity is obtained for interfering signals, $1/f$ noise and parameter drift. The dynamic range of the signals can be extended using dynamic amplifiers, dynamic dividers and special front-end amplifiers. When applied in a universal sensor interface these techniques yield a practical, high-performance device, which can be used for fast prototyping of sensor systems.

Acknowledgements

This work is supported by the Dutch Technology Foundation STW.

References

[1] F.M.L. van der Goes, "Low-cost smart sensor interfacing", PhD thesis, TUDelft, The Netherlands, 1996.
[2] F.M.L. van der Goes and G.C.M. Meijer, "A universal transducer interface for capacitive and resistive sensor elements".*Analog Integr. Circ. and Sign.Processing*, 1997, Vol. 14, pp. 249-260.
[3] A. Bakker, K. Thiele and J.H. Huijsing, "A CMOS nested-chopper instrumentation amplifier with 100 nV offset", *IEEE J. Sol. St. Circ.*, 2000, Vol. 35, pp. 1877-1883.
[4] G.C.M. Meijer and A.W. van Herwaarden, "Thermal Sensors", Bristol and Philadelphia, IOP Publ., 1994.
[5] P.C. de Jong, G.C.M. Meijer and A.H.M. van Roermund, "A 300^0C dynamic-feedback instrumentation amplifier", *IEEE J. Sol. St.Circ*, 1998., Vol. 33, pp. 1999-2009.
[6] P.C de Jong, Dutch Patent Application, No. 1002732, 1996.
[7] G.Wang, Dutch Patent Application, No.1014551, 2000.
[8] G. Wang and G. C. M. Meijer, "Accurate DEM SC amplification of small differential voltage signal with CM level from ground to VDD", *Proc.SPIE'2000*, Newport Beach, USA, 2000.
[9] F.M.L. van der Goes and G.C.M. Meijer, "A simple and accurate dynamic voltage divider for resistive-bridge transducer interfaces", *Proc. Instrum. and Meas. Techn. Conf.*, Hamamatsu, Japan, 1994, pp. 784-787.
[10] G.C.M. Meijer and V.P. Iordanov, "SC front-end with a wide dynamic range", *IEE Electronics Letters*, 2001, Vol. 37, No 23, pp. 1-2.
[11] X. Li, G.C.M. Meijer, R. de Boer and M. van der Lee, "A high-performance universal sensor interface", *Proc. Sicon/01*, Rosemount, USA, 2001, pp. 19-22.
[12] Smartec, "Universal Transducer Interface (UTI), revolution in sensor interfacing", 1996, www.smartec.nl.

IMPLEMENTATION OF NEURAL NET IN CONTINUOUS ANALOG CIRCUITRY

Frank Stüpmann, Gundolf Geske, Silicann Technologies GmbH, Rostock, Joachim-Jungius-Straße 9, 18059 Rostock, Germany

Ansgar Wego, University of Rostock, Institute GS, Dep. Electrical Engineering, Einsteinstr. 2, 18059 Rostock, Germany

ABSTRACT

It will be shown the newest results of a hardware realization of a neural net for fast decision making functions in real time. There is a digital micro core with several functions – proceeding of the learning and testing of the net, supervising of training process and computation of some calculations in pre- and post-processing. The patterns are automatically presented to the network. The heart of the classifier is a trainable integrated analog neural network structure. Because of its speed the hardware realization is able to solve real time image recognition problems The number of neurons integrated in the whole chip is 100 in the input layer, 60 in the hidden layer and 10 in the output layer. The back propagation algorithm is implemented in an analog circuit.

INTRODUCTION

The chip is meant to be used for making decision functions in real time. [1], [7] deal with the examination of existing hardware realization of neural nets. There are some analog neural net chips [2], [3], [4], [5], [6]. In [1] it was stated that previous solutions contain some disadvantages. Thus the number of the integrated neurons is small and often on-chip learning is not possible. The low complexity, not sufficient for many problems, only permits a restricted number of applications. Therefore the aim of the work was deduced to contribute to the development of an fast, complex neural integrated circuit capable of learning.

The classifier consists of the units switch, classification and control. The switch unit carries out the switching between learning vectors and input vectors requested from the unit classification in correspondence to the learning process or the working process respectively.

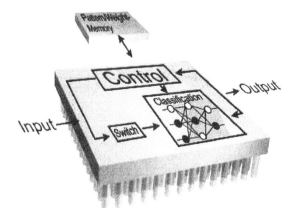

Fig.1: Structure of the whole chip

In Fig.1 the whole neural classifier is shown with its units control, switch and classification. The classifier is realized as a single chip.

It is not necessary to use a second chip for controlling the neural net itself because the fast analog realization of the net and the control function as a digital part are implemented on one chip. The switch unit is realized by an analog switch. The switch unit carries out the switching between learning vectors and input vectors that are requested by the classification unit in correspondence with the learning process or the working process respectively. It is possible to program the chip like a microcontroller but the internal processing speed is determined by a fast analog structure.

The net's topology integrated in the function classification in analog circuitry is the multi-layer perceptron. All operations for the learning and the reproduction phase including the learning in hidden neurons are implemented in analog circuitry. The learning algorithm used is the back propagation algorithm (BPA). The chip uses a SIMD-architecture. The time it takes for the data to propagate from the input to the output in the working process is 2µs. Fortunately, neural systems are more tolerant of low-accuracy components than conventional computation systems. The internal resolution is 6 bit and the resolution at the signal inputs and outputs is 10 bit. The chip has analog input/output buffers. The technology used is the 0,6 µm CMOS-technology CUP from austriamicrosystems.

Multilayer perceptron with Learning Automation

The net's topology integrated in the chip is the multi-layer perceptron. The learning algorithm used is the backpropagation algorithm (BPA), this algorithm is well known with all its advantages and disadvantages. A lot of users will apply this chip and more than fifty percent of current applications of neural networks use the BPA. In the final version it will have 100 input neurons, 60 hidden neurons and 10 output neurons. The activation of the neurons lies in the range of [0,1]. A sigmoid function where the final value is reached asymptotically very fast is used.

$$f(x) = \frac{1}{1 + e^{-\beta x}}$$

with $\beta > 0$.

In the backpropagation algorithm the changing of the weights W of the multi-layer perceptron is realized after propagation of the input pattern $i^{(p)}$ ($p \in L$) by:

$$?_n W(u, v) = ?d_v^{(p)} a_u^{(p)} \text{ with } u \in U_{i-1}, v \in U_i$$

$(2 \le i \le n)$ and $\eta > 0$, whereas

$$d_v^{(p)} = \begin{cases} f'(net_v^{(p)})(t_v^{(p)} - a_v^{(p)}) & \text{if } v \in U_n \\ f'(net_v^{(p)}) \sum_{\tilde{v}} d_{\tilde{v}}^{(p)} \cdot W(v, \tilde{v}) & \text{if } v \in U_j ; 2 \le j \le n-1 \end{cases}$$

is.

There is au(p) the activation of the unit u after propagation of the input patterns i(p) and tv(p), v? Un is the default output prescribed from the output pattern t(p) of an output unit un.

The realized net has no bias values, like the implementation of the function classification.

CONTROL UNIT

The *control* unit controls the chip. This unit is subdivided in control functions which have the following tasks during the separate oparating phases:

control functions	Function	phase, in which the unit has a meaning
pattern-control	presentation of the input- and output-patterns	learning phase
weight-control	supervision and control of initialization, update and refresh	learning, test and operating phase
error-control	supervision of the error in the learning- and test phase	learning and test phase
random-unit	random numbers for initialization and pattern presentation	learning and test phase

TABLE I
Tasks of control function

PATTERN CONTROL

The pattern control function controls the reading in of learn patterns into the pattern memory. Afterwards this function reads the patterns from the pattern memory and presents them to input and output of the function classification. On this occasion every pattern is presented separately for a short time. If all patterns from the pattern memory are presented and if the learning was successful, the test patterns are presented to the classificator. The output of the pattern elements and also of the variable learnrate is realized as 10 bit value.

WEIGHT CONTROL

The internal representation of knowledge has to be adapted to the function which has to be solved. The algorithms necessary for this purpose are realized in the function classification. The system stays in the state of learning until the error in all patterns becomes small enough and the adaptation of the classificator is successful.

Weight control controls the initialization of the weights, the state of learning, the learn rate and the outputs for the learning patterns.

Before the learning weight-control forwards the initialization values of the weights to the synapses of the function classification. Furthermore the central refresh of weights and the saving to the pattern/weight memory are realized by weight-control.

ERROR CONTROL

During the learning process the square error is calculated from the digitized value handed over by the output neurons and the expected outputs. If the calculated error is greater than the maximum permissible net error the instruction of the net has to be continued. If the error is less than the desired minimal error the learning is successful and the net has converged. If the converging fails the net has to be reinitialized and to be reinstructed with a smaller learn rate. To control the learning only the learn rate and the number of learning cycles are available. After the learning the test patterns are presented to the classificator. The observed error is summed to a summation error. This function is also solved by the error control. If greater errors appear, the net has also to be reinstructed. It is switched over to learning mode again.

RANDOM UNIT

The random function is one of the most important functions of the controller, because it is both necessary for initialization and for pattern presentation. At the beginning of learning the weights are initialized with low random values so that the values of weights are approximately zero. The derivation of the logistic activation function has a maximum in that region. It has to be guaranteed, that the values of weight are different from zero. During the learning control takes care of presenting each pattern to the net only once and in every cycle in another order. This characteristic of pattern presentation is a basic condition for the success of learning.

The controller generates the random numbers with method of additive congruence. The numbers are serially selected after that [8]. Beside the simple implementation of this algorithm the advantages are the occurrence of any possible combination in each cycle and the fast process because of the absence of complicated and time consuming multiplication and division instructions.

CONCEPT OF ANALOG STORAGE

The analog storage of the synaptic weights is a difficult task, since not all desirable parameters like precision, long time stability, fast adaption ability can be found in one circuit variant. Earlier analog implementations of neural networks used floating gates (e.g. Intel ETANN). Transistors with floating gates can change their threshold voltage and therefore represent the synaptic weight in the zero stages of multipliers. This is a very high circuit area economizing implementation. Weights remain stored up to ten years. The disadvantages are, that floating gates can not be produced in standard technology, programming takes several milliseconds and this is the reason why a weight update of a whole neural net can take up to several minutes, because programming works only sequentially. Also a high programming voltage is necessary. The reproducibility of programming is limited to $10^4 \cdots 10^6$ programming cycles.

Since high speed on chip learning should be an essential feature of the development of this chip the emphasis is put on capacitive storage in development. Capacities can relatively fast and arbitrarily often be reprogrammed (ns ... µs range). Furthermore they can be easily produced in standard technology.

However leak currents are a large disadvantage which must be overcome by suitable circuit technology. Integrated capacities are already unloaded after some milliseconds by leak currents. Not only the leak current of the capacity but also the reverse current of the source bulk junction of the pass transistor must be considered as well as the drain source leak current. The latter occurs at high drain source voltages and causes the touch of the space charge zones (punchthrough effect). Therefore a refresh must be implemented. This, however, means that the weight signal must become discrete and the precision of the weight is limited.

The punchthrough effect can be avoided by the reduction of the source drain voltage of the pass transistor. Figure 2 shows a circuit [9] for the reduction of the leak currents, called cunit. The stored voltage will loop back through an operational amplifier and a transistor M3 to the input of the pass transistor M2 so that the voltage drop over it is 0V and the leak currents are fundamentally reduced.

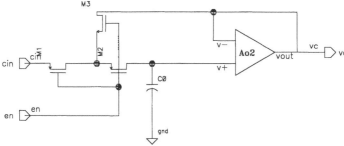

Fig. 2: storage times from a few seconds up to one minute are obtainable with the cunit using a capacity of only 1pF

The feedback transistor M3 switches inversely to the pass transistor M2 so that during loading there is no feedback. Another transistor (M1) is at the input for decoupling the input cin from output vc. An advantage of this circuit is the load independence of the output.

LEARNING WITH CAPACITIVE STORAGE CELLS

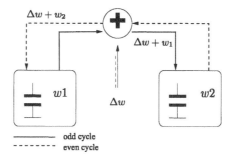

Fig. 3: Principle of the update of the weights in the weight processing unit (WPU)

How does learning with capacitive storage cells work? The BPA works in a way that at the beginning of learning all weights are initialized with coincidental values near zero. The weights are then changed so that the output vectors approach the desired training vectors. The necessary weight modification ?w are calculated by the BPA . This value must be added to the current weight:

$$w_{new} = w_{old} + \Delta w$$

Since w_{old} and $?\,w$ are in the form of voltages, w can not easily be added to in a capacity. Therefore the principle from figure 3 is used.

Two capacities are needed of which only one is active at one moment. Assuming w_1 is initialized, the first learning cycle starts by adding calculated weight modification $?\,w$ to w_1 (= w_{old}) and being saved as w_2 (= w_{new}). In the next learning cycle $w_2 = w_{old}$ and $w_1 = w_{new}$ etc. The storage units cunit are in the weight processing unit (WPU). Every single synapse contains a WPU. The update process is executed for all synapses at the same time. By this massively parallel mode of operation learning becomes very fast since the learning process does not have to be executed by a processor working sequentially.

RESULTS

For the realization of the neural structures a test chip was designed which contains single components as well as complex circuit blocks. WPUs, subtractors, operational amplifiers, gilbert multiplier, cunits (see figure 2) single synapses and neurons as well as a small neural network belong to this. The results of the cunit shall be represented here.

The circuit cunit was created and simulated with Cadence. Already in the simulations has been recognized that at a resolution of 6 bit (6 bit corresponds to a quantization level of 31.25mV) storage times in a range of seconds to minutes can be achieved. Offset of the operational amplifiers as well as geometrical dimensions of the pass transistor M2 are parameters influencing this characteristic quantity. Figure 4 shows the drift of the stored voltage over the period of time of more than one hour starting at different start points. These are already real measurement results and no simulations. For the worst case the drift is approx. 0.5 mV/s. The drift in the measurement result is positive, however, what does not mandatorily always have to be that way.

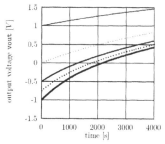

Fig. 4: measuring of the drift of one cunit (capacity 1pF)
over a periodof time of more than one hour

However not only the drift but also the voltage level difference when closing the cunit is decisive for the storage duration. This excursion results from the voltage edge (5V) at the *en*-pin and the quotient of capacities $C_{Gate\ M2}/C_0$. This voltage level excursion is desired and is +8mV on average. By this excursion the stored voltage is raised and centered more or less between two quantization levels to gain time for a refresh also at a negative drift. Figure 5 shows a zoom section for the drift of a voltage started at 1V. The switching point of *en* lies at approx. 300ms. It takes approx. 50s on average for the reaching of the next quantization level (0.96875mV). Unfortunately, a large scattering of the offset of the operational amplifiers occurred by the scattering of the transistor parameters. Normally all curves should start with -1V and then at the LH edge of *en* jump for +8mV. However the margin up to the quantization levels is strongly limited by the statistical distribution of the offset. In spite of this it is positiv that the offset did not affect the drift as strongly as assumed.

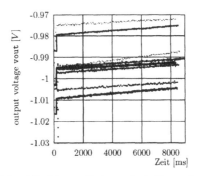

Fig.5: measuring of the drift of ten cunits (capacity 1pF) over the time of 8 seconds

CONCLUSION

The evaluation of the storage behavior of the cunit has shown that the capacitive storage method even with small capacities of only 1pF is suitable as an analog storage for the synaptic weights of a neural network. Due to this storage duration, that is for a resolution of 6 bit in a range of seconds, 10.000 synapses can be easily refreshed by a central refresh unit even if the refresh of a single synaptic weight takes up to 500µs.

The problem of the offset in the operational amplifier of the cunit occurring till now does not only impair the storage duration but also affects harmfully the BPA. The algorithm cannot converge for very small weight modifications since an offset is added to weight modification ? *w* at every learning cycle. This can be avoided by using more narrowly offset tolerated operational amplifiers. On the one hand this is to achieve by larger circuit area and on the other hand by better layout methods such as folding of zero stages.

Since the drift of the storage voltage despite high offsets is very little the prospects of future developments are promising.

[1]   Lindsey, C.S.: Neural networks in hardware: architectures, products and applications, *www.practicle.kth.se/lindsey*, 1998. Lecture at Royal Institute of Techn. Stockholm, Sweden.
[2]   INTEL: 80170NW electrically trainable analog neural network. *INTEL Information Sheet E358*, INTEL Corporation, 2200 Mission College Boulevard, Santa Clara, USA, 1990
[3]   INTEL: 80170NX Neural network technology and applications. *Technical report INTEL Corporation*, 2200 Mission College Boulevard, Santa Clara, USA, 1992
[4]   Ramacher, U. and Rückert, U.: *VLSI design of neural networks*, Kluwer, Boston, USA, 1991
[5]   Masa, P. and Hoen, K. and Wallinga, H.: A high-speed analog neural processor; *IEEE Micro-Journal*; vol. 14, pages 40-50; 1994
[6]   Hammerstrom, D.: A VLSI architecture for high performance, low-cost, on-chip-learning; *IEEE-Journal*; vol. II, pages 537-544; 1990;
[7]   Graf, H.P. and Sackinger, E. and Jackel, L.D.: Recent developments of electronic neural nets in north America; *Journal of VLSI Signal Processing*; vol. 5; pages 19-31; 1993
[8]   Pfenniger, E.: Erzeugen von Zufallszahlen mittels Schieberegister, *www.ing.pfenniger.ch/zufall.html*, 1998.
[9]   M. Kruse: Entwurf einer Synapse für eine selbstlernendes neuronales Netz als analoge VLSI Schaltung, Universität Rostock, Master's Thesis, 1997

A Demonstration System for a New Time-Triggered Sensor Network, Based on IEEE 1451

P Doyle, D Heffernan, D Duma

PEI/CSRC, E&CE Department, University of Limerick

Abstract: The IEEE 1451 family of standards defines a common interface for 'smart transducers' allowing them to be adaptable to virtually any fieldbus or control network on the market. IEEE 1451 has a wide-range of applications, but its event-driven communication is inadequate for use in the real-time deterministic systems that are becoming increasingly prevalent. This work outlines a real-time enhancement of the IEEE 1451 architecture. The time-driven (time-triggered) communication is implemented through the replacement of the defined transducer interface with the Time Triggered Controller Area Network (TTCAN) protocol. TTCAN is a new time-triggered development of the well-known CAN standard. The TTCAN protocol was co-developed by PEI/CSRC at the University of Limerick. This work briefly describes TTCAN and IEEE 1451 and presents a prototype system that introduces a real-time enhanced solution for the existing IEEE 1451 architecture. This prototype system demonstrates that a real-time, deterministic IEEE 1451 solution is both workable and feasible.

1. Introduction

The IEEE 1451 family of standards [1][2][3][4] were developed to provide transducer manufacturers and users with a common interface for 'smart-transducers' to the variety of fieldbus and control networks employed in industry. These standards define how smart transducers are mapped to a target control network based on a network-independent object model, with the aid of descriptive data sheets (TEDS), which are embedded within the transducer. Each network will have at least one Network Capable Application Processor (NCAP), which acts as a gateway between the transducer clusters and the target (backbone) network.

Although IEEE 1451 is suitable for most applications, it is not suitable for real-time and deterministic environments as it assumes an event-driven communication interface between the various transducers and the backbone network. This event-driven implementation will not satisfy the strict timing requirements that are paramount to a real-time system. Throughout the past decade, extensive research work [5][6][7][8][9] on control network scheduling has resulted in the conclusion that time driven (time triggered) scheduling is the most reliable solution for environments where safety-critical, stringently deterministic operation is necessary.

This paper introduces an enhanced architecture for the IEEE 1451 standards, where the message communication between the distributed transducer devices is scheduled in real-time, based on a time-triggered scheduling paradigm. This communication is based on the time-triggered control network, TTCAN [10]. The paper provides a brief introduction to IEEE 1451 and TTCAN, and illustrates some of the benefits of this enhanced architectural model. This paper also describes the design implementation of a prototype demonstration system that is based on this new concept.

2. IEEE 1451

The development of the IEEE 1451 standards was initiated by the NIST [11] (National Institute of Standards and Technology, USA) at the behest of transducer manufacturers for a common interface standard to give a network independent view of transducer devices in a distributed environment. IEEE 1451 proposes a standard interface for "smart transducers" that will have the intelligence to identify themselves and integrate with any given type of industrial fieldbus protocol on the market, enabling 'plug-and-play'[12] integration. The standard offers the ability to plug a new transducer into a network and have it declare its availability along with information such as its serial number, calibration data, measurement range etc.

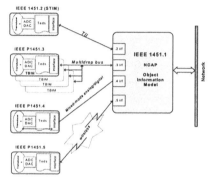

IEEE 1451 is composed of five separate standards. IEEE 1451.1 and IEEE 1451.2 have been voted upon and accepted by the IEEE, while P1451.3, P1451.4 and P1451.5 are still undergoing development. The entire

Figure 1. The IEEE 1451 Family of Standards

IEEE 1451 family of standards, which includes proposals for wireless and mixed-mode implementations, is illustrated in figure 1.

The IEEE 1451.1 standard defines a network-independent information model, enabling transducers to interface to network capable application processors (NCAP). It also provides a definition for a transducer and its components using an object-oriented model. The IEEE 1451.2 standard defines a Smart Transducer Interface Module (STIM). Within the STIM, the standard defines the

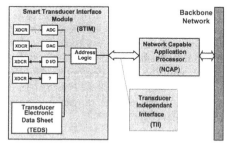

Figure 2. IEEE 1451.1 and IEEE 1451.2

data format for the Transducer Electronic Data Sheets (TEDS) and the communication protocols. The digital interface to the NCAP specifies a 10-wire electrical interface. The TEDS contains information that describes the transducers that are embodied within the STIM. The amount of information contained in the TEDS is application-specific, although critical information will always be present.

IEEE 1451.1 and 1451.2, see figure 2, are the only standards of the family used in the prototype implementation presented in this paper, as together they are sufficient to define a specification for a network of smart transducers.

3. TTCAN

The CAN protocol was developed by Robert Bosch GmbH in the mid-1980s and is now a well-established protocol in the automotive industry as well as in general industrial automation. CAN is essentially an event-driven protocol in which message latency is not constant and increases with higher bus message traffic loads. During the 1990s, much research [5][6][7][8] was undertaken to establish the optimal messaging scheduling scheme for CAN in a real-time control environment. The conclusion was to develop a new session layer for CAN, which would support the time-triggered scheduling model. The new session layer sits on top of the existing CAN two-layer network and transforms the event-driven CAN network into a time-triggered control network. The new time-triggered CAN (TTCAN) [9][10][13] protocol is now a draft ISO standard (ISO- 11898-4) and will be voted upon at the end of this year. Its intended use includes engine management systems, transmission control systems, and various chassis control systems with scope for X-by- wire [14] applications.

TTCAN is driven by the progression of time. TTCANs sense of global time is clocked by a time master's reference message, see figure 3, which is identifiable to every node through its unique identifier. The period between two reference messages is called the basic cycle. The time windows of a basic cycle can be used for periodic or spontaneous messages. All messages are based on the CAN [15] message formats. A time window for periodic messages is called an exclusive time window, while a time window for spontaneous messages is called an arbitrating time window. An arbitrating window operates according to CAN's native non-destructive bitwise arbitration scheme. The TTCAN specification builds a communication matrix, referred to as a matrix cycle, for the application's needs, where several basic cycles are configured in rows. An example TTCAN matrix cycle is shown in figure 3.

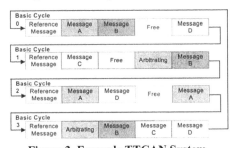

Figure 3. Example TTCAN System Matrix

4. Proposal for an enhanced real-time IEEE 1451 model

IEEE 1451.2 proposes a low-performance, point-to-point digital interface based on the well-known SPI interface [16]. In the real-time enhanced IEEE 1451 model, developed by the authors of this work, we replace the existing STIM/NCAP interface with a true time-triggered network, based on TTCAN, this model is detailed in figure 4. The benefits of an enhanced real-time IEEE 1451 model are:

- Verifiable real-time message scheduling.
- Tried and tested CAN network for the data-link and physical layers.
- Synchronised fault tolerant global clock.
- Multi-master networked interface.

- Interface can be synchronised to external clock sources e.g. GPS.
- Supports 'time-aware hot plug & play'.
- TTCAN interface is an ISO standard.
- Can support multiple NCAPs.

The implementation of IEEE 1451.1 and IEEE 1451.2 is divided into three distinct sections: the Standard Transducer Interface Module (STIM), the Transducer Independent Interface (TII), and the Network Capable Application Processor (NCAP). To avoid any confusion this project adds the hyphenated extension '-RT' to these three IEEE 1451 components to reflect the real-time enhanced architecture. Thus, the sections are referred to as follows: STIM-RT, TII-RT and NCAP-RT.

These components fulfil their IEEE 1451 descriptions, but are further enhanced to operate in real-time. The current event-driven point-to-point TII is replaced by the TII-RT (TTCAN), which will provide a fully multiplexed solution with real-time deterministic behaviour.

The STIM-RT and NCAP-RT, who are connected via the TII-RT, must now host real-time interfaces and be able to operate within a time-triggered scheduling environment. This enhanced solution will allow multiple STIM-RTs to communicate with one or more NCAP-RTs. The backbone network employed in this implementation is Ethernet hence the NCAP-RT now becomes a gateway between the TII-RT (TTCAN) and Ethernet.

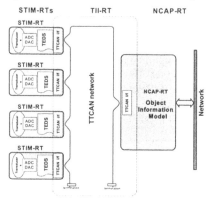

Figure 4. Real-time Enhanced 1451 Model

5. Demonstration system Prototype Design

Outlined below is a description of the prototype design implementation for the real-time enhanced IEEE 1451 solution. The demonstration system is composed of two STIM-RTs connected, via the TII-RT, to one NCAP-RT. The following is a brief description of the system components:

5.1 TII-RT

The event-driven serial TII has been replaced with TTCAN to give a time-triggered interface to interconnect the STIM-RTs with NCAP-RTs. This is implemented at each node using the Bosch TTCAN gate array chip, which is available on a sampling basis.

5.2 STIM-RT

The STIM-RT prototype is based on the 8-bit Infineon microcomputer, the SAB-C515C. The SAB-C515C is fully compatible with the standard 8051/C501 microcontroller family. It incorporates a CPU with 8 data pointers, a 10-bit ADC, a capture/compare unit, a Synchronous Serial Controller interface, a USART serial interface, I/O ports and XRAM data memory. Although the C515 has a full CAN module, a TTCAN controller is interfaced to the STIM-RT to form the TII-RT interface.

5.3 NCAP-RT

The NCAP-RT hardware prototype design is based on the Net Silicon Net+Works development board [17][18], which incorporates a Net+ARM chip with an integrated ARM 7TDMI [19] processor, a 10/100 Ethernet MAC interface, serial ports, IEEE 1284 parallel ports, memory controllers and parallel I/O ports. A TTCAN board, based on the Bosch TTCAN chip, was developed to provide the TII-RT hardware interface. The software interface was developed as a TTCAN device driver. The driver was developed in 'C', using the basic structure of an I/O device driver. The device driver is implemented on the ARM 7TDMI processor, as a formal device driver under the Multi2000© real-time operating system [20]. Through the device driver, users can dynamically manipulate messages that are scheduled in the system matrix.

5.4 System Overview

The prototype demonstration system was developed as illustrated in figure 5. It includes two STIM-RTs that are interfaced to one NCAP-RT, via the TII-RT. The NCAP-RT now functions as a gateway between the TII-RT and Ethernet. Sensors are interfaced to the STIM-RT nodes and TEDSs are written to interpret the physical sensors. The NCAP-RT TEDSs are implemented in software, based on an object-oriented model. A simple message schedule matrix consisting of

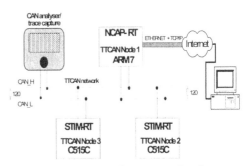

Figure 5. Prototype Demonstration System

two basic cycles was defined. The schedule was carefully timed to act in a request/response fashion to mimic the   operation of   a conventional STIM/NCAP configuration. This worked well but was a contrived effort to use the time-triggered network in an event   driven environment. A proposed modification to the IEEE 1451 standard, by the authors, will de-emphasise the inherent event-driven character of the IEEE 1451.2 standard to allow a fully time-triggered STIM-RT to be implemented.

During initialisation of the system, which is performed via the CAN protocol, a request/response communication allows all the TEDSs data to be transferred from STIM-RTs to NCAP-RT, where received data are stored locally in objects.

When initialisation is complete, the NCAP-RT orders every node on the network to commence TTCAN communication. The TII-RT interface is configured to operate at a data rate of 125Kbps. A CAN network analyser is connected, as a network node, to assess the network's live scheduling behaviour. The analyser has the capability to trace the message history. The results showed that the schedule was consistent and repeatable. Each node was configured as a potential time master, with unique priority. The highest priority time master was assigned to the NCAP-RT, making it the default time master. Testing showed the capability of the network to adapt to the late addition of the different nodes, and observed how a node of higher priority could take over the task of transmitting the reference message.

Figure 6. User Interface

In the system, the Ethernet network was connected to an Internet gateway to provide users with the capability of remote control/monitoring of the system. An embedded web server was developed on the NCAP-RT to provide this service. Simple html forms, as shown in figure 6, enable end-users to remotely interact with the system. The pages can be accessed via the NCAP-RTs IP address.

6. Results

In summary, the prototype demonstration system experimentally demonstrated some important results. Firstly, it proved that it is feasible to implement a time-triggered transducer network, modelled on the IEEE 1451 architecture. It also proved that it is possible for multiple STIM-RTs to form a transducer cluster network, which can be gated to Ethernet/TCP/IP. Testing showed that the off-line static message scheduling worked in a consistent manner and that the potential time-masters worked as expected, where the next highest priority time master replaces a faulty time master. The system also showed that the currently defined TEDS in the IEEE 1451 standard contain sufficient information fields to support basic time-triggered communication

7. Conclusions & Future Work

This paper has presented a prototype system for a real-time enhanced IEEE 1451 solution where deterministic communication between the transducer devices is guaranteed through the replacement of the TII by TTCAN. Thorough testing of the system proved that this enhancement is both workable and feasible and showed that it is possible to apply such time-triggered paradigms to general transducer interfacing environments.

Although TTCAN was the time-triggered network of choice, the concept will also work for other time-triggered control networks such as TTP/C [21] and the emerging FlexRay [14].

The system presented here is somewhat limited in that it only provides deterministic real-time communication from the transducer devices to the TII-RT interface of the NCAP-RT. Future work will concern the widening of the real-time scope to include the backbone network. This would realise a truly real-time gateway between the backbone network and islands of transducer devices. The current IEEE 1451 standards permit transducers to be 'hot swapped' into the system. This work has proposed the development of 'time-aware hot plug and play', where an applicant transducer is assessed in terms of its schedulability before being granted admission to the network.

Acknowledgements

The authors wish to acknowledge the support of PEI Circuits and Systems Research Centre and the Irish state agency, Enterprise Ireland, for their continued support and funding of the research project.

References

[1] IEEE Standard for a Smart Transducer Interface for Sensors and Actuators-Transducer to Microprocessor Communication Protocols and Transducer Electronic Data Sheet (TEDS) Formats, IEEE Std 1451.2-1997, September 26, 1997

[2] IEEE Standard for a Smart Transducer Interface for Sensors and Actuators-Network Capable Application Processor Information Model, IEEE Std P1451.1-1999, June 26, 1999

[3] IEEE Draft Standard for a Smart Transducer Interface for Sensors and Actuators-Digital Communication and Transducer Electronic Data Sheet (TEDS) Formats for Distributed Multidrop Systems, IEEE Draft Standard P1451.3

[4] IEEE Draft Standard for a Smart Transducer Interface for Sensors and Actuators-Mixed Mode Communication Protocols and Transducer Electronic Data Sheet (TEDS) Formats, IEEE Draft Standard P1451.4

[5] Tindell, K., Hansson H., Wellings A.J. (1994) "Analysing real-time communications: Controller Area Network (CAN)", Proceedings of IEEE Real Time Systems Symposium.(ISBN: 0-8186-6600-5), pp. 256-263.

[6] Cena, G. and Valenzano, H. (1997) "An Improved CAN Fieldbus for industrial applications", IEEE Transactions on Industrial Electronics", Vol 44 No 4, pp 553-564.

[7] Navet N. (1998) "Controller Area Networks: automotive applications", IEEE Potentials. Vol. 17, issue 4, pp 12-14.

[8] Zuberi, K.M. and Shin, K.G.(1997) "Scheduling messages on Controller Area Networks for real-time CIM applications". IEEE Transactions on Control System Technology". Vol. 3 No. 2 pp. 310-314.

[9] Führer, T., Muller, B., Dieterle, W., Hartwich, F., Hugel, R. and Walther, M. "Time Triggred Communications on CAN (Time Triggered CAN – TTCAN)". Seventh International CAN Conference (ICC), Amsterdam, Netherlands, Oct 24-25, 2000.

[10] Leen, G. and Heffernan, D "Time Triggered Controller Area Network" , IEE Computing and Control Engineering Journal. Vol 12. Issue 6. Pp 245-256. Dec 2001.

[11] National Institute of Standards and Technology (NIST) IEEE 1451 homepage available at http://ieee1451.nist.gov/

[12] Johnson, R.N. "Building Plug-and-Play Networked Smart Transducers", Sensors Magazine, October 1997.

[13] Leen, G. and Heffernan, D. "TTCAN: A New Time-Triggered Controller Area Network ". Microprocessors and Microsystems Journal, Elsevier. Vol. 26, Issue 2. pp. 77-94. March 2002.

[14] Shaheen, S. Heffernan, D. and Leen, G. "A comparison of emerging time-triggered protocols for automotive X-by-wire protocols". ImechE Journal of automobile Engineering. U.K. Vol. 217. Part D. Jan. 2003.

[15] ISO Standard: ISO 11898:1993 (E) "Road Vehicles, Interchange of digital information – controller area network (CAN) for high speed communications". Nov. 1993

[16] D. Wobschall and Hari Sai Prasad. Esbus - a sensor bus based on the spi serial interface Sensors, 2002. Proceedings of IEEE , Volume: 2 , Page(s): 1516 –1519, 2002

[17] NetSilicon Inc. available at http://www.netsilicon.com

[18] NET+Works for NET+ARM Hardware Reference Guide; NetSilicon Inc.; 2000

[19] Furber, S. "ARM system-on-chip architecture". Addison-Wesley Publishers, 2000 ISBN 0-201-67519-6

[20] Green Hills available at http://www.ghs.com

[21]TTP/C Specification version 0.5 edition 0.1 21 Jul 1999. TTTech Computertechnik AG: http://www.tttech.com/.

A Prototype Steer-by-Wire and Brake-by-Wire System for Educational Research Projects

Colin Ryan[1], Gabriel Leen[1], and Donal Heffernan[2]

[1] PEI/CSRC, University of Limerick, Limerick, Ireland
[2] Department of Electronic and Computer Engineering, University of Limerick, Limerick, Ireland

Abstract – X-by-wire is an advanced engineering concept which involves removing mechanical components and replacing them with electronic sensors, actuators and networks. This paper describes the implementation of a research educational prototype Steer-by-wire and Brake-by-wire system using a time-triggered network, namely TTCAN.

1. Introduction

Automotive manufacturers are currently attempting to replace many mechanical components in vehicles with ultra dependable fault tolerant electronic systems, referred to as X-by-wire systems. Mechanical components such as drive belts, water pumps, hydraulic brakes and steering columns can be replaced with electronic systems. This results in a lighter, safer, more fuel efficient and less expensive X-by-wire vehicle. In such vehicles there are fewer environmentally un-friendly fluids to contend with, systems are self-diagnosing, re-configurable, and easily adapted across vehicle platforms. X-by-wire systems allow the tightest possible integration of distributed functionality within the vehicle, in contrast to the discrete, often disjoint, operation of conventional mechanical systems. X-by-wire allows dynamic and optimum "solidarity" among system elements unlike their mechanical counterparts, which are set-up in production for a single fixed operating point for the lifetime of the vehicle. These are some of the key concepts, which are motivating the introduction of distributed control systems to eliminate several existing mechanical components. As a consequence of X-by-wire the affordability equation of advanced systems will be re-written. Features such as Electronic Brake Management (EBM) which combine the functionality of Anti-Lock Brakes (ABS), Cornering Brake Control (CBC), Dynamic Brake Control (DBC), Automatic Stability Control (ACS), and Engine Drag Force Control (MSR) will become available across several classes of vehicle and not only the luxury class as they are today. The introduction of an X-by-wire vehicle infrastructure facilitates the implementation of many active safety improvements based on advanced electronic systems. Examples include Autonomous Cruise Control, Road Recognition and Autonomous Driving. It also facilitates the integration of systems such as Active Steering, Dynamic Stability control, Electronic Brake Management and Integrated Chassis Management.

There are a number of important safety requirements of X-by-wire systems and the control networks used. The ESPRIT Open Microprocessor system Initiative (OMI) project; Time Triggered Architecture (TTA, project 23396, Dec '96 – Nov. '98) and the BRITE-EURAM "X-By-Wire" project (project BE-1329, Jan '96-Dec'98) have helped to establish the time-triggered paradigm as the most appropriate scheduling strategy for high-reliability and safety-critical embedded system communications. As a result of this work several new protocols have emerged which incorporate many of the concepts developed. These include TTP/C, ByteFlight, FlexRay, and TTCAN. For a comparison of emerging TT networks refer to [1].

The following section gives an overview of TTCAN, which is the chosen control network in this work. This is followed by a description of an experimental demonstration system for a Steer-by-Wire and Brake-by-Wire system, which has been implemented at the University of Limerick.

2. TTCAN

2.1 Introduction to CAN and TTCAN

Controller area network (CAN) was developed by Robert Bosch GmbH in the mid 1980's. In 1991 CAN was deployed for the first time in a motor vehicle; the Mercedes Benz S-Class. In 1993 CAN was standardised by the International Standards Organisation (ISO) [2]. Since then CAN has become the world's most successful in-vehicle network; with annual device sales exceeding 150 million units in 2001, and forecast sales of 315 million units in 2003 [3,4]. CAN is essentially an event-triggered protocol, and thus has a somewhat un-deterministic real-time nature. For applications such as Steer-by-wire and Brake-by-wire, CAN may exhibit excessive message latency characteristics. For a detailed description of the CAN protocol refer to [5].

Time-Triggered CAN (TTCAN) appends a set of new features to the CAN protocol through the introduction of a new session layer. It introduces the idea of a fixed, predefined schedule, known as the TTCAN system matrix to control the exchange of messages. The static schedule sequence is based on a Time Division Access (TDA) scheme whereby message exchanges may only occur during specific time slots or time windows. In order to run the schedule, all nodes on the network have a clock, which generates a local view of time. Synchronisation of time throughout the network, to achieve a sense of global time, is achieved through the use of a specific message known as the reference message.

2.2 Matrix Cycle

The system matrix correlates the message transactions and the time windows in which they occur. The matrix is composed of basic cycles (rows of the matrix) and transmission columns (columns of the matrix). Each basic cycle commences with the occurrence of a reference message, and has a duration equal to the interval between references messages, also known as the Cycle_Time. The complete matrix, consisting of all the basic cycles, is known as the matrix cycle and is illustrated in Figure 1. The matrix cycle describes the message transaction schedule, which is continuously repeated.

2.3 Time Windows

A matrix cycle is composed of time windows. The time windows of a transmission column are all of the same length (equal in the time domain) but may be of different

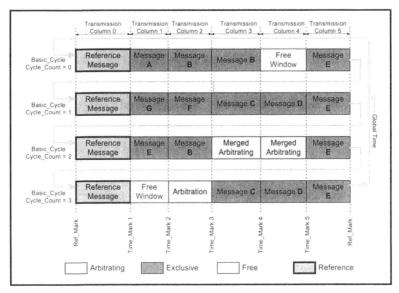

Figure 1 TTCAN Matrix Cycle

type and contain different messages (i.e. differ in the data domain). There are three basic types of time windows:

> ➢ Free time windows
> ➢ Arbitrating time windows
> ➢ Exclusive time windows

Free time windows are intervals in time where the bus is scheduled to be idle. No message transactions are scheduled to occur during these intervals, which are included to allow for future system expansion.

Within an exclusive or arbitrating time window, the transmission of a message may only commence during an interval known as the Tx_Enable window. If the bus is not idle during this initial phase of the time window, then the message will fail to be transmitted. This requirement is necessary to ensure that messages will be fully transmitted in their respective time windows, and will not infringe on the next time window in the matrix cycle.

In arbitration time windows, a number of different messages may be assigned 'permission' to transmit. Message collisions are resolved using CAN's native bitwise arbitration process. Several nodes may attempt to transmit during the Tx_Enable phase of an arbitrating time window. The automatic retransmit function of CAN is disabled and hence messages, which loose arbitration, may not re-try. The only exception to this rule exists in the case of merged arbitrating windows. Arbitrating time windows, which are consecutive in a basic cycle, may be merged to form a single large merged arbitrating time window. In this case the Tx_Enable trigger extends to the last arbitrating time window's Tx_Enable window. Retransmission of a message is allowed provided it commences within the Tx_Enable window and will complete before the end of the final arbitrating time window.

Exclusive time windows are windows in which only one message from a specific network node may be transmitted. During this window there is no competition for the medium, as the window has been assigned to a single node that may exclusively use the medium to transmit its message during this time.

2.4 Reference Message

The TTCAN reference message is used to synchronise communications across the TTCAN network. The reference message is sent by a node known as the time master and is recognised by all nodes on the network by virtue of its CAN frame identifier. It creates a reference point in time to which all other communication transactions are related. This point in time is defined by the start of frame (SOF) bit of the reference message. This bit causes the start of the Cycle_Time in all nodes on the network. The reference message contains the Cycle_Count value. This is used to index the current basic cycle of the matrix, and ensure that every node on the network is at the same point in the schedule.

All time-triggered communications on a TTCAN network are dependant on the transmission of a reference message by the current time master of the network. In the event of the time master failing another time master performs its synchronisation function by transmitting a reference message. These backup time masters are set-up as potential time masters on the network. Up to 8 nodes in a TTCAN network can act as a potential time master.

2.5 TTCAN Synchronisation and Global Time

The time masters view of time is referred to as the network's global time and all nodes try to adhere to this sense of time. There are two levels of TTCAN synchronisation, level 1 and level 2. In level 1 synchronisation all nodes on the network synchronise to the start of a basic cycle through the reception of a reference message. However the local clock rate of all nodes on the network is not tightly synchronised. In level 2 TTCAN synchronisation, the reference message contains the time masters view of time at the SOF bit of the reference message itself, known as the Master_Ref_Mark. In this way there is a distributed view of global time across the network. This is used to implement clock drift compensation in other nodes on the network running level 2 TTCAN synchronisation. For a more detained description of TTCAN synchronisation refer to [6].

2.6 Time units in the TTCAN network

All Time Marks or instances in time with respect to network transactions are counted in network time units (NTU). The NTU may be equal in duration to a CAN bit time in a level 1 TTCAN implementation or may be related to a fraction of the physical second in a level 2 TTCAN implementation.

2.7 Sending and receiving messages

Message transmissions by a node are initiated by a Tx_Trigger entity which is a register set holding the necessary information relating to when a specified message is to be transmitted. A Tx_Trigger has four component data entities:
1. A pointer to a specific single message
2. The transmission column in which the message is to be transmitted
3. The basic cycle in which the message is to be first transmitted (Cycle_Count)
4. A repeat factor, which determines at which position in the specified transmission column the message is transmitted again, provided it is repeated again, otherwise the repeat factor is not active.

The reception of messages is monitored through the use of Rx_Triggers. These are register sets similar to Tx_Triggers, which store information relating to the expected arrival time of a specific message.

2.8 Message Status Counters

The message status counter (MSC) registers provide an additional means of error detection during time triggered operation. Each MSC register is assigned to a specific message, which is either transmitted or received. Communications failure for this specific message results in the incrementing of the MSC, while correct communication will decrement the MSC. When the difference between any two MSCs is greater than 2 or when an MSC reaches 7, an error is flagged.

2.9 Time triggered and event synchronised operation

In a strictly time triggered TTCAN communications network the reference message is transmitted periodically in equidistant time windows. Once the matrix cycle has completed it simply starts again and repeats. However, TTCAN allows basic cycles to synchronise to the occurrence of an event in a time master. In this mode a time interval with no network activity can be introduced. Network activity is resumed when a time master transmits a reference message, thus announcing the start of the next basic cycle

2.10 TTCAN Realisation

A number of semiconductor manufacturer are developing TTCAN solutions. For example NEC [7] and Bosch [8] have hardware available that support the TTCAN protocol. Research work has been carried out at the University of Limerick to formally verify the correctness and behaviour of the TTCAN protocol. The TTCAN protocol is currently a draft international standard (DIS) as of 13-02-2003, and voting on the formal standard is expected at the end of this year.

3. Implementation

3.1 Network Implementation

TTCAN is used in a Steer and Brake-by-wire prototype system, which has been developed by the authors. At the time when this project was started there was no TTCAN protocol engine available in silicon. Because of this CAN driver chips were used with the Infineon C515C microcontroller and an application layer based on the TTCAN protocol with level 1 synchronisation was implemented in software. Figure 2 shows the TTCAN message matrix used, which consists of two basic cycles. Each basic cycle contains a reference message, a steering wheel position message, a brake pedal angle message, and a feedback message. The reference message is used to synchronise the network by resetting the cycle time in each node on the network. It also contains the current basic cycle count, used to ensure that all nodes are in the correct location of the schedule. The TTCAN local clock was implemented using onboard timers, and the

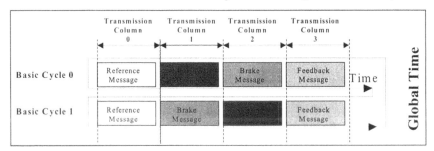

Figure 2 Message Matrix

TTCAN triggers were implemented using real-time interrupts generated by the timer overflow.

Each node monitors the transmission and reception of relevant messages. If an expected message is not transmitted or received, a corresponding message status counter (MSC) is incremented. If any MSC reaches a predefined limit then an error is flagged and appropriate action is taken. In this case the node is reconfigured and attempts to rejoin the network.

3.2 Steering Implementation

To demonstrate the steering system, an automobile steering wheel was obtained; the steering column was removed and replaced with a position sensor as seen in Figure 3a. A rotational absolute encoder is used to measure the position and thus the angle of the steering wheel. The sensor measures 128 positions per revolution, therefore has a resolution of 2.81 degrees. The sensor is connected to a microcontroller, and the position of the sensor is sent over the TTCAN network in the steering wheel position message. More accurate sensors with greater resolution were considered along with the possibility of gearing this sensor for greater resolution, however for this concept demonstration model such accuracy was not considered necessary. A real steer-by-wire system would probably require a sensor with a resolution in the order of 0.5 degrees, or better.

At the rack and pinion a simple 12V DC motor is used to change the wheel steering angle. A faster and more precise motor could be used, but again for this model it is not required. A rotational position sensor connected to the rack and pinion drive is used to measure the angle of the wheels. There are also magnetic limit switches to detect the left and right steering lock positions. Figure 3b shows the steering system used. The electronic control unit (ECU) receives the steering wheel angle from the network and moves the wheels to the required position. In a production implementation all sensors and actuators would be replicated forming redundant back-up systems.

3.3 Brake Implementation

To demonstrate the braking system, an automobile brake pedal is used. The hydraulic and mechanical links were removed and replaced with a linear potentiometer to measure the angle of the brake pedal, as seen in Figure 4a. The voltage drop across the potentiometer is converted to an 8-bit digital number and sent over the network in the brake position TTCAN message. There is a linear voltage drop difference of approximately 3V over the travel of the pedal, which results in a sensitive detection of the brake pedal position.

(a)   (b)

Figure 3 (a) Steering wheel with position sensor (b) Rack and pinion steering system

 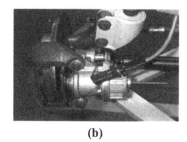

(a)   (b)

Figure 4 (a) Brake pedal with position sensor (b) Brake unit with motor

A sliding calliper brake unit with the hydraulics removed is used to demonstrate the braking of a wheel. A linear stepper motor is used to adjust the position of the brake pad, thus apply the braking force. A faster, more powerful motor could have been used, however for the purposes of demonstration this was not necessary. Figure 4b shows the brake system used. The ECU receives the brake pedal angle over the network and adjusts the position of the brake pad accordingly.

3.4 Feedback Message
No physical force feedback was implemented in this prototype. However a feedback message is sent from the rack and pinion to the steering wheel, containing information relating to the steering angle. This message was included to demonstrate that a feedback message could be easily incorporated into TTCAN's network traffic.

4. Results

The final prototype can be seen in Figure 5. The Steer-by-wire and Brake-by-wire systems were implemented on a common network. The network is run at a 250Kbaud with an NTU of 1.2µs. The time windows are 2ms long. The reference and feedback messages take 0.236ms each to be transmitted, and the steering and brake messages take 0.276ms each to be transmitted. This results in a total network bandwidth usage, excluding error frames, of 12.8%. It should be noted that in this configuration the network is only running at one quarter of its maximum speed of 1Mbaud.

5. Conclusions

This model demonstrates the 'proof of concept' of using Time Triggered networks to implement X-by-wire systems using low cost components for educational, demonstration and research purposes. For the implementation of such systems in real automobiles, a fault tolerant network is required, along with fault tolerant sensors and actuators. There is ongoing work at the University of Limerick in researching such fault tolerant networks, and specifically a fault tolerant version of TTCAN. Presently all major automotive manufacturers are either developing prototype or production X-by-wire systems. The widespread deployment of such systems is expected within the next ten to fifteen years.

Figure 5 X-by-Wire systems

The authors wish to acknowledge the support of the PEI/CSRC centre at the University of Limerick and in particular wish to thank Prof. Phil Burton for his on-going support in the area of automotive electronics research. We would also like to thank Damien O'Sullivan and Joseph Leen for their work on the X-by-wire prototype system.

6. References

[1] Shaheen, S., Heffernan, D. and Leen, G. "A Comparison of Emerging Time Triggered Protocols for Automotive X-by-wire Control Networks. ImechE, Journal of Automobile Engineering. Vol. 217, No. D1. Jan. 2003. P. 13-22.

[2] ISO 11898: 'Road Vehicles-interchange of digital information-controller area network (CAN) for high-speed communications', First edition: 1993-11-15, Reference number ISO 11898:1993(E)

[3] G. Leen, D.Heffernan, TTCAN: a new time-triggered controller area network, Microprocessors and Microsystems 26 (2002) 77-94.

[4] CAN in Automation (CiA), avalaible at: http://www.can-cia.com/

[5] FARSI, M., RATCLIFF, K., and BARBOSA, M.: 'An overview of controller area network', computing & Control Engineering Journal, 1999, 10,(3), pp113-120

[6] G.Leen, D. Heffernan, Time-triggered controller area network, IEE Computer and Control Engineering 12 (6) (2001).

[7] Warwick Control Technologies, Technical introduction to TTCAN, available at: http://www.warwickcontrol.com/TTCANfaq.htm

[8] Bosch, Time Triggered communication on CAN, avalaible at: http://www.can.bosch.com/content/TT_CAN.html

Section 9

Sensor Applications II

Mobility Monitoring using Mobile Telephony

C Ni Scanaill, G M Lyons

Biomedical Electronics Laboratory, ECE Department, University of Limerick, Ireland

Abstract: A portable system has been developed based on mobile telephony to remotely monitor the mobility of subjects as they perform their activities of daily living in their home environment. The system will continually monitor the person's mobility using accelerometers and a portable, battery-powered data acquisition unit. The mobility data is transmitted hourly, using the GSM network, to a specially developed central database. Each subject's mobility trends are monitored using custom-designed mobility alert software and the appropriate medical personnel are alerted, by SMS, if a reduction in the subject's mobility levels is detected. It is expected that this system would reduce the financial burden on health boards, by allowing elderly subjects to remain in their home environment, which is their preferred option.

1. Introduction

Mobility is a subjective quality, which has been shown to indicate a subject's stamina, strength, or psychological well-being [1]. An objective mobility assessment tool could be used to monitor health, to assess the relevance of certain medical treatments and to determine the quality of life of a patient. The need for expensive residential care or prolonged stays in hospital (estimated at €820 per patient per day [2]) could be decreased if such monitoring techniques were employed by the health services.

Existing methods for mobility measurement include observation, clinical tests, physiological measurements, diaries and questionnaires. Diaries and questionnaires require a high level of user compliance and are retrospective and subjective. Observational and clinometric measurements are usually carried out in artificial clinical environments, over short periods of time, and rely heavily on the administrator's subjectivity. Physiological techniques, though objective, have a high cost per measurement. Accelerometry provides a low cost, long term, objective monitoring technique.

In 1995, Bussmann [3] and Veltink [4] showed that two uni-axial accelerometers, one mounted on the trunk and another mounted on the thigh is sufficient to discriminate between sitting, standing and lying and movement. Several short-term (<4 hours) controlled tests, using Bussmann and Veltink's 2-sensor configuration have successfully monitored mobility, in a clinical environment.

Pilot work by the authors, using the Analog Devices family of accelerometers and the placement configuration proposed by Veltink, has been validated [5]. Medium-term (6 hours per day for 4 days) clinical trials [6], measuring the mobility patterns of 5 elderly subjects, have been completed by the authors and have showed an average detection accuracy of 95%

The system described in this paper combines accelerometry and mobile telephony to monitor the mobility of several subjects and to remotely track any changes in their recorded mobility using a central server.

2. System Overview

The system (Figure 1(a)) described in this paper consists of a portable unit, worn by each subject monitored, and a central server. The portable unit measures, analyses, and transmits the mobility data of each monitored subject to the server at hourly intervals. The mobility data is received, analysed, and stored on the server's database. This database is queried daily, and the appropriate medical personnel are informed if an alarming trend is observed in a patient's mobility status. Communication between the portable unit and the server is via the GSM network using SMS messaging.

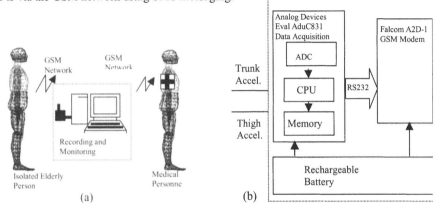

Figure 1: (a) System Overview. (b) Block Diagram of Portable Unit.

2.1 The Portable Unit

The portable unit (Figure 1(b)) houses the data acquisition unit (DAU), a GSM modem, and a battery-based power supply. Two integrated accelerometers are input to the portable unit through the analogue inputs of the DAU. The DAU carries out the following functions:

- Data acquisition of sensors
- Preliminary sensor signal processing
- Transmission of mobility summary data to the GSM modem.

2.1.1 Data Acquisition of sensor data

Accelerometers detect tilt by measuring static acceleration (gravity) and detect movement by measuring dynamic acceleration. Veltink [4] demonstrated that two uni-axial accelerometers, one mounted radially on the trunk and the other mounted radially on the thigh (Figure 2) were sufficient to distinguish between sitting, standing, lying and dynamic movement. For elderly subjects dynamic data, which indicates walking and running, can be assumed to correspond to walking. This simple configuration of accelerometers was adopted to measure the mobility status of elderly subjects and has been confirmed by the authors to deliver excellent detection accuracy [6].

The accelerometer used in this study was the Analog Devices ADXL202 biaxial integrated accelerometer. This device is a small (5mm X 5mm X 2mm), low-cost ($10.20), low power (<0.6 mA), dual-axis accelerometer with a range of $\pm 10g$, and has a typical sensitivity of 0.1V per g. The analogue output from the y-axis of the accelerometer, namely $Y_{Filt,}$ was used for this application, with the output bandlimited to 100Hz using an external capacitor.

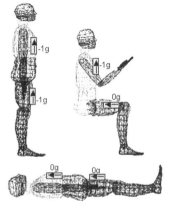

Figure 2: Posture Detection using Accelerometers.

2.1.2 Preliminary Sensor Signal Processing

Data acquisition of sensors was implemented using an ADuC831 microcontroller board, with the Y_{filt} signal of each accelerometer connected to one of the analogue input pins of the ADuC831. The accelerometers were sampled at 500Hz (200 Hz each) and converted to digital form using the microcontroller's on board ADC.

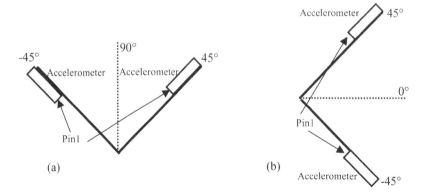

Figure 3: Posture Thresholds for vertical position (b) Posture Thresholds for horizontal position.

2.2.3 Transmission of mobility summary data to the server

The thresholds for determining upright or horizontal position of a sensor were set to ±45° to the vertical or horizontal (Figure 3) line respectively. The subject's posture was calculated every second by comparing the mean signal of each accelerometer to the predefined thresholds for sitting, standing, lying and walking and the corresponding posture counter was incremented. Every hour, the posture counters were read to calculate the percentage of time spent in each posture during the previous hour.

The mobility data, converted into percentages spent sitting, standing, lying and walking, was formatted according to a protocol shown in Table 1. A message sent according to this protocol looks like:

PATIENTSMS¦MOBIL¦001¦020¦030¦010¦040

where this is the first message sent by the microcontroller, and the values for lying, sitting, standing and walking are 20%, 30%, 10%, and 40% respectively. This protocol encrypts each patient's data to maintain confidentiality; recognises which percentage was associated with which posture; and distinguishes between a valid message and a message accidentally sent to the server.

Character	Function	Actual value
0-9	Message validity	PATIENTSMS
10-14	Message type	MOBIL
15-17	Message ID	000…999
18-20	Value Lying	000…100
21-23	Value Sitting	000…100
24-26	Value Standing	000…100
27-29	Value Walking	000…100

Table 1: The Mobility Protocol.

The formatted data and the server's phone number were inserted into the Hayes AT modem command for sending SMS messages, and sent to the GSM modem, via the microcontroller's UART. The GSM modem sent the data, as an SMS message to the supplied server phone number.

2.2 The Remote Server

The remote server is simply a PC interfaced to a GSM modem. The PC controls the modem using Hayes AT commands sent via the serial port. A C++ program was created to continually poll the modem's SIM card for new messages. The telephone number and the formatted string containing mobility data are extracted from each new message; the message is then deleted from the SIM card.

The phone number is compared to the database list to find the patient; if the number is not found in the database the message is from an invalid source and is discarded. The message string from valid source is then dissected and tested for "PATIENTSMS" to ensure message validity. The message type of a valid message shows the server that the values following it, in 3-character groups, are mobility values. The mobility values for that patient are extracted and entered into the "PatientMobilityToday" table in the database.

A relational database (Microsoft Access 2000) was chosen for this system, as it is flexible and easily extendible. Three related tables were created containing the patient's personal data, the patient's average mobility, and the patient's mobility for that day. Forms provided the user interface to the database and were used to enter, change and view data records.

Graphs (Figure 4) were used to illustrate the patient's mobility in the short, medium and long-term. The forms and graphs were created using the wizard provided, and SQL statements were written to drive them.

An appropriate mobility threshold was set for each patient in the database. The database is queried each evening for alarming mobility patterns. If a patient is found to have exceeded their threshold, the server software will format a string containing that subject's name and address and will forward it, over the GSM network as an SMS message, to the medical person listed in the database as their carer.

A web interface to the database is also available, which is a big advantage to the carer who does not have to return to the administrative centre to view the mobility data but can access the data remotely at any PC or PDA with Internet access. The database is password protected, to ensure data confidentiality and integrity on the Internet.

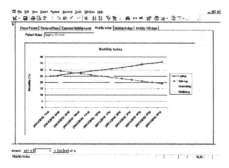

Figure 4: Access Graph for "Mobility Today".

3. Results and Analysis

Prior to fitting the equipment on the target population, namely the elderly person living alone, the system had to be thoroughly tested to confirm its performance to specification.

3.1 Bench Testing

Individual components of the developed system were bench tested in the laboratory. The microcontroller software was shown to correctly identify static postures from two accelerometer inputs and posture information was automatically inserted, according to the protocol, into an SMS message and transmitted correctly to the server.

Simulated problems to test the server's ability to cope with invalid messages, network failure, and power failure were all successfully dealt with by the software during bench testing. The mobility data from valid mobility messages were all dissected and inserted correctly into the database. The server software also detected a mobility pattern, inserted into the database, and an SMS message containing that pseudo patient's data was sent to a designated mobile phone.

3.2 Campus Test

The system was evaluated while monitoring the mobility status of a healthy subject, for 1 hour, in a controlled environment in UL. The server was located in the Biomedical Electronics Laboratory in UL and an investigator recorded the subject's postures to confirm the accuracy of mobility recording. The subject completed a sequence of sitting, standing, lying and walking. SMS messages were sent every 5 minutes to the server, to accelerate the monitoring process (normal interval 1 hour) and stored in the database. During the test the observers and the test subject noticed that the system monitored what was actually happening. When the trial was completed the investigator's recorded data was compared to the database data. Sitting, standing and lying postures were detected with 97%, 95% and 93% accuracy.

4. Conclusion

A GSM-based remote monitoring system has been developed. The system has been shown to perform well during a short trial on one healthy subject. The server system detected artificial alarm conditions and sent alert messages to a designated mobile phone.

The next stage in the project is to carry out more comprehensive mobility testing on the system initially monitoring several healthy subjects within the UL campus and then on the target population of elderly subjects living remotely in their homes.

References

[1]   Office of the U.S Surgeon General. "Physical Activity and Health: A Report of the Surgeon General. Washington D.C., U.S. Department of Health and Human Services, Public Health Service," 1996.

[2]   United States Council on Competitiveness. "Highway to Health; Transforming US Health Care in the Information Age," 1996.
        http://nii.nist.gov/pubs/coc_hghwy_to_hlth/title_page.html

[3]   Bussmann, J.B.J., et al., *Ambulatory monitoring of mobility related activities: the initial phase of the development of an activity monitor.* Eur J Phys Med Rehabil, 1995. **5**: p. 2-7.

[4]   Veltink, P.H., et al., *Detection of Static and Dynamic Activities Using Uniaxial Accelerometers.* IEEE Transactions on Rehabilitation Engineering, 1996. **4**(4): p. 375-385.

[5]   Dunne, D.M., G.M. Lyons, and P. Grace, *The feasibility of posture and physical movement detection using accelerometers.* Irish Journal of Medical Science, 2000. **169**(22).

[6]   Cullnane K., L.G.M., Lyons D., *Long-term Mobility Monitoring of the Elderly using Accelerometers in a Clinical Environment.* Clinical Rehabilitation, (submitted) 2003.

Section 9: Sensor Applications II
Paper presented at Sensors and their Applications XII, September 2003
©*2003 IOP Publishing Ltd*

A combined sensor incorporating real-time DSP for the imaging of concrete reinforcement and corrosion visualisation

G Miller, P Gaydecki, S Quek, B Fernandes and M A M Zaid

Department of Instrumentation and Analytical Science, UMIST, PO Box 88
Manchester M60 1QD, United Kingdom
Email: patrick.gaydecki@umist.ac.uk

Abstract: A new sensor, based on the principles of Q-detection and heterodyne technology is described. The sensing coil is part of a self-resonating oscillating system. When a metal target brought into the vicinity of the sensor, it changes the coil's impedance. This change is expressed as a voltage and used to generate an image. The heterodyne section enables the sensor to image surface corrosion. The sensor can detect bars down to a depth of 150mm. Clear images can also be generated to a depth of between 30 to 60mm.

1. Introduction

Concrete is used as a building material because it is easily formed, relatively lightweight and cheap. It is weak under a tensile load and strong under a compressive load. Therefore steel reinforcement bars (termed rebars) are included within the matrix during the casting process in order to increase the structural strength, which is termed *pre-stressing*. It is vital for the health of the concrete structure that these bars remain in good condition. The life expectancy of these structures has been reduced due to the presence of corrosion product and many methods have been developed over the years for the inspection, testing and monitoring. As a result the authors have reported on a number inductive sensors used in conjunction with motorised x-y scanning systems to image a steel bar mesh [1-3]. Here a new combined sensor has been developed based upon the principle of Q-detection and heterodyne technology incorporating a real time DSP system. The DSP system is used to dramatically improve the signal to noise ratio and will in-turn increase the detection range of the sensor. The sensor operates on the principle whereby a time varying magnetic field is induced around the sensing coil carrying an alternating current. When a metal or non-conducting ferrous target is placed within the vicinity eddy currents will be induced as a result of the conductivity or the magnetic flux density will increase around it depending upon the permeability. This will cause eddy currents to be induced in the sensing coil that will oppose the original flow representing a change in impedance of the coil. This change is then represented as a voltage level by the sensor and processed by an analogue-to-digital converter within a DSP system. A filtering algorithm is then applied after the data has been digitised to enable grey-scale image to be generated of the underlying steel together with the visualisation of surface corrosion.

2.   Theory

2.1   Fundamental relationships

The sensor is based upon the principle of Ampere's Law, Faradays Law and Lenz's Law. Ampere's law states that a time varying magnetic field is formed around a target carrying an alternating current, ie:

$$\oint B.dl = \mu_0 I \tag{1}$$

where μ_0 is the relative permeability of free space. This equation is related to the Biot-Savart law, which allows the magnetic flux density to be determined at a particular point p along the centre of the axis of the coil at a distance z thus:

$$B = \oint dB_z = \frac{\mu_0}{4\pi} \oint \frac{Idl \cos\theta}{a^2} \tag{2}$$

where d is the distance from the circumference of the coil to point p and a is the radius. Solving equation 2 we obtain [4]

$$B = \frac{\mu_0 I a^2}{2(a^2 + z^2)^{3/2}} \approx \frac{kINa^2}{z^3} \tag{3}$$

where k is a constant and N is the number of turns. Two observations can be made concerning equation 3: the equation can be simplified if z is much greater than a and the magnetic flux density is inversely proportional to the cube of the distance. If a metal target is placed into the vicinity of the sensing coil an EMF will be induced in the target according to faraday's law whose magnitude is proportional to the rate of change of that flux, since:

$$E = -N\frac{d\theta}{dt} \tag{4}$$

Lenz's law states that the direction of the induced EMF is such that it will oppose the change causing it, hence the minus sign in equation 4. This will cause a secondary magnetic field to be formed around the target and as a result will generate new currents in the sensing coil that oppose the original flow. This will represent a change in impedance of the coil, since:

$$Z_t = Z_r + Z_i \tag{5}$$

From equation 5 conductive targets tend to cause the resistive part to increase. This will in-turn cause the Q-factor to decrease since this parameter is inversely proportional to the resistance of the coil. The Q-factor of a self resonating oscillator is defined as the ratio of the input energy needed to maintain oscillation divided by the energy lost due to the resistance of the coil, since:

$$Q = \frac{2\pi L I^2 f}{I^2 R} = \frac{\omega L}{R} \qquad (6)$$

where L is the inductance of the coil in henrys and f is the excitation frequency. From equation 6 it can be seen that an increase in R increases the energy lost.

When a non-conducting ferrous target is placed within the vicinity of the sensing coil there will be no conductive paths for eddy currents to be induced. However due to the increase in permeability the magnetic flux density around the target will increase, since:

$$B = \mu H \qquad (7)$$

where H is the magnetic field strength in A m^{-1}. Due to this the inductance of the coil will also increase, since:

$$L = \frac{\mu N^2 A}{l} \qquad (8)$$

where N is the number of turns, A is the cross sectional area (m^2) and l is the length. As a result the resonant frequency of the tuned oscillator will decrease according, since:

$$f = \frac{1}{2\pi\sqrt{LC}} \qquad (9)$$

where C is the combined capacitance of the system. From equation 6 it can be seen that this will cause the Q-factor to increase because less energy is required to generate a given magnetic flux density.

2.2   Choice of excitation frequency and skin depth

The excitation frequency has to be determined in relation to the skin depth. At frequencies above 1MHz electromagnetic interference may be induced in the sensing system, which is not desirable. A higher excitation frequency however insures a smaller skin depth, since [5]:

$$\delta^2 = \frac{2}{\sigma\mu\omega} \qquad (10)$$

where σ is the conductivity. In the case for re-bars most corrosion product lies on the surface so the skin effect needs to be exploited to induce surface eddy currents. For example with steel having approximate electromagnetic properties $\mu_r = 600$, $\sigma = 10^7$ carrying a sinusoidal current at a frequency of 50kHz, the skin effect equals 0.029mm. At 1kHz the skin depth for the same material is 0.2mm.

3.   System overview

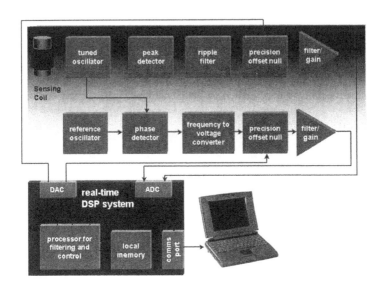

Figure 1. *Sensing system*

The design is based around a self resonating oscillator generating a alternating voltage with a constant amplitude and frequency. The Q-sensor is used to measure the change in voltage amplitude and the heterodyne sensor is used to measure the change in frequency. When the amplitude changes it is converted to a voltage level by a peak detector. The ripple filter is used to remove any ripple components that may be present in the output after which further filtering and amplification is applied. The frequency change is compared to a reference frequency using a phase detector that generates a beat frequency under null conditions. A change in frequency will cause a frequency change to occur on the output of the phase detector that is measured by a frequency to voltage converter after which amplification and filtering is then performed. The data is digitised by the ADC by a real time DSP system after which a filtering algorithm is applied based upon the process of convolution. This is used to yield a better signal to noise ratio. The data is then stored in local memory and after the scanning procedure it is transferred to a data acquisition computer. The DSP system is also used to perform a process called automatic offset correction that is performed using the DAC. This is a process whereby a negative voltage is output to a procession offset null circuit in order to counteract the positive voltages from the peak detector and frequency to voltage converter, which will zero the output of the sensor under null conditions. A DSP56309 is used to perform the latter process and digital filtering because of its high speed compared to the DSP56002 previously used.

4.   Results

Figure 2 shows the Q output verses depth for a steel bar. Real time averaging greatly improves the signal to noise ratio when the two curves are compared with the bold curve representing 700 samples per point. The maximum detection range is approximately 150mm. It can therefore be seen from the signal which involves taking only 5 samples per point that the noise limits the detection range of the sensor and not the sensitivity.

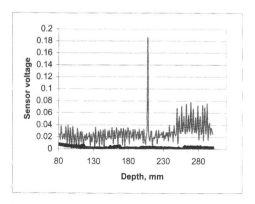

Figure 2. *Far field sensor response of a 20mm diameter bar; 5 averages (normal) and averaged (bold) traces.*

Corrosion detection is employed using the heterodyne sensor. A 20mm diameter bar was used that was corroded half way along its length using an automated corrosion tank until the corrosion layer was approximately 2mm thick. The bar was butt-joined in the middle with another of the same material and original dimension. Figure 3 shows a line scan verses distance with a sensor height of approximately 55mm above the surface of the bar. It can be seen that the signal to noise ratio is greatly improved when using 700 samples per point compared to only using 5 samples per point. The right hand side of the line scan represents the corroded region.

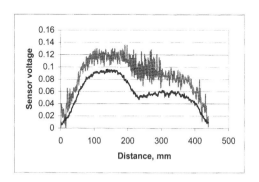

Figure 3. *Line scan of a half corroded bar; 5 averages (normal) and 700 averages (bold) traces.*

This steel bar is shown in figure 4a together with its resulting image in figure 4b and was obtained when the bar was placed under a ceramic tile with a scan height of 30mm. The corroded region is clearly visible with the butt-join indicated by the dark band. The dark band occurs due to the interruption of eddy current blow across the surface, and suggests that this sensor could also be used to image cracks.

Figure 4a. *Photograph of 20mm corroded steel bar butt-joined to uncorroded sample.*

Figure 4b. *Phase-sensitive sensor image of a bar (shown in Figure 4a), placed under a 30mm thick ceramic tile. The negative image is shown, in which the good steel section is on the right.*

5.   Discussion

A combined Q and heterodyne sensor has been developed which has been used to image the parent steel and provide corrosion visualisation. In conjunction with a DSP system is it possible to produce higher resolution images at a greater depth penetration due to the improved signal to noise ratio. This system will therefore be of great benefit when monitoring the condition of reinforced and prestressed concrete.

6.   Conclusion

A combined sensor has been designed which incorporates both Q-detection and heterodyne technology for the imaging of steel reinforcement bars and surface corrosion. A real time DSP system has been employed to improve depth penetration. The sensor works according to the theory and will provide a valuable tool for on site inspection work.

Acknowledgement

The authors wish to express their thanks to the Engineering and Physical Sciences Research Council (EPSRC) for continuing to support this work financially.

References

[1]   Gaydecki PA and Burdekin FM 1994 An Inductive Scanning System for Two-Dimensional Imaging of Reinforced Components in Concrete Structures *Meas. Sci. Technol.* **5** 1272-1280

[2]   Yu ZZ, Gaydecki PA, Silva I, Arif MI, Fernandes BT and Burdekin FM 1998 Magnetic field imaging of steel reinforcing bars in concrete using portable scanning systems *25th Annual Review of Progress in Quantitative Nondestructive Evaluation,* Utah, 18B 2145-2153

[3]   Gaydecki PA Silva I, Fernandes BT and Yu ZZ 2000 A portable inductive scanning system for imaging steel reinforcing bars embedded within concrete *Sensors and Actuators A: Physical* **84** 25-32

[4]   Archenbold W F 1976 *Engineering and Electromagnetics Using SI units* (Edinburgh: Oliver and Boyd) ISBN 0-05-002470-1

[5]   Arthur F. Kip Fundamentals of electricity and magnetism *Second edition, McGraw-Hill International Editons*

Optimization of erbium-doped fibre lasers and their sensor applications

J Mandal, S Pal, T Sun, K T V Grattan, A T Augousti[1]

School of Engineering and Mathematical Sciences, City University,
Northampton
Square, London, EC1V 0HB, United Kingdom.
[1] Figleaf IT, Kingston, United Kingdom.

Abstract: Optimization of erbium-doped fibre lasers has been carried out, with consideration of several factors such as the cavity configurations, the concentrations of erbium ions in the fibres, the length of the fibres used and the effects of pump power variations. Laser cavities were formed using thermally annealed fibre Bragg gratings ˙(FBGs) pair and pumped by a 1480nm laser diode. The characteristics of the laser systems thus created and some initial results from sensor applications are reported and discussed.

1. Introduction

Stable, spectrally narrow linewidth, linear fibre laser sources are key components to meet the increasing demand for high speed, high-capacity wavelength division multiplexed (WDM) optical systems, not only for communications [1], but for sensors and spectroscopic applications showing as they do the advantages of easy integration, alignment insensitivity, compact size, and device compatibility with basic fibre and waveguide optic systems. The developments in fibre Bragg gratings (FBGs) in recent years have further enhanced the functionality and potential of fibre laser technology, especially for sensor applications. The ability to incorporate fibre Bragg gratings into or fixed close to rare earth doped fibre with low insertion loss, low polarization sensitivity, high wavelength selectivity and good alignment insensitivity has revolutionized fibre laser technology. FBGs can be fabricated over a wide, and desirable range of reflectivities and the flexibility offered is convenient to optimize a range of fibre laser systems.

To achieve narrow-linewidth fibre lasers, several types of cavity feedback configurations have been reported, for example intra-core Bragg reflectors [2], coupled linear cavity resonators [3], Fox-Smith resonators [4] and the ring resonator configurations [5], used to promote single-mode operation. In this work, a study of the sensor potential of fibre lasers of this type, with thermally annealed FBGs pairs used for cavity feedback, fusion spliced to erbium-doped fibre as the active medium and pumped by a 1480 nm laser source was undertaken. Optimization of erbium-doped fibre laser systems was carried out with consideration of a range of contributing factors such as the reflectivities of the pairs of FBGs, the dopant levels of different erbium fibres used and the length of the gain medium, for example. The use of laser systems of this type in sensor applications is considered and discussed as, to date, relatively little work on the sensor applications of such lasers has been carried out [6]. The theoretical background to these lasers has been considered elsewhere, e.g. by Becker et al [7] and Digonnet [8] and is now discussed further here.

2. Experimental Set-Up

In this work, several types of fibre laser were designed for further investigation for sensor applications. Figure 1 shows the experimental set-up of an FBG based erbium-doped fibre laser system. A 1480nm FBG stabilized laser diode was used as a pump source, connected to one of the input arms of the 1480nm/1550nm WDM coupler used.

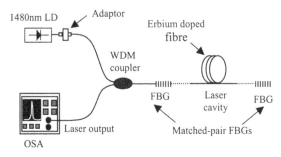

Figure 1. Schematic diagram of an FBG based erbium-doped fibre laser system.

Erbium doped fibre was chosen as the lasing medium as different samples of fibre with this dopant are most widely available and it was fusion spliced to specialized fibre Bragg gratings (FBGs) at both ends to form the laser cavity. The less reflective FBG was fusion spliced to the output arm of the WDM coupler through a length of Corning SMF-28 optical fibre. Two different types of erbium-doped fibres were used in this work, supplied by HighWave Optical Technology, France and Table 1 summarises the details of the rare earth doped fibre used in this work.

	Core compositions and their concentrations	Core diameter	NA	Absorption @1530nm	$\lambda_{cut\text{-}off}$
Fibre 2	Er_2O_3 (650-700 ppm-wt) +GeO_2 (25-28-wt%) +Al_2O_3 (3-wt%) in a silica host	3.0 μm	0.26	7 dB/m (manufactured)	1.0 μm
Fibre 1	Er_2O_3 (350-400 ppm-wt)+GeO_2 (25-28-wt%)+Al_2O_3 (3-wt%) in a silica host	2.3 μm	0.27	4 dB/m (manufactured)	0.85 μm

Table 1: Characteristics of the rare earth doped fibres used in this work

The fibre Bragg gratings (FBGs) used were fabricated on B-Ge co-doped photosensitive fibres (fibre type: PS1250/1500, mode field diameter (MFD): 9.6μm at 1550nm, single mode fibre (SMF); cladding diameter: 125 μm, *NA*: 0.14, cut-off wavelength, $\lambda_{cut\text{-}off}$: 1247 nm, supplied by Fibercore Ltd., UK) using a KrF excimer laser (wavelength = 248 nm, pulse duration = 10 ns, pulse energy = 12mJ, frequency = 200 Hz, Braggstar-500 supplied by Tuilaser AG) as a UV light source and through phase-mask techniques (phase-mask period: 1060 nm, dimension: 25mm x 3mm, zero-order: 0.94 %, supplied by O/E Land Inc., Canada) [9].

In addition, several pairs of different FBGs were tested in this work, where one FBG with high reflectivity (%R = 99.9%) was fixed as one reflector and the other FBG was fabricated with lower reflectivity, varying from 78%-95%, for the transmission "mirror" of the laser system. The laser output spectrum was monitored through the other arm of the 1480nm/1550nm WDM coupler using an optical spectrum analyzer (OSA, Model: HP-86140A) using a resolution of 0.02nm.

In each laser system investigated, a pair of thermally annealed fibre Bragg gratings (FBGs) [10] was used to form the laser cavities. Four different types of FBG pairs, creating Cavity 1, Cavity 2, Cavity 3 and Cavity 4 respectively, were used to optimize the cavity configuration design. Care had been taken to minimize the splicing loss due to the change of the cavity configuration. Each cavity was characterised in terms of the laser power observed through the less reflective FBG, while keeping the highly reflective FBG (%R = 99.9%) fixed as the other "mirror". Details of laser cavity configurations and lasing wavelength using different FBG pairs after annealing are shown in Table 2.

Laser cavities	Reflectivity of the FBG (Variable)	Reflectivity of the FBG (Fixed)	Lasing wavelength (nm)	At the lasing wavelengths reflectivity of the FBGs	
				(Variable)	(Fixed)
Cavity 1	95.15 %	99.89 %	1535.56	94.80 %	99.83%
Cavity 2	91.19 %	99.89 %	1535.64	90.52%	99.80%
Cavity 3	84.00%	99.89 %	1535.58	83.91%	99.88%
Cavity 4	79.57 %	99.89 %	1535.50	78.00%	98.00%

Table 2: Laser cavity configurations and lasing wavelength using different FBG pairs after annealing.

Previous work has shown that to achieve a highly stable performance, each of the gratings should be annealed. For this experiments, the gratings pairs were annealed at 100°C for 6 hours. Care had been taken to avoid any strain effects on these gratings while experiments were in progress. The spectrum of the gratings was monitored in the transmission mode using the OSA (using a resolution of 0.01nm), at room temperature (~23°C) before and after annealing. A decay in the grating reflectivity was observed due to the degradation of the UV-induced refractive index modulation at the annealing temperature [10] and the linewidths (FWHM) of these gratings were also seen to have decreased, with a typical example shown in Figure 2.

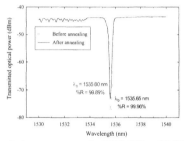

Figure 2: Transmission spectrum of a FBG (%R = 99.9%) as a function of wavelength at ambient temperature before and after annealing at 100 °C.

For each erbium-doped fibre used, the core composition was $Er_2O_3+GeO_2+Al_2O_3$ in a silica host. Aluminium oxide was added to enhance the solubility of the erbium ions in the silica, to improve the pumping efficiency of the fibre by confining erbium ions to the centre of the core of the fibre, as well as to alleviate the clustering effects [8]. The erbium-fibre length used was varied from 3 to 5 metres, in all the tests carried out. Samples designated 1 – 6, tested in this work, are indicated in Table 3.

Fibre lengths	Fibre 1 (350-400 ppm-wt)	Fibre 2 (650-700 ppm-wt)
3 metres	Sample 1	Sample 4
4 metres	Sample 2	Sample 5
5 metres	Sample 3	Sample 6

Table 3: Characteristics of different samples used in this work

3. Experimental results on the effect of Erbium-doped fibres on laser performance

A number of tests was carried out to evaluate the performance of the various laser systems used, as discussed below. A higher concentration of the ion dopant is expected to offer a higher level of gain, this being further confirmed in the results of this work. Figure 3 shows a typical result obtained from samples 1 and 4. However sample 6 showed different behaviour, likely due to the saturation of the absorption of the fibre.

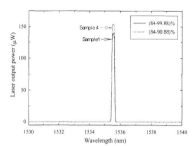

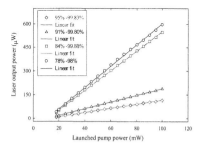

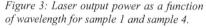

Figure 3: Laser output power as a function of wavelength for sample 1 and sample 4.

Figure 4: Laser output power as a function of launched pump power for sample 3.

The laser output power was found to be linearly proportional to the pump power after lasing occurs [8] and this is clearly shown in Figure 4. Similar results were obtained for other samples. The laser wavelength was consistent with the wavelength of the less reflective FBGs for all the cavities of the six samples tested and was shown in Table 2.

As an example of the capabilities of such systems, the laser output power was also tested as a function of erbium fibre length for cavity 4 (78%-98%) for fibre 1 (350-400 ppm-wt) and fibre 2 (650-700 ppm-wt) respectively, with a pump power of 100mW and the result is shown in Figure 5. The figure shows that the maximum laser output power was obtained for the longest length of the fibre, due to the higher absorption of the pump power along the length of the fibre.

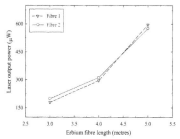

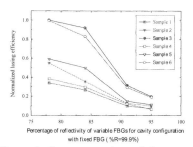

Figure 5: Laser output power as a function of erbium fibre length for cavity 4 (78%-98%) for fibre 1 (350-400 ppm-wt) and fibre 2 (650-700 ppm-wt).

Figure 6: Cross-comparison of the normalized lasing efficiencies as a function of different cavity configurations for fibre 1(350-400 ppm-wt) and fibre 2 (650-700 ppm-wt).

Figure 6 shows a cross-comparison of the normalized lasing efficiencies with respect to different cavity configurations for samples 1 - 6. The normalized lasing efficiencies quoted here were calculated from the linear fit to the laser output power with respect to the pump power, for all the cavities and samples quoted. The highest overall relative efficiency among the samples tested was obtained from sample 3 and cavity 4 due to the optimization of laser output power and the high pump efficiency of the fibre, the details of which are listed in Table 4.

Lasing Efficiencies (η)									
Fibre 1	Sample 1			Sample 2			Sample 3		
Cavities	(% η)	*COD	*SD	(% η)	COD	SD	(% η)	COD	SD
Cavity 1	0.047	>0.99	0.72	0.066	>0.99	0.62	0.133	>0.99	0.84
Cavity 2	0.066	>0.99	1.00	0.084	>0.99	1.13	0.213	>0.99	0.99
Cavity 3	0.18	>0.99	2.81	0.24	>0.99	2.64	0.62	>0.99	2.94
Cavity 4	0.23	>0.99	3.44	0.37	>0.99	2.94	0.68	>0.99	5.18
Fibre 2	Sample 4			Sample 5			Sample 6		
Cavity 1	0.043	>0.99	0.87	0.075	>0.99	1.35	0.126	>0.99	0.70
Cavity 2	0.080	>0.99	1.58	0.099	>0.99	1.83	0.197	>0.99	1.99
Cavity 3	0.20	>0.99	3.86	0.34	>0.99	4.98	0.56	>0.99	4.20
Cavity 4	0.26	>0.99	4.59	0.40	>0.99	6.52	0.67	>0.99	2.86

* COD - coefficient of determination,   * SD-standard deviation

Table 4: Details of the lasing efficiencies with respect to different cavity configurations and different samples for fibre 1 (350-400ppm-wt) and fibre 2 (650-700ppm-wt) respectively

4. Fibre optic laser-based temperature sensor

Figure 7 shows the experimental set-up of a laser-based sensor system for temperature monitoring. The laser cavity was formed using a *chirped* Bragg grating (GR$_2$) and a normal grating (GR$_1$). The normal grating, GR$_1$ written in Ge-doped fibre [9], had a reflectivity of 99.4% (after annealing at 320°C for more than six hours) and was mounted loosely in a Carbolite tube oven (MTF12/38/400) to allow it to respond to different applied temperatures. The remaining part of the experimental set-up was exactly the same as that used for the laser configuration with Fibre 1 (erbium concentration of 350-400 ppm-wt) where a length of 5 metres was used as the gain medium for the laser-based sensor.

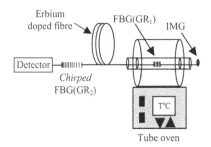

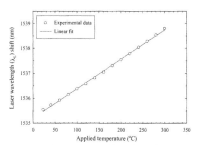

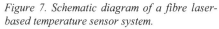

Figure 7. Schematic diagram of a fibre laser-based temperature sensor system.

Figure 8: Laser wavelength shift with applied temperatures.

Figure 8 shows the calibration of the laser sensor formed, in terms of the laser wavelength shift with temperature. This was raised from room temperature to 300°C in incremental steps of 20°C and the reading was monitored using the OSA (using a resolution of 0.02nm) after the stabilization at each temperature. The response of laser wavelength shift with temperature was linear and highly repeatable over the measurement range. This measurement range was limited by the bandwidth (~ 6nm, $\%R = 75$ (±5) %) of the *chirped* Bragg grating (GR_2). With the appropriate choice of the *chirped* Bragg grating (GR_2) and the normal grating (GR_1), the upper limit of the temperature range is likely to be around 500°C, this being chosen to give a realistic lifetime for the gratings in the sensor. The laser wavelength (λ_L) shift with temperature can be determined using the formula shown in equation (1) below:

$$\lambda_L = 1535.21 + 0.011785\ T \tag{1}$$

where λ_L is the laser wavelength and T (°C) is the temperature applied to the normal grating (GR_1). The sensitivity of the laser wavelength was found to be 11.8 pm/°C, giving a precision of ± 1.7 °C over temperature range up to 300 °C. This result is very relevant to the calibration of a laser-based probe using this system design.

5. Conclusion

The contributing factors to the laser performance such as the fibre length, erbium ion concentration, the use of matched-pair gratings, pump power, etc. were evaluated and optimized. The laser output power has been maximized by optimizing the fibre length, erbium concentration and cavity configuration. The highest lasing efficiency was obtained for cavity 4 (~78%-98%) and sample 3 (fibre 1, 350-400 ppm-wt, 5 metres fibre length) among the six tested samples owing to its high pump efficiency of the fibre. Fibre 1 (350-400 ppm-wt of erbium) was used with a small effective cross-sectional area ($A_{eff} = 4.2\ \mu m^2$ with a fibre core diameter of 2.3 μm) of the fibre core compared to that of fibre 2 (650-700 ppm-wt of erbium, $A_{eff} = 7.1\ \mu m^2$ with a fibre core diameter of 3 μm) and also optimizing the laser output power. The overall performance of this device was found very stable at the operating temperatures considered, regardless of the differences of cavity configurations and samples. This is one of the requirements of prime importance for effective communication and also sensor applications. The approach has been illustrated in the creation of a simple, yet highly reliable and cost effective fibre optic temperature sensor system.

6. Acknowledgements

The authors are pleased to acknowledge the support of EPSRC through various schemes including the INTErSECT Faraday Partnership. JM acknowledges Figleaf IT and the Worshipful Company of Tin Plate workers alias Wire Workers for support and Khulna University, Khulna, Bangladesh for study-leave. SP acknowledges Commonwealth Scholarship Commission for support and CEERI, Pilani, India for study-leave.

References

[1]   Miniscalco W. J. Erbium-Doped Glasses for Fiber Amplifiers at 1500 nm. J. Lightwave Technology, Vol. 9, No. 2, 1991, pp. 234-250.

[2]   Ball G. A., Morey W. W. and Glenn W. H. Standing-Wave Monomode Erbium Fiber Laser. IEEE Photonics Technology Letters, Vol. 3, No. 7, 1991, pp. 613-615.

[3]   Gilbert S. L. Frequency stabilization of a tunable erbium-doped fiber laser. Optics Letters, Vol. 16, No. 3, 1991, pp. 150-152.

[4]   Barnsley P., Urquhart P., Millar C. and Brierley M. Fiber Fox-Smith resonators: application to single-longitudinal-mode operation of fiber lasers. J. Opt. Soc. Am. A, Vol. 5, No. 8, 1988, pp.1339-1346.

[5]   Cowle G. J., Payne D. N. and Reid D. Single–frequency Travelling-Wave Erbium-Doped Fibre Loop Laser. Electronics Letters, Vol. 27, No. 3, 1991, pp. 229-230.

[6]   Kim B.Y. 'Fiber lasers in optical sensors' in "Optical Fiber Sensor Technology", Vol. 2: Devices and Technology, edited by Grattan K.T.V. and Meggitt B.T., Chapman and Hall, London, UK, pp. 99-115, (1998)

[7]   Becker P. C., Olsson N. A. and Simpson J. R. "Erbium-Doped Fiber Amplifiers Fundamentals and Technology", Academic Press, San Diego, (1999)

[8]   Digonnet M. J. F. "Rare Earth Doped Fiber Lasers and Amplifiers", M. Dekker, INC. New York, (1993)

[9]   Sun T., Pal S., Mandal J. and Grattan K. T. V. Fibre Bragg grating fabrication using fluoride excimer laser for sensing and communication applications. in CLF (Central Laser Facility, Rutherford Appleton Laboratory), pp. 147-149, Annual Report 2001/2002.

[10]   Pal S., Mandal J., Sun T. and Grattan K. T. V. Analysis of thermal decay and prediction of operational lifetime for a type I boron-germanium codoped Fiber Bragg grating. Applied Optics, Vol. 42, No. 12, 2003, pp. 2188-2197.

The development of a robust, autonomous sensor network platform for environmental monitoring

L Sacks[1], M Britton[1], I Wokoma[1], A Marbini[1], T Adebutu[1,] I Marshall[2], C Roadknight[2], J Tateson[2], D Robinson[2]

[1] Department of E&EE, University College London, UK.
[2] Btexect Technologies, UK.

Abstract: This paper describes an approach to the approaches being explored for a Sensor Network platform being developed for the DTI/NextWave technologies programme. The approach being adopted is to develop the system as a community of devices which use self-organising techniques to provide key functions. The devices are, largely, based on commodity technologies, thus providing a low cost basis. We give an outline of the approach and project and illustrate the techniques being developed with specific functions for: control, management, data retrieval and data quality control. The target application is off shore sea shelf monitoring; but the techniques being developed may be applied to a range of problems.

1. Introduction

Developing complete sensor systems for environmental monitoring in harsh environments is currently a complex and expensive exercise. There is a trend towards leveraging commodity technologies to enable the construction of cost effective sensor platforms, enabled by recent growth in wireless communications and microprocessor controlled consumer devices. This leads to small, low cost, limited functionality devices working in a cooperative networked mode, to form a single system. This model brings cost advantages and extra science value. By constructing a *network* of sensors, a greater area may be covered allowing better spatial resolution. Further, redundancy in components may be introduced thus providing more robust and longer lived operations. These features are highly advantages in applications covering large, hostile environments such as: glaciers, volcanoes or off-shore sea beds.

This paper presents the development of such a sensor network (SN) platform for the monitoring of environmental impact on a coastal sea bed of a wind farm. Wind farms are seen as a key feature of the governments' sustainable energy policy. It is critical to ensure that their placement does not produce negative, environmental impacts. The complex interplay between the: oceans currents; wind; coast line and off shore features (e.g. sand banks) is poorly understood; but it is critical that natural costal defences such as sand banks are not disturbed or destroyed inadvertently. Thus there is a great need for continuous measurement of off shore features in such a context.

This approach and application requires trade-off between design and capability. Using low cost devices entails limited functionality and performance. Target environments considered here are, by their nature, highly turbulent which entails that the sensor platforms have to be able to adjust to the environment on a continuous, though hard to predict, basis. These problems have led us to explore a new approach to the system engineering. In this approach the system is composed of a *peer-to-peer* system, supporting intrinsically networked algorithms; which derive their full functionality from the symmetric cooperative and self-organisation. This contrasts to current approaches of confining the functionality to individual nodes and considering the sensor network as a slave device. To achieve this, approach is to model the algorithms on behaviours found in natural biological systems. This paper presents this overall concept and some example details of the algorithm design.

2.   Application Context & Motivation

The work reported has been carried out in preparation for implementation under the DTI NextWave [I] technologies program, for the project SECOS [II], part of the "Centre for Pervasive Computing in the Environment". The wind farm development located on the Scroby Sands [III] sand bank provides the key requirements capture for the SN development. The location of wind turbines on sand banks can produce highly complex turbulent flows both locally, producing sand shift and scouring of the turbine bases; and on a wider scale having impacts on alongshore sand banks and sea barriers. Measuring the ocean in this environment is currently expensive. Typical landers will cost £100-200K to build including: sensors, acoustic releases etc. and have high deployment and retrieval costs (as the lander must be precision located). A single lander can only measure the environment at a single point; it is vulnerable to destruction by storms, burial by moving sand waves and accidents from trawlers – as such it is a 'one shot' proposal. In science terms, the key problem is that the current technologies do not give good coverage in either time or space. For most of each deployment little happens so power and data storage is wasted; operational cycles are always compromises between the need for high resolution measurements during the active periods (usually storms) and the uncertainties in the frequency, timing and severity of the active periods. Typically, most monitoring systems undergo fixed sleep-sample duty cycles which are good enough for monitoring regular features (e.g. tides) but cannot reactively monitor transit effects. Turbulence is far from regular through space and there is clear need for high special sampling to measure features such as sand: drift, scouring and build up.

The approach we are developing is to develop a SN platform consisting of a large number (30-50) of low cost (<£1000) nodes. Each node has basic functionality; a small µ-processor, low rate communications and sensing capabilities. Sensing capabilities will focus on features such as, optical backscatter – a measure of sediment load in the water – pressure, and temperature. In addition to the general sensor nodes, there will be a few master packages which can measure other parameters (e.g. surface pressure) and for communications with the science base-stations. For deployment the nodes could be scattered from a boat, if they are capable of determining their own location. Key amongst the technical requirements is the need for the nodes to be able to communicate between them selves on a nearest neighbour basis. This can be achieved using surface-floating tethered buoys supporting low cost, low power radio communications such as 802.15.4. operating in the IMS spectrum with low data rates (250kbs) and high node densities (~250) and low power demands (*e.g.* 30µW for a 1000/1 sleep/transmit cycle). For more general applications sonar communications are a viable alternative.

This approach brings its own set of functional requirements. The nodes must be able to configure them selves for monitoring frequency in space and time, to be able to condense data and communication it back to the base station, to be able to substitute for each other in case of failure, and be able to locate them selves. The nodes must respond to local changes in the environment (both the monitored environment and each other) as autonomously as possible. Thus there is a need for a general, localise control system with the whole system manageable by the scientist / operator. It is not possible for the operator to control the each node individually so the system must be manageable at a high level.

3.   A Kind of Operating System

From the above discussion it is clear that nodes loaded with classical micro kernel Operating Systems (OS) is inappropriate for this kind of application. Such OSs take substantial resources and their functionality is generally confined to single nodes. In this context, we require a light weight service kernel, supporting distributed algorithms. The best models we currently have for such as systems can be found in the biological world in which systems at all scales (from inside the Cell, through to social systems) can be seen to derive their functionality as much from how components interact as from the intrinsic functionality of the components them selves; in a. fully dispersed architecture. The idea that the 'algorithms' which a system is performing is derived as much from the way things interact as from a specific algorithm is key behind the concept of Emergence. That concept of control occurring through dispersed cooperative action is the key notion behind evolving self-organising systems [IV]. Hence the OS framework should be based on concepts of emergent, self-organisation.

The engineering strengths of this approach are two fold. Firstly, by constructing the system out of a number of independent self-organising components in which much of the information and functionality is in the connectivity, we can build a system which is robust against failure or partial functioning of individual components. Secondly, the work performed by individual nodes is small, only being a partial part of the full algorithm, thus the processing & memory demands on individual nodes can be small. We refer to the operating system based on these attributes as the kOS.

We illustrate the functionality of the kOS through the development of some of the key functional components. The components provide the following functionalities, typically found in operating systems in one form or another: Synchronisation between nodes for coordination of communications and work loading; User management through policies; Location based Identity; Persistent Data storage; Work scheduling & partitioning; Control Optimisation. In order to give a feel how the approaches describe here can be used to provide these functionalities; initial explorations of mechanisms are described below.

3.1.   Synchronisation

On any system it is important that the individual components can synchronise their operation. In the SN platform the components not only need a degree of synchronisation but need to be able to adjust the sync-time base to enable energy conserving operation. The synchronisation mechanism has to be increase frequency to able to adjust to transient changes in the system – for example when a user management command is entered – and return to the base state. To provide this functionality we have developed a system, loosely based on the concept of fire-fly synchronisation. Fireflies are known to emit flashes at regular intervals when isolated but in

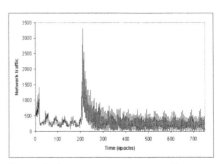

Figure 1. Synchronisation across a Transient event.

a group they entrain the pulsing of their lights to converge upon the same rhythm as that of other fireflies in the group until synchronicity is reached. A non-deterministic distributed mechanism is used in the system where messages are passed between individual fireflies to achieve their coordination[V] which is based on a network of pulse-coupled integrate-and-fire elements[VI]. Each node adjusts its flash intervals through a concave control surface which enables synchronisation lock. In the SN platform, nodes exchange messages in line with the flash interval and 'lock' onto the message exchange phase in order to achieve synchronisation. An illustration of the synchronisation phase is shown in Figure 1 which shows the rate of message exchange before and after a transient event which has cause in frequency increase in the flash interval.

3.2.   System Management

Any operating system has to allow the operator to control and configure the system. In this context, controlling an individual node is not meaningful as nodes will exchange roles for a range of operational reasons (power conservation, destruction *etc.*). The management system we have explored is based on the concept of a Policy. Policies are high level operational rules which are meaningful to the operator and it is up to the individual components to adjust their behaviour to fulfil these policies. Policies in this context might

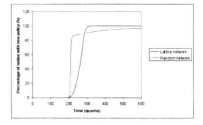

Figure 2. Policy Distribution

determine such operational requirements and the length of the experiment or the resolution of the data required by the scientists. This approach has been extensively explored in the world of network and service management [VII] including its use within self-organising systems [VIII]. In this context, the key problem is to ensure that all nodes in the system are aware of the introduction of new policies. The typical procedure for this would be to send the data to each node individually. This would required that the system support a complete routed network infrastructure such as those found on ad-hoc networks. Many ad hoc routing protocols have been devised. Some of the most widely known are DSDV[IX], TORA[X], DSR[XI] and AODV[XII]. A comparison of the performance of these protocols [XIII] has shown widely differing results in the size of routing overhead and can become unacceptable for a network size of 30 nodes. The main problems using these in a SN are the size of processor and memory required and the protocols are not energy usage aware. As an alternative, we have adopted the approach of a modified gossip protocol called Time-Stamped Anti-Entropy (TSAE) [XIV] which provides weak-consistency based information distribution [XV]. The combination of TSAE and fire-fly protocols has been demonstrated in the context of policy based management for fixed networks[XVI]; the key point is that this never depended on specific routing or communications models, indeed, it performance on random networks was seen to be substantially better than on normally routed networks is illustrated in Figure 2.

3.3. Persistent storage / Data acquisition

The key use of persistent data storage in this context will be for data acquisition. In the context described here, each node will not be able to store much data locally; rather they will send the data back to base stations for archive and analysis[XVII, XVIII]. As discussed above, using classical ad-hoc networking technologies is extremely expensive and not necessary for control and management purposes. Thus the task is to devise a data retrieval system that does not depend on node identity. The approach taken here [XIX] is based on a least resistance flow concept and platforms on the basic synchronisation capabilities described above. Data is, effectively broadcast from each node to its nearest neighbours who, in turn, re-broadcast it based on a cost function. The cost function its self is developed

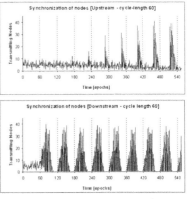

Figure 3. Up and Down Stream Synchronisation

through a form of back propagation from the base stations – who understand that they are the appropriate destination of the data. Figure 3 illustrates the forward and abackward developent of the data flow synchronisation through the system.

3.4. Data Quality selection.

With communications overheads being a major resource cost, it is essential that data is filtered at or close to the sensing nodes so that only changes in measurements are transmitted. The environment being measured is, however, turbulent, with high fractal components in the dynamics. Thus it is hard, if not impossible, to use classical control systems to track the sampling frequencies; a situation that is exacerbated by the fact that the nodes are likely to

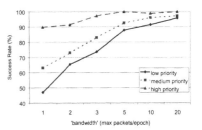

Figure 4. Adaptive Sensing Efficacy

be 16bit, fixed point arithmetic devices which will struggle with floating point computations. For these reasons we have been exploring the use of adaptive control algorithms. The one described here is based on the evolution and exchange of short plasmid strings representing the key sensing operations – more or less like a distributed genetic algorithm. Each node continuously evolves its notion of sampling and exchanges its best guess representation with neighbouring nodes. The discreet operations are: Sense, Forward, Delete, Compressing and Idle - each operation being performed for a variable amount of time. The ratio of these operations is determined both through the evolutionary process and locally under Policy control (in line with the nodes local objectives). These rules are applied for each measured quantity individually or in combination (as part of a local data fusion approach) with differential priorities being given to each measurement – under policy control. This approach has been modelled for a scenario with 3 kinds of measurement and a network of 30 nodes. Figure 4 illustrates how each measurement can be differentiated under varying (policy) requirements on retrieved data.

3.5.   Location Based identity.

In the applications considered here, the important form of reference is functional in the sense that the key issues are the science values at spatial locations; independently of which node is actually sampling that location. Indeed, it is possible that through the lifetime of an experiment the node(s) responsible for sampling some part of the environment will change due to failure or movement. In our architecture, a node is principally identified functionally, by its position and thus it is necessary that nodes can determine where they are. It will be possible to determine some nodes location absolutely by equipping them with GPS systems or attaching them to wind turbine masts. For the majority of nodes it will be necessary that they can locate them selves relatively to other nodes through techniques such as sonar ranging or – for floating buoys – radio ranging. Nodes can then exchange measurements of relative position with nearest neighbours to locate them selves in absolute term. Some approaches to this have been explored else where [XX,XXI]. Current developments [XXII] focuses on robust operations.

3.6.   Work partitioning

Most operating systems will have some means of multi-tasking, allowing the system to perform a range of tasks simultaneously. In this context individual nodes will have a single tasking kernel and the multi-tasking aspect of the work will be achieved through partitioning of work between the available nodes. To do this in a self-organising manor, it is necessary that the nodes agree locally, between themselves, which work will be undertaken by which nodes. To develop this approach, we have explored the use of Quorum Sensing [XXIII]. Quorum Sensing is a form of intercellular signalling that regulates gene expression in response to cell-population density. In this process, each cell is able to individually sense when a quorum of bacteria, a minimum population, has been achieved and results in a behavioural change in the bacteria. An example of this is bioluminescence which occurs in the organs of deep sea fish due to the bacterium, Vibrio fischeri. Having established a core quorum forming system, the sensor nodes will be able to from them selves into groups dedicated to sampling at required rates; performing data fusion, hibernating and so on.

3.7.   Practical Implementation

The implementation of the kOS in software is driven by a requirement for only minimal OS-like functionality. As the nodes self-synchronise their functional operation, we have no requirement for concurrent task operation and we do not therefore require a multi-tasking system. We can then have run-to-completion tasks that are queued in a simple FIFO manner-greatly simplifying system operation and minimising the code footprint. This also results in predictable task execution and removes any need for inter-task communication. One non-standard feature required for the kOS is in-application re-programmability, where new sensing instructions are installed in non-volatile memory areas of nodes from remote workstations.

The choice of microprocessor is an important design decision. In most modern microprocessors, processing speed and programmable non-volatile memory is cheap in terms of cost and power consumption, and is generally more than adequate for our purposes. Also, many microcontrollers have features we require such as analogue-to-digital converters (ADCs) and several timers or counters. Driving requirements are therefore for low power consumption devices (including enough low-power modes of operation) and suitable communications and peripheral interfaces, as both networking and sensing are core features of the system. As we have only a simple memory organisation and speed is not a primary concern we also find that a Von Neumann (Princeton) architecture is adequate.

4.   Related Work

Researchers at the University of California Berkeley have worked for several years on the design and implementation of a self-contained, millimeter-scale sensing and communication platform in the Smart Dust project [XXIV]. The aim is to design a sensor device that will have some processing ability, wireless communications and a battery supply, while being inexpensive enough to deploy in large numbers. A number of such "motes" have been implemented, and a minimal component-based operating system TinyOS [XXV] has been written specifically for the platform. Other related projects include Manatee [XXVI], involving Bluetooth-based communications and environmental sensing. The University of California LA's Centre for Embedded Networked Sensing [XXVII] conducts a considerable amount of research in this area.

5.   Conclusion

This paper has described our approach to the developments of a sensor network platform based on, biologically inspired, self-organising algorithms. At this stage we have established the viability of implementing key functions using this approach. Critical work is now underway to implement the system in suitable packaging and with appropriate technologies for the target applications. This overall approached hold out the possibility of low cost sensor applications for a range of science tasks; and with low management overheads and hardware costs.

References

[1]   http://www.nextwave.org.uk/
[2]   SECOS: "Self-Organising Collegiate Sensor Networks"
         http://www.adastral.ucl.ac.uk/sensornets/secos/
[3]   http://www.offshorewindfarms.co.uk/sites/scroby-sands.html

[4]   I.W.Marshall and C.Roadknight, "Adaptive management of an active services network", British Telecom. Technol. J., 18, 4, pp78-84 Oct 2000

[5]   U. Wilensky, K. Reisman, "Learning Biology through Constructing and Testing Computational Theories – an Embodied Modelling Approach".

[6]   X. Guardiola, A. Diaz-Guilera, M. Lias, C. J. Perez, "Synchronisation, diversity, and topology of networks of integrate- and-fire oscillators", Physical Review E, October 2000.

[7]   M. Sloman, "Policy Driven Management for Distributed Systems", Journal of Network and Systems Management, 1994

[8]   L. Sacks, O. Prnjat, I. Liabotis, T. Olukemi, A. Ching, M. Fisher, P. Mckee, N. Georgalas, H. Yoshii, "Active Robust Resource Management in Cluster Computing Using Policies", Journal of Network and Systems Management, Special Issue on Policy Based Management of Networks and Services, 2003.

[9]   C Perkins and P Bhagwat, Highly Dynamic Destination-Sequenced Distance-Vector Routing (DSDV) for mobile computers, Proceedings of the SIGCOMM '94

[10]   VD Park and MS Corson, A Highly Adaptive Distributed routing Algorithm for Mobile Wireless Networks, Proceedings of INFOCOM '97, pages 1405-1413, April 1997

[11]   DB Johnson, Routing in Ad Hoc Networks of Mobile Hosts, Proceedings of the IEEE Workshop on Mobile Computing Systems and Applications, pages 158-163, December 1994

[12]   C Perkins, Ad Hoc On Demand Distance Vector (AODV) Routing, Internet-Draft, draft-ietf-manet-aodv-04.txt, October 1999

[13]   J Broch, DA. Maltz, DB. Johnson, Y-C Hu, J Jetcheva, A Performance Comparison of Multi-Hop Wireless Ad Hoc Routing Protocols, Proceedings of Mobicom '98

[14]   C. L. van Eijl, C. Mallia, "The time-stamped anti-entropy protocol- a weak consistency protocol for replicated data".

[15]   R. Golding, D. D. E. Long, "Modelling replica divergence in a weak-consistency protocol for global-scale distributed data bases", http://www.cse.ucsc.edu/~golding/

[16]   I. Wokoma, I. Liabotis, O. Prnjat, L. Sacks, I. Marshall, "A Weakly Coupled Adaptive Gossip Protocol for Application Level Active Networks"; IEEE/Policy 2002, Monterey, CA, USA, June 2002.

[17]   Tian H., Stankovic J., Chenyang J., "SPEED: A Stateless Protocol for Real-Time Communication in sensor networks", International Conference on Distributed Computing Systems (ICDCS 2003).

[18]   Deepak G, Krishnamachari B, Woo A, Culler D, Estrin D and Wicker S, "Complex Behavior at Scale: An Experimental Study of Low-Power Wireless Sensor Networks", UCLA/CSD-TR 02-0013.

[19]   A. E. Gonzalez-Velazquez, L.E. Sacks and I. W. Marshall, "Simple spontaneous mechanism for flexible data communication in wireless ad hoc sensor networks", The London Communications Symposium 2002

[20]   Savvides, A. Park, H. Srivastava, M. 'The bits and flops of the N-hop multilateration primitive for node localization problems' In Proceedings of 1st ACM International Workshop on Wireless Sensor Networks and Applications, Sept, 2002

[21]   Savarese, C. Rabaey, J. Langendoen, K. 'Robust Positioning Algorithms for Distributed Ad-hoc Wireless Sensor Networks' USENIX Technical annual Conference, pgs 317-28, Monteray, CA, June 2001.

[22]   D. P. Robinson and I. W. Marshall "An Iterative Approach to Locating Simple Devices in an Ad-Hoc Network" The London Communications Symposium 2002.

[23]   "Quorum Sensing in Bacteria", Melissa B. Miller and Bonnie L. Bassler, Annual Review in Microbiology, pp.165-99, 2001.

[24]   K. S. J. Pister, J. M. Kahn and B. E. Boser, Smart dust: Wireless networks of millimeter-scale sensor nodes, 1999, Highlight Article in 1999 Electronics Research Laboratory Research Summary.

[25]   J. Hill, R. Szewczyk, A. Woo, S. Hollar, D. Cullar and K. Pister, System Architecture Directions for Networked Sensors, ASPLOS 2000, Cambridge, November 2000. http://webs.cs.berkeley.edu/tos/papers/tos.pdf

[26]   Manatee project web page. http://www.distlab.dk/manatee/

[27]   The Centre for Embedded Networked Sensing web page. http://cens.ucla.edu/

Section 10

Water Quality Monitoring

A feasibility study into differential tensiography for water pollution studies with some important monitoring proposals

[1]A C Bertho, [1]D D G McMillan, [1]N D McMillan, [1]B O'Rourke, [1]G Doyle, [2]M O'Neill and [3]M Neill

[1] Institute of Technology Carlow, Kilkenny Road, Carlow, Ireland
  Phone: 353 503 70400 Fax: 353 503 70500
[2] Carl Stuart Ltd., Tallaght Business Park, Whitestown, Dublin 24, Ireland
[3] Regional Water Laboratories, Environmental Protection Agency, The Butts, Kilkenny, Ireland

Principal Author: email: berthoa@itcarlow.ie

Abstract A differential tensiographic method for the fingerprinting of liquids has been developed for tensiography, based on a visual graphical assessment of the differences between tensiotraces. The most useful analytical application in water science will perhaps be for the detection and quantification of trace-level differences. The extreme sensitivity in this method is illustrated by this work, which introduces the new technique by way of a feasibility study and as a consequence this should find application in water pollution studies and other areas. The technique has been applied to two organic pollutants and for the monitoring of colour and turbidity in water. The paper concludes with an outline of the future work to be done in the ongoing research project with suggestions on how to analyse trihalomethanes (THMs).

1. Introduction

The development of double-beam detection began in the late 1850s and was given wide publicity by the 1861 Bakerian Lecture with the first discourse on environmental monitoring [1] by the Carlow born physicist John Tyndall. He is further remembered for demonstrating liquid jet experiments with intense light beams to show for the first guided light beam, which today has led to the fiber optic [1]. The current paper was led by a Carlow team and involves differential development of tensiography, drops and fiber technologies, all of which of course were pioneered by Tyndall. The development of drop UV-visible spectroscopy has been dealt with at length in a paper entitled 'Drop spectroscopy [2]. The development of modern double-beam detection systems developed from the late 1950s with the introduction of chopped beam techniques [3]. The advantages of double-beam null-balance (or ratioed) systems were such that these instrumental techniques are today becoming multiplexed with the development of CCD technology in HPLC systems and elsewhere [4]. The initial development of tensiography from the early 1990s was by of necessity a single-beam Amplitude Modulated Fiber Optic Sensor (AMFOS) technique [5]. The tensiograph drophead does not readily lend itself to modifications involving double-beam fiber paths because this requires a doubling of the instrumentation with two dropheads. The instrumental development described here is therefore based on the development of software techniques [6] for generating differential signals from two tensiotraces recorded sequentially.

2. Theory

The tensiotrace is primarily the result of a series of total internal reflections (TIRs) inside the droplet and provides a unique fingerprint of the liquid being studied. Figure 1(a) shows the convention for heights and time of peaks used in the tensiograph analysis. Various characteristic features of the tensiotrace provide information about the physical properties of the liquid. Details of how to obtain these parameters are available elsewhere [7]. For measurements of colour and turbidity, it is the tensiograph and rainbow peaks heights that are used as the measurands. The rainbow peak is recorded in the early stage of drop development with a drop shape that is usually not too far removed from the remnant drop shape. The tensiograph peak is obtained from light couplings in the fuller pendant drop. These higher-order couplings are generally associated with reflections from an elongated drop base. The rainbow peak height is determined in the first instance by the refractive index of the liquid under test (LUT), whilst the tensiograph peak shows a strong dependence on the absorbance of the liquid. Both peaks however depend on the refractive index and absorbance of the LUT.

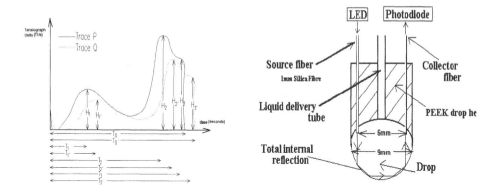

Figure 1(a): *Convention for heights and times of peaks used in the tensiograph analysis*
Figure 1(b): *Diagram of the drophead*

The fingerprinting of a liquid is based on matching (M) and subtraction (S) values. One of the approaches integrated into the multianalyser tensiograph analysis package is trace features ratioing using M-functions. The software produces the M and S functions (S = 1-M) from a reference trace and a test trace. The software separates each feature i.e. drop period, rainbow peak period etc. and the match between both traces is determined. The M - Value for a peak height is given by

$$M = f(H_m) = H_m/H_{m'} \text{ or } H_{m'}/H_m \ (0<M<1) \tag{1}$$

In order to determine the tensiographic absorbance, it is convenient to assume that the tensiograph peak maximum is in an equivalent position for each trace. The tensiographic absorbance is then given by

$$A_T = \log_{10} \frac{H_m}{H_{m'}} \; or \; \frac{H_{m'}}{H_m}$$

$$(2)$$

where the peak height is that of the tensiograph peak.

It has been shown that both the tensiographic absorbance and turbidity obey the Beer-Lambert Law [8] very closely, and hence the following equation applies

$$A = \varepsilon c \int \partial l = \varepsilon c l_{av} \qquad\qquad A(t) = (\tau.c)\int \partial l = \tau c l_{av} \qquad\qquad (3)$$

where ε is the tensiograph molar absorptivity (A-units mol^{-1} cm^{-1}), and c is the concentration in mol l^{-1}. The integral represents the sum of the path lengths connecting the source and detector fibers at the tensiograph peak maximum where τ is the tensiograph turbidity coefficient (A-units mol^{-1} cm^{-1}).

3. Apparatus

The work presented here uses a concave drop head, for which drawings are shown in Figure 1(b). The drop head employs standard 1 mm polymer or silica fibers. The diameter of the head is 9mm with the fibers separated at 6mm. The fibers are positioned to give a standard tensiotrace and protrude a small distance from the concave base. The liquid is delivered through a HPLC capillary glued into the centre of the head. Ideally the head should be designed such that it wets (i.e. the suspended liquid covers the entire lower surface of the drop head) and exhibits all the characteristic tensiotrace peaks.

4. Experimental:

4.1   Application for trace level pollutant detection

Two organic pollutants have been studied to assess the performances of differential tensiography in distinguishing similar traces. The pollutants are anthracene ($C_{14}H_{10}$) and naphthalene ($C_{10}H_8$). They are both PAHs (Polyaromatic hydrocarbons) and are found in industrial waters. The usual method of analysis is via liquid chromatography or optical detection. As a consequence of the two or three aromatic rings these molecules strongly absorb in the UV region, but neither absorb in the visible range.

Although both anthracene and naphthalene are very insoluble in water because of their strongly absorbing properties, even a little amount of these substances can be detected. The greater solubility of the molecules is in organic solvents such as alcohols and benzene. Methanol has been used to make up standard solutions. Anthracene samples were prepared by making up water based solutions. The solution was then filtered before being analysed with the multianalyser. The molarity of this sample was calculated from the calibration plot of a range of known concentration of anthracene standards solutions in methanol, run on a UV-Visible spectrophotometer at $\lambda_{Max \; C14H10}$ = 252.1nm and found to have a concentration of $1.3915.10^{-5}$ M. The tensiographic analysis was carried out at 470nm, although this sample has very little absorbance at this wavelength. Figure 2(a), shows the tensiotraces for this solution showing measurable differences, making it visually impossible to quantify these differences. The tensiotraces were then analysed with the TraceMiner software and the differential trace was obtained as seen in Figure 2(b).

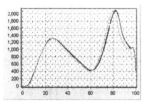

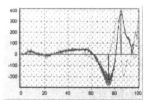

Figure 2(a): Traces of water and anthracene (1.391.10⁻⁵M) recorded at 470nm and visualised with the TraceMiner software

Figure 2(b): Differential trace of anthracene (1.391.10⁻⁵M) at 470nm. Smoothing chosen = 101.

The differential of these two traces, i.e. the calculation trace of water *minus* trace of anthracene taken point by point shows large visual differences with peaks that importantly can be quantified. The same approach to analysis was carried out for naphthalene. Again, the same protocol as described above was employed. The absorbance of the sample was measured at $\lambda_{Max\ C10H8}$ = 220.3nm and gave a concentration for the naphthalene of 1.3935.10⁻⁵ M. The sample was then analysed by tensiography at 470nm and produced very little visual difference with that of water as can be seen from Figure 3(a).

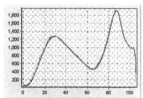

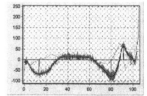

Figure 3(a): Traces of water and naphthalene (1.393.10⁻⁵M) recorded at 470 nm and visualised with the TraceMiner software.

Figure 3(b): Differential trace of naphthalene (1.393.10⁻⁵M) at 470nm, smoothing chosen = 101

Differential tensiography produces a characteristic trace for naphthalene, which is different from that of anthracene. It appears from our initial feasibility study that the form of these differential tensiotraces are indeed a characteristic of the priority pollutant and furthermore the trace height increases in the initial stages with concentration, maintaining the form of the differential tensiotrace. The author's believe that this technique is really only qualitative for low concentrations of pollutants. It is thought to be unlikely that all pollutants have unique differential tensiotraces, although at this point no two such matches have been observed after six or seven studies. We feel that there is a very good chance that the differential trace of pollutants can give in many cases useful qualitative results. The vexed question then of mixtures of pollutants would obviously need great experimental scrutiny. What is proposed in the next phase of the work is to investigate natural waters to see if the addition of pollutants gives a characteristic change in the differential tensiotrace as this would allow qualitative monitoring with the analysis of water continually done with requisite time delay between measurements to pick up a pollution incident.

4.2   Application for Colour and turbidity monitoring in water

The standards for colour measurements in water are solutions of chloroplatinic acid in the presence of cobalt. The colour value in Hazen Units is numerically equal to the platinum concentration of the solution in mg per litre. Turbidity in water usually comes from the presence of very finely suspended solid matter such as clay, organic and inorganic materials, but may also be due to the presence of crystspiridium, a pathogenic protozoan that can cause severe gastrointestinal illness. The standard spectroscopic technique for measuring the turbidity in water samples is the Attenuated Radiation Method (ARM), which is carried out at 860nm.

Ten replicate tensiotraces of each standard were generated at 470nm and at a temperature of 20°C, with a PEEK drophead. The average values of rainbow and tensiograph peak heights were calculated, along with the standard deviation for each parameter. Each of the ten test tensiotraces was compared to the ten reference tensiotraces obtained for distilled water giving 100 M and S values. This approach obviously greatly improved the statistical validity of the results. The M and S values were calculated for each of the water traces with each of the test traces for all of the solutions. The differential of each tensiotrace were taken with water as a reference.

The important conclusion, anticipated from theory, is that the M and S values of both colour and turbidity contributions are additive. This result was confirmed by analysing all data-points of the test traces. It was shown that the addition of the S-values of 'two' samples (one coloured and one turbid) matched that of a 'single' mixture with the corresponding concentrations of the same two components. This result is shown in Figure 4(a) where the S value for the mixture solution of concentration 50HU+5NTU are plotted for every point of the data set (in black on the chart) and compared with the addition of the S value for the 50HU solution and S value for 5NU solution (in grey on the chart). This graph shows a good overlap of the data sets i.e. the S values for the 'single' mixture 50HU+5NTU equals the addition of the 'two' solutions of 50HU and 5 NTU solutions.

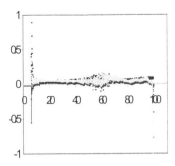

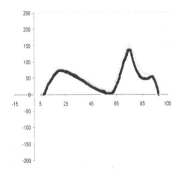

Figure 4(a): *S values traces*   **Figure 4(b):** *Differential traces*

The same additive relationship, it should be noted, exists also for differential tensiography. Figure 4(b) shows a graph representing the differential trace of the mixture solution (50HU+5NTU) in black and also the addition of the differentials traces of 50 HU and 5 NTU: Again, these two traces overlap, showing colour and scattering are both measured by absorbance, and can not be differentiated using tensiography, which is also the case in UV-visible spectrophotometry. It is conceivable that surface chemistry may however be used in tensiography, to develop different drop shape changes with regard to the two components, and give different differential tensiotraces. Such methods are worth pursuing.

5. Conclusions

This paper has shown the potential of differential tensiography as a qualitative tool in the detection of trace pollutants. The feasibility study shows the approach offers some advantages; it is both very sensitive and fast. In the next phase of work, careful quantitative studies are to be done to quantify these effects and demonstrate that the detection limits make the method useful for water monitoring. The above experiment on anthracene and naphthalene has shown that differential tensiography is useful for the detection of trace levels amount of pollutant in a water sample. Secondly, the possibility of using differential tensiography as a method for qualitative analysis has been investigated for monitoring colour and turbidity in water. The detection levels are acceptable, but at high concentrations the errors increase in these measurements for the turbid solutions to effectively limit the measurement range to below 75 NTUs. We anticipate this problem will be overcome in future work. Other work, which has not here been described in this paper, has shown that water drops absorb very strongly methanol vapour to produce a characteristic tensiotrace for the adsorbing species. It is felt that THMs could be monitored in this way by using a suitable set-up to pass the extracted vapours and gases from the water over a drophead delivering a drop with a suitable liquid. The liquid should adsorb strongly the THMs and not the other possible components of the vapour/gas extracts to give a good discrimination in this measurement. Furthermore, the application of Dual Wavelength spectroscopy to differential tensiography has to be investigated [9].

6. References

[1]   Tyndall J, On some phenomena connected to the motion of liquids, Proc. Roy. Inst., 19 May 1854.
[2]   McMillan N.D., O' Neill M., O'Mongain E., Neill M., O'Rourke B., Riedel S., Smith S., Wüstneck N., Wüstneck R., Quantitative drop spectroscopy: An experimental and theoretical investigation into the instrumental advantage and utility of a fiber optic drop-head spectrophotometer, In preparation.
[3]   Currell G., Analytical Instrumentation, Wiley, 72, 2000.
[4]   *ibid* 152
[5]   McMillan N.D., Finlayson O., Fortune F., Fingelton M., Daly D., Townsend D., McMillan D.D.G., Dalton M.J. and Cryan C., A fiber drop analyser: A new analytical instrument for the individual, sequential, or collective measurement of the physical and chemical properties of a liquid, Rev. Sci. Instrum. 63(6), 216-227, 1992.
[6]   Mcmillan N.D., Riedel S.M., O'Rourke B., Hammond J., Doyle G., Murthagh F., Kököer M., Whyte N., O'Neill A., McMillan D.G.E., Beverley K., Augousti A., Mason J., Bertelsen H.S., Asbjørnsen S., Chemometrics and intelligent laboratory systems (Amsterdam: Elsevier), In Press.
[7]   McMillan, N. D., Lawlor, V., Baker, M., and Smith, S., (1998), Drops and Bubbles in Interfacial Research, D. Mobius and R. Miller (Ed's), Elsevier Press, 593.
[8]   *op.cit.,* Note 2.
[9]   Burgess C. and Frost T., Standards and best practice in absorption spectrometry, Blackwell Science, 214-215, 1999.

Ultrasonic tensiographic measurements on liquid drops for liquid fingerprinting

A Augousti[1], J Mason[1], H Morgan[1], and N D McMillan[2]

[1] School of Life Sciences, Kingston University, Surrey, KT1 2EE, UK
[2] Carlow Institute of Technology, Kilkenny Road, Carlow, Ireland

Abstract: A new tensiographic system modality based on the use of ultrasound has been developed. The ultrasonic tensiograph consist of a constant-head liquid reservoir that supplies liquid to a drop-head, where drops are formed. The drops are monitored using ultrasound during drop growth, and the signal that is produced, which is characteristic for a particular liquid, is termed a tensiogram when plotted. The system is interfaced to a PC, and it is controlled by a program written in Visual Basic. The system has been used to test a range of liquids, and results are reported here for samples of de-ionised water, sodium chloride solutions ranging from 0.001M to 0.1M and pure methanol. The results show that the system is capable of differentiating liquids easily, and that the resulting tensiogram is independent of the flow rate to within a factor of nearly 2.

Keywords: liquids, fingerprinting, drops, tensiography,

1. Introduction

The subject of tensiography, defined as the study of liquid characteristics based on the observation of liquid drops, has been pioneered over the last few years by the work of groups in Carlow Institute of Technology and Kingston University [1-7]. The work to date has concentrated on the use of two independent modalities, one using light to monitor drop growth [1,3,5,6], and more recently using capacitance [2,4,7]. This paper reports the development of a new modality based on the use of ultrasound, which extends the techniques reported previously, and which provides additional benefits in comparison with the two earlier modalities.

In common with the optical and capacitive modalities, the new system consists of a liquid delivery system, which can be either constant-head or pump-driven, delivering liquid to a drop-head where pendant drops are formed. In this instance the drop growth is monitored using two ultrasonic transducers, one for emission and one for detection, and the signal that is received during drop growth is recorded and displayed. The detailed construction and operation of the system is described below, and results on samples of de-ionised water, a range of concentrations of sodium chloride solutions, and pure methanol are presented here. Results have also been reported elsewhere on the capability of this system to distinguish between different lagers and admixtures of lagers following the application of advanced feature extraction algorithms [8,9].

2. Construction and operation

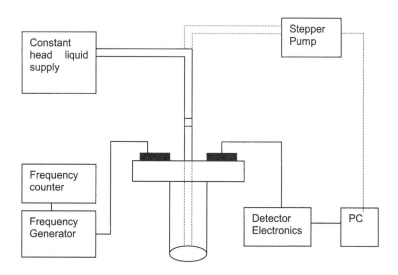

Figure 1: Experimental set-up of the ultrasonic system

The constant head reservoir consisted of a cylindrical copper crucible fitted with a hypodermic needle which connected to translucent vinyl tubing (ID of 1.00 mm and OD of 2.00 mm), which directed the liquid to the drop-head. The drop-head was also fitted with a second hypodermic needle, which coupled to the other end of the tubing. The height of the liquid level in the constant pressure head was used to control the flow rate of the liquid, and was chosen so that in the case of de-ionised water, for example, a typical lifetime for drop formation and disconnection was of the order of 180s. The liquid level depth was exactly 2cm in this case, and flow rate based on depths of 3cm and 4cm were also investigated. The stepper pump shown in the diagram here using dotted lines indicated that stepper-pump delivery of liquid under computer control has also been investigated, although those results are not presented here.

The temperature of liquid was monitored using a standard laboratory mercury-bulb thermometer positioned within the liquid. Two ultrasonic transducers are soldered on the upper surface of the probe head, which was constructed of brass, and the dimensions of the construction are presented in more detail in Figure 2.

The emitting transducer was connected to Farnell signal generator operating at approximately 521kHz and a Racal-Dana 9910 VHF frequency counter. The other was connected to a custom electronics interface to detect and amplify the signal amplitude. The signal phase relative to the driving signal may also be monitored, although those results are not presented here.

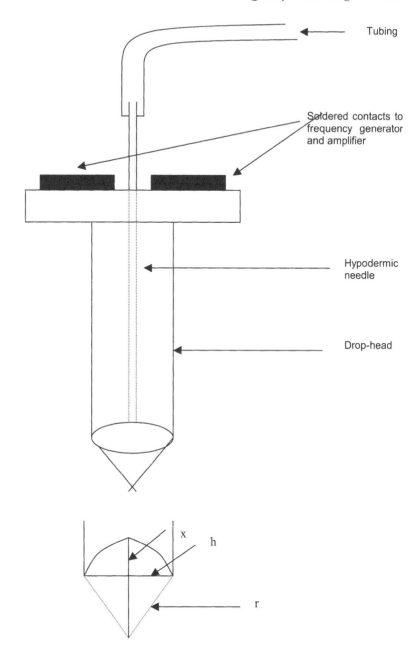

Figure 2: schematic diagram for the drop head x=2mm approx., h=9mm

3. Results and discussion

Sample results for de-ionised water are shown in Figure 3.

Deionised water sample No. 13 at height 4 cm date 26/02/03

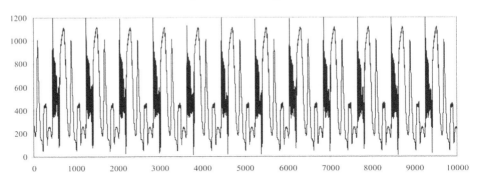

Figure 3. Sample tensiogram for de-ionised water. The x-axis scale is labelled with data point number, and corresponds to an overall time of approximately 184 seconds, and the y-axis scale is in arbitrary units.

The first point to note from the tensiogram is the excellent repeatability of the signal during drop formation. Although it is not clear at this scale, the repeatability in terms of the signal features is very high even at magnified scales. The tensiogram shows a very high modulation depth, indicating the existence of strong features, and the tensiogram for a single drop is feature-rich, pariculary when compared with those arising from the two previous other modalities. The optical modality tensiogram usually contains two peaks at most [1], and the capacitive modality tensiogram is usually monotonic [4].

In order to establish the dependence on flow rate, tensiograms were obtained at three different flow rates, corresponding to variations in flow rate of approximately 90%, i.e. within the same time interval of 184 seconds, 13, 10 and 7 drops were recorded respectively. A comparison of the tensiograms is shown in Figure 4. The x-axes of the three tensiograms have been scaled in order to compare the shape of the tensiogram for a single drop. It is evident that there is no significant variation in the shape over a wide variation of flow rate.

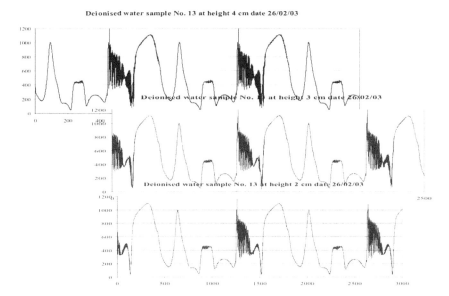

Figure 4 A comparison of the tensiograms for deionised water at different flow rates. The horizontal axes have been scaled so that the tensiograms for a single drop are easily compared.

Given the frequency of the ultrasound and the acoustic impedance of water at this frequency, one may calculate the wavelength of the acoustic waves, which is very close to 3mm, comparable to the dimensions of the drop itself. It also follows from considerations of the acoustic impedance of air (<0.0005) and water (*ca.* 1.5) that reflections at the drop boundary are very nearly 100%. The system is therefore operating as an acoustic interferometer with a complex transfer function, and a significant finesse, in contrast to the optical modality in particular, which is an amplitude-based incoherent wave system.

4. Conclusion

A new modality for tensiography has been developed. The system shows extreme sensitivity, and is useful for studying a range of liquids, such as opaque ones, that are inaccessible to the optical modality in particular. The resulting tensiograms are extremely feature-rich, and it has been demonstrated that the form of the tensiograms is largely flow-rate independent, at least for flow rates that are not so slow that Laplace forces arising from drop surface tensions become comparable to the pressure head producing the flow [7]. The system has also been applied to the characterisation of beverages such as lagers, and has been successful, when combined with advanced feature recognition techniques, at distinguishing lagers and admixtures of lagers [8,9].

The dependence of the form of the tensiogram on temperature (<0.1°C) and frequency (<5 Hz in 520kHz) excursions appears to be very high, as would be expected from an interferometric system. Currently tensiogram reproducibility remains an issue at least until accurate system temperature and frequency stability is established.

Acknowledgement:

The authors would like to thank the EU for support (EVK1-CT2000-00006) which was carried out at Kingston University.

References

[1]   The fibre drop analyser: a new multianalyser analytical instrument with applications in sugar processing and for the analysis of pure liquids **N D McMillan, O Finlayson, F Fortune, M Fingleton, D Daly, D Townsend, D D G McMillan and M J Dalton** *Meas. Sci. Technol.* **3** 746-764 1992

[2]   Capacitive Tensiography: A new modality for the fiber drop multianalyser **A.T. Augousti, J. Mason, ND McMillan and CH Wang** *Applied Optics and Optoelectronics 1998* **KTV Grattan (Ed)** p193-198 IOP Publishing 1998

[3]   The search for technological advantages and commercial success in sensor applications: lessons of industrial trials for fiber multianalyser technology **N.D.McMillan, M.Baker, M.O'Neill S.Smith, A.Augousti, J.Mason, B.Ryan and R.A. Ryan** (Process Monitoring using Fiber Optic Sensors, Boston, November, 1998

[4]   The Capacitive Drop Tensiometer – a novel multianalyzing technique for measuring the properties of liquids **CH Wang, AT Augousti, J Mason and N.D. McMillan** *Meas Sci Technol* **10(1)** p19-24 1999

[5]   A Hough Transform inspired technique for the rapid fingerprinting and conceptual archiving of Multianalyser tensiotraces **N.D. McMillan, S. Riedel, M.O'Neill, N. Whyte, J. McDonald, A. Augousti and J. Mason** p330-346 IMVIP 99 (Irish Machine Vision and Image Processing), 8-9th September 1999, Dublin, Ireland

[6]   The development of a commercial multivariate laboratory and process control measurement system with industrial and medical applications **N.D. McMillan, S.M. Riedel, A.T. Augousti and J. Mason** *Sensors and Their Applications XI* **K.T.V Grattan and S. H. Khan (Eds)** p191-196 IOP Publishing 2001

[7]   Investigation of capillary flow forces using capacitive tensiography **A.T. Augousti, J. Mason, N.D. McMillan and D. Zhang** *Sensors and Their Applications XI* **K.T.V Grattan and S. H. Khan (Eds)** p197-202 IOP Publishing 2001

[8]   Wavelet- and Entropy-Based Feature Extraction: Application to Distinguishing Mixtures of Beverages **M. Kököer, F. Murtagh, A. T. Augousti, J. Mason and N. D. McMillan** *Optical Metrology, Imaging and Machine Vision* SPIE 4877 Galway, Ireland 5-6th September 2002

[9]   A Wavelet, Fourier and PCA Data Analysis Pipeline: Application to Distinguishing Mixtures of Liquids **M. Kököer, F. Murtagh, N. D. McMillan, S. Riedel, B. O'Rourke, K. Beverly, A. T. Augousti, J. Mason and** *J Chem Inf and Comp Sci* http://pubs3.acs.org/acs/journals/toc.page?incoden= jcisd8&indecade=0&involume=0&inissue=0 ASAP Web release date: 23rd January 2003

A Dual Element Optical Fibre Water Contamination Sensor Utilising Artificial Neural Network Pattern Recognition Coupled With Digital Signal Processing Techniques

D King, W B Lyons, C Flanagan and E Lewis

ECE Dept., F1 Foundation Bldg., University of Limerick, Limerick, Ireland.

Abstract: A multipoint optical fibre sensor system is reported which is capable of detecting ethanol in water supplies. The sensor system comprises of two U-bend sensor elements, which are incorporated into a 1km length of polymer-clad silica (PCS) optical fibre. The sensor system is interrogated using Optical Time Domain Reflectometry, OTDR and it is proposed to apply artificial neural network pattern recognition to the OTDR output data to accurately classify the sensor test conditions. Signal processing techniques are applied to the resulting OTDR data with the aim of reducing the required ANN computational resources i.e. reducing the number of required nodes.

Keywords: Optical Fibre Sensor, U Bend, Measurement, Optical Time Domain Reflectometry, Discrete Fourier Transform, Pattern Recognition, Artificial Neural Networks.

1. Introduction

Optical fibre sensors possess a number of advantages over conventional electronic sensing techniques, which make them attractive for use in a wide range of application areas. These advantages include safety in chemically hostile environments, immunity to electromagnetic interference, electrically passive operation, sensitivity, weight and versatility. As a result of this, the use of optical fibre sensors for the purpose of environmental monitoring has expanded rapidly over the past decade [1-4]. In this investigation two sensors were incorporated into a 1Km length of 62.5 μm core polymer-clad silica (PCS) optical fibre. In order to maximize sensitivity, a U bend configuration was used for each sensor where the cladding was removed and the core exposed directly to the measurand.

The interrogation of such a sensor system uses a technique known as Optical Time Domain Reflectrometery, OTDR. OTDR is capable of detecting attenuation as a function of distance along the fibre and therefore is able to locate position and changes in the sensor signals along the fibre [5-6]. As a result of this, OTDR has found many applications in both single and multipoint sensors where the OTDR instrument is used to monitor the fluctuation in the optical fibre attenuation caused by an external parameter induced by a measurand.

Optical fibre sensor signals can often be complex and cross coupling of signals from external parameters e.g. temperature (the true measurand) and strain or microbending (interfering parameters in this case), adds to the difficulty interpreting data from such systems. It has been proposed that for many applications of optical fibre sensors, artificial neural network pattern recognition techniques may be used to resolve the problems arising from cross-sensitivity to other parameters [7].

Previous work by Lyons *et al* [8], in which data from a single U-bend optical fibre sensor was used initially to train a single layer perceptron and then a multi layer perceptron, and a comparative analysis of their results showed that a multi layer perceptron was required to adequately classify the data. This provided a means of validating the technique, which has been extended using signal processing techniques in this work.

The signal processing technique has been designed to optimise the artificial neural network adopted in the existing sensor system. In this investigation the Discrete Fourier Transform (DFT) using a Fast Fourier Transform (FFT) algorithm is used and it is shown that its application leads to an improvement in efficiency of the neural network i.e. minimising the computing resources. Initial investigations have reported that it is possible to train a neural network based on the frequency domain response of a single U-bend sensor element [9]. In this investigation, the sensor system has been expanded to include two U-bend sensor elements and the test conditions for both sensors in this investigation are combinations of air and 50% ethanol.

2. Experimental Setup

In this investigation the sensors were incorporated into a 1Km length of 62.5 μm core polymer-clad silica (PCS) optical fibre. In order to maximise sensitivity, a U bend configuration was used for the sensor where the cladding was removed and the core exposed directly to the measurand.

In order to minimise optical power loss, the sensors were fabricated from the same fibre as that used to transmit light through the system (i.e. 62.5μm PCS), removing the need to splice the sensor into the transmitting fibre and the resulting power losses.

The operation of the sensors is based on the modulation of the light intensity propagating in the fibre by the measurand as a result of the interaction with the evanescent field penetrating into the absorbing measurand. Much experimental work has already been reported [10-11] for a single U-bend sensor detailing resulting sensitivity gains from evanescent wave increases from the curving of the sensing fibre [12]. It has been shown by *Gupta et al* that the sensitivity of the sensor increases with decreasing bend radius of the probe and also with the increase in refractive index of the fluid under test [10-11].

3. U-Bend Sensor Fabrication

In order to produce the U-bend in the fibre, the buffer and cladding were chemically removed from a 2cm length section located at 665m and 756m from the launch end of the fibre. In order to shape the fibre to the desired sensor configuration, the exposed fibre was cleaned using acetone and was then slowly bent into a U shape using heat from a flame. The bending procedure was controlled using an in house developed fixture to improve the repeatability of the sensor manufacturing and hence improve the reproducibility characteristics between successive sensors. The final bend radii were measured to be 1mm using a conventional optical laboratory microscope. A schematic and a photo of a manufactured sensor are shown in figure 1.

4. Measurement System Configuration

The system configuration for this investigation is shown in figure 2. It comprises of the optical fibre sensors, an EXFO IQ7000 (0.85μm) OTDR and a Pentium MMX 200 MHz PC running LabVIEW [13] Virtual Instrument (VI) programs for data capture and pre-processing. The LabVIEW VI programs were developed by the authors specifically for this investigation, and the resulting data output were made available for analysis by the signal processing technique and the artificial neural network implemented using SNNS V3.2 (Stuttgart Neural Network Simulator) [14].

5. Results

In order to train and test an Artificial Neural Network (ANN) pattern recognition system, it is necessary to obtain a large number of patterns. For this reason, numerous OTDR readings were taken for each of the sensing conditions under test as shown in table 1. In order to allow the sensor to fully stabilize, the first OTDR reading taken for each test condition is ignored and the remaining fifty patterns are used in the training and testing of the artificial neural network implemented in SNNS.

The sensing area of interest on the OTDR trace forms a relatively small part of the overall trace (approx. 256 points out of a total of 12,000, see figure 3) and therefore to maximise efficiency of the computer algorithm, it was necessary to design an in-house LabVIEW VI that would locate the sensor peak, select the required window width and save this data for analysis by the signal processing and the SNNS software. A typical output measurement trace obtained from the OTDR when the sensors are exposed to combinations of air and 50% ethanol is shown in figure 4.

Table 1: Sensor Test Conditions Investigated

Sensor 1	Sensor 2	No. Of OTDR Readings Taken
Air	Air	51
Air	50% Ethanol	51
50% Ethanol	Air	51
50% Ethanol	50% Ethanol	51

6. Signal Processing Analysis

Previous work by *Lyons et al* [15] has shown it is possible to train a multi layer perceptron using the data obtained from a U bend sensor interrogated by an OTDR. In this investigation it has been proposed to apply a digital signal processing technique to the obtained sensor data, prior to the training of the artificial neural network, with the aim of reducing the computational resources of the implemented artificial neural network, i.e. fewer nodes required in the input and hidden layers.

It can be observed from the extracted OTDR peaks shown in figure 4, that it is relatively low frequency information that is of interest in this classification. Due to the low frequency nature of the information, a discrete Fourier transform (DFT) using an FFT algorithm can be directly applied to the OTDR peaks without having to apply any windowing transform. The low frequency nature of the information also reduces concerns over filtering and aliasing in the frequency domain [16].

Prior to application to the signal processing technique, the OTDR sensor data is normalised between +1 and –1 using the standard LabVIEW Scale 2D array VI. The signal processing analysis in this investigation is performed using an in house developed MATLAB program. Once the extracted OTDR sensor peaks are inputted into MATLAB, a 256-point DFT of the peak is calculated using an FFT algorithm and from the resulting Fourier transform the power spectral density (PSD) of the OTDR output is calculated.

The resulting PSD plots are shown in figure 5. As anticipated, the main PSD area of interest is located in the low frequency region. As a result of the application of the discrete Fourier transform, the OTDR peak information is now more explicit and easier for the user to access in comparison to time-domain based results which require all of the extracted OTDR peak data points. An empirical decision was made to select the first six points of the PSD plot as the main area of interest on the trace and these six points form the input layer of the ANN implemented in SNNS.

7. Artificial Neural Network Analysis

Using SNNS, a three layer feed forward neural network was implemented, figure 6, consisting of a six node input layer, a three node hidden layer and a four node output layer. The latter includes one node to represent each sensor test condition. To determine the optimal size of the hidden layer used, multiple trials using hidden layers consisting of one node up to five nodes were performed. In this investigation, a hidden layer of three nodes was found to perform best. The use of a feed forward network is a robust way of building a non-parametric classifier.

The application of the discrete Fourier transform to the extracted output peak from the OTDR has achieved it aim as it has significantly reduced the computational resources of the artificial neural network in comparison to time domain based results. The network resources required with the DFT applied were an input layer of six nodes, a hidden layer of three nodes and an output layer of four nodes in comparison to previous work by *Lyons et al* which required ANN resources of (i) an input layer of thirty one nodes, a hidden layer of five nodes and an output layer of four nodes [17] and (ii) an input layer of 59 nodes, a hidden layer of ten nodes and an output layer of five nodes [18] based on time domain OTDR results for a single U bend sensor.

A total of 160 result patterns were used to train the network – 40 for each condition listed in Table 1. For training the feed-forward network, 500 cycles were required using a back-propagation algorithm. The algorithm used in this classification is listed in the SNNS learning functions as *BackPropMomentum*. The network was initialised with randomised weights and trained with a "topological order" update function [14].

In order to test the operation of the trained network, an independent set of data to that which had been applied in the training of the network was used. This was generated from the remaining unused 10 result patterns for each condition listed in table 1. The resulting 40 patterns were applied to the trained network and were all classified correctly. Table 2 shows a sample of the results obtained when the test pattern was applied to the trained network.

Table 2: ANN Analysis Test Results

Test Conditions		Expected Results				Observed Results			
Air	Air	0	0	0	**1**	0.010	0.001	0.088	**0.901**
Air	50% Ethanol	0	0	**1**	0	0.096	0.000	**0.904**	0.000
50% Ethanol	Air	0	**1**	0	0	0.036	**0.964**	0.000	0.000
50% Ethanol	50% Ethanol	**1**	0	0	0	**0.911**	0.087	0.000	0.002

From the results obtained in table 2, it can be seen that the application of the discrete Fourier transform has achieved its aim. The computational resources of the ANN have been significantly reduced by the application of the DFT, i.e. fewer nodes used, in comparison to previous work by *Lyons et al* [17-18] without affecting the accuracy of the ANN's classifications. The ANN, using the frequency domain response of the sensor as its input layer, accurately classifies all four of the sensor's test conditions. Based on the results shown in table 2, the example of the ANN shown in figure 6 shows sensor 1 detecting 50% ethanol and sensor 2 detecting air.

8. Conclusion

A reliable measurement system using optical fibre sensors and OTDR interrogation for fluid detection has been presented. A dual element U-bend optical fibre sensor system, on a 1km continuous length of fibre, has been investigated and proven to be capable of detecting the presence of combinations of air and 50% ethanol, at the sensing points of interest using OTDR techniques. The U-bend evanescent wave absorption sensors were developed with 62.5μm polymer-clad silica fibre, which had its cladding removed in the sensing region. Although the length of the fibre used in this investigation was 1 km, longer or shorter lengths may be used as required.

The application of a discrete Fourier transform, utilizing an FFT algorithm, to the extracted OTDR peaks has resulted in a significant reduction in the artificial neural network resources in comparison to previous work by *Lyons et al* [17-18]. Results have shown that it is possible to train an artificial neural network to recognize the frequency domain response of exposure to air and immersion in 50% ethanol.

Artificial neural network techniques have allowed the resulting OTDR signals to be accurately determined using pattern recognition. The ANN implemented in SNNS successfully classified each sensor test condition correctly based on the frequency domain response of the sensors. Due to the application of the Fourier transform to the OTDR peaks, refinements have been made to the resources used by the feed forward ANN, with a result of a reduction in the training time of the ANN.

Figure 1: Fabricated U-Bend Optical Fibre Sensor

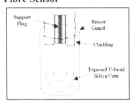

Figure 2: Measurement System Configuration

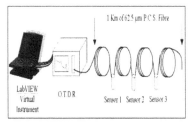

Figure 3: OTDR Output Trace

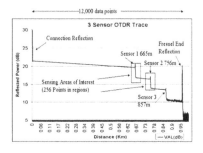

Figure 4: Extracted OTDR Peaks

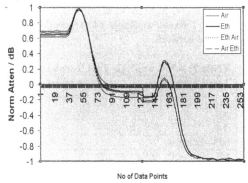

Figure 5: PSD Traces

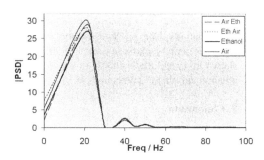

Figure 6: ANN Implemented in SNNS.

References

[1] Kersey A.D., "A Review of Recent Developments in Fiber Optic Sensor Technology", *Optical Fiber Technology* **Vol2**, pp291-317 1996.
[2] Udd E., "An Overview of Fiber Optic Sensors", *Rev.Sci.Instrum* **66(8)**, pp4015-4030 1995.
[3] Grattan K.T.V., Sun T., "Fiber Optic Sensor Technology: An Overview", *Sensors and Actuators A* **Vol. 82**, pp40-61, 2000.
[4] Kersey A.D., Dandridge A., "Applications of Fiber-Optic Sensors", *IEEE Transactions on Components, Hybrids and Manufacturing Technology*, **Vol. 13(1)**, pp137-143, 1990.
[5] Barnoski, M.K., Jensen S.N., "Fiber Waveguides: a Novel Technique for Investigating Attenuation Characteristics", *Applied Optics* **15**, pp2112-2115, 1976.
[6] Neumann E.G., "Optical Time Domain Reflectometer: Comment", *Applied Optics* **17(11)**, pp1675, 1978.
[7] Lyons W.B., Lewis E., "Neural networks and pattern recognition techniques applied to optical fibre sensors" *Transactions of the Institute of Measurement and Control* **22 (5)** pp385-404 2000a.
[8] Lyons W.B., Ewald H., Lewis E., "An Optical Fibre Distributed Sensor Based on Pattern Recognition", *ICPR-15: Manufacturing for a Global Market*, Limerick, Ireland August 1996b.
[9] King D., Flanagan C., Lyons W.B. and Lewis E. "An Optical Fibre Water Sensor Utilizing Signal Processing Techniques and Artificial Neural Network Pattern Recognition", *Proc. IEEE International Conference on Sensors (IEEE) Sensors*, **1 Vol. II**, Orlando, Fl., USA, pp1374-1378 2002.
[10] Gupta, B.D., et al, "Fibre optic evanescent field absorption sensor based on a U-shaped probe" *Optical and Quantum Electronics,* **28**, pp1629-1639 1996.
[11] Khijwania, S.K., Gupta B.D., "Maximum achievable sensitivity of the fibre optic evanescent field absorption sensor based on a U-shape probe" *Optics Communications,* **175**, pp135-137 2000.
[12] Soichi Otsuki et al, "A novel fibre optic gas-sensing configuration using extremely curbed optical fibres and an attempt for optical humidity detection" *Sensors and Actuators* **B53**, pp91-96 1998.
[13] LabVIEW Function and VI Reference Manual, P/N 321526B-01, Jan 1998.
[14] Stuggart Neural Network Simulator (SNNS), User Manual, Version 4.1, Report No. 6/95, 1995. http://www-ra.informatik.uni-tuebingen.de/SNNS/
[15] Lyons W.B., Lewis E., Ewald H., "An Optical Fibre Multipoint U Bend Sensor Based on Pattern Recognition", *Proc Irish Signals & Systems Conference (ISSC) 2000, UCD, Dublin, Ireland*, pp412-417 2000.
[16] Lynn P.A., Fuerst W., "Introductory Digital Signal Processing", John Wiley and Sons, July 1994.
[17] Lyons W.B., Ewald H., Flanagan C., and Lewis E., "An Optical Fibre Multipoint U Bend Sensor Based On Artificial Neural Network Pattern Recognition" *Proc Artificial Neural Networks in Engineering Conference 2000 (ANNIE 2000)*, St. Louis, USA, pp663-670 2000.
[18] Lyons W.B., Ewald H, Flanagan C, Lochmann S, Lewis E, "A Neural Network Based Approach for Determining Fouling of Multipoint Optical Fibre Sensors in Water Systems", *Meas.Sci.Techol* **12**, pp958-965 2001.

Paper presented at Sensors and their Applications XII, September 2003

Novel NIR Water Sensor based on a Doped Glass Bead

John R Donohue[1], Hugh J Masterson[2], Kieran O'Dwyer[1], Gwenael Mazé[3], Brian D MacCraith[1]

[1] School of Physical Sciences, National Centre for Sensor Research, Dublin City University, Glasnevin, Dublin 11, Ireland
[2] Boulder Nonlinear Systems, Lafayette, CO 80026, USA
[3] Le Verre Fluoré, Campus Ker Lann, F-35 170 Bruz, Brittany, France

Abstract: The detection and measurement of vapour-phase or liquid-phase water is important in many industrial and chemical processes. Water exhibits strong absorption bands compared to other substances in the near infrared (NIR), and for this reason NIR spectroscopy is especially well suited to moisture determination.

We have developed a modulatable IR source for use in a moisture sensor. In the system, the luminescent emission from optically pumped rare earth doped glasses is used.

Results illustrating the effectiveness of the novel IR source in a sensor platform to measure trace amounts of water vapour are presented.

1. Introduction

A lack of suitable sources in the NIR has impeded the application of optical sensors to water detection. Many systems rely on the use of incandescent sources in conjunction with narrow bandpass filters to produce emission bands corresponding to the absorption bands of water. Disadvantages associated with this type of system include the temperature dependence of the filters and limited lifetime of the incandescent source. In addition, such sources can pose a fire hazard in environments where a need for moisture detection exists, such as in gas mains.

LED sources that are available for water or humidity sensing are generally unstable, have insufficient output power, and are not well matched spectrally to the relevant absorption bands of water. Recent studies have shown that rare earth doped glasses are particularly suitable as alternative sources of NIR radiation for moisture detection[1 - 4]. Thulium doped zirconium fluoride glass ($ZBLAN:Tm^{3+}$) was the material chosen as the basis of the development of a moisture sensor. The material fluoresces over a range of wavelengths which overlaps significantly with the main NIR absorption band of water.

Design considerations for the sensor have centred around low power consumption (the use of a laser diode for the pump source), stability in detecting small levels of moisture (employing drift referencing using a dual detector approach), and simplicity of construction. The design and construction of an industrial prototype instrument are discussed and results are presented.

2. Source Development

ZBLAN:Tm^{3+} glass was chosen for development of the sensor based on its luminescence properties. Moisture detection is achieved by monitoring the optical absorption of NIR radiation by the 1.93μm water absorption band. A spherical bead of ZBLAN:Tm^{3+} is used as the IR fluorescence can be more easily collimated than with an arbitrary shaped piece of doped glass. The glass bead is pumped by a 60mW 685nm laser diode, which coincides with a strong absorption line of the doped glass. Figure 1 shows the optical absorption spectrum of the bead. Figure 2 shows the emitted NIR radiation centred at 1.83 μm, superimposed on the NIR absorption spectrum of water. The emitted radiation corresponds to the $^{3}F_{4}{\rightarrow}^{3}H_{6}$ crystal field transition of the thulium in the ZBLAN glass matrix. Although the optimum pumping wavelength is at 685nm, pumping may also be achieved at other more commercially available wavelengths, in particular, 670nm, and 780nm where the absorption, though weaker, is significant.

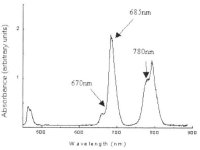

Figure 1. Absorption spectrum of ZBLAN: Tm^{3+} glass.

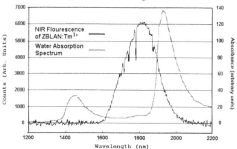

Figure 2. NIR fluorescence of ZBLAN:Tm^{3+} glass when pumped with 685nm diode laser. The absorption spectrum of water is superimposed.

3. Sensor Configuration

An industrial prototype for moisture detection has been designed and constructed for field use, based on the principle of matching the doped glass fluorescence with the NIR absorption band of water. The construction of the prototype was modelled using TracePro, an advanced ray-tracing package, and the dimensions and characteristics of the optical components were chosen accordingly. The tolerances, however, are very wide, given the nature of the emitted IR energy, which is quasi-collimated.

Depending on the application, the system either operates by direct transmission through a sample liquid cell for water detection in a liquid, by direct transmission through a gas flow cell for relative humidity detection in gases, or by reflection from a moisture bearing surface. The sensor uses a reference beam and a probe beam to monitor changes in absorption due to moisture or water vapour. Measurement of the presence of moisture or water vapour is achieved by comparing the reference and probe wavelengths simultaneously (685nm and 1.83μm, respectively).

Optimisation of the system involved reduction of the linear dimension of the sensor as much as possible to ensure maximum NIR energy arriving at the detector, and to minimise the effects of beam divergence.

The system employs dual detector referencing with a Cal Sensors lead sulphide (PbS) quad detector. Although the detector has four sensing elements (each with independent output channels) only two channels are required. IR-transparent anti-reflection coated silicon (LWP-1000, transmittance > 90% in the range 1.80 – 2.60μm, zero transmittance at wavelengths below 1.0μm) was installed over two detecting elements, while IR-opaque KG-3 glass (transmittance of 60% at 685nm, <1% at 1.85μm) was installed over the other two elements. The filters ensure that the NIR probe beam can be detected without interference from the visible reference beam, and vice versa, using the single detector. As the PbS quad detector elements were manufactured from the same die, mutual detector drift is minimised.

The prototype is illustrated in Figure 3. A cylindrical housing covers the electronics situated on the bottom side of a platform while the top side serves as an optical bench for the laser diode, ZBLAN: Tm^{3+} bead and parabolic reflector, and the PbS quad detector.

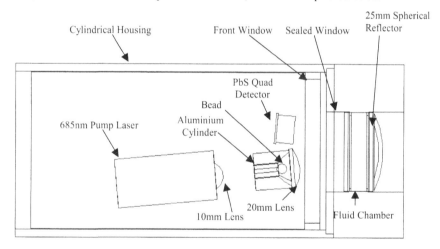

Figure 3. Schematic view of the industrial prototype with the liquid head attached.

Moisture bearing samples are interrogated by reflection through a fluid chamber: a 10mm gap in the case of liquids and a 50mm cavity in the case of gases. Two interchangeable heads have been constructed to facilitate measurement in each of these cases. Behind the fluid chamber in each case is situated a mirror, which is spherical to aid in refocusing the IR radiation back into the detector. The liquid head employs a 25mm focal length mirror while the gas head employs a 50mm focal length mirror. An f = 20mm lens is employed in front of the parabolic reflector containing the ZBLAN: Tm^{3+} bead to counter the tendency of the beam to diverge and an f = 10mm lens is attached to the laser diode aperture. The latter element is used to focus the laser diode light into the bead through a 2mm diameter opening in a small aluminium cylinder which slides into the back of the reflector. Maximum luminescence occurs when the laser diode energy is focused inside the bead. The laser diode-lens combination is therefore positioned at a precise distance from the reflector bead combination.

The bead itself is epoxied to the front of the small cylinder, which is situated to position the bead at the focus of the parabolic reflector. A 5mm diameter bead doped with 12% thulium is used in this case and the reflector is a Carley Lamps parabolic reflector of 19 mm diameter (reference no RF 1936). This source optics linear combination is situated on the platform close to the front window of the housing, such that a parallel beam emitted from the reflector is incident at 5° to the central axis of the back reflecting spherical mirror. The detector is situated at 5° from the central axis on the side opposite the source optics. The reference beam follows the same path as the probe beam.

4. Results and Discussion

The performance of the industrial prototype as a humidity sensor for gases was investigated. The presence of water vapour in gas was measured through the use of the Gas Flow Sensor Head, to which nitrogen was introduced with a known moisture content referenced against dry nitrogen. The apparatus and experimental set-up used to characterise the response of the sensor when measuring the presence of water vapour in nitrogen is shown in Figure 4. A mass flow controller (MFC 1) allows nitrogen (water content <5ppm) to be fed into a humidifer, while another mass flow controller (MFC 2) facilitates mixing of set amounts of the humidified nitrogen (Wet N2) and the dry nitrogen (Dry N2). The moisture content inside the Gas Flow Sensor Head is therefore strictly controlled.

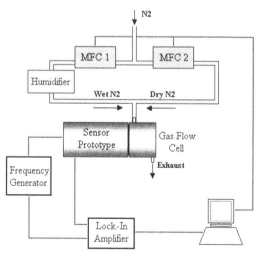

Figure 4. Experimental set-up used for measuring the response of the laboratory-based prototype and the industrial prototype by measuring the relative humidity of a gas in a flow cell.

Figure 5(a) illustrates the response of the sensor to relative humidity (RH) levels of 0% and 100% in nitrogen. Aside from a short settling time, the sensor exhibits excellent repeatability in sensing humidity levels. Note that the negative y-axis is a result of the electronic configuration of the detector circuitry, and is not significant. The very small drift with time at 0% relative humidity is due to temperature effects on the PbS dual detector. The dual detector has an in-built single-stage Peltier cooler, enabling it to be thermoelectrically cooled to temperatures as low as –15° Celsius. At such low temperatures, the sensitivity of the detector is significantly increased, thus enabling it to monitor smaller changes in the IR Probe signal than would be possible were the detector at room temperature. However, this results in the detector being very sensitive to changes in ambient temperature in its surrounding environment. In addition, any tiny fluctuation or drift in the thermoelectric cooler power supply has a knock-on effect on the response of the detector. A built-in thermistor allows the temperature of the detector to be monitored. Small changes in the resistance of the thermistor were observed during the experiment, corresponding to small variations in the detector temperature.

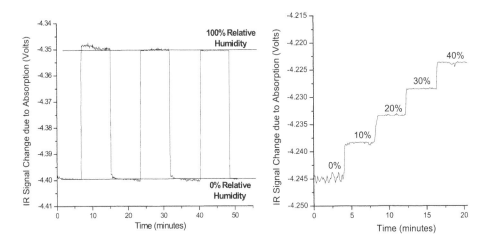

Figure 5(a). IR absorption due to the presence of water vapour in nitrogen.

Figure 5(b). IR absorption due to changing moisture content of nitrogen (0 – 40%).

In Figure 5(b), the response of the sensor to small humidity changes in nitrogen can be seen. The IR signal decreases as the humidity level is increased in steps of 10%, corresponding to the increasing NIR absorption. The temperature of the PbS dual detector is monitored with the in-built thermistor. In spite of some initial instability (corresponding to small temperature fluctuations of the detector), the ability of the sensor to monitor small relative humidity changes in nitrogen is apparent. Currently, the lower limit of resolution is 2%, due to the minimum flow control of the MFCs.

It is noted that the IR probe signal is sufficient to provide stable moisture readings without the need for referencing it with the reference signal. The reference beam is not utilised, as it is not fully optimised with respect to the IR probe beam at present.

It is clear from the preliminary results that the sensor has the capability of detecting moisture levels in gas down to at least 2% RH. Further work needs to be done to minimise the temperature effects. The industrial prototype system will also be used to detect moisture in solvents.

5. Conclusion

We have shown that thulium-doped zirconium fluoride glass can be used as a novel NIR source. When used in conjunction with a 685nm diode laser as a pump source, the

ZBLAN:Tm^{3+} glass bead emits a modulated and stable signal that is very suitable for moisture sensing. It has been shown that the ZBLAN:Tm^{3+} can be incorporated into a modular sensor platform and that gaseous moisture content can be monitored. Using the industrial prototype, full-scale (0 – 100%) RH levels in nitrogen can be measured repeatedly, as can smaller levels in the range 0 – 40%. The resolution of the sensor is currently limited by the mass flow control system to 2% RH.

Other applications for moisture sensing that we will look into in the near future include a study of the suitability of the sensor in monitoring trace amounts of water in solid surfaces by diffuse reflection. This is an important avenue of investigation for us, as the development of a non-contact in-situ online moisture detector would prove beneficial to certain industries, e.g. paper manufacturers. The capabilities of the industrial prototype as a sensor for moisture in liquids will also be investigated.

Future work in the further development of the sensing system itself will involve a study of how alternative pumping wavelengths may be used to pump the doped glass (e.g. a 670nm diode laser is less expensive and more readily available than a 685nm diode laser). Also of interest is how the use of other fluorescent sources, such as holmium, may be used to tune the NIR fluorescence of the doped glass bead to improve the spectral match between the fluorescence and the absorption spectrum of water. Finally, there is the issue of eliminating temperature-induced effects on the PbS detector itself. One proposed solution is the incorporation of a PID controller into the sensor, using the thermistor on the PbS dual detector in a feedback role.

Acknowledgements

The authors would like to thank the staff of the Optical Sensors Laboratory at the National Centre for Sensor Research in Dublin City University for all their help and assistance. We would also like to thank Gwenael Mazé (Le Verre Fluoré, Campus Ker Lann, F-35 170 Bruz, Brittany, France) for supplying us with thulium-doped ZBLAN glasses. Finally we would like to thank Enterprise Ireland, who supported this work under contract PRP / 00 / OPT / 08.

References

[1] F. McAleavy, J. O'Gorman, J.F. Donegan, B.D. MacCraith, J. Hegarty, G. Mazé, "Narrow Linewidth, Tunable Tm^{3+}-Doped Fluoride Fibre Laser for Optical-Based Hydrocarbon Gas Sensing", *IEEE Journal of Selected Topics in Quantum Electronics*, **3** (4) 1103-1111, 1997.

[2] F. McAleavy, B.D. MacCraith, "Diode-pumped thulium-doped zirconium fluoride fibre as a fluorescent source for water sensing", *Electronic Letters* **31** 1379-1380, 1995.

[3] M. Yokota, T. Yoshino, "Optical fibre water droplet sensor using absorption of fluorescence of Tm^{3+}:YAG", *Applied Optics* **37** 2526-2533, 1998.

[4] M. Yokota, T. Yoshino, "An optical-fibre water-concentration sensor using Tm^{3+}:YAG fluorescent light", *Measurement Science Technology* **11** 152-156, 2000.

Section 10: Water Quality Monitoring
Paper presented at Sensors and their Applications XII, September 2003
©2003 IOP Publishing Ltd

A New n-Layer mRDG Ellipsoidal Model of Light Scattering for Rapid Monitoring Tests of Coliforms in Potable Water Samples

G Chliveros‡, M A Rodrigues, and D Cooper

Computer Vision Pattern Recognition & Artificial Intelligence Group, Computing Research Centre, Sheffield Hallam University, Howard Street, Sheffield S1 1WB, UK

Abstract. This paper presents a new model for fast calculation of light scattering by inhomogeneous ellipsoids with multiple layers. Explicit expressions for the scattering amplitude are derived using the modified Rayleigh-Debye approximation. The suitability of the model for the identification of coliforms in water based environments is tested through simulation algorithms in which scattered light intensity is estimated from asymmetric cell populations. Sensitivity analysis is performed on the ratio of the scattering matrix elements regarding changes in axial ratio and appearance of additional layers. It is shown that the monitoring of scattering ratio and local maxima allows the identification of a sample's geometrical morphology and internal structure, leading thus, to fast monitoring of coliforms in water samples.

Keywords: Optical Sensors, Light Scattering, mRDG, Water Quality Sensing, Bacteria

1. Introduction

Methods for rapid estimation of bacteriological content in water samples is a recognised need of the water industry. The principal indicator of potable water suitability is the coliform group (genera *Enterobacter*, *Citrobacter*, *Klebsiella* and *Escherichia*) as their presence renders unsuitable and potentially unsafe use. European standards require that 95% of potable water samples contain no coliforms and no samples should contain faecal coliforms. Thus, rapid detection of coliforms would call for further investigation to determine whether faecal coliform is present or not. While there is considerable interest in the analysis of wastewater, more emphasis is placed on rapid analysis of potable water due to its implications to public health. Also, the monitoring of coliforms can provide trend estimate of many other pathogenic bacteria.

It has been indicated that many techniques based on laser light scattering are sensitive and fast enough to permit rapid detection and can be brought to comply with international standards [4]. Optical data obtained from a circular array of photo detectors, Figure 1, can be used to characterise suspended cells. However, any proposed computer algorithm must not significantly add in time to the detection process. This means that the required computational time has to be within the bounds of what is considered to be a rapid test. Computational methods such as the Discrete Dipole Approximation have shown some success [2] but the computational time requirements for scattering calculations may exceed the bounds of rapid testing. Earlier methods based on simpler theories such as the Rayleigh-Debye-Gans approximation (RDG) have been investigated [11] and are still used for rapid monitoring [5]. However, due to the constraints of the method its validity is questionable for sizes greater than $1.2\mu m$. To overcome this limitation, several extensions and modifications to the RDG theory have been proposed [9]. An interesting application of the modified RDG (mRDG) is the characterisation of blood cells by generalising from a homogeneous sphere to a radially inhomogeneous two layer model [10].

‡ To whom correspondence should be addressed (g.chliveros@shu.ac.uk)

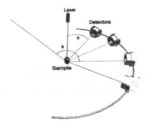

Figure 1. Optical sensors' configuration for light scattering measurements.

In this paper, using the mRDG expansion we derive and extend the theory for an ellipsoidal cell with an arbitrary number of layers. The new proposed model assumes particle's symmetrical coatings with corresponding thickness and relative refractive indices. We also investigate the effects of multiple layers on the ratio of the Müeller Matrix elements, S_{33}/S_{11}, and propose a method for identifying ellipsoids as opposed to spheroids.

2. The n-Layer Ellipsoidal Model

Under the physical basis of RDG scattering a particle can be divided into a collection of points of infinitely small volume that scatter independently as long as $|1 - m| \ll 1$, where m is the relative refractive index. This condition has been re-evaluated in several publications (e.g. [6]) and the general consensus is that if one accepts for a 10% error in the results then $|1 - m| < 1$. This condition is valid for waterborne bacteria since 70-86% of their content is water ([8], p24). In the picture of Figure 1, photodetectors are in the same plane which is defined from the origin by an angle ϕ while θ is the scattering angle. As such, for a particle of arbitrary shape an element of volume dV scatters with amplitude $dS_\perp$ for perpendicularly polarised incident light so that

$$dS_\perp(\theta, \phi) = \frac{jk^3}{2\pi} \left(m(r, \theta, \phi) - 1 \right) e^{j\delta} dV \tag{1}$$

where the phase difference δ within the volume elements is measured with respect to a fixed origin of r at the scattering observation (θ, ϕ). In Equation (1), $S_\perp(\theta, \phi)$ is a complex number and j denotes $\sqrt{-1}$, whilst k is the propagation constant in the water medium and m denotes the relative refractive index. By assuming that the particle's region consists of small slices perpendicular to the bisectrix (v) of the scattering observation's supplement, then the total scattering amplitude of the particle can be found by integration over the volume elements of these slices. Hence,

$$S_\perp(\theta, \phi) = \frac{jk^3}{2\pi} \int_{-v}^{v} \left(m(v) - 1 \right) A(v) \exp(j\delta) dp \tag{2}$$

where A is the area of the cross section on the bisectrix at v. It has be shown that under the RDG conditions $\delta = 2kp \sin(\theta/2)$, where p is the projection of r on the bisectrix. However, it is assumed that the applied field inside the particle equals that in the medium. Hence, the propagation constant in and out of the particle's region is unchanged. In [9] the RDG applicability is extended by accommodating for the contributions resulting from the field inside the particle. As a result, the particle's refractive index is taken into account so that now k is replaced by $km(v)$ resulting in $\delta = 2km(v)p \sin(\theta/2)$. Equation (2) becomes

$$S_\perp(\theta, \phi) = \frac{jk^3}{2\pi} \int_{-v}^{v} \left(m(v) - 1 \right) A(v) \exp(j2km(v)p \sin(\theta/2)) dp \tag{3}$$

In several microbiology manuals coliforms and in particular *E.coli* are described as rod-shaped bacteria. However, electron microscopy images reveal an almost ellipsoidal like shape. Thus, adopting this as a general structure with semi-axes a, b and c, in cartesian coordinates

$$\left(\frac{x}{a}\right)^2 + \left(\frac{y}{b}\right)^2 + \left(\frac{z}{c}\right)^2 = 1 \tag{4}$$

Let the direction cosines of the bisectrix be l, m and n. It can be shown that for a plane normal to this direction and tangent to the ellipsoid at a point on its surface, say (x_1, y_1, z_1), then

$$\frac{xx_1}{a^2} + \frac{yy_1}{b^2} + \frac{zz_1}{c^2} = 1 \quad \text{and} \quad lx + my + nz = v \tag{5}$$

Equations in 5 are equivalent. It follows that $x_1 = la^2/v$, $y_1 = mb^2/v$ and $z_1 = nc^2/v$. Therefore substituting into Equation (4) results in

$$v = \sqrt{(la)^2 + (mb)^2 + (nc)^2} \tag{6}$$

The area of a slice as defined in Equation (2) will be $A = \pi abc(1 - p^2/v^2)/v$, where in the current context p is the distance from the center of the ellipsoid to this plane measured along the bisectrix. It follows from Equation (3) that

$$
\begin{aligned}
|S_\perp(\theta, \phi)| &= \frac{k^3}{2\pi v} \int_{-v}^{v} (m(v) - 1) \left(\pi abc(1 - \frac{p^2}{v^2})\right) \exp(j2km(v)p\sin(\theta/2))dp \\
&= \frac{k^3 abc}{2v} \int_{-v}^{v} (m(v) - 1)(1 - \frac{p^2}{v^2}) \exp(2jkm(v)p\sin(\theta/2))dp
\end{aligned} \tag{7}
$$

The integral of Equation (7) would be evaluated over each of the homogeneous regions $[-v_1, v_1]$ to $[v_{n-1}, v_n]$. That is assuming n symmetrically placed coatings. For the internal part of the cell and substituting by $u_1 = 2km_1v_1\sin(\theta/2)$, $w = p/v$ we have that

$$
\begin{aligned}
[|S_\perp(\theta, \phi)|]_{-v_1}^{v_1} &= \frac{k^3 abc}{2}(m_1 - 1) \int_{-1}^{1} (1 - w^2) \exp(ju_1 w)dw \\
&= k^3 abc(m_1 - 1) \int_{0}^{1} (1 - w^2) \cos(u_1 w)dw \\
&= k^3 abc(m_1 - 1) \frac{\sqrt{\pi u_1}}{u_1^2} J_{3/2}(u_1)
\end{aligned} \tag{8}
$$

where $J_{3/2}$ is the Bessel function of order $3/2$. The remaining terms that need to be evaluated in Equation (7), correspond to each of the n-layers. Following the reasoning of the RDG physical basis but for the cell's region described by the mRDG, a layer can be thought of as a collection of elements that will scatter light independent of any other collection of elements of another layer. It is therefore a natural consequence that the kth layer of the cell will scatter light proportional to that of a homogeneous ellipsoid of corresponding v_k and m_k, by subtraction of the contribution arising from a $(k - 1)$ homogeneous ellipsoid of corresponding v_{k-1} but having the same refractive index (m_k). Bearing this in mind and for the kth layer we write:

$$
\begin{aligned}
\frac{1}{2}[|S_\perp(\theta, \phi)|]_{v_{k-1}}^{v_k} &= [|S_\perp(\theta, \phi)|]_0^{v_k} - [|S_\perp(\theta, \phi)|]_0^{v_{k-1}} \\
&= k^3 abc(m_k - 1) \left(\frac{\sqrt{\pi u_{k,k}}}{u_{k,k}^2} J_{3/2}(u_{k,k}) - \frac{\sqrt{\pi u_{k,k-1}}}{u_{k,k-1}^2} J_{3/2}(u_{k,k-1}) \right)
\end{aligned} \tag{9}
$$

where $u_{i,\ell} = 2km_iv_\ell\sin(\theta/2)$. Generalising for an n-layered ellipsoid cell using Equation (9), Equation (7) will become

$$|S_\perp(\theta,\phi)| \approx \sum_{i=1}^{n} K_iG_{i,i} - K_iG_{i,i-1} \qquad (10)$$

where $i \in \mathbb{N}^*$ and

$$K_i = 2k^3abc(m_i - 1) \quad , \quad G_{i,\ell} = \frac{\sqrt{\pi u_{i,\ell}}}{u_{i,\ell}^2}J_{3/2}(u_{i,\ell}) \qquad (11)$$

with initial conditions

$$G_{1,1} = \frac{\sqrt{\pi u_1}}{2u_1^2}J_{3/2}(u_1) \quad , \quad G_{1,0} = 0 \qquad (12)$$

To conclude this mathematical model, we return to the geometrical relationships that resulted in Equation (6). Taking into account the assumptions we have made about the cell's geometry and with respect to the x-axis of the ellipsoid if β is the angle of the bisectrix and α is the angle of the incident beam, then it has been shown [7] that they relate to the scattering observation (θ, ϕ) by $\cos\beta = -\cos\alpha\sin(\theta/2) + \sin\alpha\cos(\theta/2)\cos\phi$.

3. The Scattering Relation

For cells that appear alone, that is, when there is no binding of cells together, and at low concentrations ($\lesssim 10^5$ cells/ml), where multiple scattering effects can be ignored, the average scattering pattern can be calculated using a size distribution. The term *size* in the current context should be interpreted as length of the semi axis of the ellipsoid form denoted by s. Using a probability density function $\tilde{P}(s)$ for the size, and assuming that we have N size ranges with midpoints $s_1, s_2, \ldots, s_N$, then the relative frequencies of the cell samples in the ranges are approximated by the density function at the mid-points, so that the mean of the scattering matrix elements S_{IJ} would be calculated by

$$\langle S_{IJ} \rangle = \frac{\displaystyle\int_0^\infty S_{IJ}P(s)\mathrm{d}s}{\displaystyle\int_0^\infty P(s)\mathrm{d}s} \simeq \frac{\displaystyle\sum_{k=1}^{N} S_{IJ}(s_k)P(s_k)}{\displaystyle\sum_{k=1}^{N} P(s_k)} \qquad (13)$$

where S_{IJ} are defined in p65 of [1] whilst for incident light of parallel polarisation (hence the subscript $\|$), $|S_\|(\theta,\phi)| = |S_\perp(\theta,\phi)|\cos(\theta)$, where $|S_\perp(\theta,\phi)|$ is defined in Equations 10–12. We have to note that the mRDG is an expansion of the RDG and as such only the matrix elements S_{11}, S_{12}, S_{33} and S_{34} are of relevance. The density function for the size can be selected to be any of the known statistical probability functions but for the reasons outlined in [11] we use the distribution proposed therein with 30% variation as in natural environments.

4. Results

Computer simulations were implemented in Matlab for estimation of the light scattered intensity from asymmetric bacteria populations. In general bacteria cells present a structure that consists mainly of the cell wall, cytoplasmic membrane, cytoplasm and nucleoid. Other morphological characteristics may also appear such as a slime layer (capsule) external to the cell wall or spore

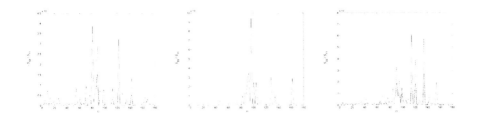

Figure 2. The angular dependence of the scattering ratio for an n-layer ellipsoid. Left: n=3, a three layer ellipsoid model with different axial ratios (dashed line: $t < 1$, solid line: $t = 1$, dotted line: $t > 1$). Centre: n=4, a four layer ellipsoid model with different axial ratios. Right: n=6, a six layer ellipsoid model with different axial ratios.

inclusions in the cytoplasmic area. As such, the use of an n-layered model as defined here is justified since a more accurate representation of the cell can be achieved when each morphological characteristic is modelled as a separate layer. In the present study it is assumed that $n_{\max} = 6$. Following the findings of many studies we investigate for relative refractive index values that have been found to be true for waterborne biological cells; that is, in the range $1 < m < 1.3$.

In order to better report and present our results we have parametrically defined the geometry of the cell with $b = c = s$ with $a = ts$ where a, b and c are those of Equation (4), s is a sampled value s_k as defined in Equation (13) and t denotes the axial ratio. It follows that

$$v = s\sqrt{t^2 \cos^2 \beta + \sin^2 \beta} \tag{14}$$

The cytoplasm dominates the internal part of the cell and so it is assumed that it will always have a size value $v_1 = 0.8v$. The remaining layers will have a value $v_{i+1} = fv + v_i$ so that f is a uniform random number where $f < 0.2$ and relates to the proportion of s allocated for the specific layer thickness, whilst for the outermost layer's compartment it follows that $v_n = v$. To assess the applicability of our models we have concentrated on the examination of the pattern resulting from the ratio of the population averaged scattering matrix elements, $-|\langle S_{33}\rangle|/|\langle S_{11}\rangle|$. First we investigate the effect of altering the axial ratio (t in Equation (14)) on this ratio. That is to say for $t < 1, t = 1$ which corresponds to spherical morphology, and $t > 1$.

Figure 2, left, presents results for a three layer model. Note that for the plot of $t > 1$ (doted line) there are two prominent peaks at angles $\sim 90^{\circ}$ and 130°; whilst for $t < 1$ (dashed line) there is a prominent peak at $\sim 142^{\circ}$. The peaks of $t > 1$ at $90^{\circ},130^{\circ}$ is unique, as no other pattern presents such a significant maximum in magnitude. Consulting the results of additional layers in the model, that is for four layers in Figure 2, left, and six layers on the right, we verify that there is a consistent appearance of large peaks for each of the patterns for different t values. For example for n=6 the peaks for $t > 1$ appear at $\sim 90^{\circ}$ and 133°, whilst a magnitude peak for $t < 1$ appears at $\sim 136^{\circ}$. These results indicate that there will always be a significant difference in the patterns of $-|\langle S_{33}\rangle|/|\langle S_{11}\rangle|$ allowing for the morphological characterisation or identification of bacteria in water samples by simply monitoring this ratio.

This statement is further supported by examining particular peaks of Figure 2. When we have a three layer model a prominent peak appears at $\sim 90^{\circ}$. By adding one layer to the model, that is for populations of four layer ellipsoid cells (n=4), the magnitude peak now appears at $\sim 105^{\circ}$. For populations of six layer ellipsoid cells (n=6) a further angular shift on the magnitude peak of $t > 1$ results in a maxima at $\sim 133^{\circ}$. Similar arguments can be made for particular maxima of the remaining patterns. As a result we can now generalise the statement made above not only to the axial ratio but also to the number of layers that suffice to describe a bacterial cell. Our results indicate that by monitoring the ratio $-|\langle S_{33} \rangle|/|\langle S_{11} \rangle|$ and for a particular local maxima (peak), a significant difference appears in these patterns that allow for identification of geometrical morphology (spheres or ellipsoids) and also their internal structure (number of layers).

Clearly, and returning to our initial question, coliforms with ellipsoidal morphology will present unique patterns that are significantly different from those found for cocci or sporulating bacilli (rods - spheres). As such, we would easily alert for their appearance in water samples by monitoring the scattering ratio as defined in Equation (13) much faster than using other theories or approximations. The latter is supported by the results in [3], which indicate that the mRDG is comparable in performance but much faster than competitive theories.

5. Conclusions

In this paper we have derived a new model for the n-layer ellipsoid problem based on the mRDG approximation and used it to simulate light scattering phenomena in bacteria cell populations. The proposed models can be used for screening for coliforms in water samples. The models have been implemented and tested by computer algorithms developed in Matlab. It has been demonstrated that the method can be used for the identification of cell morphology and internal structure. All simulations have been conducted using sizes and relative refractive indices in accordance with values found in bacteria cells.

It appears that the patterns generated for the scattering ratio $-|\langle S_{33} \rangle|/|\langle S_{11} \rangle|$ are suitable for real time identification of coliforms. Our findings indicate that the all important issue of characterising bacteria morphology in water's suitability rapid testing can be performed by using the sensor array of Figure 1 in conjunction with the n-layer mRDG ellipsoid approximation as presented in this paper. The consistency of the magnitude peaks throughout results for multiple layers on the ellipsoidal form and for different values of axial ratios support this claim.

Further research is required for the examination of these approximation patterns in the presence of noise, i.e. dead bacteria cells, particles other than bacteria and so on. Even though we have reported a uniqueness in morphology patterns, there is still a need for the investigation from true light intensity scattering patterns from benchmark prokaryotic cells.

References

[1] Bohren C F and Huffman D R, 1998. *Absorption and Scattering of Light by Small Particles*. New York: Wiley
[2] Bronk B V, Druger S D, Czege J and Van de Merwe W P 1995. Measuring diameters of rod shaped bacteria in vivo with polarised light scattering, *Biophysical Journal*, 69, 1170–1177.
[3] Chliveros G, Rodrigues M A and Cooper D 2003. Modelling Populations of Prokaryotic Cells: the n-layered mRDG Approximation, *SCS-Europe Proc. 17th European Simulation Multi-Conference* [BioMed], Nottingham (UK), June2003, 338–344
[4] Clesceri L S, Greenberg A E and Eaton A D (editors) 1998. *Standard Methods for the examination of Water and Wastewater*. Washington: American Public Health Association.
[5] Conville P, Witebsky F G and MacLowry J D 1994. Antimicrobial Susceptibilities of Mycobacteria by Differential Light Scattering, *Journal of Clinical Microbiology*, 32(6), 1554–1559.
[6] Farias T L, Koylu U O and Carvalho M G 1996. Range of validity of the Rayleigh-Debye-Gans theory for optics of fractal aggregates, *Applied Optics*, 35(33), 6560–6567.
[7] Gucker F T and Egan J J 1961. Measurement of the angular variation of light scattered from single aerosol droplets, *Journal of Colloid Science*, 16, 68–74.
[8] Schlegel H G 1997 (7th ed.). *General Microbiology*. London: Cambridge University Press
[9] Shimizu K, 1983. Modification of the Rayleigh-Debye Approximation, *J.Opt.Soc.Am. Letters*, 73, 504–507.
[10] Sloot P M A, Figdor C G 1986. Elastic Light Scattering from nucleated blood cells: rapid numerical analysis, *Applied Optics*, 25(19), 3559–3565.
[11] Wyatt P J 1973. Differential Light Scattering Techniques for Microbiology. In: JR Norris and DW Ribbons (eds.) *Methods in Microbiology*, Volume 8. Academic Press: 183–263.

Poster Programme

Poster Programme
Paper presented at Sensors and their Applications XII, September 2003
©2003 IOP Publishing Ltd

Period Dependent Temperature and Ambient Index Effects on Long Period Fibre Gratings

K P Dowker[1*], Z F Ghassemlooy[1], A K Ray[1], F J O'Flaherty[2] and P S Mangat[2]

[1] Electronics Research Group, School of Engineering, Sheffield Hallam University, City Campus, Sheffield S1 1WB, UK.
[2] Centre For Infrastracture Management, , Sheffield Hallam University, City Campus, Sheffield S1 1WB, UK.

Abstract. The ambient index and temperature effects on the spectral profiles of two sets of long period gratings (LPGs) of different periods were investigated. The shorter period LPGs were found to be more sensitive than the longer period LPGs over identical ambient index ranges but less sensitive over identical temperature ranges. The coupling wavelength shifts due to temperature are also seen to be linear and in opposite directions in each set of LPGs and unlike the ambient index shifts there seems to be no obvious modal dependency with respect to sensitivity in any individual LPG. The conclusion to this investigation is that it may be possible to design an LPG of such a period that parts of the spectral profile are unaffected by temperature whilst maintaining a reasonable ambient index sensitivity.

1. Introduction

In recent years, fibre gratings have been extensively investigated for their applications as a variety of sensors [1,2]. The transmission spectra of long period gratings (LPGs) have been shown to be sensitive to changes in index of refraction of the medium surrounding the fibre cladding (ambient refractive index) in the vicinity of the LPG [3-7] and also to changes in the ambient temperature [8,9]. Changes in the coupling wavelengths from the core modes to the various cladding modes due to changing ambient refractive indices using index matching gels have been the subject of many reports [10,11]. This investigation examines the effect of the grating period on the temperature profile of an LPG. LPGs with 2 different grating periods (450μm and 700μm) written into the cores of identical monomode fibres have been utilised and comparisons between the 2 sets of temperature profiles and ambient index profiles have been made in order to ascertain the possibility of achieving temperature immunity in a single refractive index sensing LPG.

---

* Present address: Health and Safety Executive, Broad Lane Sheffield S3

2.   Experimental Procedure

The six fibre gratings utilised were of the same material and optical and physical parameters before the gratings were written into the respective cores. The grating period of fibres numbered 1 to 4 is 700µm and numbers 5 and 6 is 450µm. All spectral profiles were observed and recorded using a HP 86142A Optical Spectrum Analyser (OSA), the light source used being the 1300nm and 1550nm internal EELED sources of the OSA. The ends of each fibre were connected into the light source and detector inputs of the OSA using bare fibre connectors. The transmission spectra of all of the LPGs were recorded in air at room temperature before any experiments were undertaken. Spectral profiles of each fibre grating were taken using a range of index matching gels. Temperature spectral profiles were taken by placing the grating section under slight tension in a purposely designed heat chamber and recording the spectra at selected temperatures from room temperature to a maximum of around 80°C.

3.   Results and Discussion

The transmission spectra at room temperature in air are shown in Figure 1 for gratings with periods of 700µm and 450 µm respectively.

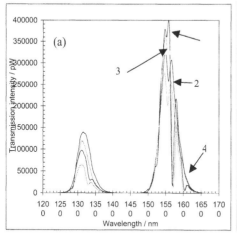
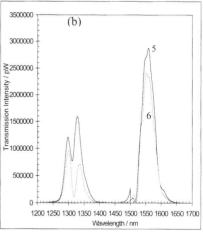

Figure 1: Spectral Profile of (a) 4 LPGs with Periods of 700mm (numbered 1-4) and (b) 2 LPGs with Periods of 450mm (numbered 5-6) at 300K.

The wavelengths at which coupling will occur is written in the form [5]

$$\lambda_{(i)} = \left(\beta_{core}^{(i)} - \beta_{clad}^{(i)} \right) \frac{\Lambda}{k} \qquad (1)$$

where the propagation constants (β) are quoted as functions of the mode number i and coupling wavelength λ, k is the wave number ($2\pi/\lambda$) and Λ is the grating period. Coupling was evident in each fibre at the different wavelengths as summarised in Table I.

Table I: Optical and physical parameters and coupling wavelengths of all six LPGs

core refractive index (n_{core})		cladding refractive index (n_{clad})			core radius (r_{core})		cladding radius (r_{clad})
1.4499		1.4441			3.9µm		62.5µm
LPG No	Λ/ µm	Coupling λs / nm					
1	700				1552.75	1570	
2	700	1285.5		1525.5	1556.5	1572	
3	700		1335		1558.5		1604
4	700		1329.5		1556		1601.5
5	450		1311.5	1516	1550		
6	450		1317.5	1520	1558		

Thus, identical fibres underwent identical writing procedures but the wavelengths at which coupling occurs differs in each grating of the same period. The propagation constants of the cladding modes will change with the ambient refractive index around the cladding, resulting in a shift in coupling wavelengths, the sensitivity increasing with mode number [12]. The ambient refractive index spectral profiles of all 6 fibres shown in Figure 2 showed shifts to shorter wavelengths for each modal coupling as the ambient refractive index was increased. All showed the degree of shift in coupling wavelength to increase as the ambient refractive index approached that of the cladding. The coupling wavelengths showed a positive shift to values greater than in air as an ambient refractive index of 1.452 was exceeded. It has been previously suggested that the process of writing the grating in the core has increased the refractive index of the cladding around the grating from its original value of 1.4441 to at least 1.452 [5]. The core index must also have increased to a value greater than 1.452 for guided modes to exist in the core, which will have a considerable effect on the initial core and cladding modes. The fibres in the period range used in most previously reported experiments of 450µm period, react as predicted by theory. The longer wavelengths coupling to the higher order cladding modes are more sensitive to ambient index changes than the shorter wavelengths. Two of the fibres containing the longer period of 700µm, LPG 1 and LPG2, responded as expected with the longer wavelengths generally showing greater sensitivity to the ambient index changes, however, LPG3 and LPG4 showed the shortest coupling wavelength to be around 6 times more sensitive than the longer ones. The high sensitivity of this wavelength suggests that this may be a higher order of diffraction (-2 order) of the grating period as reported by [9] and is actually coupling to a much higher cladding mode than the wavelengths in the 1500 to 1600nm region.

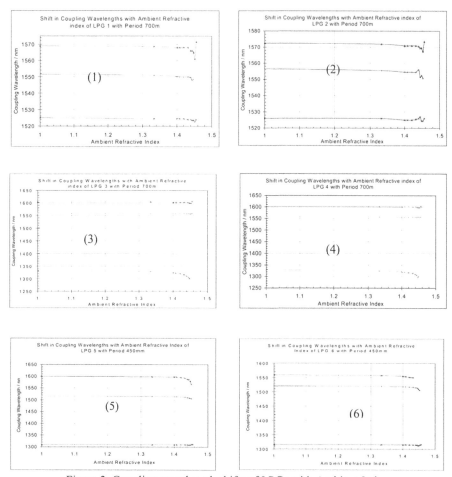

Figure 2: Coupling wavelength shifts of LPGs with Ambient Index

In the gratings with the 450µm period the cladding modes into which the respective wavelengths are coupling will be a higher order than for the gratings with the 700µm period. This would explain the greater sensitivity of the shorter period gratings to the ambient refractive index, as sensitivity is known to increase as the coupling mode increases.

The shifts due to the change in temperature for all 6 fibres are shown in Figure 3. The main feature of this investigation is the fact that all of the LPGs with 700µm periods have a negative temperature profile whilst all of the LPGs with 450 µm periods have a positive temperature profile. Differentiating equation 1 with respect to temperature shows the combined period and refractive index dependency on temperature.[8]:

$$\frac{d\lambda}{dT} = \frac{1}{k}\left[\frac{d\lambda}{d(\delta\beta)}\left(\frac{d\beta_{core}}{dT} - \frac{d\beta_{clad}}{dT}\right) + \Lambda\frac{d\lambda}{d\Lambda}\frac{1}{L}\frac{dL}{dT}\right]$$ (2)

where L is the length of the grating. The increase in temperature will cause expansion of the fibre and hence an increase in the grating period. This will result in smaller diffraction angles of all wavelengths and an increase in the wavelength which will match the coupling conditions for the first mode, and thus subsequent modes. It follows therefore that this effect will tend to cause the coupling wavelengths in each cladding mode to undergo a red shift. The directional shift of the coupling wavelength due to the change in refractive indices of the core and the cladding caused by the temperature change is not as obvious from Equation (1). Firstly, the increase in grating period due to the temperature increase would result in coupling to longer wavelengths. As these new wavelengths have their own individual modal core and cladding paths and as the propagation constants are dependent on the modal path this factor will itself cause different propagation constants in both media to be considered.

For the LPGs used in this investigation it is shown that an increase in the ambient index from that of air to values approaching that of the cladding will cause the core to cladding mode coupling wavelengths to undergo a blue shift. The magnitude of these blue shifts increase with cladding mode order and the and the sensitivity increases to an exponential approximation in all modes as the ambient index approaches that of the cladding. Compensating methods have been reported [13,14] including a combination of a number of LPGs or by the combination of LPGs and other sensing mechanisms, each of which has inherent limitations. Compensation systems employing only LPGs include the use of a 2nd fibre with an optically identical grating as a control to reduce the unwanted effect, and fibres containing 2 identical gratings in the same core. However, the production of 2 identical gratings is very difficult, as is shown in this investigation. Other limitations relate to the exact positioning of the 2nd LPG to receive the identical effects of the unwanted measureand in the first system and the fact that the radiation coupled out of the core into the cladding at the first grating may be coupled back into the core at the second grating in the second system. It will therefore have a much reduced intensity loss at the output of the fibre.

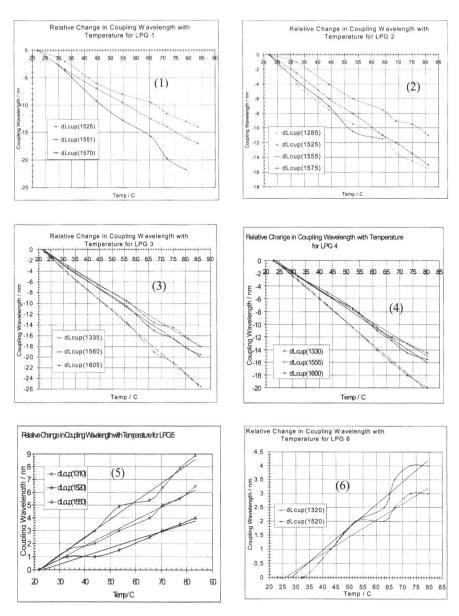

Figure 3: Coupling wavelength shifts of LPGs with Temperature

4.   Conclusion

The direction of the temperature gradient of an LPG is found to depend on the period of the grating, for fibres with identical optical and physical properties. If the fractional increase of the period is sufficient the temperature effects on the period become more significant and may overcome the effects of the refractive index changes. The possibility of using a single temperature immune LPG for ambient index sensing has been described. Further work will investigate an optimum period length at which temperature will have negligible affect on the optical profile of one or more of the coupling modes, but is still of a period length which is sensitive enough to detect small changes in ambient indices.

Acknowledgements

The authors are grateful to the Engineering and Physical Science Research Council for the award of a grant to develop all optical sensors for use in the construction industry (ref: GR/R19151) and TQ Environmental plc, Wakefield for participation in a CASE studentship.

References

[1]   Peng PC, Tseng HY, Chi S 2002 *Electronics Letters* **38 (24)** 1510-1512

[2]   Zhao DH, Shu XW, Zhang L, Bennion I 2002 *Electronics Letters* **38 (7)** 312-313

[3]   Erdogan, T. 1997*Journal of Lightwave Technology*, **15 (8)** 1277-1294,.

[4]   Bhatia, V. and Vengsarkar, A.M. 1996 *Optics Letters*, **21(9)** 692-694.

[5]   Patrick, H.J., Kersey, A.D. and Bucholtz, F. 1998.*Journal of Lightwave Technology*, **16** 1606-1607.

[6]   Lee, B.H., Liu, Y., Lee, S.B., Choi, S.S., and Jang, N.J 1997 *Optics Letters*, **22(23)** 1769-1771.

[7]   Bhatia, V., Vengsarkar, A.M. 1996 *Optics Letters*, **21(9)** 692-694.

[8]   Bhatia, V. 1999 *Optics Express*,.**4(11)** 457-460.

[9]   Shu, X., Zhang, L., Bennion, I. 2002 *Optics Communications*, **203** 277-281.

[10] Shu, X., Huang, D. 1999 *Optics Communications*, **171** 65-69.

[11] Shu, X., Zhu, X., Jiang, S., Shi, W., and Huang, D. 1999 *Electronics Letters*, **35(18)** 1580-1581

[12] Patrick, H.J., Kersey, A.D., Bucholtz, F., Ewing, K.J., Judkins, J.B., and Vengsarkar, A.M. 1997 "Chemical Sensor Basesd on Long Period Fiber Grating Response to Index of Refraction", in Proc. Conf. Lasers Electro-Opt., l.11,.420-421.

[13] Qin, L., Wei, Z.X., LI, H.P., Zheng, W., Zhang, Y.S., and Gao, D.S. 2000 *Optical Materials,* **14** 239-242.

[14] Harumoto, M., Shigehara, M., Kakui, M., Kanamori, H., and Nishimura, M. 2000 *Electronics Letters*, **36(6)** 512-513.

Poster Programme
Paper presented at Sensors and their Applications XII, September 2003
©*2003 IOP Publishing Ltd*

IR Gas Sensors of a Liquid Crystal Type For Environmental Monitoring

K U Nazarava and V I Navumenka

Scientific Research Laboratory of Environmental Physics, Department of Physics and High Mathematics, International Sakharov Environmental University, Minsk, Belarus

Abstract: When one aspires to obtain spectroscopic data about different ingredients in the atmosphere or indoor air a whole arsenal of complicated and cumbersome instruments and refined techniques are necessary. One of the most interesting spectroscopic information is contained in the shapes of spectral bands and the instruments intended to obtain such details must be multichannel or scanning. In that case, the difficulties above grow by many times.

Unfortunately, the spectrometric sensors of types known up today have some essential drawbacks that restrict roughly their wide applicability and do them poor-suitable for field, mobile and high-altitude jobs. Those drawbacks are the construction complexity, cumbersome embodiments, difficulty of the control, and (the main) obligatory presence of the moving parts (mirrors, prisms or gratings). Meanwhile, there could be applications where any vibrations caused by moving parts were strictly inadmissible.

We propose a new method that combines in some aspects the derivative approach and simplicity of non-scanning instruments. In other words, we developed a method to investigate spectral band shapes without prisms, gratings, interferometers and moving parts in the spectral instruments. An instrument of the type proposed is very small, even miniature. The main element in the instrument (sensor) of the type proposed is a liquid-crystalline cell possessing marked dichroism of optical absorption in some spectral range.

One of the most important tasks in the environmental monitoring field is a detection and concentration measurement of different ingredients in natural and artificial mixtures of substances. According to the measuring of value of IR absorption of different gases one can determine their concentration in the environment. For example, it is known a lot of devices (sensors) of optoacoustical type, which are used for this purpose. Moreover, the modulation of external optical radiation is usually organized with the help of some interrupting device. Interrupting devices are as a rule multiwaved in spectral sense (for example, mechanical) or wideband (for example, electrooptical). Although, application of narrowband modulators, tuned to the absorption range of determined gas, increases sensitivity of the method [1].

The construction complexity, cumbersome embodiments of the sensors on the basis of interrupting devices and their other disadvantages are well-known. That is why there is a great practical and on principle fundamental interest to investigate an opportunity of making the devices which could be used for measuring the absorption of external gases in IR optical range and also would not have the disadvantages of optoacoustical sensors [2].

New opportunities can be found out by using liquid crystals. As it is well-known, in liquid crystals one can observe several electrooptical effects, which are accompanied by changing the colour. But changing the colour is a spectroselective transformation of incident optical radiation. If one can control this transformation, in particular, by gradually varying the magnitude of electric field applied (see Fig. 1), so it will be spectral scanning (or spectroselective modulation). In addition, liquid crystal (LC) cells have miniature sizes and they are produced now by well-developed technologies. So, LC cells are reliable, long-lived and very cheap.

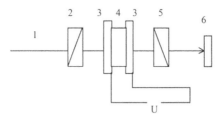

1 – incident light

2 – polarizer

3 – transparent electrodes

4 – oriented LC layer

5 - analyzer

6 - receiver

Figure 1. *The Scheme of electrically controlled LC cell.*

The main element in a sensor of the type we are proposing [3] is a liquid-crystalline cell possessing marked dichroism of optical absorption in some spectral range. By fitting the absorbance and alignment parameters of the LC cell, it is possible to create the effect of the absorption edge shift (see Fig. 2) when varying the magnitude of electric field applied (line 1 corresponds to U_1 magnitude of applied electric field and line 2 corresponds to U_2 magnitude of applied electric field). This leads to some kind of wavelength modulation of an external optical radiation in the pre-determined spectral range. Adding an appropriate light detector to the system described, we obtain a miniature scanning spectral instrument (sensor) of the new type without any moving parts, without using of dangerous high voltages and without other drawbacks mentioned above.

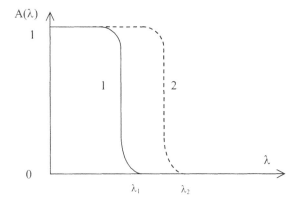

Figure 2. *The Absorption Edge Shift.*

So, it meets requirements for well functioning gas sensors - high selectivity, high sensitivity, high stability, small sizes and high signal to noise ratio.

It is well-known that one of the problems of environmental monitoring is detecting minor components of the atmosphere. Among these minor components there are nitrogen oxide, methane, and some other hydrocarbons. Their absorption spectra are situated in the near IR band (see Fig. 3), but these molecules have different spectral shapes and we can detect them easily.

That's why we have investigated several LC cells with absorption bands in spectral range near 3 micrometers where some atmospheric hydrocarbons absorb. Some results gave very hopes. One example of realisation of the idea was a LC cell with thickness near 20 micrometers. This LC cell gives the absorption edge scanning shift of about 100 nm in this region under field about 3 V. The absorption near 3 micrometers is due to CH bond stretching vibrations and these bonds and corresponding transition dipole moments orient in very different ways relative to long axes of the LC molecules. Nevertheless, even this shift can be used we consider for the wavelength (wave number) modulation.

A response from a photodetector placed after the LC cell is a little similar to the original external spectrum, but can be reduced to the original spectrum with the help of some computational operations (differentiation, iteration). We achieved good results in revealing the external spectral band shapes with the help if iterations. Here is no place to demonstrate how one can reduce the true external spectrum after recording such a LC cell-modulated signal. Computer simulation and direct experiments showed good results. The technique for reducing true external spectrum from the signal detected after the modulation process is described in detail in [4].

Once more about advantages: we already said about the absence of any moving parts and low controlling voltages. Additive advantages are miniature sizes of such an instrument: LC cell is one or several cm^2 in area and one or two mm in thickness. An appropriate attached photodiode, for example, is also of small size. Moreover, LC cells are produced now by well-developed technologies and so are reliable, long-lived and very cheap. An electronic part of the instrument is another task and has little inventional interest.

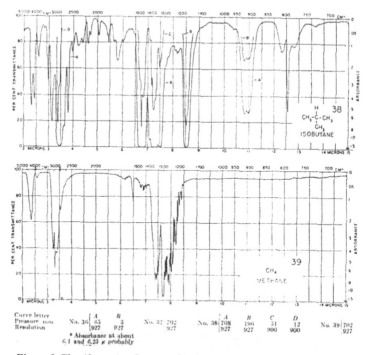

Figure 3. *The Absorption Spectra of Isobutane and Methane.*

References

[1] Knapp K T Handbook of Air Pollution Technology ed. By S.Calvent and H.M.Englund «John Wiley & Sons»

[2] Navumenka V I, Ljaonava I G, Lapanik V I, and Samcou M P 1999 Abstracts of European Conference on Liquid Crystals 99 P1-001.

[3] Navumenka V and Nazarava K 2002 Materials of the 19th International Liquid Crystal Conference P619.

[4] Navumenka V and Ljavonava I 2000 Materials of the 18th International Liquid Crystal Conference Japan 25D-54-P.

Poster Programme
Paper presented at Sensors and their Applications XII, September 2003
©*2003 IOP Publishing Ltd*

Micromoulded Elastomer springs for biomedical sensors using micromachined silicon moulds.

Vincent Casey†, Jacha I van Hout‡and Jörg Scheurer‡

‡ Physics Department, University of Limerick, Limerick, Ireland
† Institut für Mikrotechnik Mainz GmbH, Carl-Zeiss-Straße 18-20, D-55129, Mainz, Germany

E-mail: vincent.casey@ul.ie

Abstract. Elastomer deformation is used as the basis for sensing many mechanical quantities such as force, pressure and acceleration. Elastomer shape, characterised by the so called 'shape-factor' is an important index in optimising the elastic properties of such structures. In this work we exploit the characteristic shapes of trenches that arise in anisotropically etched silicon to form micro-spring structures. The etched silicon structures are used as moulds to form elastomer microsprings with carefully controlled shape-factors designed to optimise spring mechanical performance parameters such as linearity and hysteresis. Preliminary stress-strain data for PDMS microsprings is presented.

1. Introduction

A wide range of micromechanical sensors require some form of deflection structure to serve as the primary sensing element. In biomedical interface pressure measurement direct mechanical contact between the sensor and body parts or pressure/force applying parts is required. Large errors can arise through coupling artifacts associated with the body sensor interface[1]. Simple capacitor structures where a very thin elastomer microspring, structured, in order to optimize mechanical properties, is sandwiched between two electrodes[2], figure (1), have been used in many studies concerned with biomedical interface pressure measurement since such devices are inherently planar and of low profile. Thick film technology is often used to form the elastomer structures. However, it does not allow the degree of structure control required in order to optimise the deformation properties of the elastomer.

Channels anisotropically etched in [100] silicon wafers are inherently trapezoidal in cross section. It should be possible, therefore, to machine rib- and pyramid-shape structures in silicon which could in turn be used to form or mould mechanically stable replica elastomer structures with controlled shapes. However, precise mask alignment to the crystallographic orientation of the silicon wafers is essential in order to preserve rib dimensions[3] which can be corrupted by etch undercutting. A technique similar to that developed by the Uppsala group[4] was used to obtain precision alignment and

Micromoulded springs.

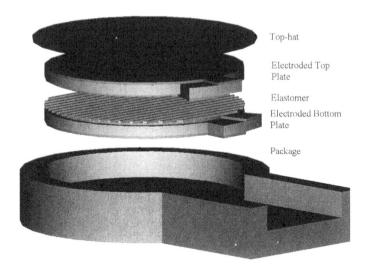

Figure 1. Planer biomedical sensor and package configuration.

thereby maintain close geometrical control over etched features, particularly trenches, which extended continuously across almost the entire die.

2. Theory

The problem of large elastic compression of finite rectangular blocks has no exact solution (see [5] however) and much of the engineering literature is based on load-deflection relations of the type emanating from the linear theory of elasticity but couched in terms of the shape-factor S which is defined as the ratio of one loaded surface area, i.e. top surface, to the load-free surface area, i.e. sides, for a device having bonded metal plates on its upper and lower surface and free to bulge at the sides. For a rectangular block of length L, breadth B, and thickness T the shape factor has the form

$$S = \frac{LB}{2(Lt + Bt)} = \frac{S_o}{\lambda} \tag{1}$$

where t is the strained thickness of the block, and S_o is the unstrained shape factor. For compression the resulting stress-strain relation has the form

$$\sigma = \frac{F}{LB} = E_a \frac{\Delta T}{T} = E_a[1 - \lambda] \tag{2}$$

where E_a is the apparent Young's modulus which takes the form

$$E_a = E_h + ECS^2. \tag{3}$$

with C a constant which has the value 4/3 when $L >> B$ and 7/3 for the situation where $L = B$, $\lambda = 1 - \Delta T/T$ is the compression ratio. E_h is known as the 'homogeneous'

Micromoulded springs.

compression modulus which has values of approximately $4/3E$ and E for the limiting cases of infinitely long block (extended rib) and square block (flat topped pyramid) respectively. Analysis of the problem using this formulation yields the following stress-strain relation:

$$\sigma = E[C((\frac{1}{\lambda^2} - 1)S_o^2 - \log \lambda)]. \qquad (4)$$

According to this approach, a plot of the nominal stress versus the strain term (square brackets) appropriate to the geometry of the experimental situation should yield a straight line.

3. Experimental

The silicon wafers used in this study were $125mm$ diameter with (100)-orientation. Wafers were thermally oxidized to obtain a $500nm$ thick SiO_2 layer to serve as a stable mask during the final anisotropic KOH-etching of silicon to form the moulds. An alignment fork similar to that used by Vangbo and Backlund[4], was used in order to establish accurate alignment with the crystallographic orientation of the silicon wafers. The silicon was etched using 20% KOH at $80°C$. Variation of the etching time of the silicon for a given wafer yielded a series of feature depths ($5\mu m$, $10\mu m$, $20\mu m$, $30\mu m$ and $40\mu m$). The remaining SiO_2 mask was removed using a buffered HF solution.

Channel patterns precisely aligned to the crystallographic orientation of the silicon wafer will have channel walls tapered at an angle of $54.7°$. Flat bottomed channels may be produced by ensuring that the ratio of channel width to depth is greater than 1.4, otherwise V-grooves will result. Of course the moulded elastomer will have complimentary form to the silicon moulds. The shape factor for moulded structures produced using these micromoulds ranged in value from 0.03 to 0.70 for ribbed structures and was of the order 0.45 for square pyramid structures. A range of structures was incorporated into the micromould CAD design. These ranged from channel (rib) structures with widths from 25 to $64\mu m$ and which extended over the entire length of the die, to shorter channels (same width range) which were either staggered or aligned (see figure(2)). The square pyramid openings were all $84 \times 84\mu m$.

Microspring samples were formed on polyester film (Melinex from Autotype Ltd., $75\mu m$ thick) which had an adhesion promoting pretreatment on both sides. A small amount of elastomer prepolymer mix (Sylgard 184, Dow Corning) was placed onto the substrate. The silicon micromould die was pressed onto the elastomer. The combination was placed in a vacuum desscicator which was connected to a rotary vacuum pump and degassed for 5 minutes. Samples were cured in an oven at $100°C$ for 1 hour. Demoulding of the polymer structures was simply done by pealing/prizing the substrate away from the die.

A custom built test jig, was assembled in order to perform a preliminary mechanical characterisation of the microspring arrays. A PosiCon 150/1 (Piezomechanik GmbH) was used for position control of a piezoelectric stack actuator (PSt 150/7/20) with in-

Micromoulded springs.

built strain gauge feedback control. It was possible to adjust the stack tip position over the range $0 - 20\mu m$ with $0.01\mu m$ step size. A miniature load cell (LPM530, Cooper Instruments UK) was fixed in the test gig with a sample holder plate attached. It was possible to load the sample by adjustment of the position of the actuator head. The position data was used to determine the strain and the load cell provided a direct measurement of the force.

4. Results

SEM images of some mould structures etched in silicon are shown in figure(2). Structure dimensions showed excellent correspondence with design feature size and the integrity of the structures was preserved across individual die and indeed across the entire wafer. PDMS (Sylgard, Dow Corning) microsprings were formed on flexible but relatively imcompressible polyester substrate using the silicon micromoulds, figure(2). Mould release agent was not required as the PDMS detached readily from the silicon. Anticipated problems with the moulding of pyramid structures were not realized as the PDMS filled the mould and demoulded with ease.

Simple stress versus strain plots, corrected for load-cell deflection, for ribbed and pyramid elastomer structures, displayed pronounced non-linear (second order polynomial) behaviour. The hysteresis level was similar in magnitude to the value obtained for the load-cell on its own and so is probably largely due to the load-cell. Unfortunately, it is therefore not possible to obtain a reliable estimate of the spring hysteresis with the test jig as currently configured.

When the spring elements are compressed by the applied load, they will spread and thereby increase in area. For a rectangular rib, the loaded edge length would increase by a factor $1/\lambda$ as would the loaded area (conservation of volume principle). However, the ribs used here are of trapezoidal cross section. Even without bulging of the spring elastomer, the simple geometry of the trapezoid would give a rib width increase with increasing compression given by, $\Delta W = 2\Delta T / \tan(54.7°)$ where ΔT is the spring deflection. Taking these changes into account in the area calculations for the stress for both rib and pyramid geometries and plotting the data according to the shape-factor formulation represented by equation(4) yields the stress-strain characteristics shown in figure(3). Excellent straight line fits result for both the ribbed (slope, $279kPa$. $R^2 = 0.986$) and the pyramid (slope, $149kPa$, $R^2 = 0.997$) sets of data. Experimental values for the Young's modulus for each of these structures may be obtained from the slopes of these plots once it is remembered that the elastomer structure constitutes a solid fraction of about 0.5 of the volume loaded (0.45 for ribbed structure; 0.5 for pyramid structure). Thus the effective Young's modulus for the ribbed structure is $560kPa$ whereas the pyramid structure has the lower value of $300kPa$. Values for E quoted in the literature range from about 300kPa to 2MPa for bulk PDMS.

Micromoulded springs.

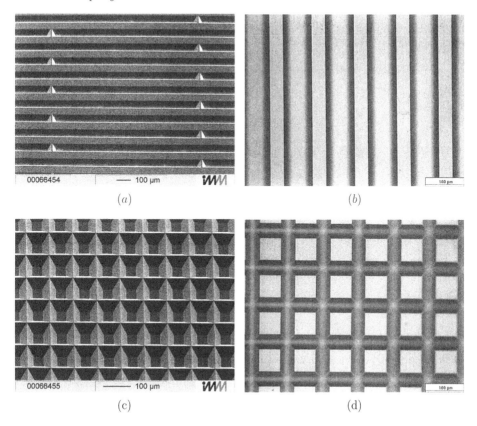

Figure 2. Silicon moulds and moulded PDMS microsprings. (a)Staggered trapezoidal channel. (b)Moulded Continuous Rib. (c)Pyramid array mould. (d)Moulded PDMS Pyramid.

5. Conclusions

An etch alignment tool was used for precision mask alignment to the crystallographic axis of (100) silicon wafers to carry out anisotropic etching of silicon to produce micromould features which were highly faithful to the design layout and which in some cases extended as one continuous feature to the edges of the die. The etched silicon die were used to mould PDMS microsprings on flexible polyester sheet. A preliminary investigation of the mechanical properties of the microsprings was carried out. The simple stress-strain characteristics of the microsprings were non-linear with the stress following a second order polynomial dependence on strain. However, excellent agreement between experimental data and a standard 'shape-factor' formulation for large scale deformation of bonded rectangular blocks was found for two distinct structures studied. Further work will focus on quantifying the hysteresis of these structures in order to

Micromoulded springs.

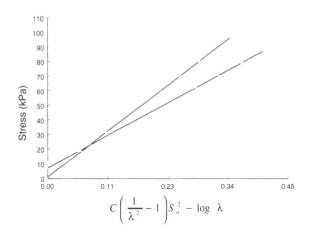

$$C\left(\frac{1}{\lambda^{2}}-1\right)S_{o}^{2}-\log\ \lambda$$

Figure 3. Plot of stress corrected for strained trapezoidal area versus strain for PDMS ribbed (∇) and pyramid ($\triangle$) springs.

identify optimum structures for incorporation in biomedical pressure sensors.

Acknowledgments

This project was funded by the European Commission's Improving the Human Potential and Socio-Economic Knowledge Base Programme, EMERGE, EU-contract HPRI-CT-1999-00023

References

[1] V. Casey, S. Griffin and S. B. G. O'Brien, An investigation of the hammocking effect associated with interface pressure measurements using pneumatic tourniquet cuffs, Medical Engineering and Physics, 23/7 (2001) 511-517.

[2] Casey, O'Sullivan, Nagle, Planar transducer for measuring biomedical pressures, World Patent.WO0001296, European Patent, EP1094747, Canadian Patent, CN1299247T, Australian Patent, AU8238898, priority/filing date 3rd July 1998.

[3] V K Dwivedi, R Gopal and S Ahmad, Fabrication of very smooth walls and bottoms of silicon microchannels for heat dissipation of semiconductor devices, Microelectronics Journal 31 (2000) 405-410.

[4] M Vangbo and Y Bäcklund, Precise mask alignment to the crystallographic orientation of silicon wafers using wet anisotropic etching, J Micromech. Microeng. 6 (1996) 279-284.

[5] J M Hill, A review of partial solutions of finite elasticity and their applications, Int J Non-Linear Mechanics, (36) (2001) 447-463.

Poster Programme
Paper presented at Sensors and their Applications XII, September 2003
©2003 IOP Publishing Ltd

Sensor properties of porous polymer composites

L Quercia*, F Loffredo, G Di Francia

ENEA, Centro Ricerche Portici, 80055 Portici (NA), Italy

Abstract. Conductive polymer composites are widely used as vapor sensors because their resistance changes when polymer matrix swells in presence of organic vapor. It is very interesting to investigate methods to improve the response of these composites for example increasing porosity of polymer matrix. In this work we have prepared polymer composites adding a porous agent in the polymer matrix (Poly(methyl-methacrylate)) to increase film porosity. Different mass fraction of porous agent added to the polymers allow to prepare conductive composites with different permeability to organic vapors. The morphology and sensor properties of these porous composites films will be studied by optical and electrical characterizations.

1. Introduction

Polymer composite gas sensors are characterized by an insulating polymer whose electrical properties are modulated by a conductive "filler". These conductive composite thin films have already shown their potentialities as sensing elements in commercial electronic noses [1,2]. The simplicity of signal transduction and the low material cost of the detectors make them attractive even as specific application single sensors, pushing for a significant improvement of their performances. Changes in selectivity have been obtained using different polymer matrixes [1-3], while sensitivity can be strongly improved working on the "filler" characteristics [4,5]. In general, the thin film morphology affects critically the sensing device performances but the morphology of these polymer composite films prepared by spinning is quite complex and difficult to be adjusted. In order to significantly increase the thin film porosity we have chosen to investigate the approach of increasing the composite blend with another phase extracted at the end of the process. This approach allows a better control of film porosity as has already been showed in the preparation of polymeric membranes [6].

In this work we have prepared polymer composites adding a porous agent (Poly(propylene-glycol)) in the polymer matrix (Poly(methyl-methacrylate)) to increase film porosity. A porogen is a non-reactive species that is added to a polymer solution prior curing or casting. Once the film has been formed a solvent which is selective towards the porogen is used to extract the porogen from the film. This leaves behind a series of void spaces in the place of the porogens. Different mass fraction of porous agent added to the polymers allow to prepare conductive composites with different

Corresponding author: tel.+390817723302, fax.+390817723344, e-mail: quercia@portici.enea.it permeability to organic vapors. The morphology and sensor properties of these porous composites films will be studied by optical and electrical characterizations.

2. Experimental

2.1 Preparation composite thin films

A commercial carbon black was used as filler to obtain conductive composites and sensor devices: Black Pearls 2000. It was furnace carbon black grade donated by Cabot Co with a 1500 m²/g surface area and 12 nm particle size.

The solvents (reagent grade), Poly(propylene-glycol) (PPG) and the Poly(methyl-methacrylate) (PMMA) used in this study were purchased from Sigma-Aldrich and were used as received.

In first step we prepared suspensions of carbon black by dissolving PMMA in tetrahydrofuran (1.8% by weight of polymer) and dispersing the filler (approximately 25% by weight) in the solution by ultrasonic bath for 90 minutes. Subsequently various amounts of PPG were added to the suspension, then composite films were obtained by depositing the final suspension by spin coating on the glass substrate. Films based only on PPMA/CB have been prepared as reference. Generally, film thicknesses are in the sub-micrometer range as measured by Tencor P10 surface profiler. The summary of experimental conditions used to prepared composite films is shown in Table 1.

Composite films were soaked in ethanol for 6h or 12h at 45°C to extract PPG. Films morphologies obtained before and after extraction have been investigated by optical microscopy (Reichert-Jung) and scanning electronic microscopy. SEM images were taken with a LEO 1530 after metallic covering with gold.

Table 1. Experimental conditions used to produce PPG/PMMA/CB composite thin films.

Sample name	PPG(g) / PMMA(g)	PPG (mg)	PMMA (mg)	CB (mg)	THF (ml)
PMMA	0	0	80	20	5
PPG20/PMMA	0.25	20	80	20	5
PPG50/PMMA	0.50	40	80	20	5
PPG70/PMMA	0.75	60	80	20	5

2.2 Sensing devices.

Sensing devices were realized simply depositing two fingers of silver paste separated by a gap of 5 mm. Fig.1 shows the scheme of the chemiresistor with the electrical measurements in a controlled environment. The detailed characterization of sensor response to VOC has been performed using a Gas Sensor Characterization System (GSCS) already described [5]. An airtight gas sensor test stainless steel chamber is placed in a climatic chamber to keep constant the temperature (20°C). In order to obtain a controlled concentration of volatile organic compounds (VOC) and humidity the flow-through saturating method has been used [7]: the carrier gas flows through a condenser in the climatic chamber placed after a bubbler filled by the liquid of interest and kept at constant temperature by a thermostatic bath (25°C). At the outlet of the condenser the flow is saturated with organic vapors that can be diluted to the desired concentration (up to 1/500 of vapor pressure Pv(T)) adding carrier gas by Mass Flow Controllers (MFC) . The GSCS is interfaced with a computer for data acquisition, storage and analysis.

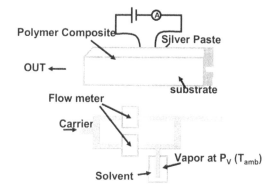

Figure 1. Schematic view of the sensing devices and the electrical measurement of sensor response to vapors under controlled environment.

3. Results and discussion

A series of composite films containing PMMA/PPG/CB have been prepared using different amount of PPG [Table 1]. Every film composite was soaked to ethanol for different times to extract PPG. Morphologies and vapor sensors responses of samples before and after treatments were studied.

Optical microscopy that can be easily performed on every sample (see Fig. 2) can give a rough idea of the complex composite film type morphology. In figure 2 is reported an example of two areas of PPG70/PMMA sample after PPG extraction.

A lot of void characterized by different size are visible in the sample especially next carbon aggregates. Moreover it is possible to observe also some white particles near bigger carbon aggregates (figure 2B). This particles have been observed only in samples treated with ethanol (both in PMMA and in PPG/PMMA composites). This seems to suggest that ethanol treatment influences also PMMA phase.

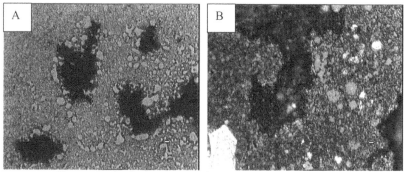

Figure 2. Optical micrograph at 500 magnification of PPG70/PMMA composite after long ethanol treatments to extract PPG.

SEM analysis has pointed out large differences between samples prepared with a different amounts of PPG (Fig. 3). For instance, PPG20/PMMA shows a more homogeneous distribution of particles than PPG70/PMMA. Anyway, a further glimpse into the films microstructure can be obtained by an accurate characterization of the sensor device to vapors.

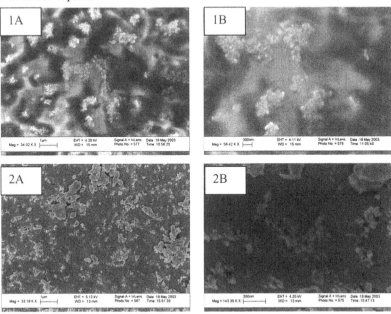

Figure 3. SEM images of PPG/PMMA composites films after ethanol treatments for PPG extraction. 1A) and 1B) PPG20/PMMA surface morphologies ad different magnifications. 2A) and 2B) show PPG70/PMMA morphologies.

We have carried on a first screening of all sensor types electrical characteristics (PMMA, PPG/PMMA and PPG/PMMA after extraction) in presence of different vapors (ethanol, n-pentane, acetone). Moreover PPG/PMMA sensor type has been submitted to different time of extraction to evaluate the influence of this parameter.

In Fig. 4 the dynamic responses of all sensors types are compared while in Table 2 the relative response times are reported. It is very interesting to observe that all sample types with PPG are clearly faster than the PMMA reference sensor type, in agreement with the aim of this work. Furthermore the PPG50/PMMA shows even an improvement in sensitivity however before extraction, suggesting that the optimized ratio of PPG and PMMA is 0.5. Fig. 5 compares the acetone sensitivities for all sensor types spanning the solvent concentration range from 0.2% to 70% of its vapor pressure at T=20°C. Apart from PPG50/PMMA type sensors, all the others show lower sensitivity than the reference PMMA type for high solvent concentrations (>20%) due to a very large linearity range. Further work is needed to understand in detail the influence of PPG addiction on the final sensing devices.

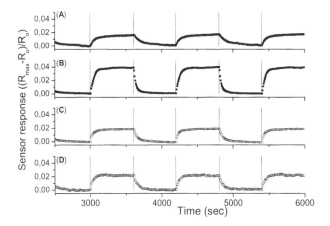

Figure 4. Dynamic responses to acetone flow (total flow=500sccm; T=20°C) of all sensor types: a). PMMA; b) PPG50/PMMA; c) PPG50/PMMA after 6h of extraction; d) PPG70/PMMA after 12h of extraction. Acetone concentration is at 6% of its vapor pressure at T=20°C. The acetone flow in and out is shown in the figure with a vertical line.

Table 2. Response times obtained by Fig. 4 relative to all sensor types.

Sample	Response time (s)
PMMA	> 254
PPG50/PMMA	93
PPG50/PMMA extracted for 6h	108
PPG70/PMMA extracted for 12h	69

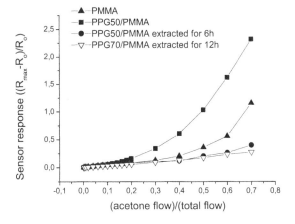

Figure 5. Sensor response to acetone flow (total flow=500sccm; T=20°C) of all sensor types.

4. Conclusions

We have shown that is possible to improve sensor performances working on the "porosization" of polymer matrix.

In particular, response and recovery times can be strongly improved adding PPG to the polymer matrix. The optimized PPG/PMMA weight ratio seems to be around 0.5.

The role of PPG extraction still needs further work of investigation to be fully understood, however a clear trend towards higher linearity, shorter responses and recovery times and lower sensitivity is suggested.

Acknowledgements

This work has been partially funded by CAMPEC project.

References

[1] Burl M.C., Sisk B.C., Vaid T.P., and Lewis N.S. 2002 Sensors and Actuators B87 130.

[2] Koscho M.E., Grubbs R.H. and Lewis N.S. 2002 Anal. Chem. 74 1307.

[3] Doleman B.J., Sanner R.D., Severin E.J., Grubbs R.H. and Lewis N.S. 1998 Anal. Chem. 70 2560.

[4] Chen X., Jiang Y., Wu Z., Li D. and Yang J. 2000 Sensors and Actuators B66 37.

[5] Quercia L, Loffredo F., Alfano B., La Ferrara V. and Di Francia G. submitted to Sensor and Actuators B.

[6] Nikpur M., Chaouk H., Mau A., Chung D.J. and Wallace G. 1999 Snthetic Metals 99 121.

[7] Endres H.E., Hildegard D.J. and Gottler W. Sensors and Actuators 1995 B23 163

Poster Programme
Paper presented at Sensors and their Applications XII, September 2003
©2003 IOP Publishing Ltd

Sensor Studies for Non Destructive Examination of Spot Welding

A M A El-Rasheed, J D Cullen, A Shaw, A I Al-Shamma'a, J Lucas

Department of Electrical Engineering and Electronics, University of Liverpool, Liverpool L69 3BX, UK.

Abstract: Spot-welding is a widely used welding method in industry. However, little advancement of on-line non-destructive examination (NDE) has been achieved to date. Traditional spot welding quality control has been carried out using either a chisel test, or a peel test, in a production environment; this is undesirable, since this approach is inherently expensive due to material wastage and time consumption. Through this study some promising techniques for NDE of resistance spot-welding are analysed. Particular attention is placed on electrical parameters such as current, voltage, and dynamic resistance, as well as, infrared emissions and acoustic emissions. Set-up and results pertaining to each technique are presented and discussed.

1. Introduction

Spot welding is an important welding technique extensively used in industry that has been long established [1]. The main quality control tests are the destructive chisel test and peel test, which are carried out on a randomly chosen welds obtained from the production line of the product [2]. The failure rates are then detected, examined and evaluated and if there is substantial increase, then the whole batch of the products is rejected as faulty. The labour costs incurred while performing this type of testing also make this approach unattractive. Industry rules, demand increase of the productivity and reduction of the production costs. This fact points to the direction of inventing other test procedures, which will be mainly, based on the NDE methods and techniques. Conventional resistance welding process control has been based upon monitoring the voltage and current, or their derivatives, power and resistance [3,4]. While these sensors work well under ideal conditions, surface contaminants or metal impurities can cause under strength, or under sized welds to be formed while retaining the same voltage and current values as for welds conforming to the ideal standards [5]. The answer to this problem begins with a look at how the electric current creates welds in resistance spot welding. According to Joule's law:

$$H \text{ (cal)} = \frac{I^2 R T}{J} \tag{1}$$

Where J is the electrical equivalent of heat in Joule, T is the temperature in Kelvin, the electrical resistance in ohm and I is the electrical current in Amp.

From equation (1) it is simple to understand that one must know the electrical parameters and the temperature reached in order to determine if enough heat has been generated for weld formation. However, as experts in industry would testify, simply having the right conditions for weld formation does not mean that a weld has been formed. Therefore, other parameters must be examined as well in order to give a complete picture of the quality of weld formation [6]. Figure 1 shows the weld formation as a result of the destructive 'peel test.' Along with the 'chisel test', the 'peel test' is the most reliable method of weld inspection, but the greatest draw back, is that they are destructive tests. Using these destructive tests weld quality is determined through the measurement of the average weld diameter (nugget size) by (d1+d2)/2. Figure 1 also illustrates the parameters analysed in this paper.

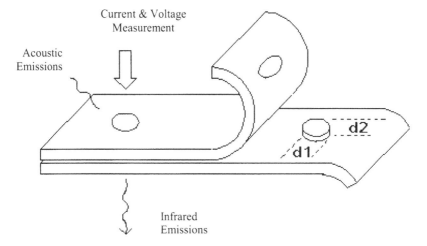

Figure 1: Diagram of weld formation and the parameters inspected in this study.

2. Experimental Set-Up

There are many types of spot welders available [7]. For this study the pedestal type spot welder is used; in the pedestal type there is a fixed vertical pedestal frame and integral transformer and control cabinet. The bottom arm is fixed to the frame that is stationary during welding, and takes the weight of the work piece. The top arm may be hinged to move down the arc of a circle, or it may be moved down in a straight line. Pivoting arms are adjustable so as to have a large gap between the electrodes, the arms are easily adjusted in the hubs and various length arms are easily fitted giving easy access to difficult joints.

The sensors needed are placed in relevant positions for most effective detection. The current sensor is a coil from Rogowski Coil manufacturers. This sensor gives a voltage output proportional to the current induced through the arms of the spot welder. The voltage sensor is, simply, two leads connected to each spot welder electrode. This signal is then passed through a buffer/amplifier circuit and into a computer. The infrared sensor is a photoelectric infrared-diode placed at 45° from the vertical and at a distance of 2cm from the electrode tips, also passed through a buffer/amplifier circuit and then into a computer. Finally, the acoustic sensor is a microphone placed at the same position as the infrared sensor. Figure 2 is a labelled photograph of this set-up.

All, these sensors have their readings captured using relevant data acquisition boards and recorded using a personal computer. The data-acquisition programme is specifically written for this project in the Pascal based Delphi environment. The signals are sampled at 1000 Hz each for current, voltage, and infrared, while the acoustic signal is sampled at 44100 Hz. All the data is recorded simultaneously for 1 second.

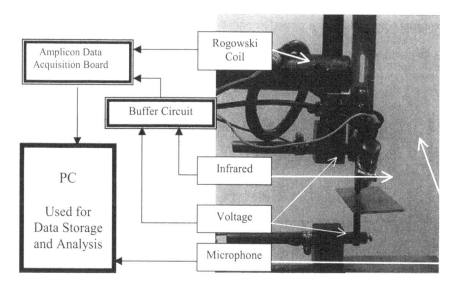

Figure 2: The Experimental set-up

3. Results and Analysis

3.1 Current Sensor

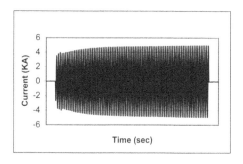

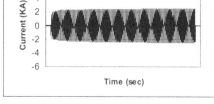

Figure 3: Current values for a good weld.　　　*Figure 4: Current values for a poor weld.*

From figures 3 and 4 a good weld can be easily identified from a poor weld by the amount of current induced through the welding circuit. However, the problem of identifying a good weld is not as simple as a current level threshold. There can be other electrical contacts on the weld plates (through impurities or deformations) through which the current can flow. So, although the current induced through the weld circuit is an important factor, it is by no-means the only factor one should look at.

The current sensor is an already available technology that has many advantages including remote sensing, robustness, and reliability. This sensor is an available technology meaning that it has been perfected and is therefore accurate, robust and reliable. These characteristics make this sensor ideal for factory environments except that it is also relatively expensive. Also, as the sensor measures the electromagnetic field generated by such high currents, therefore it is a non-impinging sensor that does not interfere with the welding circuit.

3.2 Voltage Sensor

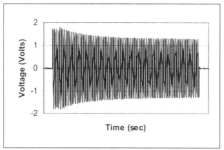
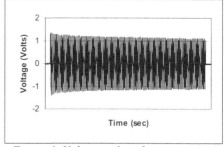

Figure 5: Voltage values for a good weld.   *Figure 6: Voltage values for a poor weld.*

Figures 5 and 6 show the difference between the voltage readings of a good weld and a very poor weld. As can be seen the difference between them is not too great. This verifies that the voltage readings from resistance spot-welding are not particularly useful on their own. However, when trying to calculate the dynamic resistance these readings are necessary. In spite of this, it was established in previous studies [8] that the integration of the voltage readings over the period of the weld could give an indication of the heat generated. Yet, this study finds that the infrared sensor is far more accurate for this purpose.

3.3 Infrared Sensor

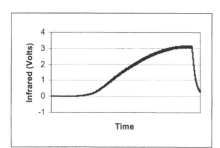
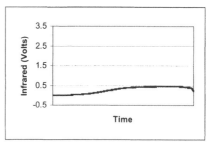

Figure 7: Infrared values for a good weld.   *Figure 8: Infrared values for a poor weld.*

Thermoelectric voltage measurements between the work piece and the electrode have been tried previously [9]. This technique has a problem associated with metal purity, contaminants and the quality of metal preparation regarding the reproducibility of results. The infrared sensor has been used and the readings show the heat radiated from the tip of the welding electrodes. Figures 7 and 8 show a good weld and a poor weld respectively. This sensor is relatively straight forward in that the data it provides, one can through a lengthy study identify the amount of heat needed to generate a specific weld sizes. The photoelectric infrared diode used is relatively inexpensive, robust, and small in size. All these factors draw a picture of the perfect sensor, however, there are draw backs to this sensor also. The infrared readings vary greatly depending on its position with respect to the welding electrode tip. Also, the heat radiated by the electrode tips is not necessarily proportional to the heat generated at the weld.

3.4 Acoustic Sensor

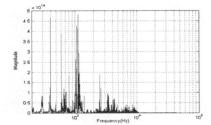 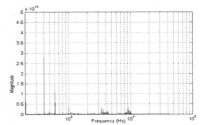

Figure 9: Frequency analysis for a good weld.   *Figure 10: Frequency analysis for a poor weld.*

Finally the acoustic sensor it is small, robust, inexpensive, and as shown in figures 9 and 10 the sensor can distinguish between a good and a poor weld. This sensor's readings can indicate the amount of electric power used for a weld, it can indicate the amount of vibrations by the electrode tips, and it can record the occurrence of splash (or splatter) during welding. Again, one finds a promising sensor able to give an indication of weld quality, but this sensor also has its disadvantages. Firstly, the acoustic sensor by its nature is vulnerable to noise interferences. Also, this sensor because of its high sampling rate (44.1kHz) has a large amount of data to record and analyze which would incur extra expenses.

4. Conclusions and Future Work

The results of these tests show that these four sensors give accurate indication of weld quality in a laboratory environment. It is important to note that although the raw readings show great accuracy in weld quality prediction, this accuracy becomes even more precise when the raw readings are analyzed. So, integrating the current readings over the period of the weld, combining the readings of the voltage sensor and the current sensor to produce dynamic resistance, finding the average infrared reading and its integral over the weld period, and Fourier analysis of the acoustic sensor readings all combine to make this out-put of these sensors even more accurate.

One must comprehensively resolve any queries as to the accuracy, robustness, reliability, and cost effectiveness of these already investigated sensors. All four mentioned sensors must undergo tests that subject them to noise. This noise will be electric, electromagnetic, and environmental. Under these stresses, the four sensors will need to show similar accuracy because through these tests factory level robustness is determined.

Figure 11 shows the possible future work of this study. A system that reads in the data from these sensors, analyses them, and gives one final verdict on the quality of the weld. As seen in this sample graphical user interface, the user would be able to use any combination of sensors required, the programme would record the data of all the welds as well as give a decision of the weld quality. The final aim is to make this kind of system factory worthy, thereby suitable for industry.

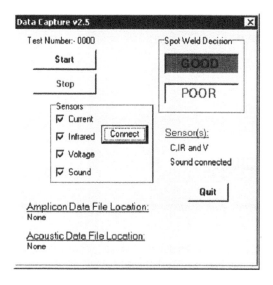

Figure 11: Future work based on this study includes an automated quality assurance system.

References

[1] Larson C.F., "Bibliography on Resistance Welding", 1950-1971, Welding Journal, Vol. 51, 1972.

[2] Krieger R.J., Wenk S.A., McMaster R.C., "Non destructive test methods for inspection of welded joints", Welding Journal, Vol. 33, 1954.

[3] Savage W.F., Nippes E.F., Wassel F.A., "Dynamic contact resistance of series spot welds", Welding Journal, Volume 57, February 1978.

[4] Livshits A.G., "Universal Quality assurance method for resistance spot welding based on dynamic resistance", Welding Research Supplement, September 1997.

[5] "Specification for resistance spot welding of uncoated and coated low carbon steel", British Standards, BS1140:1993.

[6] "Spot welding system and method for sensing welding conditions in real time", Matsuyama K., US Paten No. 6,506,997 B2, January 2003.

[7] "Tecna S.p.A (Welder Manufacturers) website," http://www.tecna.net/ , 2003.

[8] Ablewhite J. D., "The Application of Artificial Intelligence Techniques to the Control of Spot Welding", University of Swansea, 2000.

[9] Faber W, Uhimann M, "Temperature Measurement in resistance welding", Welding Research Abroad, December 1994.

Online Cost-Effective Sensor System for Microwave UV Lamp

I Pandithas, J D Cullen, A I Al-Shamma'a, J Lucas

The University of Liverpool, Department of Electrical Engineering and Electronics, Liverpool L69 3GJ, UK.

Abstract: This paper is concerned with the development of an online cost-effective sensor system, which calculates the efficiency of the microwave ultraviolet (UV) lamp. Nowadays, the commercial technology provides UV lamp systems for water treatment, curing applications and food processing which are either bulky or difficult to maintain. The novel microwave UV lamp, overcomes the limitations of the current technology. As a part of this project, a simple cost-effective online sensor was developed to monitor the UV intensity and evaluate its performance. The sensor system is based upon basic principles of optics and light sensing. The system setup and the results obtained are thoroughly explained.

1. Introduction

UV radiation is a form of energy, which can be absorbed by and can bring about structural changes of systems [1]. The exposure of microbiological systems to UV radiation, within the wavelength range from 220nm to 300nm can dissociate the DNA, which are vital to metabolic and reproductive functions and thus inactivating the microorganisms. The germicidal function of the UV light is important for water and food processing. Additionally, the interaction of deep UV spectral lines (185nm) with the oxygen content of the air helps produce ozone as well.

The current technology offers UV lamps that are based upon the electric field sustained by two electrodes on either side of the lamp [2]. There are practical limitations to this approach that reduce the efficiency of such lamps and restrict them lengthwise in terms of output UV power.

The microwave UV lamp does not use any electrodes and microwaves of 2.45GHz of frequency are coupled into a resonant cavity, which sustains a TE_{010} propagation mode at resonance [3]. The lamp is supported by the cavity where the microwaves resonant cavity. The result is a more efficient ultraviolet light source that does not suffer any of the restrictions of the commercial technology. The importance of introducing a new sensor technology is outlined by the fact that there is always the need for lamp monitoring for UV processing installations that should not only indicate the state of the lamp but calculate their efficiency, as well. For industrial applications the sensor also has to meet certain specifications, such as durability, small size, high fidelity and low cost [4].

2. Experimental Set-Up

2.1 Design Considerations

The most common light sensor is the photodiode. However, their technology has advanced to the level that specialised types are available for commercial use, such as the UV sensitive photodiode, available from RS components, UK [5]. This is the base for the sensor system and its characteristics are shown in table 1.

TABLE 1. UV photodiode Characteristics

Active area		Responsivity Amp/Watt (typical)			Peak Responsivity
cm^2	cm	190nm	245nm	340nm	(Typical)
5.8×10^{-2}	0.24 x 0.24	0.12	0.14	0.19	950nm

For the case of the microwave lamp the approach of utilising a PID UV sensitive photodiode has proven to be sufficient and cost effective at the same time. It also gave the opportunity to devise a simple MOSFET transimpedance amplifier of variable gain to couple out the UV signal levels to the form the sensor to the digital storage devices, which is shown in figure 1.

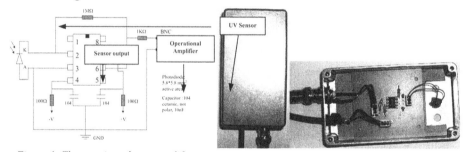

Figure 1: The transimpedance amplifier circuit

Figure 2: The sensor box.

The transimpedance amplifier circuit raised the detector output at sufficient levels. A feedback resistor of 10^6 Ohms provided the amplification factor. The basic component of the circuit is the TL071 JFET integrated operational amplifier from Texas Instruments. The choice of the specific operational amplifier was based upon the fact that the TL071 has low harmonic distortion of the input signals, low noise and high slew rate. Consequently, TL071 does not suffer from the glitches of the common operational amplifiers such as the CA314 or similar common op-amps. The above characteristics of the TL071 would also ensure the clarity of the signal under all experimental conditions and low distortion levels under pulsed lamp operations. The detector and the sensor circuit were enclosed in a metallic box. Additionally, as a precaution, all the cabling used was shielded and the equipment used was checked for grounding to avoid any noise induced by the microwaves. The sensor box is illustrated in Figure 2. Prior to any measurement, the photodiode [5] has to employ a UV bandpass filter of the desired wavelength that has to be measured. For the purposed of this study, two filters were used, the one for the 254nm emission spectrum and the other for the 185nm emission spectrum, obtained from Coherent/Ealing corporation [6]. Their characteristics are presented in Tables 2 and 3.

TABLE 2. 254nm Filter characteristics

Centre Wavelength (nm)	Bandwidth FWHM (nm)	Centre Tolerance (nm)	Peak Transmission (%)	Source	Catalogue Number
185	27.5	2.5	15	Hg	37-2968

TABLE 3. 185nm Filter Characteristics

Centre Wavelength (nm)	Bandwidth FWHM (nm)	Peak Transmission (%)	Effective Index	Number of Cavities	Catalogue Number
254	7	10	1.45	2	35-2807

Finally, the UV lamp envelope material is quartz. The grade of the quartz affects the transmission efficiency of the material [6] and a coefficient can be derived from the glass manufacturer for the relevant wavelengths the lamp operates, according to the tube wall thicknesses.

2.2 The Experimental Apparatus

The experimental apparatus is illustrated in figure 3. It comprises of the microwave UV lamp, the sensor box equipped with the filter and the data acquisition system, which can be a PC-based data acquisition system or a digital storage oscilloscope. This depends upon the application or the experimental requirements.

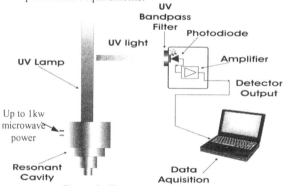

Figure 3: The experimental setup

Since the parameters of the system are known (Tables 1 to 3) an expression that correlates the detector output and the UV light strength can be derived. Let V be the detector readout (in Volts) and P the UV radiation (expressed in $\mu W/cm^2$). The detector signal is affected by the transmission efficiency of the filter at the specified frequency for this case either, the sensitivity of the detector and the detector area and finally, by the amplification factor of the detector circuit (which is expressed as the value of the feedback resistor used on the operational amplifier of the circuit) as illustrated in table 4.

TABLE 4. Calculation Parameters

	Detector Sensitivity	Filter Transmission	Detector Area	Glass Transmission
256nm	0.14 (A/W/cm^2)	0.13	5.8 mm^2	0.9
185nm	0.12 (A/W/cm^2)	0.17	5.8 mm^2	0.53
Feedback Resistor for Sensor circuit: $10^6\Omega$ or $1M\Omega$				

For the 254nm emissions, it is:

$$P = \frac{10^6}{0.14 \times 0.13 \times 5.8 \times 10^4 \times 0.9} = \frac{10^6}{950.04} \times V = 1052 \times V \tag{1}$$

Using the same method, a similar approach is applied for the 185nm, where:

$$P = 1594.6 \times V \tag{2}$$

Equations (1) and (2) are also the base of the lamp efficiency calculations and the general form is given by equation (3).

$$P = a_{sensor} \times V \tag{3}$$

a_{sensor} takes the value of 1052 for the 254nm emissions and 1594.6 for the 185nm emissions.

3. Efficiency Calculations

The calculations assume a uniform emission lamp model as shown in Figure 4a, where 'd' is the lamp-sensor distance from the detector and 'l' is the lamp length. The important parameter that needed to be resolved was the solid angle viewpoints induced by the sensor finite area. Figure 4b shows the experimental setup followed to resolve the issue. From the experimental measurements, using a simple trigonometric approach, the solid angle of the sensor was found to be 30.4 degrees from the isosceles triangle formed by the sensor and the "viewable" lamp, which was 12cm of the 40cm lamp length that was used. The experiment lamp-sensor distance was set at 10cm.

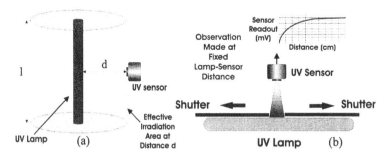

Figure 4: (a) The lamp emission model (b)The solid angle experimental setup

The coefficient derived from the solid angle (a_{solid_angle}) shows the observable lamp length. It is:

$$a_{solid\_angle} = \frac{2d\sin(30.4°)}{l} \qquad (4)$$

The integral of the detector area for the cylinder formed by the 'visible' lamp length, is given by:

$$a_{transmission} = \frac{2\pi d(2d\sin(30.4°))}{5.8\cdot10^{-2}} \qquad (5)$$

The correction factor for the solid angle (a_{sa}), allows for the whole lamp area to take place in the emissions calculation, which is:

$$a_{sa} = \frac{1}{a_{solid\_angle}} \qquad (6)$$

This leads to a generic formula for the lamp emissions, which is a revised version of equation (3):

$$P_{germicidal} = a_{sensor}\cdot a_{transmission}\cdot a_{sa}\cdot V_{sensor} \qquad (7)$$

The same method follows for the 185nm emissions using the appropriate coefficients.

Assuming that the microwave power is coupled into the cavity resonator via a coaxial cable that suffers losses, then the coefficient derived for cable losses of D decibels is:

$$a_c = \left(10^{\frac{D}{10}}\right)^{-1} \qquad (8)$$

The net microwave power can be now calculated from the incident microwave power P_i, the reflected power P_r and the cable losses a_c:

$$P_{net} = P_i\cdot a_c - \frac{P_r}{a_c} \qquad (9)$$

Then, the efficiency of the microwave lamp can be calculated by referring to the net power, as shown in Table 5.

TABLE 5. Efficiency calculations

254nm UV Efficiency	185nm Efficiency	Total Losses
$\eta_{germicidal} = \left(1 - \dfrac{P_{germicidal}}{P_{net}}\right)\cdot 100\%$	$\eta_{ozone} = \left(1 - \dfrac{P_{ozone}}{P_{net}}\right)\cdot 100\%$	$\eta_{total} = \left(1 - \dfrac{P_{germicidal} + P_{ozone}}{P_{net}}\right)\cdot 100\%$

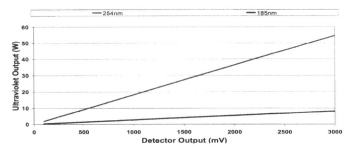

Figure 5: 254nm and 185nm UV radiation Calibration chart for the case of a 64Watts, 32cm microwave lamp

4. Results and Discussion

Using the equations above, the software has the ability to calculate the lamp efficiencies under different experimental conditions using the sensor values. Furthermore, calibration charts specific to a microwave lamp can be derived, which help to decide the system operating conditions, depending upon the sensor values, as shown in Figures 5 and 6 for the case of a 32cm lamp run under 64Watts of net microwave power. Figure 7 shows the experimental results obtained from a 32cm, 53.8W microwave lamp as they appear from the data logging device.

To conclude, this implementation fulfils the requirements for a compact cost-effective sensor system that calculates the microwave lamp efficiencies under different experimental conditions. The implementation however, has ample room for improvements, such as the implementation of parameters concerning the commercial lamps, and probably a software version that provides real-time responses or an implementation to an optical fibre sensor system for UV emissions [8]. Previous research [2] has outlined the need for such a sensing system that can be highly competitive to the current ultraviolet sensors, being accurate and low cost, for industrial applications.

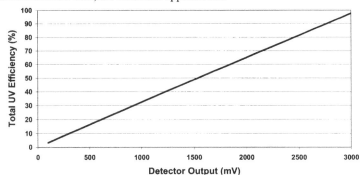

Figure 6: UV lamp efficiency Calibration Chart for the case of a 64 Watts, 32cm microwave lamp

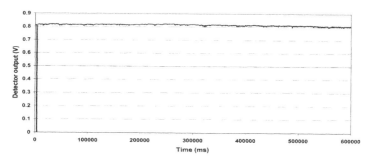

Figure7: Data logger response for a 32cm 53.8W net powered microwave lamp

References

[1] Harm W., "Biological Effects of Electromagnetic fields", Cambridge University Press – ISBN: 0521-2221-8, (1980).

[2] Al-Shamma'a A.I., Pandithas I. and Lucas J., "Low Pressure Microwave Plasma UV Lamp for Water Purification and Ozone Production", J. Phys D: Appl. Phys. Special Issue, Vol. 34, No. 14, pp.2775-2781, (2001).

[3] Pandithas I., Al-Shamma'a A.I., Cullen J.D., Lucas J.
"Low Pressure Industrial Microwave Plasma UV Lamp for Germicidal Applications and Ozone Generation", Proceedings of the XIV International Conference on Gas Discharges and Their Applications, Vol. 2, ISBN 0 9539105 1 2, pp: 124-127, (2002).

[4] Senior J.M., "Optical Fibre Communications, Principle and Practice", Prentice Hall, International Series in Optoelectronics, Second Edition – ISBN: 0-13-635424-6, (1992).

[5] RS Components Datasheet, "Photodiodes", Issued July 1998 (Order Code 298-4562)

[6] Ealing Catalog Inc., 3845 Atherton Road, Suite #1Rocklin, CA 95765, USA.

[7] General Electric Quartz Inc.,"Lamp Tubing" – technical information document, General Electric Quartz website (www.gequartz.com), Willoughby, OH 44094, USA, (2002).

[8] Fitzpatrick C., Lewis E., Al-Shamma'a A.I., Pandithas I., Cullen J., Lucas J., "An Optical Fibre Sensor Based on Cladding Photoluminescence for High Power Microwave Plasma Ultraviolet Lamps Used in Water Treatment", Optical Review, Vol. 8, No. 6, pp 459-462, (2001).

Poster Programme
Paper presented at Sensors and their Applications XII, September 2003
©2003 IOP Publishing Ltd

Sensing of the Microwave High Velocity Thermal Spraying

P Mavromatidis[1], C Fitzpatrick[2], A I Al-Shamma'a[1], J Lucas[1]

[1] Department of Electrical Engineering & Electronics, University of Liverpool, Liverpool, L69 3GJ, UK.
[2] Department of Electronic & Computer Engineering, University of Limerick, Ireland.

Abstract: This research program investigates the use of microwave plasma, at 2.45GHz frequency, to preheat nitrogen gas/coating powders when using a high velocity spray technique. The heating process of this novel technique is obtained by the combination of dielectric and microwave plasma heating by using a long interaction cavity. As a result the thermal efficiency of the process will be as high as 50% compared with the present commercial plasma heated device which is in the order of 5% enabling temperatures of 1000°C to be obtained. The paper presents a preliminary result of an optical fibre sensor operating in harsh microwave environment. The design and construction of the novel microwave thermal spraying system will be described.

1. Introduction

The thermal spraying process evolved in the early 20th century and for many years on was used for heavy-duty components, bridges, tank cars, ships and water tanks. With the evolution of emerging technologies in the global market the process of thermal spraying has been addressed at a larger scale of applications. Today it is broadly applied to a variety of areas, like aerospace, automotive material processing and many others to provide longer life expectancy, durability, mechanical strength and protection from corrosion. As thermal spray coatings have been widely spread in many areas they have received considerable industrial and academic research in the effort to produce new processes for higher efficiency coatings and extended variety of coating materials.

This research program investigates the novelty of using 2.45 GHz microwaves to preheat a mixture of gas and coating powders using a long heat interaction between the mixture and the heat source to increase the adhesion efficiency of the projected particles and the surface. In most spraying methods, the sensing mechanism of monitoring the stability and the power transfer to the coating process is poor because it is mainly based on theoretical calculations. However, in the case of microwave thermal spraying the use of microwaves raises problems of non-linearity, EM interference and high cost equipment. As a result, the microwave market is dictated by very expensive sensors that require constant calibration and are not always as reliable. In this research program a new methodical sensing system is implemented to the specific spraying technique to overcome the hazardous environment of high temperatures and microwaves. This sensing mechanism [1] is based on the principle operation of a microwave electric field optical sensor probe.

2. Microwave spraying system

The spraying technique introduced in this project is a direct, continuous melting spray technique using a high efficiency-heating source to melt the powder and only then accelerate it to high velocities to be projected onto the substrate surface. Most of the existing spraying processes mainly rely on a small heat interaction zone to rapidly melt the powder particles concentrating on high velocities to deposit the coating on the surface. The microwave thermal spraying technique, instead, uses a long interaction path to gradually and uniformly distributes the temperature to the particles reaching at slightly above the melting temperature of which use a high kinetic energy to propel them onto the surface. The microwave configuration uses a WG9A waveguide system shown in figure 1, operating at 2.45 GHz.

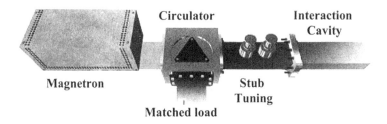

Figure 1: The microwave system

The system-heating source is that of an electrodless, low-pressure gas discharge lamp. When the low-pressure gas is interacted with microwaves its atoms are ionised creating a plasma discharge, which evolves heat and light [2]. Unlike commercial UV lamps that use electrodes to propagate power and produce plasma, this lamp is made of dielectric material allowing a dense discharge and exposure to larger powers. In order to absorb the temperature emitted, a spiral flow configuration was constructed surrounding the UV lamp. This structure will absorb the temperature allowing the heat transfer from the UV lamp to the gas powder mixture and at the same time provide a long interaction path by keeping the mixture near the heat zone for a longer period of time. Figure 2 depicts the experimental setup of the heat source and gas flow structure.

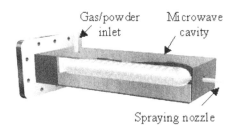

Figure 2: The heat source

The nitrogen gas is selected based up on allowing the temperature transfer from the heat source to the powder particles and that will further propel them with the necessary kinetic energy. The advantage of nitrogen is its neutral behaviour that will prevent oxidization of the particles during their heating process and its low cost. When the molten particles are projected on the surface they form "splats" adhering onto one another and the surface. Coating thickness is determined by controlling the speed of movement of the surface (or the spray gun), the distance of the spray gun and the surface, the number of spraying passes, the gas flow and pressure of the carrier gas and the amount of powder mixed with the gas flow controlled by a gas-powder mixer [3]. Figure 3 depicts two examples of the coating thickness. Both coatings are tin powder sprayed with the microwave thermal spraying technique on a steel surface at 30 litre/min. Although tin melts at around 230°C a higher temperature of the carrier gas was used in order to fully melt all powder.

Steel surface   Tin coating

Figure 3: Tin coatings on steel surface

3. Commercial microwave sensors

The harsh environment of microwaves and extreme temperatures make the sensing of the microwave spraying process parameters complex and difficult to obtain. In principle the operation of a microwave system, the microwave propagation and field distribution can be fully described using Maxwell's equation to obtain a theoretical model. If the transmission guide was to be uniform in cross section and non-dissipative, the relative values of the fields could be determined by the cross sectional dimensions of the waveguide and the magnitude and direction of power flow [4]. However in practice this is not always the case.

In the microwave thermal spraying process the presence of dielectric materials of the heat source, the plasma discharge inside the cavity and the adjustment of the stub tuner will change the resonance of the cavity making theoretical calculations complex and non-linear. This resorted in the use of empirical and practical methods for an indirect measurement of microwave parameters. These methods are based on the use of semiconductors or thermally sensitive devices. The most widely spread methodology is the use of semiconductors such as the Schottky diode [5]. The principle of its operation relies on a small loop antenna inside the electromagnetic field. When the electromagnetic fields cross the wire loop a proportional electric current is induced inside the wire, which is used as a measurement reference. The problem however with semiconductors is their non-linear behaviour and their vulnerability to microwave leakage. As a result constant calibration is required. Thus they can be used only for low power measurements and are unsuitable for high temperature systems such as the microwave spraying process. As the most common and reliable power measurement technique, the calorimeter is distinguished which is part of the matched load [6].

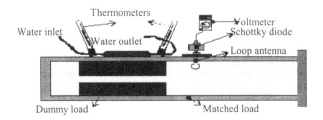

Figure 4: Matched Load Cross-section

The fundamental calorimeter technique is based on the use of a load that will dissipate heat in the absorption of microwaves. The temperature rise of the load will allow a physical measurement of the microwave power absorbed. The disadvantage of the calorimeter method is can mainly be used for high power measurements. The temperature rise is measured by two mercury thermometers, see figure 4, adjusted at the cooling inlet and outlet of the matched load [3]. This makes it an unreliable method for low powers as the specific heat capacity of water will allow small increases of temperature (~0.1 °C), which is difficult to monitor.

4. Microwave spraying sensor

The initial monitoring technique of the microwave thermal spraying process faced crucial problems in microwave power and temperature measurement. At first the propagated microwave power was measured using a Schottky diode adjusted to a loop antenna inside the matched load to measure the reflected power and thus deduce the power dissipated in the heat source. This proved to be rather impractical and unreliable since the diode was often subject to leakage radiation and frequent calibration was required. Another technique to use the calorimeter method, which proved to be inaccurate since the operating powers of the spraying system are less than 1 kW and the monitoring temperature, was limited to a 0.1°C resolution. Further sensing involved the use of a thermocouple placed at the outlet of the spraying nozzle monitoring the temperature of only nitrogen flow, the carrier gas. When the critical melting temperature was reached, the thermocouple had to be manually removed and only then was the powder fed to the gas stream to be coated to the surface. This empirical methodology raised concerns since the temperature during the spraying process would remain uncertain and any critical disturbance or variation of the microwave propagation would result in failure of the coating. The advantage though of using a low gas discharge lamp is that no matter what the form of the emitted temperature is the dominating one; there will always be a visible light emission proportional to the microwave power. Thus, allowing the monitoring of the heat source through the intensity of the plasma discharge. When the temperature of a gas increases electrons in the molecules are raised to higher energy states and with a variety of energy transitions taking place, photons are emitted. The higher the temperature, the greater is the function of these photons at higher energies. The gas pressure is chosen such that the heat emission to be the dominate one limiting UV emission to the region of UVA, typically 400-320 nm. This eliminates the requirement of using phosphor filter layers for UV to visible light conversion. So the dominate energies released from the discharge lamp is heat and light. This allows the use of an optical fibre to measure the intensity and relate it to temperature and power absorption. Furthermore the use of quartz as the lamp construction material does not only serve the purpose of high temperatures (melting temperature of quartz is around 1700°C) but also excellent transmission properties for the ease of the intensity measurements.

5. Experimental results

For the purpose of intensity measurement of the heat source, an aperture is created at the top of the cavity facing the UV lamp. The aperture is covered by a metallic mesh to prevent microwave leakage. The optical fibre is placed at a distance above the aperture of the cavity as shown in figure 5 and uses a long length cable before being coupled to a photodiode. Preliminary investigations have been carried out to use the optical fibre sensor to measure the system intensity. These measurements have lead to the predictions of the system total power and temperature. Argon and nitrogen were individually tested for the plasma discharge inside the UV lamp. The pressure of argon gas was set at 0.11 mbar and the microwave system incident power was set at 80 Watt incremented to 300 Watt. Figure 6 depicts the direct proportional of temperature and intensity versus incident power. In order to enhance the cavity Q factor (i.e. highly resonated) a tuning effect can be observed at 200 Watt of which the system temperature has increased from 140°C to 200°C.

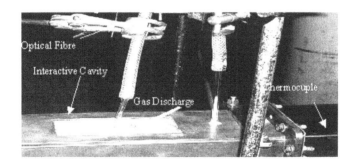

Figure 5: Experimental setup for optical measurements

Same system setup has been used with the nitrogen gas discharge. Figure 7 displays the temperature and intensity response, again showing an almost linear response with the increase of microwave power.

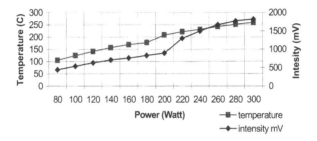

Figure 6: Temperature and intensity of argon discharge

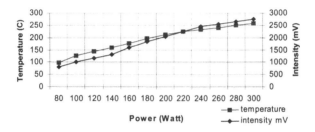

Figure 7: Temperature and intensity of nitrogen discharge

6. Conclusions

A very simple and low cost optical fibre sensor has been designed, constructed and implemented successfully in microwave environment. The preliminary measurements have proved the applicability of such a sensor with the response of intensity being proportional to temperature and microwave power. Further development will accomplice a reliable and accurate sensing mechanism.

References

[1]   Fitzpatrick C., Mavromatidis P., Lewis E., Lucas J., Al-Shamma'a A.I. "Experimental investigation into low pressure gas discharge in microwave electric field optical sensor probes", Sensor review journal, Volume 23 · Number 1 · 2003 · pp. 44–47.
[2]   Mavromatidis P., Al-Shamma'a A.I., Lucas J., Stuart R.A., "Microwave plasma high velocity thermal spraying", XIV International Conference on Gas Discharges and their Applications, 1-6 September 2002,
[3]   Mavromatidis P., Al-Shamma'a A.I., Lucas J., Lucas W., "Computer control of a microwave spray system" , 50th International conference on welding technologies, Sydney, Australia 26th-30th August 2002.
[4]   Ginzton E.L., "Microwave Measurements", McGraw Hill, 1957.
[5]   Laverghetta T.S., "Modern Microwave Measuremends and Techniques", Artech House, 1988.
[6]   Bailey A.E., "Microwave Measurement", IEE, 1985.

An Innovative Approach to Tyre Pressure Monitoring

N Arvanitis, J D Cullen, A I Al-Shamma'a, J Lucas

The University of Liverpool, Department of Electrical Engineering and Electronics, Liverpool L69 3GJ, UK.

Abstract: In recent years the automobile industry has been adopting ever-higher safety standards. Under this light, the development of a tyre pressure monitoring system that aims to overcome the difficulties faced by the already present systems becomes even more important. In order to meet that specific need, a system capable of relaying the status of tyre pressure from a sensor at the wheel to a display unit in the vehicle cab was constructed. This paper describes the Controller Area Network (CAN) bus networking technique, the design and implementation of the system and the results of in-field testing will be presented.

1. Introduction

Low tyre pressure can adversely affect fuel consumption and vehicle handling. It can also be the sign of a slow puncture, in which a sharp object pierces the tyre and then remains on the tyre, in some cases for up to 250 miles, resulting in a gradual loss of pressure, which even further reduces the response of the tyre in aggressive situations, for example during an emergency stop. For trucks this is also a problem, but also a similar condition can arise where there are two wheels on either end of an axle. In such a scenario, if a blow-out occurs in one wheel while the other wheel is under inflated, resulting in an overloading of the remaining tyre, then the chances of failure of the remaining tyre are increased. Previous pressure monitoring methods traditionally relied on the presence of some form of powered unit on the wheel, typically a transmitter; sometimes a microcontroller is required as well. This resulted in a need to have regular maintenance to change the power supply and to ensure the transmitter is still working. Another problem with this type of system may occur if several vehicles are in close proximity to each other, where the transmitter signal may cancel each other out. Therefore a system was chosen that allows the measurement to be taken off the wheel with no power being used. This is based up on using capacitance to provide the measurement [1]. This was chosen because of the non-contact nature of capacitance plates allowed a pickup system to be constructed to get the data off the rotating wheels onto the non-rotating vehicle body.

2. The CAN Bus

The CAN (Controller Area Network) bus is an ISO defined serial communications bus that was originally developed during the 1980's for the automotive industry [2]. Its basic design specification called for a high bit rate, high immunity to electrical interference and an ability to detect any errors produced. Not surprisingly due to these features the CAN serial communications bus has become widely used throughout the automotive, manufacturing and aerospace industries. CAN is an asynchronous serial bus system with one logical bus line. It has an open, linear bus structure with equal priority bus nodes. A CAN bus consists of two or more nodes [3], with nodes being able to be added or removed without affecting the remaining nodes. This allows easy connection and disconnection of bus nodes. The bus lines are called "CAN_H" and "CAN_L". They are driven by the nodes with a differential signal. The twisted wire pair is terminated by terminating resistors at each end of the bus line. The maximum bus speed is 1 Mbaud with a bus length of up to 40 m. The maximum number of nodes on a CAN bus is typically 32.

3. The Detectors and Capacitance Ring

The pressure detector being used is a specially constructed semiconductor based switch. This switch is either open if the pressure is suitable, or shorted out if the pressure drops below a preset level. The switch part of the detector operates in the same way as a standard single pole - single throw switch. A high resistance (1 MΩ) associated with the switch if a relatively small voltage (< 1 Volt) is applied across the switch. This is due to the resistance of the P-N junction, which is very high when the breakdown voltage has not been reached [4]. This necessitates including this resistance in the equivalent circuit. The sensor is then connected to the capacitance pickup system, which consists of two segmented rings, one fixed on the non-rotating part of the wheel assembly and the other placed on the spinning wheel. The two rings face each other creating a capacitor. The system is illustrated in figure 1.

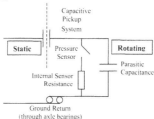

Figure 1: Electrical Equivalent Circuits for the pressure sensor

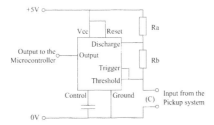

Figure 2: The 555 timer astable mode circuit

As can be seen that, the right hand side is on the spinning wheel, while the left hand side is not moving. The circuit includes the components of parasitic capacitance, which is due to the lengths of cable used to connect up the pressure sensor to the capacitance pickup system. In most electronic systems the presence of parasitic capacitance is an undesired element, but in this type of system the parasitic capacitance allows the system to work. Indeed, without the parasitic capacitance present on the wheel, this system would not function. To be able to perform measurements of this value the sub-system is then connected into a system that takes the capacitance as the input variable, in this case, a 555-timer chip is used [5]. The frequency of the 555 timer at astable mode is governed by three components, *Ra*, *Rb* and *C*. The equation for the output frequency is:

$$f = \frac{1.45}{(Ra + 2Rb)C}$$

(1)

By fixing any two of these components, the output frequency can be made to vary with the third component. The natural component to adjust is the capacitance, *C*, because the frequency is then inversely proportional to the capacitance. The output of the 555 timer fed into the microcontroller, where the frequency can be measured and relayed to the display unit to warn the driver.

4. The Microcontroller

The Microcontroller used in this application is an Infineon C505CA Microcontroller. This device is used because it has the relevant electronics for driving the CAN Bus built into the controller, therefore eliminating some of the external components required. The only external components required for this microcontroller, are the timing circuit, reset circuitry, and the CAN line driver, which takes the single ended CAN output from the microcontroller, and converts it into the differential voltage required by the CAN bus, and also takes the differential voltage present on the CAN bus and converts it into the single ended CAN input for the microcontroller. A circuit board used to develop the code for the C505. Figure 3 shows the developed board, which has the 555 timers used for the capacitance measurement mounted on the board. The two different types of microcontroller code required to operate this application are shown in overview in figure 4. When the power on the vehicle is turned on, all the nodes start to initialise themselves, with the master node (display unit) initialising its CAN controller, LCD display and serial port, while the slave nodes (detection units) initialise their CAN controller and their own board ID.

Figure 3: Final Microcontroller Board

5. In-Field Trials

The vehicle, which was used for the in-field trials, was a fire engine and an outline of the system installation is provided in figure 5. The recording of all measurements and results during the trials was carried out using a laptop PC directly connected through its serial communications port to the display unit. The PC was equipped with logging software implemented in the C programming language. During the trials, the top station microprocessor situated in the drivers cabin would take turns requesting information from each detection unit, wait for an answer, display the answer and repeat for the next wheel. This process was arranged to last for 4.8sec in order to provide enough time for the pulse count to be read by the operator. Although the test was conducted for just two of the wheels, the system was designed and built having four wheels in mind. That leads to a total time of 19.2sec between two continuous samples from the same wheel. Once the system was installed in the vehicle, there were two kinds of tests performed: A static test and a driving test. The purpose of this test was first to insure that the system was producing the desired response. Once the proper function of the system was established, then the static test could provide a valid reference point against which the performance of the system during a driving test could be compared and evaluated.

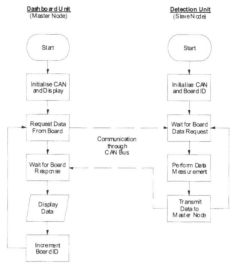

Figure 4: Outline of software operation

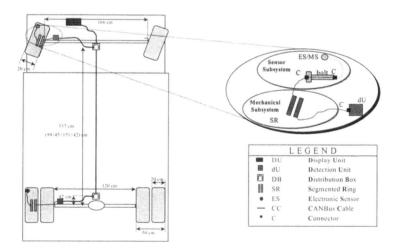

Figure 5: System Overview

6. Static Test

During the static test one wheel was lifted so it could freely rotate. The wheel was then marked in regular 22.5° intervals into 16 segments. For each one of these 16 points two sets of measurements were taken, one for each state of the pressure switch. The change of the effective area is the only factor that governs the change of the capacitance during the wheel rotation in a static test, since both the distance between the two rings and the dielectric between them have constant values. Figures 6 and 7 show a sample of static measurements and the minimum and maximum count values for wheels 1 and 2 for each switch state (open or shorted). It appears as if the behaviour of the quality factor is contradicting the theoretical predictions.

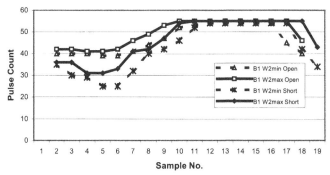

Figure 6: Static pulse count values for front wheel configuration, where B1 W2 is the front sensor board for the right wheel (open and short switch).

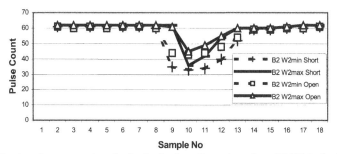

Figure 7: Static pulse count values for back wheel configuration where B2 W2 is the back sensor board for the right wheel (open and short switch).

7. Driving Test

The trial consisted of a test drive of the vehicle during which the response of the system was recorded. The test drive was contacted over a rough terrain with a speed between 40-60Km/hr. This was done in order to subject the capacitance rings in a fair amount of vibration so their stability can be tested, as well. The results are shown in figure 8. The front wheel configuration (B1) seem to be more stable while the response of the back wheel configuration (B2) shows more fluctuation. That can be attributed to a less stable fitting and to the fact that the back wheels of a loaded back wheel drive vehicle, like the truck used for the test, are subjected to greater vibration, which undermines the mechanical stability of the rings. That fluctuation only comes into effect for the minimum count values, since the maximum count values are less prone to fluctuation because they're only affected by the parasitic capacitance of the system, which is constant.

8. Conclusions

The in-field trials, although not extensive enough to produce firm conclusions, indicated that the system displays the required stability to perform efficiently in a "real life" situation. One of the main objectives of the tests was to investigate in which degree the performance of the electronics subsystem would be affected by the working environment. It appeared that the presence of heat, vibration and dust did not hinder the operation of the system since it operated as it did in the laboratory tests. Keeping in mind that the casing and installation of the electronics were done using rather simple equipment and materials, it only points to the fact that this is a rather reliable and robust system. Overall, the performance of the system can be deemed satisfactory.

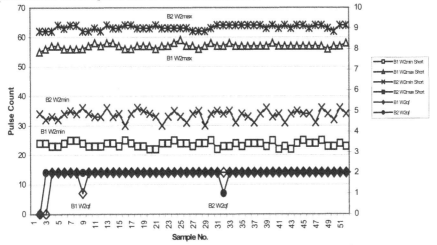

Figure 8: Driving pulse counts values and quality factor (qf) for both wheels (short switch).

References

[1]   Tsagas N., "An Indicator of the Abnormal and Normal Pneumatic Tyre pressure During the Driving and Stopping of a Vehicle", European Patent Number EP0550701, 1996, Worldwide Patent Number WO 92/13730, 1992.

[2]   "ISO 11898: 1992, Road Vehicles - Interchange of Digital Information Controller Area Network (CAN) for High-Speed Communication", International Standards Organisation, 1992.

[3]   "CAN Physical Layer for Industrial Applications, DS102 V2.0", CAN in Automation, http://www.can-cia.de, 1994.

[4]   "Semiconductors and Electronic Devices", Adir Bar-Lev, Prentice Hall, 1984, ISBN 013806265x.

[5]   555 Timer Data Sheet, RS Components, http://rswww.com.

Poster Programme
Paper presented at Sensors and their Applications XII, September 2003
©2003 IOP Publishing Ltd

Development of the Diagnostic Module - an optical Measurement Device for non-invasive determination of haemoglobin content in human blood

J Kraitl[1], H Matz[2], H Ewald[3], H Gehring[4]

[1] STS Systemtechnik GmbH, Bremen, Germany
[2] Institute of Biomedical Engineering, University of Luebeck, Germany
[3] Institute of General Electrical Engineering, University of Rostock, Germany
[4] Clinic of Anesthesiology, University of Luebeck, Germany

Abstract: It is well-known that pulsatile changes of blood volume in tissue can be observed by measuring the transmission or the reflection of light. This method is called photo-electric plethysmography (PhEP or PPG).
The development of the Diagnostic Module is based on the realisation of a photo-plethysmography measurement device developed by STS Systemtechnik GmbH Germany, for the German/Russian manned space flight on board the former MIR station in 1997.
One of the devices tasks was the logging of fluid shifts in the circulating intravasal volume and the evaluation of dynamic changes in the skin micro-circulation, both under micro-gravity conditions and during orthostatic disturbances (Lit. 1).
The objective of a later further developed device described here is the non-invasive continuous measurement of light absorbent blood components in the arterial blood of the human finger, such as oxidised and reduced haemoglobin for the calculation of S_aO_2 and determination of relative haemoglobin concentration change.

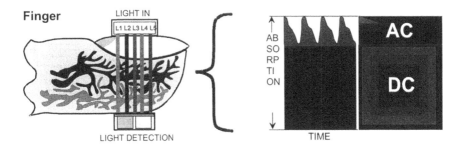

Figure 1: Principle of transmission pulsatile blood volume changes detection and signal-form

Materials and Methods

The method used by the Diagnostic Module is based on the radiation of monochromatical light, emitted by lasers diodes in the range of 600nm to 1400nm, through an area of skin on the finger.

After interaction with the tissue the transmitted light is detected non-invasively by photo-diodes (figure 1). The method makes use of the intensity fluctuations caused by the pulse wave and the ratio of relative changes of the pulse sizes for different wavelengths and its dependence on the optical absorbality characteristics of human blood.

Four of the five laser diodes emit light in the wavelengths range of 600 – 1000nm (675, 808, 905 and 980nm). This is the region in which the blood absorption is dominated by the haemoglobin derivatives. At 980nm besides the haemoglobin absorption exists a weak absorption band for water. Additionally there is a 1310nm integrated laser diode, at this wavelength the absorption of water is dominant (figure2).

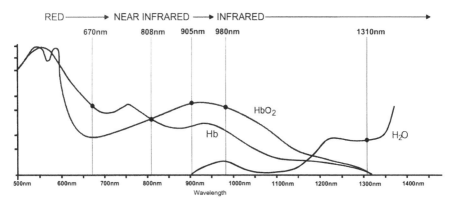

Figure 2: Progression of light absorption for different blood components in principle

Deferrals between the proportions of haemoglobin and water in the intravasal volume should be detected photo-electrically by signal-analytic evaluation of the signals. The computed coefficients are used for the measurement and calculation of S_aO_2 and the relative haemoglobin concentration change.

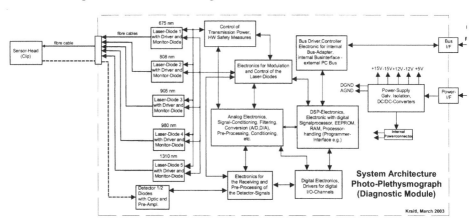

Figure 3: Block diagram of Diagnostic Module

Inside the measurement device the laser diodes are integrated together with the required control electronics. The laser light is conducted to a special optic by means of optical fibres inside the sensor head. Photo-detectors are also contained in the sensor head together with the required pre-amplifiers, the sensor signals detected here will be processed inside the measurement device.

The device electronics consists of the required components for signal amplification and digitalisation and the triggering of the laser diodes, which operates in a pulse mode.

A further component of the measurement device is a high performance DSP-system with a floating-point processor, flash and memory.

This enables the possibility for a DSP software-controlled time-multiplexed laser operation respectively and the control of the appendant receiving channels.

The evaluation of the data and the operation of mathematical algorithms for pre-processing, e.g. digital filtering and averaging, is achieved by using the DSP-software.

The data viewing and storing is achieved via the serial RS232 I/F connection on a Laptop or personal computer. The application software is LabView® programmed.

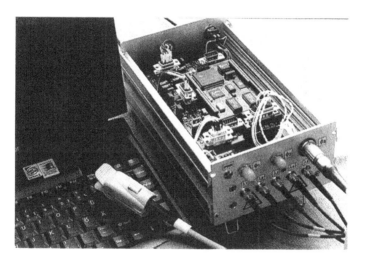

Figure 4: Photo Diagnostic Module with sensor clip and Laptop

The measurement method of Diagnostic Module was tested on patients during two clinical studies:

1. On haemodialysis patients with a terminal kidney insufficiency during dialysis sessions.

By using a dialyser (haemofilter) the patient has dialysat (prevailing water) distracted. This deferral means a fluid reduction for the patient during the ultra-filtration. The change in blood volume involves a change of the haematocrit status, which is monitored continuously by the extra-corporal circuit of the dialysis apparatus (Crit-Line III™, In-Line Diagnostics Corp., Kaysville, UT, USA). In these studies the continuous recorded data of the diagnostic module was compared with the haematocrit changes during the dialysis sessions.

2. With urologic patients which undergo surgery, were it was to presume that with the apparent loss of blood during surgery a change of haemoglobin concentration occurs. Thereby the recorded data of the diagnostic module was compared with the data of the blood-gas-analysis from the A. radialis (arterial oxygenic saturation - S_aO_2 in percent and haemoglobin concentration – cHb in g/dl).

The required approval came beforehand from the ethics board responsible.

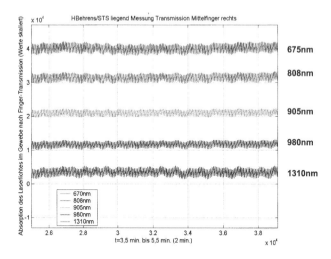

Figure 5: Diagnostic Module photo-current time signals of the transmission-light after passing through the finger of a healthy and dormant proband (scaled values)

Preliminary Results

Figure 5 shows the time signal photo-currents of the silicon- and the InGaAs-photo diode after the passage of the laser light through the middle finger. The measurements was performed with a proband lying calmly in a horizontal position at room temperature.

The transmission values of the pulse waves processed are constant with breath dependent periodical oscillations for all laser wavelengths. Figure 6 shows an overview of the recorded measurement values from a patient during a dialysis session.

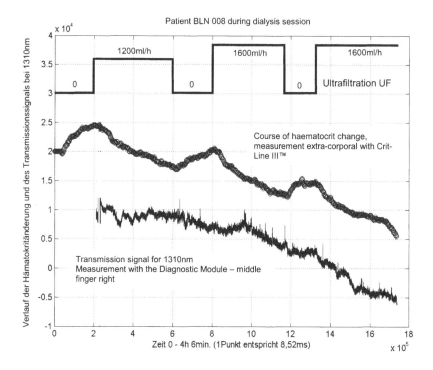

Figure 6: Time signal of photo current for the 1310nm laser diode light transmission after its passage through the right middle finger of a patient during a dialysis session is the lower curve of the figure.
Ultra-filtration profile and rate of the dialysis is represented by the upper curve.
The change of haematocrit measured in the extra-corporal circuit of the dialysis device with the Crit-LineIII™ device is displayed by the middle curve.
The time signal of the transmission value for 1310nm shows an analogue course similar to the haematocrit change during the dialysis of the patient.

Previous measurements of the transmission signals of the five wavelengths showed an apparent variation of the arterial pulse. The signal quality was sufficient to analyse the signal components and to calculate relative attenuation coefficients of the arterial blood. With regard to the components at 1310nm an evaluation of the relative portions of haemoglobin and water in the blood should be possible.

The data recording from 10 urologic patients and 5 dialysis patients is completed. At this time a further redesign period for the diagnostic module is foreseen and we are working on the optimisation of the used evaluation algorithms and statistical analyses.

References

[1] A. Samel et. Al., "Langzeitmonitoring psychophysiologischer Größen in der Flugphysiologie", Internist Vol38: p755 – p769, 1997

Poster Programme
Paper presented at Sensors and their Applications XII, September 2003
©*2003 IOP Publishing Ltd*

Unobtrusive Transducer Augmentation of Everyday Objects for Systems with Dynamic Interactivity

Kieran Delaney, Stephen Bellis, Achilles Kameas[1], Irene Mavrommati[1], Martin Colley[2], Anthony Pounds-Cornish[2]

NMRC, Ireland, [1]Computer Technology Institute, Greece, [2]University of Essex, UK.

Abstract: This paper reports upon recent research in creating effective transducer networks for Ambient Intelligence (AmI). Techniques for unobtrusively embedding high granularity transducer networks into everyday objects have been investigated as part of a systemic approach to implementing "plug-and-play" objects, achieved recently within the research project "Extrovert Gadgets". Analysis has shown that the nature of the object is an inherent factor in the network development. This is a materials issue, as the object becomes part of the 'packaging' of the network devices, and a scaling issue, as the size of the network is linked to its effectiveness as an AmI system.

1. Introduction

The future of information technology systems will be driven by the vision of Ambient Intelligence [1]. In this vision, Ambient Intelligence will surround us with proactive interfaces supported by computing and networking technology platforms that are everywhere; for instance, its systems would be embedded into everyday objects such as furniture, clothes, vehicles, roads and even in decorative materials like paint, wallpaper, etc. They would be unobtrusive, often invisible. They will provide a seamless environment of computing, advanced networking technology and specific interfaces. The systems will be aware of the specific characteristics of human presence and personalities, and will take care of needs. It will be capable of responding intelligently to spoken or gestured indications of desire, and could engage in dialogue. Interacting with ambient intelligence would be relaxing and enjoyable for the citizen, and not involve steep learning curves.

In Europe, a focal point of this research is the "Disappearing Computer" initiative [2]. The goal of this program is to explore how everyday life can be supported and enhanced through the use of collections of interacting artefacts. The initiative has three inter-linked objectives:

- *Develop new methods for the embedding computation in everyday objects.*
- *Research on how new functionality can emerge from interacting artefacts.*

- *Ensure people's experience is coherent and engaging in space and time.* The program consists of 17 projects [3], each addressing aspects of the above objectives. Among them is the "Extrovert-Gadgets" project [4], which aims to provide a conceptual and technological framework that will engage and assist ordinary people in composing, (re)configuring or using systems of computationally enabled everyday objects, which are able to communicate using wireless networks. It does this by adopting a dynamic "plug-and-play" approach with tangible, transducer-enhanced objects, in a manner that has much in common with the process where system builders design software systems out of components.

1.1 Extrovert Gadgets and Gadgetworlds

The objective of this project is to develop a method of diffusing information technology into everyday objects and settings, and assisting ordinary people in composing, (re)configuring or using groups of these objects. To achieve this, the project has developed and validated a *Gadgetware Architectural Style (GAS)*. This constitutes a generic framework shared by both gadget designers and users for consistently describing, using and reasoning about families of augmented objects (or eGadgets). Generally speaking, eGadgets are everyday tangible (physical) objects enhanced with sensing, acting, processing and communication abilities. This may entail "intelligent" behaviour, which can be manifested at various levels. In effect, eGadgets are 'GAS-aware' artefacts, which are used as the building blocks of Gadgetworlds.

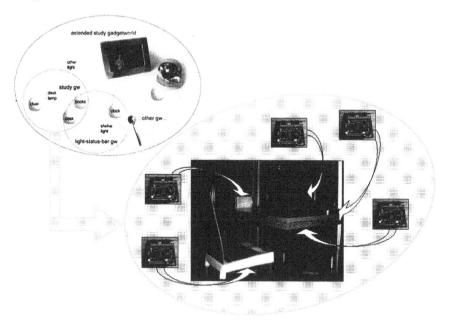

Figure 1: The construction of Study Gadgetworld

1.2 Gadgetworlds

A Gadgetworld is a distinguishable, specific configuration of associated extrovert gadgets (figure 1), formed purposefully by a user, which communicates and / or collaborates in order to realize a collective function. This can be achieved through the use of "plugs", which are software classes, defined at the GAS layer, that abstract the gadget capabilities to make them visible to people and to other objects. Plugs have a direct relation to the sensors / actuators and the gadget functions as implemented by its manufacturer. Making associations between compatible plugs form what are known as synapses. Once a synapse is established, the involved gadgets interact on their own, independently and transparently of the existence of plugs. A set of such synapses defines interactions between gadgets and thus forms the Gadgetworld. All eGadgets come with two types of plug, a Tplug (one per gadget) that describes the physical properties of the gadget, and a set of Splugs that describe services (and which depends on the set of sensors/actuators of the gadget).

1.3 Gadgetworld Construction: The Augmented Study

A test system was constructed to demonstrate the Gadgetworlds concept [5]. It was based upon a user-created study environment in which desk-lamps and floor lamps responded automatically to user needs by, for instance, illuminating a desk-space when the user sits down to study. Subsequently, it maintains a constant, comfortable level of light on the surface of the study book even as it is move around the desk by the user. A sample section of this Gadget world is composed of four gadgets (Desk, desk-lamp, book, chair), which can be broken up into two main classes: sensing eGadgets (eBook, eDesk, eChair), and actuating eGadgets (eLamp). The plugs of each gadget are shown in Figure 2. Once formed, the Gadgetworld becomes operational and remains so until explicitly disassembled by the user. A Gadgetworld should be considered as being always on. When switched on, each gadget constantly attempts to re-establish its Synapses. Each Synapse within a Gadgetworld may be mandatory or optional.

Desk	Lamp	Book	Chair
T-plug O	T-plug O	T-plug O	T-plug O
Weight O	On/off I	Open/close O	Occupy O
Proximity O	Intensity I	Luminosity O	

The Synapses required are:
Desk (weight=1) -> Lamp (on/off=on)
Chair (occupancy=1) -> Lamp (on/off=on)
Book (open/closed=open) -> Lamp (on/off=on)
Book (luminosity) -> Lamp (intensity).

O= output plug, I=input plug

Figure 2: The Plugs of Sample Gadgets

The overall Gadget world function can be described simply as:

When the CHAIR is NEAR the DESK
AND
the BOOK is ON the DESK,
AND
SOMEONE is sitting on the CHAIR
AND
The BOOK is OPEN
THEN
ADJUST the LAMP INTENSITY according to the book LUMINOCITY.

2. Gadgetworld System Implementation

2.1 Extrovert Gadget Architecture

The high-level architecture of an eGadget is shown in figure 3. According to it, an eGadget contains gadget *management software* (GadgetOS), responsible for providing access to resources (e.g. the RF unit, any sensors or actuators etc), *collaboration logic* for service discovery, and *computation logic* to implement the intrinsic gadget functions. GAS-related services are implemented by plug and synapse management services, and agent implemented intelligent mechanisms. GAS-OS is the middleware that runs on every gadget and implements GAS concepts. Its services include *plug discovery and advertising, synapse establishment – disestablishment, and synapse management.* GAS-OS software is composed of the **eComP**, which handles the networking communication between gadgets, the **message factory,** responsible for encoding and decoding GAS-OS messages, **the Gadget-OS interface** for handling the communication between the GAS-OS and the gadget-OS, the **Agent Proxy,** which collaborates with the Intelligent Agent, and the **GAS-OS kernel**, the central module of the GAS operating system.

2.2 Extrovert Gadget Hardware

The detailed eGadget architecture, which represents the way eGadgets are being implemented, is shown in figure 4. Current eGadgets are made of a matrix of sensors and actuators, conditioning and aggregation boards to implement communication between the sensor/actuator boards and the processor, and a (processor + memory + wireless) module itself. The transducer data aggregation portions of the hardware modules are based around field programmable gate arrays (FPGAs), which give flexibility for implementing many different types of eGadgets due to their reprogrammability.

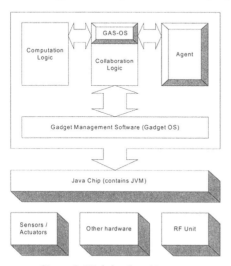

Figure 3. High-level architecture

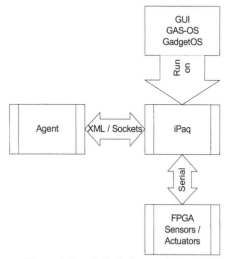

Figure 4. Detailed eGadget architecture

Figure 5: FPGA Aggregation module

The FPGA module, in figure 5, uses the Xilinx Spartan XC2S300E-7FG456C FPGA. The circuit board includes an on board EEPROM to retain configuration after the power is turned off. The board also contains fast ZBT SRAM for data storage. There are approximately 200 digital I/Os available for sensor input. In the current scenario only a small percentage of these are used, the table representing the maximum usage of 30 LDR inputs. There are also 10 analogue inputs available which use the Texas Instruments TLC549 A to D converter.

2.3 Sensing eGadgets

The sensors used are pressure pads, light dependent resistors (LDRs), bend sensors, ultrasonic transducers and tilt switches. Pressure pads are used to detect the presence of a substantial weight such as a person sitting on the eChair, while the LDRs have a dual purpose; firstly for detecting a bigger range of weights such as an object on the desk and secondly to detect a range of light levels, the luminosity of the eBook for example. The eBook has bend sensors embedded into the spine to detect whether it is open or closed. An ultrasonic transmitter and receiver pair is used to detect the proximity between the eTable and eChair. Infrared (IrDA) was also used to send eBook identification to the table. Signal conditioning logic is used to interface the sensors with a Field Programmable Gate Array (FPGA) circuit, which does some initial pre-processing on the sensor data before sending it serially to the iPAQ or Laptop. The iPAQ/laptop is used to host the GAS-OS java programs, which respond to the serial sensor data. Wireless local area network PCMCIA cards are then used to interconnect the eGadgets.

2.4 Actuating eGadgets

The actuating eGadget in the current Gadgetworld is the Desk-eLamp. The eLamp receives status information from synapsed eGadgets over the WLAN and this data is processed by the GAS-OS residing in the IPAQ or the laptop. Information on the required dimmer level is sent serially to the FPGA PCB which implements pulse width modulation (PWM) to produce an analogue signal whose voltage level is proportional to the desired light level. The analogue signal is the input to a voltage-controlled dimmer, which is required to provide the higher voltages necessary to dim the eLamp.

2.5 Effectiveness of System Functionality

The current system is fully effective in demonstrating that the concept of dynamic reconfigurability of component systems can be applied to everyday objects. Groups of eGadgets can be "plugged" together to give numerous types of functional Gadgetworlds. In addition, the system shows an encouraging level of longevity in its performance. From the perspective of transducer function, the system operates acceptably within the confines of the dedicated scenario. However, there are relationships between the context of use of the Gadgetworlds and the nature of their transducer networks that requires further study.

3. Applying Transducer Networks to Augmented Objects

3.1 Unobtrusiveness of Transducers

One of the main design objectives in the production of E-Gadgets is that systems should be as unobtrusive as possible and that they do not interfere with the visual appearance or primary function(s) of the artefact. As a result, transducers (primarily sensors) are embedded into the material of the artefact or are located inside the artifact so that they are effectively invisible to the eGadget user. In this regard, the function of the sensors is only as effective as the quality of its interface with the object. For direct physical user interaction, as in the case of pressure sensing, this will be the adhesive bond to the object, as relative motion will be a significant attenuating factor. Adhesivity will also be a major factor in embedding impact and acoustic sensors, as poor quality will vastly affect performance, and provide significant variability in the network performance (performance variations in the transducers can be represented as having components due to the embedding method and to device variation). A significant factor in embedding optical sensing arrays into everyday objects will be vertical alignment. For ultrasonic and infrared sensing, typical integration techniques tend to limit field of view, thus, multiple receivers/transmitters are sometimes applied. In such cases, the 3-D alignment and bonding techniques employed are vital factors in achieving high quality performance.

3.2 Relationship between sizes of the gadget and the size of the transducer:

An integral part of achieving unobtrusive systems is through miniaturisation [7,6,8]. In general, the smaller the gadget the more difficult it is to effectively embed sufficiently dense transducer granularity levels for good performance. However, typically this is not due to the size of the transducer. It is rather due to the protective encapsulation materials, and interconnect in use to provide traditional solutions in handling, connectivity, and deployment. Device miniaturisation is in part solving this problem, however, there is a tradeoff between device size and system scale, as the complexity of the conditioning and aggregation layers tends to increase significantly, and current high density interconnect solutions are not easily designed to supply connectivity over (in relative terms) large surface areas, and more importantly are extremely difficult to deploy effectively.

3.3 Relating Gadget function and Interaction to Network Scale

The sensing capability of an eGadget is a function of the number of sensors and the purpose of those sensors. In terms of effective performance, this is inherently linked to the context of use, which will change depending upon the type of interaction. In generic terms, it is possible to define three main classes of interaction [1] *User to eGadget*, [2] *plain artefact to eGadget* and [3] *eGadget to eGadget*.

In the User to eGadget type, scaling is affected by the how the person uses the eGadget. For instance, on the eTable there are a multitude of personal interactions that could take place requiring differing degrees of sensing scale and granularity. Very few sensors with a large field of view (or interaction interface) could determine user proximity to the eTable, whereas a significant quantity would be needed to determine hand movements across the table. Additionally, detecting the fingertip of a baby or small child would require sensors with an even smaller field of view and higher granularity, making such a dimension a high imperative granularity target for fine context extraction.

Interaction between plain artefacts and eGadgets are potentially significantly difficult. The scope of legitimate artefacts that could interact with an eGadget makes the definition of scale, granularity, field of view, and device size a very difficult task. On an eDesk, a sparse sensor array could detect book-sized objects, but not a moving pencil-tip; this would be a fine-grained sensing task, even more so than tracking a child's fingertip.

Gadget-to-Gadget sensing can be more effective because the potential for collaborative sensing can be exploited. For example, mobility and location sensing in an ePen, coupled with pressure sensitivity of the eTable, more accurately determine relative motion across eTable surfaces. The combination of two eGadgets, working coherently within a Gadgetworld, gives virtual sensor granularity higher than that of the individual granularity of an eDesk used with an ordinary pen, and inherently scales the level of system capability.

4. Conclusion

Techniques for unobtrusively embedding high granularity transducer networks into everyday objects are emerging. In the "Extrovert Gadgets" project, it has been shown that newly developed tangible "plug-and-play" systems of gadgets can provide effective collaboration in sensing networks, increasing granularity, and coherence in determining user behaviour. Investigation has also shown that the physical nature and component materials of the objects affect the transducer networks behaviour to the extent that it becomes part of the 'packaging' of the network devices, and should be considered as such during development.

References

[1] IST Advisory Group (ISTAG) Scenarios for Ambient Intelligence in 2010: http://www.cordis.lu/ist/istag-reports.htm.
[2] The Disappearing Computer initiative: http://www.disappearing-computer.net/
[3] K.Delaney, J.Barton, "The Challenge of Globally Embedding Functional Electronics into Everyday Artefacts", Conference on Smart Packaging, November 14, 2001, Institute of Materials, London, UK.
[4] Extrovert Gadgets (eGadgets) website: http://www.extrovert-gadgets.net
[5] A. Kameas, S. Bellis, I. Mavrommati, K. Delaney, A. Pounds-Cornish and M. Colley, "An Architecture that Treats Everyday Objects as Communicating Tangible Components", IEEE International Conference on Pervasive Computing and Communications (PerCom2003), Dallas-Fort Worth, Texas (March 2003).
[6] J. Barton, K. Delaney, C. Ó Mathúna, J. A. Paradiso, "Miniaturised Modular Wireless Sensor Networks" UbiComp 2002, September 29th – October 1st, Göteborg, Sweden.
[7] G. Kelly, A. Morrissey, J. Alderman and H. Camon,"3D Packaging Methodologies For Microsystems", IEEE Transactions On Advanced Packaging, Volume 23, No.4, November 2000, pp. 1-8.
[8] Cian O'Mathuna, Kieran Delaney, John Alderman, "MEMS packaging for the Intelligent Environment", Invited presentation at the IMAPS Advanced Technology Workshop, San Jose, Nov 9-12, 2001.

Poster Programme
Paper presented at Sensors and their Applications XII, September 2003
©*2003 IOP Publishing Ltd*

Investigation of a Luminescent based Technique for Surface Temperature Measurements

M McSherry, C Fitzpatrick, E Lewis
Department of Electronics and Computer Engineering, University of Limerick, Ireland.

Abstract: An extrinsic optical fibre temperature sensor has been investigated and preliminary results are presented. The active coating used consists of a mixture of two phosphor powders. Thermal quenching dominates the luminescent characteristics of one phosphor at high temperatures while the other more stable phosphor is used as a reference. Ratios of their emission intensities are calculated and analysed. Preliminary results show a decreasing ratio for increasing temperature. This paper describes initial testing of the coating and its application, as part of a temperature sensor, for monitoring surface temperatures in electromagnetically harsh industrial environments.

1. Introduction

Optical fibre temperature sensors have been widely reported [1][2][3]. This paper presents preliminary results of a phosphor based temperature sensor. The coating for the sensor is a combination of two phosphors. Both phosphors absorb incident 470nm visible light, which causes them to fluoresce. This results in emissions from both phosphors occurring at separate wavelengths. The emissions are coupled via an optical fibre to a spectrometer for spectral analysis. The emission intensity of one phosphor is subject to thermal quenching while the intensity of the second phosphor remains unaltered and therefore acts as a reference. Ratios of their intensities are obtained which correspond to various changes in temperature.

Phosphor powder is commercially available. The phosphors were purchased from Phosphor Technology and are currently used in light emitting diodes. They are relatively inexpensive as only a few grams are used for any application. Phosphors are employed in printing inks or dispersed in plastics to avoid counterfeiting problems of bank notes, limited edition prints, works of art, admission tickets etc. This process provides security and avoids reproducibility of these products. The phosphors fluoresce under UV light and so are not seen by the naked eye in visible light. It is our aim to use this process to print phosphor onto materials of interest as part of a temperature sensor.

Optical fibres offer versatility of design to this temperature sensor. They are small in size, robust, lightweight with adjustable lengths. There are no electrical parts used but more importantly optical fibres offer immunity to electromagnetic interference. Compared to existing electrical thermometers, this optical fibre temperature sensor can be used in water and in applications where microwave radiation or high-voltages are present. The 470nm LED is widely obtainable and compared with infrared pyrometry whose infrared sources are expensive. The fibre optic temperature sensor is highly reproducible and the materials are inexpensive and commercially attainable. Due to its flexible design the fibre optic temperature sensor can be used to measure various temperatures in many applications. These applications include surface temperature measurements of samples in microwave material processing applications and in environments where size is a factor and low cost is a priority.

2. Sensor Specifications

2.1 Coating characteristics

The coating consists of a mixture of two phosphors, EQD25/N-U1 and QUMK58/F-D1 commercially available from Phosphor Technology. The excitation spectra of the two phosphors have common characteristics at 430nm resulting in two emissions at different wavelengths. This is shown in Figure 1. EQD25/N-U1 emits light at 660nm while QUMK58/F-D1 has a peak emission at 570nm.

Under normal temperature conditions the phosphors absorb energy from incident UV or blue light and emits a portion of this energy as visible light at a higher wavelength. This process is known as photoluminescence. At higher temperatures thermal quenching dominates and the intensity of the luminescence decreases [4]. Thermal quenching affects different phosphors at different temperatures. The intensity of phosphor EQD25/N-U1 is unaffected by variations in temperature in the investigated range and so acts as a reference. The luminescent emission characteristics of the second phosphor QUMK58/F-D1 change with increasing temperature and so are subject to quenching in the temperature range. This results in a linear decrease in its intensity relative to changes in temperature. A ratio is established between the different phosphor intensity values from which the temperature can be extrapolated.

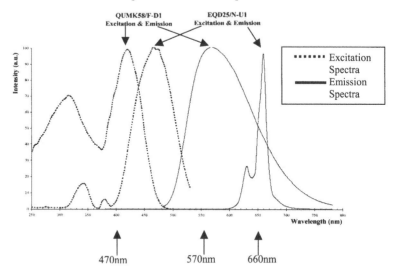

Figure 1 : Combined Phosphor Spectra

2.2 Fibre Characteristics

The optical fibre used for the sensor is from Ceramoptec. It has a 1mm pure silica core and a hard polymer cladding. The surrounding Tefzel jacket can withstand temperatures from –40°C to 150°C and is specific to temperature applications. It is 3 metres in length with an SMA connector on one end with the other end polished. The fibre transmits light from 300nm to 1200nm with high transmission of visible light.

The availability and versatility of the sensors materials provide various designs for different applications. This fibre optic temperature sensor can be used as a remote sensor as shown in Figure 2(a). For a remote sensor, the mixture of phosphor is coated onto the material of interest. An optical fibre transmits excitation light to the coating. As the phosphor mixture fluoresces a second optical fibre captures the fluorescence and transmits it to a detector for analysis. Another application of this sensor is shown in Figure 2(b). In the case the temperature of a liquid is required. The base of a beaker of liquid is coated with the phosphor mixture. Two optical fibres are used to transmit the excitation light and receive the phosphors emissions. By analysing the resulting intensity ratios the temperature of the liquid is identified.

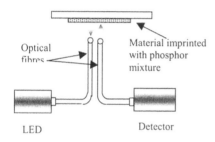

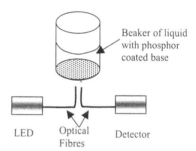

Figure 2(a) : Non contact temperature sensor sensor

Figure 2(b) : Application of temperature

3. Experimental Set-up

The experimental set up is shown in Figure 3. One gram of each phosphor powder are mixed together on a glass slide for initial testing of the coating. The phosphors and glass slide are placed on a hotplate. The hotplate is used to simulated heat for preliminary testing of the coating but is not an ideal heating solution. A laboratory-circulating fan oven is ideal as temperature is controlled and evenly distributed unlike the hotplate. The temperature of the hotplate is increased in steps of 10°C from room temperature to 120°C. A digital thermometer monitors the temperature of the hotplate.

An optical fibre transmits light from a 470nm LED to the phosphors in order to stimulate them. A second optical fibre transmits the resulting visible light from the phosphors to a spectrometer. The S2000 fibre optic spectrometer from Ocean Optics offers a detector range from 200nm-1100nm and a 2048-element linear silicon CCD array for detecting light. Each of the 2048 pixels represents a wavelength of light that strikes it. The spectrometer is connected to a PC to observe the resulting spectrums, to log the required data for each emission and to calculate the resulting ratio using Labview software.

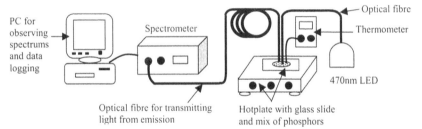

Figure 3 : Experimental Set up

4. Results

Figure 4 shows three distinct sets of peaks. The first set of peaks represent the reflected light from the LED light source at 470nm. The light intensity of the LED varies slightly but has no affect on the ratio of the phosphors intensities. The second set of peaks is the emission of visible light from phosphor QUMK58/F-D1 when it is stimulated by light at 470nm. The third set of peaks is the emission of visible light from phosphor EQD25/N-U1 when it is stimulated by light at 470nm.

Figure 4 also shows five varying waveforms at five different temperatures. The temperature of the hotplate is increased in steps of ten. The calculated ratios of the phosphors intensities are shown in the Table 1, which represent all ten temperatures. It is evident from the ratio values that a dependency exists between temperature and the intensities of the phosphors. As temperature increases there is a definite decrease in the ratio of the phosphor intensities as shown in Table 1.

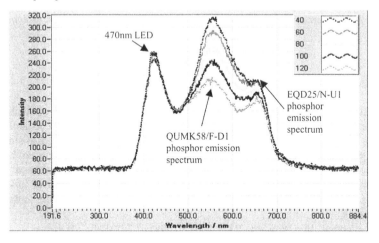

Figure 4 : Spectrum of Led with two phosphor intensity peaks

Temp(°C)	30	40	50	60	70	80	90	100	110	120
Ratio	1.49	1.47	1.43	1.39	1.38	1.34	1.32	1.28	1.21	1.17

Table 1 : Temperature with corresponding intensity ratio values

This experiment is repeated 10 times in order to show a constant relationship between the intensity ratios and temperature. The results are presented in Figure 5. Figure 5 shows an inverse proportionality between temperature and the intensity ratios. An average value of the intensity ratio is calculated at each temperature. Y error bars are shown which represent a maximum error of 3.85% and a minimum error of 3.62%. The temperature of the hotplate used in this experiment is unstable and uncontrollable which leads to slight inconsistency in the results at certain temperatures but it is evident that there is a dependence of the intensity ratios on variations in temperature. This ratio technique makes the system immune to intensity variations in the stimulation, as they will affect both peaks identically.

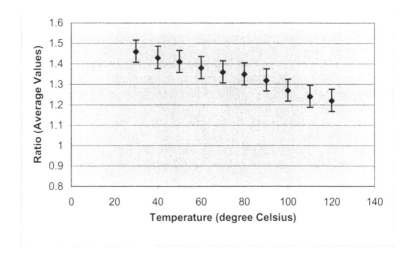

Figure 5: Variation of the ratio of phosphor intensities with temperature

5. Conclusion

The experiments have shown a dependence of the intensity ratio of the phosphors emissions on temperature. An increasing temperature corresponds to a decreasing ratio of the phosphors intensities. Tests have also shown that the phosphor mixture is reproducible from 3.85%→3.62% over the temperature range from 30°C to 120°C. These preliminary results provide a basis for more research with a view to developing a versatile and cost efficient fibre optic temperature sensor.

There are many applications for this type of fibre optic temperature sensor especially in industry where a harsh environment determines the lifetime and usability of an electronic sensor. This temperature sensor offers a technique whereby it can be used to monitor surface temperature in microwave and high voltage production line environments.

References

[1] Wickersheim K.A., Alves R.B., *" Recent advances in optical temperature measurement"*, Industrial/Research Development 1979 pp 82-89

[2] Grattan K.T.V., Zhang Z.Y., Sun T., Shen Y., Tong L., and Ding Z., *"Sapphire-ruby single-crystal fibre for application in high temperature optical fibre thermometers: studies at temperatures up to 1500°C"*, Measurement Science and Technology, Vol 12 2001 pp 981-986

[3] Wickersheim K.A., Sun M.H., *" Fiberoptic thermometry and its applications"*, Journal of Microwave Power, 1987 pp 85-94

[4] Simons A.J., McClean I.P., Stevens R., *"Phosphors for remote thermograph sensing in lower temperature ranges"*, Electronics Letters Vol 32 (3) 1996 pp 254-254

Poster Programme
Paper presented at Sensors and their Applications XII, September 2003
©*2003 IOP Publishing Ltd*

Model to predict the response of a correlation spectroscopy system for the detection of oxygen gas

P Chambers, E A Austin and J P Dakin

Optoelectronics Research Centre, University of Southampton, United Kingdom, SO17 1BJ.

Abstract: We present the first comprehensive model for the detection of a gas by absorption based correlation spectroscopy, and show the practically useful results obtained. Predictions of sensor response for a fibre-optic-coupled O_2 detection system have been made, based on gas absorption data from the HITRAN database. Fractional absorption-change values were predicted, and system S/N values were derived, taking into account effects of fundamental photon noise limitations and thermal noise contributions. A new theoretical procedure for determining the optimum optical filter width and central wavelength for best modulation depth and best signal to noise (S/N) ratio is introduced. Using typical parameters associated with available optical sources, practically achievable modulation depths, S/N ratios and O_2 detection limits have been derived.

1. Introduction

Absorption based correlation spectroscopy offers an attractive method to detect several gases of industrial importance. To-date no complete theoretical analysis of such a system has been presented and we now present such an analysis. As a case study, a correlation spectroscopy based O_2 sensing system was modelled, using published gas absorption data from the HITRAN database (http://www.hitran.com). Fractional changes in the detected response expected when gases are introduced into a correlation spectroscopy sensor system (we shall call this fractional change the modulation depth/index) were predicted and system signal to noise (S/N) values were derived, taking into account the effects of fundamental photon noise limitations and thermal noise contributions. We have shown that the model agrees well with experimental data which we have previously published.

1.1 Principle of Operation

There are several methods of detecting gases using correlation spectroscopy [[1]-[13]]. An attractive absorption based method is the correlation spectroscopy Complementary Source Modulation (CoSM) method [[14]-[16]]. This involves the alternate on/off switching of two light sources in anti-phase, passing light from one of these sources through a reference cell containing the target gas (or gases) of interest for detection, and then combining this now-partially-absorbed light beam with a fraction of unaffected light from the other source, in a proportion such as to give no net intensity modulation over an appropriately optically filtered bandwidth. This combined beam is then used to probe for the target gas. As the beam component which has passed through the reference gas sample, now has less available optical energy lying within the narrow spectral regions of the target gas spectral absorption lines, a net intensity modulation of the balanced combined beam will be re-established when it passes through a measurement cell containing the gas of interest. This induced intensity modulation is near-proportional to the target gas concentration, as only differential absorption between the two beam components can contribute to a signal modulation. A fibre optic based schematic of the absorption based correlation spectroscopy gas sensing system is shown in figure 1. The input signal detector is used to ensure balanced light power and the output signal detector is used to resolve any intensity modulation resulting from the test gas in the measurement cell.

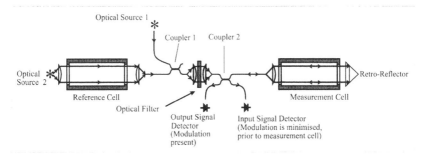

Figure 1: *Schematic of a fibre optic based implementation of a absorption correlation spectroscopy system.*

2. Modelled system parameters

It is important to be able to model the correlation spectroscopy system, not only to predict performance, but also to make the very important selection of what type of optical filter (or what type of LED or super-luminescent optical fibre source in terms of optical spectral output) to use to obtain best detection performance or best selectivity. We have defined, and analysed, the parameters that, we believe, dictate the system performance.

2.1. Modulation Depth

The first practically important system parameter that needs to be modelled is the modulation depth present at the output detector. This is defined as the peak to peak AC output signal variation, which arises when gas is introduced into the measurement cell, divided by the mean DC detected level. This is effectively a measure of optical "contrast".

Assuming that a spectrally-flat optical source is used and that the optical intensities at the reference detector are equalised, the Modulation Depth (MD) may be expressed as an optical intensity ratio, in terms of the optical transmissivities of the reference and measurement gas cells ($T_{Ref}(\lambda)$ and $T_{Meas}(\lambda)$ respectively) and the transmission spectrum of the optical filter ($F(\lambda)$):

$$MD = 2\left(\frac{\int T_{\mathrm{Re}f}(\lambda)T_{Meas}(\lambda)F(\lambda)d\lambda \int F(\lambda)d\lambda - \int T_{\mathrm{Re}f}(\lambda)F(\lambda)d\lambda \int T_{Meas}(\lambda)F(\lambda)d\lambda}{\int T_{\mathrm{Re}f}(\lambda)T_{Meas}(\lambda)F(\lambda)d\lambda \int F(\lambda)d\lambda + \int T_{\mathrm{Re}f}(\lambda)F(\lambda)d\lambda \int T_{Meas}(\lambda)F(\lambda)d\lambda} \right) \quad (1)$$

If desired, the source spectrum can also, of course, be taken into account by using additional spectral functions in equation 1.

2.2 Other parameters that dictate system performance

To obtain maximum spectral contrast, the optical filter returning maximum modulation depth (equation 1) should be chosen. This would give maximum "contrast", thus reducing sensitivity to environmental effects such as dust in sensing cells, or mechanical vibration. However, for good selectivity and to improve the S/N ratio, it is also desirable to cover many gas absorption lines, so a different filter may be required to obtain optimum S/N ratio.

3. Numerical Model Results

Below, optimal filters for modulation depth and S/N ratio are derived for an O_2 gas sensor using the absorption band at 760nm.

3.1 Detection of oxygen gas (O_2) using 760nm absorption band.

Oxygen gas (O_2) exhibits an electronic based absorption band at approximately 760nm, the spectrum of which is shown in figure 2. Although just beyond the visible region, LED and tungsten optical sources and commonly available optical components function well here. Therefore, it is relatively straightforward to implement optically based detection systems in this spectral region.

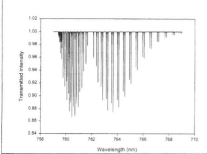	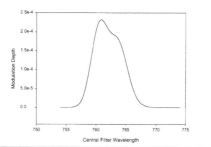
Figure 2: *Spectral transmission of 100% O_2 in the near IR region, over a 1m path at STP. (Source is HITRAN database)*	**Figure 3:** Theoretical variation of modulation depth as the centre wavelength of the Gaussian-shaped optical filter is changed. *Reference and measurement cells contain 100% O_2 at STP and the Gaussian-shaped optical filter has a 3nm bandwidth.*

3.2 Dependency of Modulation Depth on the central wavelength and bandwidth of the optical filter.

Modelling the optical filter as having a Gaussian shaped transmission function, we determined the dependency on output signal modulation depth of varying the centre wavelength and the full width at half maximum bandwidth of the Gaussian-shaped optical filter.   The result, shown in figure 3, shows that, with a 3nm Gaussian-shaped optical filter bandwidth, an optimum modulation depth (or optical contrast) is achieved at a filter centre wavelength of 760.9nm.

We then examined the effect of broadening the Gaussian-shaped optical filter bandwidth on the output signal modulation depth. To illustrate this, we performed calculations at wavelengths of 760.9nm and 760.8nm, the first wavelength being midway between two lines the second lying on an O_2 absorption line. Under the same conditions, with the both cells again containing 100% O_2, figures 4(a) and 4(b) show the expected modulation depth at 760.9nm and 760.8nm, respectively, as the bandwidth of the Gaussian-shaped optical filter is increased from 0nm to 10nm. The modulation depth in figure 4(a) actually starts at a minimum, as the centre wavelength is between absorption lines, whereas that in figure 4(b) starts at a maximum as it is on a line. A maximum in modulation depth is achieved in both plots at a Gaussian-shaped optical filter bandwidth of approximately 0.3nm. We will show that a significantly larger Gaussian-shaped optical filter bandwidth, such as 3nm, is more realistic when S/N considerations are included.

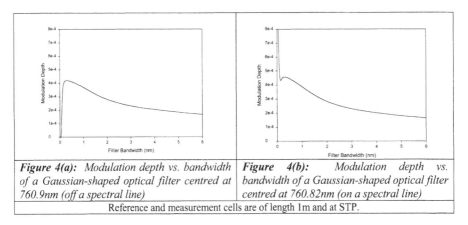

Figure 4(a): *Modulation depth vs. bandwidth of a Gaussian-shaped optical filter centred at 760.9nm (off a spectral line)*	***Figure 4(b):*** *Modulation depth vs. bandwidth of a Gaussian-shaped optical filter centred at 760.82nm (on a spectral line)*
Reference and measurement cells are of length 1m and at STP.	

The dependency on modulation depth of varying measurement cell O_2 concentration, when using a system with a 3nm bandwidth Gaussian-shaped optical filter, centred at 760.9nm, is shown on both linear and logarithmic scales in figures 5(a) and 5(b) respectively. These plots clearly show the almost linear relationship between modulation depth and measurement cell O_2 concentration over a wide range of O_2 concentrations in the measurement cell.

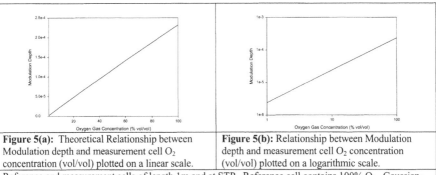

Figure 5(a): Theoretical Relationship between Modulation depth and measurement cell O_2 concentration (vol/vol) plotted on a linear scale.	**Figure 5(b):** Relationship between Modulation depth and measurement cell O_2 concentration (vol/vol) plotted on a logarithmic scale.
Reference and measurement cells of length 1m and at STP. Reference cell contains 100% O_2. Gaussian-shaped optical filter has a bandwidth of 3nm and is centred at 760.9nm.	

3.3 Gas Sensitivity Optimisation with respect to S/N Ratio.

Based on measurements from our earlier experimental results [[16]], we modelled the spectral power intensity at the output detector as being 10nW/nm. Our detection system was a silicon detector diode followed by a FET based transimpedance amplifier, with a 10MΩ feedback resistor ($R_{Feedback}$).

The total noise contribution of the system was modelled by considering contributions of both photon and thermal noise on the output detector, given by equations 2 and 3 respectively, where q is electronic charge, k is the Boltzmann constant, T is absolute temperature (modelled as 293°K), I_{sig} is the DC output current of the output detector and B is the detection bandwidth (modelled as 0.1Hz).

$$I_{Photon\ noise} = \sqrt{2qI_{Sig}B} \quad (2) \qquad\qquad V_{Thermal\ noise} = \sqrt{4kTBR_{Feedback}} \quad (3)$$

Figure 6 shows the voltage noise at the detector (left axis, solid line) and the increase in the AC amplitude of the measurement signal (right axis, dashed line) as the Gaussian-shaped optical filter bandwidth is increased.

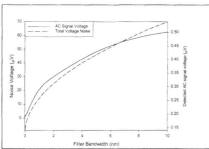

Figure 6: Graph of expected AC signal modulation (dashed line, right Y axis) and calculated noise voltage (solid line, left Y axis) against Gaussian-shaped optical filter bandwidth.	**Figure 7:** Predicted variation in S/N ratio as the bandwidth of the Gaussian-shaped optical filter (centred at 762.5nm) is changed. This is shown for four detected output optical power levels.
For the above plots, both reference and measurement gas cells contain 100% O_2 at STP.	

Figure 7 shows that a maximal S/N ratio is attained with a Gaussian-shaped optical filter bandwidth of approximately 7nm, whereas best contrast was achieved with a bandwidth of approximately 0.3nm. We consider that a 3nm bandwidth offers a reasonable compromise between these two extremes. Figure 7 also shows that increasing the signal light intensity level has a relatively small effect on the bandwidth of the Gaussian-shaped optical filter needed to get an optimised S/N ratio. Due to the near-linear relationship between O_2 measurement cell concentration and modulation depth, the S/N values are also representative of the minimum detectable concentrations of O_2 in the measurement cell.

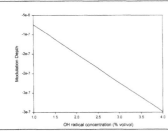 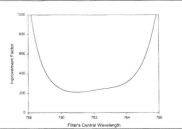

Figure 8: Near-linear dependence of Modulation Depth on concentration of gaseous OH contamination in measurement cell (using a Gaussian-shaped optical filter centred at 760.9nm of 3 nm bandwidth).	**Figure 9:** Rejection factor by which an absorption correlation spectroscopy technique offers improved selectivity when compared to a simple optical absorption method using an identical Gaussian-shaped optical filter of 3nm bandwidth.

The graphs above assume a correlation spectroscopy system based on a reference cell, at STP, containing 100% concentration O_2 gas and the measurement cell containing various concentrations of the OH radical.

3.4 Cross sensitivity

We examined the cross sensitivity of a correlation spectroscopy based system to contaminant gases. The correlation spectroscopy method, as described by equation 1, has the property that where a contaminant gas is present the cross sensitivity to it is small, even when it has optical absorption in the same spectral region as that of the target gas. We illustrate this by modelling the response of an O_2 detection system to cross talk with the OH radical, characteristic of water vapour, which is one of the few likely gaseous contaminants which are likely to absorb significantly around 760nm.

Figure 9 shows that an absorption based correlation spectroscopy system is approximately 200 times less sensitive to the OH radical than a broadband simple absorption based sensing system would be.

3.5 Use of model to predict experimental results

We demonstrated earlier the experimental operation of an O_2 absorption based correlation spectroscopy system [[16]]. There, we used a Gaussian-shaped optical filter, centred at 762.6nm, with a bandwidth of 3nm, and a 1m reference cell containing 100% O_2 and a 0.8m measurement gas cell, or 1.6m when a retro-reflector arrangement was used, containing air (21% O_2) at STP. By inserting the relevant data into equation 1, we compared the experimental results that were obtained earlier with these new theoretical results we have calculated.

Theoretical comparison with experimentally derived predictions of modulation depth.			
Reference Cell pathlength	Measurement Cell pathlength	Experimental	Predicted
1m	0.8m	2.8x10-5	3.2x10-5
1m	1.6m	5.6x10-5	6.4x10-5
Temperature and Pressure = STP	Reference cell O_2 concentration = 100%		
	Measurement cell O_2 concentration = 21% (vol/vol)(Air)		
Table 1: Theoretical comparison with experimentally derived predictions of modulation depth.			

The experimentally obtained detection limit of less than 1% (vol/vol) O_2 concentration, in the early work, compares favourably with a theoretical derived predicted value of 0.6% (vol/vol) (obtained from our S/N analysis above).

4. Conclusions

We have reported new theoretically modelled results for gas detection using the CoSM scheme for absorption based correlation spectroscopy. This includes simulated responses of signal modulation depth, predictions of the S/N ratio that may be achieved, an analysis of the system's selectivity in the presence of contaminant gases and a comparison of our predictions with our previously published experimental data. In this work we have developed a scheme that allows us to determine which light sources, optical filters and detectors would enable the optimal detection of gases using absorption based correlation spectroscopy.

This work has shown that the modulation depth response of an absorption based correlation spectroscopy system is strongly dependant of the type of optical filter chosen. The optimum bandwidth and centre wavelength of a Gaussian filter required to maximise the modulation depth in our O_2 sensing system was 0.3nm at 760.9nm. We have shown that the optimal response of the system when other key aspects, such as achieving a good S/N ratio, are taken into account, may be better when using an optical filter of somewhat higher bandwidth. For our O_2 sensing system, the best optical filter bandwidth for optimum S/N ratio was found to be 7nm. Fortunately, the modulation depth does not drop very rapidly beyond the peak at 0.3nm, so choosing wider filter bandwidths to improve selectivity and S/N ratio will not lead to a too great reduction in modulation depth (or measurement contrast).

5. Acknowledgements

The authors thank the Engineering and Physical Sciences Research Council (EPSRC) for research funding and support from industrial sponsors, via EPSRC's Faraday/INTErSECT research initiative, vis: Accurate Controls Ltd., BOC Edwards, Corus, Health & Safety Laboratories, Kidde plc, National Grid, and NPL. P. Chambers thanks the EPSRC for a student research grant.

References

[1] Hardwick, A. Berg, and D. Thingbo, *A fibre optic gas detection system*, Proc. 9th Int. Conf. on Optical Communications, 'ECOC 83' Geneva (1983), pp. 317.

[2] T. Kobayashi, M. Hirana, and H. Inaba, *Remote monitoring of NO_2 molecules by differential absorption, using optical fibre link,* Appl. Opt., **20** (1981), pp. 3279.

[3] K. Chan, H. Ito, and H. Inaba, *An optical fibre-based gas sensor for remote absorption measurements of low-level methane gas in the near-infrared region,* J. Lightwave Tech., **LT-2** (1984), pp. 234.

[4] S. Stueflotten *et al.*, *An infrared fibre optic gas detection system,* Proc. OFS-94 int conf., Stuttgart, 1994, pp. 87.

[5] R. Goody, *Cross-correlation spectrometer,* J. Opt. Soc. of Am., **58** (1968), pp. 900.

[6] H. O. Edwards and J. P. Dakin, *A novel optical fibre gas sensor employing pressure-modulation spectroscopy*, Proc. OFS-90, int conf. , Sydney, Australia 1990, pp. 377, Dec 1990.

[7] J.P. Dakin and H.O. Edwards, *Progress in fibre-remoted gas correlation spectroscopy*, Special issue of Optical Engineering, **31** (1992), pp 1616-1620

[8] H. O. Edwards and J. P. Dakin, *Correlation spectroscopy gas sensing compatible with fibre-remoted operation*, Sens. actuators, B, Chem, **11** (1993), pp. 9.

[9] J. P. Dakin, H. O. Edwards, and B. H. Weigl, *Progress with optical gas sensors using correlation spectroscopy*, Sens. actuators, B, Chem, **29** (1995), pp. 87.

[10] J. P. Dakin (**invited**)., *Evolution of highly-selective gas sensing methods using correlation spectroscopy*, Advances in Optoelectronics for environmental monitoring, Erice, Sicily, Nov 1998.

[11] J. P. Dakin, H. O. Edwards, and W. H. Weigl, *Latest developments in gas sensing using correlation spectroscopy,* Proc. SPIE Int. Conf., Munich, July 1995, paper 2508.

[12] J. P. Dakin *Sensor for sensing the light absorption of a gas,* UK Patent Application GB2219656A.

[13] H. O. Edwards and J. P. Dakin, *Measurements of cross-sensitivity to contaminant gases, using highly-selective, optical fibre-remoted methane sensor based on correlation spectroscopy,* Proc. SPIE Int. Conf "Chemical, biochemical & environmental fiber sensors"., Boston, Sept. 1991, SPIE Vol **1587** paper 33.

[14] J. P. Dakin, M. J. Gunning, P. Chambers and E. A. Austin, *Fibre optic LED-based correlation spectroscopy for O_2 detection, Proc* OFS-2002 int conf., 2002 Portland, Oregon 6-10 May 2002.

[15] J. P. Dakin, M. J. Gunning and P. Chambers, *Detection of gases and gas mixtures by correlation spectroscopy*, Europt(R)ode VI, Manchester, 7-10 Apr 2002.

[16] J.P.Dakin, M.J.Gunning, P.Chambers and Z.J.Xin, *Detection of gases by correlation spectroscopy*, Sensors & Actuators B: Chemical, accepted for publication.

Optimisation of inductive sensors using mathematical modelling

Hartmut Ewald and Andreas Wolter

University of Rostock, Department of Electrical Engineering and Information Technology, Albert-Einstein-Str. 2, D-18051 Rostock , Germany

Abstract: Over the last few years, there has been an increase in the application of inductive sensing in many areas. Presently, for many applications there are standard solutions whereby the selection of the probe coils and the parameterisation of the devices is carried out using empirical values. In part the signal evaluation adapts itself to the respective test conditions automatically.

Although this approach will succeed it does so only in the boundaries that are set by the actual sensor (the probe coil). With the help of a mathematical modelling approach it is possible to optimise all constructive and physical parameters of inductive sensors for the various settings of a task. In order to realise this approach, the engineer can make use of different numerical procedures and software packages. In this paper the boundaries of the analytical method are discussed and the possibilities of modern numeric procedures for the computer-aided draft of inductive sensors are shown.

1.   Introduction

For many years inductive sensors have been broadly used for many different tasks, for example as distance and approximation sensors in automatic control, or in non-destructive testing and quality protection for the touchless recording of material properties and defects. For each application one mostly chooses the coil model according to the setting of a task and determines the coil form and the core as well as the operating frequency empirically. If necessary experimental investigations can be used for optimisation.

While the inductive sensors (which are usually of an absolute or transformer-coupled air or ferrite core coil type) can be built simply, the mathematical modelling proves to be quite complex even for the simplest calibration set-ups. With the aid of mathematical modelling, the expected probe coil impedance effects can be estimated, and both the measuring frequency and the reactance coil geometry including the field profile can be optimised according to the problem. In order to optimise the sensor effect and the drop-in suppression, the operating parameters of the frequency and field strength, as well as the geometrical measurements must be optimal. Utilisation an experimental optimisation approach, it usually proves difficult and sometimes impossible to vary only one influence parameter.

2.   Fundamentals of the mathematical modelling of inductive sensors

The modelling of inductive sensors mathematically is based upon Maxwell's equations
of the electrodynamics in the form

$$\text{rot } \underline{H} = \underline{S} + \delta \underline{D}/\delta t \qquad (1)$$

and

$$\text{rot } \underline{E} = - \underline{B} \, \delta/\delta t \qquad (2)$$

with the material equations

$$\underline{S} = \chi \underline{E} \qquad (3)$$
$$\underline{B} = \mu \, \underline{H} \qquad (4)$$
$$\underline{D} = \varepsilon \, \underline{E} \, . \qquad (5)$$

(χ = electrical conductivity, μ = magnetic permeability, ε = dielectric permittivity)

Here $\underline{E}$ and $\underline{H}$ are the electrical and magnetic field strength, $\underline{S}$ and $\underline{D}$ are the electrical
and displacement current density and $\underline{B}$ is the magnetic flux density.
If the effect of the displacing-currents is neglected, the equation (1) can be transformed
by the introduction of the vector potential '$\underline{A}$' for linear, uniform and isotropic materials
and by using the material equations in:

$$\Delta \underline{A} + \chi \underline{A} = - \mu \, \underline{I} \qquad (6)$$

(Helmholtz Equation of the magnetic vector potential $\underline{A}$, I exciting current).

With the help of the vector potential 'A' one is able to calculate all electrical and
magnetic parameters of interest, for example the Eddy current density in the test piece
and the coil impedance. The time-dependent Helmholtz Equation (Eq. 6) can be solved
with different methods (boundary value problem).
Usually the boundary value problem is solvable only for the simplest of coil
arrangements and linear material properties, for example, a rotationally symmetrical air
coil over a conductive half-space (Figure 1, pick up coil) or for the tube probe coil
represented with uniform infinitely long core.
There are two fundamental approaches for the calculation of electro-magnetic fields: the
Analytical Solution and the Numerical (or approximated) Solution, the former being
refered to as the Analytical model, and the latter the Numeric model. A complete
modelling of the inductive sensors succeeds only with the help of numerical (discrete)
methods.

2.1   Analytical modelling of inductive sensors

One of the first analytical solutions for the modelling of an inductive sensor
arrangement (such as that which is typically used in non-destructive testing (NDT), was
developed by SOBOLEV [1] and later by DODD/DEEDS [2].
These initial approaches involved a symmetrical probe coil arrangement that was
computed from the layer thickness, conductivity and the remote sensing distance
involved. With these models for air core coils, the device parameters (measuring
frequency, probe coil geometry), the influence of the material properties on the probe
coil impedance could be calculated.

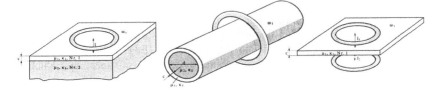

Fig. 1; Typically probe coil arrangements in non-destructive testing for analytical solutions

Core and shielded probe coils prove difficult to find an analytical solution for, even for very simple problems encountered in defect detection, i.e. the calculation of the probe coil impedance for a crack and the behaviour of the sensors at edges [3].

Non-linear material properties can only be integrated into a model, if one accepts strong simplifications [4]. Such practical arrangement important for the probe design, are only accessible with the help of numerical (discrete) model approach [5].

2.2   Numeric modelling of inductive sensors

The numeric procedures involved in the modelling of inductive sensors are based upon the discretization of the field area to be computed utilising a normal or irregular grid/mesh (1, 2 or 3 dimensional). In this case, the solution of Maxwell equations is now limited by the discrete (finite) grid points (eg. area or volumes). Different methods can used to find the solution in this case. One can divide the numeric procedures into four groups:

- the Finite difference method (FDM),
- the Finite element method (FEM),
- the Boundary integral method (BEM) and
- the Finite integration technique (FIT).

Initially the Finite difference method and the Finite element method were applied, with the Boundary integral method and Finite integration theory been successfully employed and reported recently [4],[5],[6]. In practical applications the eddy current probe axial symmetry allows for the reduction of the 3-D problems to a two dimensional problem.

The **FDM** is based on a reduction in a time-dependent Maxwell's (differential) equation to a difference equation for every point in a uniform grid/mesh present in the solution field. This results in an algebraic system of equations whose solution supplies the potentials searched for in the grid points. The relevant physically parameters are then ascertained (for example) by method of deviation from the note potentials (e.g. the y-component of the flux density: $B_Y = -\delta A/\delta x$).

The disadvantage of this procedure compared with the FEM is that the number of discrete points is much greater (especially when dealing with three dimensional Eddy current problems), whereas the accuracy is the same.   For boundary regions specific difference terms must be indicated (boundary conditions).

The **FEM** is based on the calculation of variation. The variation procedures can be distinguished according to the kind of the formulation of the variation integral (e.g. the energy functional) or according to the kind of the method of the minimization (Ritz- or Galerkin procedure). For the modelling of inductive sensors, linear and quadratic approach functions are sufficient to get the potential changes. The physical parameters are computed the same way as in the FDM.

The boundary **integral equations** are based on the 2nd Green's theorem, and involves the transformation of Maxwell equations to an integral equation (Fredholm's Integral of 1st or 2nd kind). Due to the fact that it isn't possible to solve these resulting integrals analytically, the field is subdivided into finite elements (Edge or surface). This step transfers the integral equation to an algebraic system of equations, for which the (discrete) solution shows the distribution in the finite elements searched for.

Whereas the analytical approaches reviewed so far (FDM, FEM, BEM) (for example the Laplace equation) could be discredited problem-specifically, the Finite-Integration-Technique utilises another approach. Its starting point is the discretization of the space with the help of two dual grids that are orthogonal to each other. If Maxwell equations in their integral-form are applied to this lattice structure, a complete system of consistent so called 'Maxwell-Grid-Equations' arises [7]. This discrete linear system of matrix equations of the discretized fields represents a universal basis and can be used for efficient numerical simulations on computer. The relevant matrix equations are numerically solvable according to the known method for large linear systems equations.

Due to the enormous evolution in computer engineering over the last 10 years, software packages have been developed that can be used with a standard personal computer, which are able to implement three-dimensional modelling for inductive sensors. Most frequently FDM- and FEM-programs are used. OPERA-2D/3D [8], Maxwell [9] or FEMLAB [10], are examples of FEM software packages. BEM Software worth mentioning would be VIC [11], and FIT is represented by the MAFIA program system [12] (for examples).

In the next section, the potential of the numeric modelling of inductive sensors with the FIT is demonstrated with the help of a simple example of an inductive metal detector.

3.   Optimisation of an inductive sensor

A standard difference coil probe – as illustrated figure 2 – is investigated. Such a probe design is utilised in industry for the detection of the metallic parts/pieces in volume flows, e.g. sand stream by suction dredger.

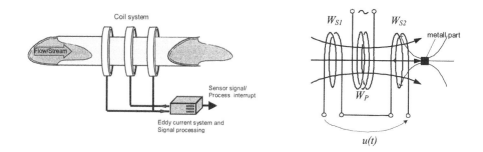

Fig. 2: Differential coil sensor system for metal detection

The magnetic field which is generated by the primary coil W_P at first induces equal voltages in the symmetric secondary coils (W_{S1}/W_{S2}). Without an electrically or magnetically conductivity metallic parts, this set-up results in a Zero voltage difference. If a metallic parts 'occurs' within the volume of the coil, a voltage difference arises from the resulting 'distortion' of this symmetry. Usually in non-destructive testing, such systems work with one or several discrete frequencies.

The investigation was carried out for a specific sensor problem (suction dredger) and involved finding the optimal set-up utilising the test frequency, geometry of the coil arrangement to that of the default inside diameter of the volume flow, the range and the material properties of the metal parts. The investigations was carried out using the FIT simulated on the MAFIA software package [11].

4. Results

The field distributions were computed for a difference probe coil order similar to that illustrated in figure 2.

Primary and second probe coils were modelled as single turns up to a diameter of 600 mm. The figures 3[a-d] show different parts of the calculated field distribution/area in the crossed R-Z coordinate system.

It can be seen from figures 3 a-d that the symmetry of the magnetic B field is disturbed by the metal part which results in a voltage arising from between both difference secondary coils.   With aid of these calculated field distributions or the computed physical values, an optimisation of the geometry and other influence parameter can be carried out.

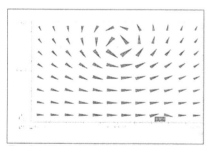

Fig. 3a: Magnetic flux density (Phase to exciting primary current: 0°)

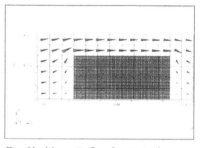

Fig. 3b: Magnetic flux density in the near of a metallic material

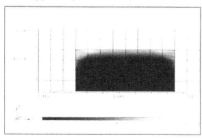

Fig. 3c: Inducted eddy current distribution in the metallic material

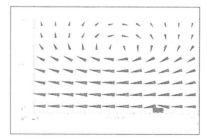

Fig. 3d: Magnetic flux density generated by the eddy current in the metallic material

Fig.. 3 a-d Magnetic flux density in a coil due to presence of a metallic material

For example, the effect on the voltage difference as a function of distance of the secondary coil was investigated, and the associated results are illustrated in figure 4a.

The results of varying the frequency are illustrated in figure 4b. The electrical and magnetic properties of the metal parts to be detected were also varied over a wide range.

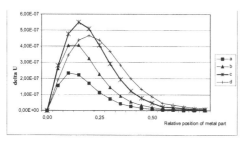
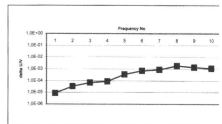

Fig. 4a: Difference voltage versus the position of a metallic material for different distances of the secondary coils a...d

Fig. 4b: Difference voltage versus frequency

The series of figures 4.a-b show that one can achieve the optimum set-up required by utilising some of the influential parameters, but if the process distortion cannot be suppressed, then a multi parameter optimisation procedure using more than one measurement frequency is necessary.

5. Summary

The mathematical modelling of inductive sensors can be calculated with both analytical and numeric methods. The impedance of an air core coils are computable utilising analytical models on condition of linear electrical and magnetic material properties. Inductive sensors with bore materials and non-linear material characteristics are modelled preferably with numeric procedures.

The finite element methods, BEM and the Finite Integration Theory currently represent the strongest procedure for the calculation of 2-D and 3-D magnetic field problems. Materials with non-linear properties can be computed, and an arbitrary form of the exciter signal response can be approved.

The numeric modelling, in conjunction with the available software packages, represents a strong "tool" for the engineer in the computer-aided design and the optimisation of inductive sensors. The optimisation of the measuring frequency, the bore material, and winding geometry can be investigated, as well as the influence of any disturbance signals on the measurement values and the possibility of the separation out.

References

[1] Sobolev, V.S.: K teorii metoda nakladnoi katuski kontrole vichrevymi tokami. -IZV AN SSSR, (1963)2,-S.78-88 (in Russian)
[2] Dodd, C.V.; Deeds, W.E.: Analytical Solution to Eddy-Current Coil Problems. Oak Ridge National Laboratory, 1967, Report ORNL-TM-1987
[3] Suchorukov, V.V.: Matematiceskoe modelirovanie elektromagnitnych polej v provadjascich sredach. Uzd. Energija, Moskva 1975 (in Russian)
[4] Russel, T.J.; Schuster, V.E.; Waidelich, D.L.: The Impedance of a Coil Placed on a Conducting Plane. -Bulletins of the University of Missouri, Columbia 1962

[5] Ida, N.; Lord, W.: Simulating Electromagnetic NDT Probe Fields. IEEE Comp. Graphic and Appl., Vol. CGA (1983)5, pp.21-28

[6] Ewald, H.; Neumann, R.: Simulation and design of inductive sensors using Finite Element-Method. Annual conference of the German Nondestructive Society (DGZfP), Proccedings of the conference, paper P35, Fulda 1992 (in German)

[7] Clemens, M.; Weiland, T.   Discrete Electromagnetism with the Finte Integration Technique. Progress In Electromagnetics Research, PIER 32, pp. 65-87, 2001

[8] OPERA-2D/3D: *Software Description, Version 8.0, Vector Fields Corp. , Oxford, GB, Spring 2002*

[9] Maxwell-3D: *Software Description, ANSOFT Corp., Pittsburgh (PA), USA, 2002*

[10] FEMLAB Toolbox for MATLAB: *Version 2.3, COMSOL AB, Stockholm, Sweden, 2002*

[11] VIC-3D: *Software Description, Version 3.0, Sabbagh Associates, Inc., Bloomington, (IN), USA*

[12] MAFIA: *Software Description, Version 4.0, CST Darmstadt, Germany, 1997-2001*

Poster Programme
Paper presented at Sensors and their Applications XII, September 2003
©2003 IOP Publishing Ltd

Wavelet-based Removal of Sinusoidal Interference from a Sensor Signal

Lijun Xu and Yong Yan

Centre for Advanced Instrumentation and Control, School of Engineering, University of Greenwich at Medway, Chatham Maritime, Kent ME4 4TB, UK.

Abstract: Presented in this paper is a wavelet-based filtering approach to the recovery of a sensor signal contaminated with sinusoidal interference of a single frequency component or its harmonics. A key step in the approach is to threshold the detail coefficients by utilizing a level-dependent threshold estimator. Unlike the use of a notch and comb-shaped filter, the new approach does not attenuate the signal at interference frequencies and is especially suited for the removal of sinusoidal interference that is located within the signal spectrum. Evaluation of the approach has been conducted using a simulated piecewise continuous signal and a real, rapidly changing, sensor signal. Results obtained have demonstrated that both the sinusoidal interference and the Gaussian noise in the signal can be removed effectively without notable attenuation to the signal at the interference frequencies.

1. Introduction

Apart from the Gaussian noise, a sensor signal is often contaminated with sinusoidal interference of a single frequency and its harmonics due to inappropriate design or shielding of the sensor or any other reasons [1-3]. Once sinusoidal interference appears in the observations of a sensor signal, subsequent data analysis or measurement can lead to erroneous results. Research into the elimination of a single frequency component or its harmonics interference has been reported in the past [4-7]. One of the approaches that have been proposed is to use a notch filter (or a comb filter) [4-6]. If the spectrum of the sensor signal does not overlap any of the interference frequencies, the interference can be removed completely. However, if the signal contains components at any of the interference frequencies, such a filter will become ineffective to remove the interference. Yan [7] developed a so-called dynamic Levkov-Christov method for the subtraction of mains interference from an ECG signal, in which identified interference values of non-linear segments were replaced with a preceding linear segment.

Signal processing techniques based on orthogonal wavelet decomposition may provide an alternative solution to this problem. By selecting a suitable wavelet, the wavelet transform of the observation sequence can suppress the main profile of the signal and retain the sharp transition details, sinusoidal interference and other noisy components (if any). The main profile of the signal is preserved in the approximate coefficients of the signal. If one of the interference frequencies is within the detail frequency subband at a certain scale, the interference is represented by the detail coefficients at that scale. Unlike in the frequency domain appearing as a sharp peak or multiple peaks, the transform coefficients of the sinusoidal interference spread throughout the time axis and appear sinusoidal. Therefore, denoising techniques based on detail coefficients thresholding [8-9] may be used to cancel the coefficients associated with the sinusoidal interference and the Gaussian noise but keep those in relation to the details of the signal. The thresholded coefficients are then used to reconstruct the signal. With this method, both the sinusoidal interference and the Gaussian noise can be removed simultaneously without significant attenuation to the signal at interference frequencies.

2.   Fundamentals

The observation model of a signal x contaminated with sinusoidal interference r and zero-mean Gaussian noise e can be described as

$$y = x + r + e \tag{1}$$

The sum of r and e is referred to as noisy components. In orthogonal wavelet decomposition, each of the detail coefficients of y at level j consists of three items, which are the transform coefficients of x, r and e, i.e.

$$d_j[n] = dx_j[n] + dr_j[n] + de_j[n] \tag{2}$$

where j is the decomposition level. In this context d_j ($j = 1, 2, \ldots$) are referred to as *detail coefficients* whilst dx_j, dr_j and de_j as the *transform coefficients* of x, r and e respectively.

If the signal x is contaminated by the Gaussian noise e only, the noise can be removed using the denoising method based on detail coefficients thresholding [10]. The transform coefficients of Gaussian noise remain Gaussian in the time-scale domain. Through orthogonal wavelet decomposition with an appropriate wavelet [10] the principal profile of the signal, of low frequency, is preserved in the approximate coefficients, and the sharp transition details together with noisy components are represented by the detail coefficients. If the signal-to-noise ratio (SNR) is large enough, the coefficients of the sharp transition details are generally larger than that of noise. Therefore, the noise can be cancelled by thresholding the detail coefficients with a suitable threshold. The expected signal can then be reconstructed from the approximate coefficients and the thresholded detail coefficients.

It can also be seen from the following that the sinusoidal interference appears sinusoidal too in the time-scale domain. This spreading characteristic can be used to remove the sinusoidal interference.

2.1 Characteristics of transform coefficients of the sinusoidal interference

Fig.1 illustrates a sinusoidal waveform of 50Hz sampled at a rate of 420Hz while shows The corresponding power spectral density (PSD) of the waveform as shown in Fig.1 was estimated using the Welch method [11] where a hamming window was applied to the waveform to reduce spectral leakage. The orthogonal wavelet decomposition of the signal into level 4 using Daubechies wavelet 'db2' [12] is depicted in Fig.1(c). The wavelet decomposition coefficients have been put in series, where d_j denotes the detail coefficients at level j and a_4 denotes the approximate coefficient at the deepest level $J=4$. In this case, $d_j = dr_j$. The coefficients d_1, d_2, d_3, d_4 and a_4 cover the frequency bands of 105-210Hz, 52.5-105Hz, 26.25-52.5Hz, 13.125-26.25Hz and 0-13.125Hz respectively. It can be seen that the sinusoidal signal remains sinusoidal after wavelet transform. In addition, although 50Hz is in the frequency band 26.25-52.5Hz covered by d_3, its energy spreads over to the other bands due to spectral leakage of the wavelet function. The leaked part is also of sinusoidal nature within that band.

In fact, the sinusoidal interference can be regarded as coloured 'noise'. The spreading nature of the sinusoidal interference in the time-scale domain is similar to that of Gaussian noise, hence the de-noising method based on detail coefficients thresholding can be applied to remove sinusoidal interference.

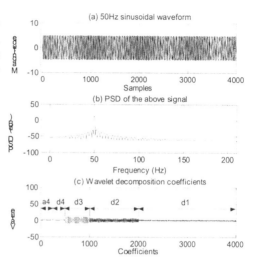

Fig.1. A sinusoidal signal (50Hz) and the orthogonal wavelet decomposition into level 4 using daubechies wavelet 'db2'.

2.2. Filtering approach

To threshold the detail coefficients of the sensor signal, the soft thresholding operator [8, 13, 14] is adopted here:

$$Th(d) = \begin{cases} y - T \cdot sign(d), & if \ |y| > T \\ 0, & if \ |y| \leq T \end{cases} \tag{3}$$

where $sign(\cdot)$ is the sign operator and T is the threshold. Since the noisy components are not white, T needs to be estimated level by level. The threshold T for decomposition level j can be estimated by:

$$T_j = \lambda \sigma_j \sqrt{2\log(N/2^j)} \tag{4}$$

where N is the length of the signal observations; σ_j denotes the standard deviation of the noisy part in the detail coefficients in level j; λ is a relaxation factor that can adjust the magnitude of the threshold so that compromise can be made between noise rejection and signal preservation. As the coefficients in relation to the sharp transition details of the signal x do not belong to noisy part, it needs to be taken out to estimate the standard deviation.

In summary, the following procedure should be followed to remove sinusoidal interference from the observations of a sensor signal:

(1) Select an appropriate wavelet and decomposition level J and obtain the orthogonal wavelet decomposition of the signal into level J;

(2) Estimate the standard deviation $\{\sigma_j\}_{0<j\leq J}$ and obtain the thresholds $\{T_j\}_{0<j\leq J}$;

(3) Threshold the detail coefficients according to equation (3) and obtain the thresholded coefficients $\{\tilde{d}_j\}_{1\leq j\leq J}$;

(4) Reconstruct the signal x with the thresholded coefficients $\{\tilde{d}_j\}_{1\leq j\leq J}$ and the approximate coefficients a_J.

3. Results and discussion

A simulated piecewise continuous signal and its corresponding PSD are shown in Fig.2. The signal was sampled at 1000Hz.

Fig.3 depicts the signal contaminated with 50Hz sinusoidal wave and its harmonics as well as Gaussian noise. The Signal-to-Noise Ratio (SNR) is 4.71dB. The orthogonal wavelet decomposition of the signal using Daubechies wavelet 'db4' is also plotted in Fig.3. Fig.4 shows the thresholded coefficients, from which the filtered signal is reconstructed, and the

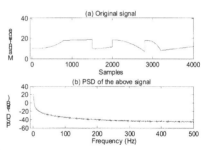

Fig.2 A piece-wise smooth signal and its PSD

waveform of the reconstructed signal and its PSD. The SNR after filtering is 17.40dB. It can be seen that the new filtering approach, unlike a notch filter or a comb-shaped filter, does not attenuate significantly the signal at the interference frequencies.

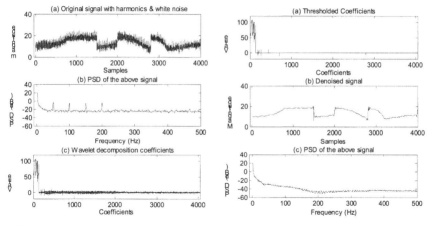

Fig.3 A piecewise smooth signal contaminated with four harmonics of 50Hz and Gaussian white noise and the corresponding orthogonal wavelet decomposition using Daubechies wavelet 'db4'.

Fig.4 Thresholded version of the wavelet coefficients in Fig.3(c) and the reconstructed signal after filtering.

It can also be seen from Fig.4 that some details of the signal has been slightly distorted due to the low SNR. This is very similar to the case of Gaussian white noise cancellation using thresholding method [8-10]. The degree of this distortion depends on the SNR. The stronger the interference, the more serious the distortion. Therefore, a compromise has to be made between the removal of the interference and the restoration of the signal details.

Fig.5 shows a real signal derived from a differential-pressure sensor. The sampling frequency was 330Hz. It is known that the signal is contaminated with 50Hz and 150Hz interference. The PSD of the signal and its corresponding orthogonal wavelet decomposition coefficients are also plotted in Fig.5. The new filtering technique is used to threshold the coefficients and the results obtained are given in Fig.6. Although the SNR can not be obtained in this case as the magnitudes of the harmonics and Gaussian noise are unknown in advance, it is still evident that the 50Hz and 150 Hz interference and the Gaussian noise have been effectively removed.

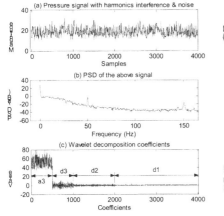

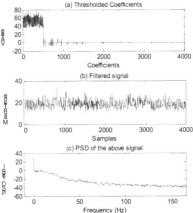

Fig.5 A real differential-pressure signal contaminated with 50Hz and 150Hz interference and Gaussian noise

Fig.6 Thresholded version of coefficients in Fig.14 (c) and the reconstructed signal in both the time and frequency domains.

For the purpose of a direct comparison, 50Hz and 150Hz notch filters were applied to filter the contaminated signal. The results obtained are shown in Fig.7. It is clear that the inherent 50Hz and 150Hz components of the signal have been notably attenuated.

In practice, the sinusoidal interference might be time-varying both in amplitude and frequency. Looking back to the characteristics of the sinusoidal interference in the time-scale domain, one finds that the spreading nature of the transform coefficients of the

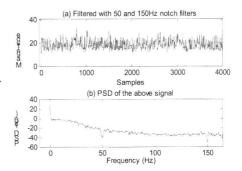

Fig.7 Signal filtered by notch filters of 50Hz and 150Hz in both the time and frequency domains.

interference is unaffected by its time-varying nature. The new filtering approach is, therefore, applicable to the removal of time-varying sinusoidal interference.

4. Conclusion

A new wavelet-based filtering approach has been presented to recover a signal contaminated with sinusoidal interference. With this approach both the sinusoidal interference and the other noisy components such as Gaussian noise can be effectively removed. Unlike the use of a notch or a comb filter, the new approach does not cause notable attenuation to the signal at the interference frequencies. This advantage makes the new approach particularly applicable to signals containing the frequency components of the sinusoidal interference. In addition, the new approach is immune to any changes in amplitude and frequency of the sinusoidal interference.

References

[1]   Kerth D A and Piasecki D S. "An oversampling converter for strain gauge transducers", IEEE Journal of Solid-State Circuits, 27(12), pp.1689-1696, (1992)

[2]   Verbeek J, Rolain Y and Pintelon R. "Leakage reduction in measurement data contaminated with 50 Hz mains and harmonic frequency components", IEEE Instrumentation and Measurement Technology Conference, 2, pp.1054-1058, (1999)

[3]   Huhta J C and Webster J G. "60-Hz interference in electrocardiography", IEEE Trans. Biomed. Eng. BME-20, pp.91-101, (1973)

[4]   Zhang H, Dai X and Shi X. "Pole-zero design method for Power frequency eliminating notch filters", Journal of Data Acquisition and Processing, 16(1), pp.68-71, (2001)

[5]   Glover J R. "Adaptive noise cancelling applied to sinusoidal interference", IEEE Trans. Acoust. Speech Signal Process. ASSP-25, pp.484-491, (1977)

[6]   Jang Y K and Chicharo J F. "Adaptive IIR comb filter for harmonic signal cancellation", International Journal of Electronics, 75(2), pp.241-250, (1993)

[7]   Yan X G. "Dynamic Levkov-Christov subtraction of mains interference", Medical and Biological Engineering and Computing, 31(6), pp.635-638, (1993)

[8]   Donoho D. "De-noising by soft-thresholding", IEEE Transactions on Information Theory, IT-41(3) 613-627, (1995)

[9]   Krim H, Tucker D, Mallat S and Donoho D L. "On denoising and best signal representation", IEEE Transactions on Information Theory, IT-45(7), pp.2225-2238, (1999)

[10]   Mallat S. A wavelet tour of signal processing. San Diego: Academic, 1998

[11]   Stoica P and Moses R L, Introduction to Spectral Analysis. Englewood Cliffs (NJ): Prentice-Hall, 1997.

[12]   Daubechies I. "Orthonormal basis of compactly supported wavelets" Comm. Pure. Applied Math. 41, pp.909-996, (1988)

[13]   Donoho D L and Johnstone I M. "Ideal spatial adaptation by wavelet shrinkage", Biometrika, 81, pp.425-455, (1994)

[14]   Donoho D L, Johnstone I M, Kerkyacharian G and Picard D. "Wavelet shrinkage: asymptopia", Jour Roy Stat Soc, Series B. 57(2), pp.301-369, (1995)

Poster Programme
Paper presented at Sensors and their Applications XII, September 2003
©2003 IOP Publishing Ltd

Emission Monitoring Using Optical and Acoustic Sensors Combined with Neural Networks

K Chua, L Xu and Y Yan

Centre for Advanced Instrumentation and Control, School of Engineering, University of Greenwich at Medway, Chatham Maritime, Kent ME4 4TB, UK.

Abstract: A simple and low-cost prototype emission monitoring system via flame detection using a combination of optical and acoustic sensors has been developed. Signals from the sensors are extracted using digital signal processing techniques to obtain features which represent flame signatures in relation to emission data. These features are applied to a neural network for training and subsequently for the monitoring of combustion emission. Preliminary results obtained on a combustion test facility show good correlation between reference emission data and those from the monitoring system. The relative error between the predicted emission and the expected value is less than 1.6%.

1. Introduction

Furnaces and boilers in industries are constantly pursuing for more efficient and economical combustion control systems in order to comply with increasingly stringent environmental regulations and requirements for cost saving. With the increasing use of low-quality fuels and biomass materials, there is a growing concern in the power generation industry over emissions such as excessive release of NOx and CO2 and associated operational problems. New techniques for advanced emission control have hence become desirable. One of such new techniques proposed is to upgrade flame detectors, which are available on most burners for safety reasons, so that they can be used to predict the emissions by analysing the flame signals. Emissions control is traditionally performed by adjusting burner's operating parameters based on the signals from a flue gas analyser installed at the exhaust of a combustion process. Real time sampling is often difficult to achieve due to inherent long time delay from the combustion chamber to the analyser at the exit of flue stack. This delay can lead to unstable control if the combustion condition is highly dynamic. By utilising a flame detector as an alternative to the conventional emission monitoring, on-line prediction of the emissions can be achieved and control loop can become more responsive, leading to improved control system performance and ultimately reduced emissions.

Various methods have been investigated in the past to optimise combustion conditions and monitoring emissions through advanced flame monitoring. Khesin *et al* [1] has used a flame spectra analyser to balance air-fuel ratio to achieve a better control of combustion emissions. Optical sensor in conjunction with neural network has been used by Von Drasek *et al* [2] to predict highly non-linear relationship of flame spectrum with combustion parameters. Acoustic sensors were also employed in investigation of combustion behaviour in various experimental studies [3-4]. Investigations conducted by M. M. Miyasato *et al* and G. S. Samuelsen *et al* [5] utilising an acoustic sensor to determine the correlation between the acoustic signal and NOx/CO emissions in a gas turbine combustor has shown positive results. This study, based on the successful implementation of optical and acoustic sensors for flame detection in previous research, explores the potential advantage of combining the two detection techniques to predict pollutant emissions. As a combustion process always emits electromagnetic radiation and acoustic energy, a combination of optical and acoustic sensors has the advantage of capturing both visual and audible characteristics of the flame. These characteristics can be identified by analysing the flame signals derived from both sensors in time and frequency domains. If quantitative relationships between the flame characteristics and the emission levels can be established, such relationships may be used to predict emissions. Although emissions level cannot be measured accurately utilising this method, the predictions may provide useful information so that rapid control actions can be taken to reduce emissions.

In this study, a prototype flame detector has been developed by combining acoustic and optical sensors in conjunction with an artificial neural network to predict combustion emissions. Digital signal processing techniques are applied to extract features from the acoustic and optical signals to establish their relationships with the emission levels. As such relationships are very complex, non-linear, the incorporation of a neural network is necessary. Neural networks have the ability to learn and store any non-linear, complex relationships between its inputs and outputs. To develop the neural network as part of the emission monitoring system, the features extracted from acoustic and optical signals and the corresponding emissions data are used to train the neural network, which is then applied to predict emissions. To evaluate the effectiveness of the system, a series of experimental programme was carried out and preliminary results obtained are reported in this paper.

2. Methodology

2.1. Sensors

Figure 1 shows a schematic diagram of the emission monitoring system.

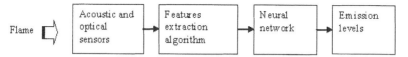

Figure 1 Schematic diagram of the emission monitoring system

The sensors used are an omni-directional electret microphone and an infrared silicon photodiode. The microphone is preferred to other acoustic sensors here because of its excellent response at the low frequency region (below 13kHz) where most acoustic emission from a combustion process normally takes place. The microphone possesses good linearity over its response range from 50Hz to 13kHz with a sensitivity of -60dB and a maximum sound pressure level of 120dB. The infrared sensor has a daylight filter to desensitise interference from non-flame source. This type of sensor was selected due to its high sensitivity to a wider range of flame radiation as compared to UV and visible light sensors. The response of the infrared sensor used covers the electromagnetic radiation wavelength from 800 to 1000nm with peak sensitivity at 900nm. Signals from both sensors are amplified through a low-noise, high input-impedance amplifier with an adjustable gain up to 120dB. These sensors generate analogue output signals the features of which are related to combustion emissions.

2.2 Features extraction

Feature extraction provides a means to quantify the flame characteristics – a process similar to finger-printing. Digital signal processing algorithms have been developed to extract these features from the sensor signals in the time and frequency domains. The time-domain features consist of DC mean, AC variance, standard deviation, kurtosis, skewness while the frequency-domain features include entropy and shape factor of the power spectral density (PSD) distribution. These features together with the corresponding emissions data can be used to train a neural network to establish the quantitative relationship between the features and emission levels. Once the network has been properly trained, it can be applied to predict emissions.

2.3 Artificial Neural Network

The neural network developed was a Multi-Layer Perceptrons (MLP) trained using backpropagation algorithm. Previous research [6] has shown that MLP can accurately represent any continuous non-linear function. This type of network is hence believed to be most suitable for the emissions monitoring purpose where non-linear relationships are expected between the flame properties and emissions.

The architecture of the network consists of one input layer, two hidden layers and one output layer. Each hidden layer has five neurons configured with non-linear transfer function known as log-sigmoid (LOGSIG) and the output layer comprises one neuron with a pure linear (PURELIN) transfer function which gives a continuous output. The learning function selected for the network is the backpropagation gradient descent with momentum weight and bias learning (LEARNGDM) and its training performance function uses the differentiable performance function of Mean Square Error (MSE). The network is trained in a supervised mode with a set of inputs and their targets provided for the network to establish the relationship between the inputs and the outputs.

3. Experimental Results and Discussion

3.1 Experimental set up

A series of experiments was carried out on a laboratory-scale combustion rig operating at atmospheric pressure, as shown in Figure 2. The system consists of a 26mm diameter butane-air premixing Bunsen burner and a cylindrical combustion chamber with a diameter of 0.2m and a height of 0.5m. The air-fuel ratio, determined from the flow rates of air and gas measured using calibrated rotameters, was varied from 10 and 44. Under each condition, five sets of data were collected to arrive at average values. Water-cooled stainless steel tubing with 6mm diameter was used to sample the combustion emissions at the exhaust of the rig continuously. The samples were pumped into a gas analyzer which determines relevant combustion emissions. In the initial studies, only CO_2 is monitored due to the limitation of the current experimental set-up.

The flame detection system consists of a sample probe installed on the side of combustor directed at the flame. The sample probe was connected to a metal enclosure which contains a microphone to detect acoustic emissions from the combustion radiation and an infrared sensor to detect flame radiation.

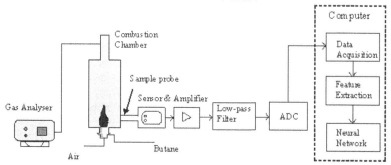

Figure 2 Experimental set up to study emission using flame monitoring system

The signals from the sensors were amplified and then passed through low-pass filters with a cut-off frequency of 10kHz to reject unwanted higher frequency noise. Both signals were fed into a 12-bit analogue-to-digital converter (ADC) and simultaneously digitised at the rate of 20kHz. The subsequent signal processing and data analysis were performed as described in Section 2.

3.2 Results and Discussion

Figure 3 shows the PSD distribution of the acoustic sensor from 0 to 1kHz for three different air-fuel ratios. It is clear that the PSD distribution varies with the test condition with the larger amplitude in lower spectral band at higher air-fuel ratio. This increase is expected as a result of the greater heat release rate and stronger acoustic energy when combustion process shifted from incomplete combustion regime towards complete combustion zone with more air supplied. For the infrared sensor, as illustrated in Figure 4, the PSD becomes smaller and smoother between 0 to 500Hz when the air-fuel ratio increases. This is believed due to the incomplete combustion at lower air-fuel ratio, resulting in a mainly orange and yellowish flame containing strong infrared energy. However, as the air-fuel ratio increases, the radiation spectrum shifts from the infrared region towards the ultraviolet zone and the flame becomes more transparent with less infrared energy.

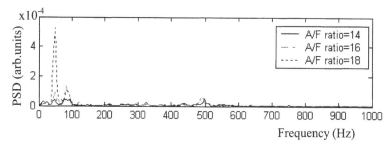

Figure 3 Power Spectral Density of the acoustic signal for different air-fuel ratios

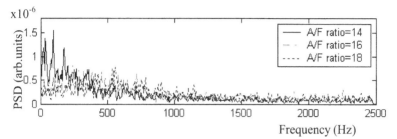

Figure 4 Power Spectral Density of the IR sensor for different air-fuel ratios

From the power spectral analysis of the optical and acoustic signals, it can be seen that they respond differently under each test condition which indicates the presence of useful information that can be extracted to yield flame features. As illustrated in Figure 5, these features were quantified using the developed digital signal processing algorithms. These features were then used to train the neural network to establish their correlations with the combustion emission, as shown in Figure 6. The trained neural network can then be applied to predict emissions under similar combustion conditions.

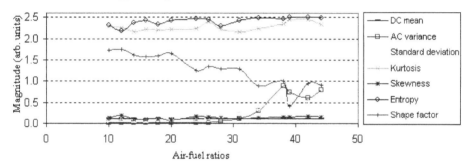

Figure 5   Flame features for various air-fuel ratios

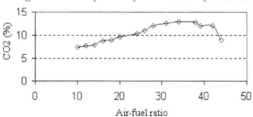

Figure 6 CO_2 emission for various air-fuel ratios

Figure 7 shows the emission level predicted by the neural network when it was tested with unseen data. It is evident that the neural network has performed well with the predicted emissions very close to the actual measurement. Figure 8 illustrates the performance of the network by comparing its predicted outputs with the expected data. The ideal results are indicated by a dashed line (45° line) and the best fitted line of the predicted data points (circles) is shown by a solid line. It can be seen from Figure 8 that the emission prediction achieved by the neural network is consistent with the expected value. The relative error between the predicted emission and the expected is no greater than 1.6%.

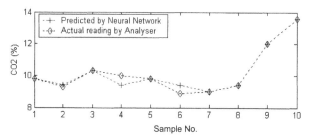

Figure 7 CO₂ emissions as predicted by the neural network and measured by gas analyser

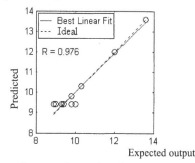

Figure 8 Comparison between the predicted and expected CO₂ emissions

4. Conclusions

A new emissions monitoring system using a combination of acoustic and optical sensors with a trained neural network has been developed. Experiments conducted on an experimental-scale combustion rig have demonstrated that the system has the potential to predict combustion emissions as expected. Results obtained are encouraging.

The microphone and infrared photodiode sensors provide a low-cost and effective solution for the system to identify the flame features in relation to combustion emissions. The neural network developed plays an important role in the establishment of the complex non-linear relationships between the flame features and emissions and in the subsequent application to emissions monitoring.

References

[1] Khesin M, Girvan R and Quenan D, "Demonstration tests of new burner diagnostic system on a 650 MW Coal-Fired Utility Boiler", Proceedings of Instrumentation, Control and Automation in the Power Industry, Vol.40, pp.127-135, (1997).
[2] Von Drasek, W., Charon, O., Marsais, O., "Industrial combustion   monitoring   using optical sensors" .SPIE Meeting on Industrial and Environmental Monitors and Biosensors, Boston, Nov. 2-5, (1998).
[3] Zukowski W. "Acoustic effects during the combustion of gaseous fuels in a bubbling fluidised bed". Combustion and Flame, Vol.117, No.3, pp.629-635, (1999).
[4] Xu Y., Liu Y., Tang X., Yang H., Gu F., "Study on properties of acoustic emissions in the coal-fired fluidised bed", Procs of 1997 Int. Symp. on Multiphase Fluid, Non-Newtonian Fluid and Physico-Chemical Fluid Flows, ISMNP'97, pp2.88-2.93, (1997).

[5] M. M. M Miyasato and G. S. Samuelsen, "Reaction chemiluminescence and its relationship to emissions and stability in a model industrial burner", American Flame Research Committee (AFRC) 1999 Fall Int. Symp., San Francisco, Oct.3-6, (1999).

[6] Cybenko, G., "Approximations by superpositions of a sigmoidal function", Mathematics of control, signals and systems, Vol.2, pp.303-14, (1989).

Poster Programme
Paper presented at Sensors and their Applications XII, September 2003
©2003 IOP Publishing Ltd

In situ and remote controlled units which activate marine systems

R Jaskulke, B Himmel

Department of Electrical Engineering and Information Technology, University of Rostock, Germany.

Abstract: Versions of an event-controlled sampling system are reported. These are designed to respond to relevant processes in the marine environment.

The *in situ* version is based on a measuring system that drives peripheral devices in respond to sensor data, but only in the case of defined events. If there is no event during a determined time, a time control takes over.

The remote controlled version is a conventional marine system as, for example, a filtration sampler or a bottom water sampler without a measuring system. A radio communication link is used to start specific sampling procedures. The identification of real events takes place in a laboratory on land. Determination of what constitutes an event is made by, inter alia, specific weather and hydrodynamic conditions.

1. Introduction

In marine research, the investigation of processes of particle exchange and transport is an important task. Filtration samplers, bottom water samplers, sediment traps or other equipment are typically used in such investigation. The development and realization of an event-controlled system will be explained, for example, for an *in situ* controlled and a remote controlled filtration sampler of the Baltic Sea Research Institute, Warnemünde (Germany) [1]. The special filtration samplers are used to collect suspended matter near the sediment-water interaction zone for a period of several months. Both systems are designed for a water depth of maximal 6000 m and they are equipped with 21 and 6 filter places, respectively.

It is estimated that relevant physical, chemical, or biological hydrodynamic processes are not recognizable when time-triggered autonomous systems are being used. There is an unusual correlation between the time of the measurement and the process. An example of such a process is the correlation of salinity with the rate of water exchange between the North Sea and the Baltic Sea. At irregular time intervals, water from the North Sea with high salinity flows into the Baltic Sea. This event is very important for the whole balance of this semi-enclosed brackish water ecosystem. Vice versa, predominantly surface water with a noticeable lower salinity flows back into the North Sea [2].

2. Problem solution

Our approach is the implementation of an intelligent control system based on relevant data and predefined decision rules for the identification of real events. The following situations are of special interest concerning event-controlled sampling:

• combination of features, e.g. the current is flowing, on the average, in a certain direction
   and the salinity simultaneously reaches an interesting range.

• crossing of given absolute limits, e.g. very strong currents give useful information about
   resuspension of sediments

• after longer periods of continuously drifting parameters, e.g. increasing or decreasing salt
   contents of water (dynamic programming)

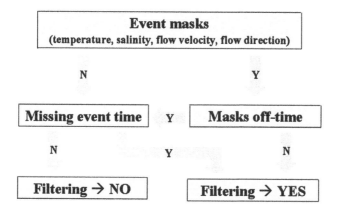

Figure 1.. Diagram for making decisions of event-driven sampling

The decision to filtrate will be determined by a comparison of the filter criteria given over programmable masks with the data stored in the memory of the control device (Fig. 1). If such a criterion is fulfilled, filtration starts and the corresponding mask is closed for a defined period (mask off time). Thus, the same criterion (that means the same event) is explicitly prevented from causing filtration processes in very short sequences. With missing events, one filtration process starts at a predefined time interval in each case (missing event time).

3. Results

The intelligence for the identification of real events can be transmitted to the filtration system from a Master PC in any laboratory on land. The *in situ* event-control developed for a filtration sampler is shown in fig. 2 as a block diagram.

The micro-controller MSP 430337 is the central component of the control unit [3,4]. One EEPROM is used for storing calibration data and status information, as well as, one flash memory which is used for storing the data measured by the sensors that are available to the MSP as external memory. Both non-volatile memories guarantee protection of all relevant data in the case of malfunction. A RS232 interface combined with a multiplexor circuit allows the MSP to communication with the master PC, the sensors for hydrological data acquisition, and an optional acoustic modem for bi-directional data exchange.

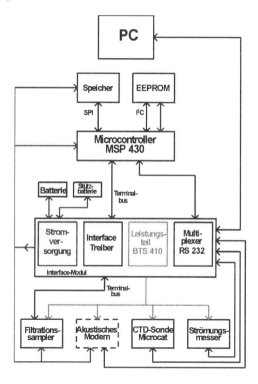

Figure 2: Block diagram of the in situ controlled system

The electric current requirement of the overall electronics assembly averages approximately 1 mA during the idle time between single measurements. Using a battery of 70 Ah, this extremely small power demand is a precondition to realise a length of application of at least one year.

The MSP is programmable with the necessary control data (filtering times, allowed masks for each filter place, rinsing time, starting time, interval of measurement, missing event interval, masks off-time interval) via a PC with a user-friendly programming interface [5,6]. The masks (e.g. the limiting levels for the individual sensors) can be set in a separate window.

The SBE 37-SI MicroCAT was selected as a CTD-detector (conductivity, temperature, and depth). An inductive current meter equipped with a magnetometer is used for measuring velocity and direction of water current. Upgrading with additional sensors to measure, inter alia, turbidity, fluorescence, oxygen, wind, and waves is ongoing, as is the provision of an optional GSM modem for bi-directional data exchange.

The sensors and parts of the filtration system are separately supplied with power by the use of circuit breakers. Thus, we have the possibility to connect only the systems and units which are needed at a given moment.

The remote control developed for other filtration samplers is shown in fig. 3 as a schematic diagram.

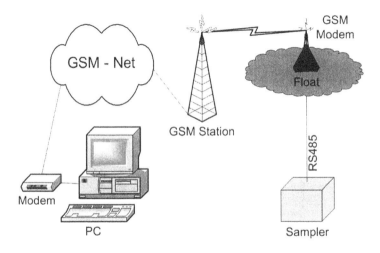

Figure 3: Schematic view of the remote controlled system

A Master PC in any laboratory on land determines if an event takes place or not. If an event occurs, the sampler receives action commands over the radio link. The Master PC receives the necessary sensor data over satellite from the operation area of the sampler with the help of a permanently installed measuring systems. Not only actions can be activated over the radio, but also the current state of the samplers, such as, the amount of filtered liquid, the error status, the state of the power supply, and other data can be requested. The embedded circuit in the sampler is again the MSP 430 because of its low power requirements. The overall power consumption of the system in this mode is drastically decreased, as no on-board sensors need be activated (all event determination is performed at the base station). Furthermore, the expense of the hardware is also reduced. The biggest advantage, though, of this kind of event control is found in the user's ability to intervene at all times in the filtration processes from the Master PC somewhere in a laboratory. Therefore, through the Master PC, it is possible to initiate filtration processes at any time. This is not possible with an autonomously working systems. However, sensor data have to be available from the area of analysis at every time. To come to a decision if an event takes place or not in a laboratory on land, the sampler in the computer receives commands via the remote radio communication.

4. Conclusion

This paper has described flexible approaches to the control of marine sampling systems. Both an *in situ* control approach and a remote approach are presented, and the advantages of each outlined. The opportunities opened up for marine research by the provision of flexible, intelligent and automated sampling systems are exciting and should encourage new research programmes in the fields of marine science and technology.

Both the hardware system and the software control system can be adapted without difficulty to other problems with requirements for autonomous operation over more extended periods.

References

[1] http://www.io-warnemuende.de/

[2] Fennel, K. and Neumann, T.: A coupled physical-chemical-biological model for the western Baltic. In: Computerized modeling of sedimentary systems. Ed. by J. Harff, W. Lemke and K. Stattegger: Berlin: Springer 1999: 169-181

[3] Company publication:"MSP430x33x - Mixed Signal Microcontrollers, SLAS163-February 1998", Texas Instruments Inc., Dallas, 1998

[4] Jaskulke, R. and Himmel, B.: Ereignisgesteuerte Probennahme auf der Basis des ultra-low-power Microcontrollers MSP430, 5. Seminar der Strömungssensorik an der Universität Rostock, 2000, S. 7.1-7.4

[5] Himmel, Bernhard; Jaskulke, Rainer: Event-driven sampling on the basis of relevant hydrological parameters, Proceedings, MAT2001 International Conference, Nürnberg 08.05.-10.05.2001, Proceedings p. 335-339

[6] Jaskulke, R. and Himmel, B.: Eventgesteuerte autonome Systeme, Landestechnologieanzeiger Mecklenburg-Vorpommern 2/2001, S. 7/8

Poster Programme
Paper presented at Sensors and their Applications XII, September 2003
©*2003 IOP Publishing Ltd*

Toward a better understanding of optical fibre Bragg grating sensors

S A Wade, C Rollinson, N M Dragomir, A Roberts[1], S F Collins and G W Baxter

Optical Technology Research Laboratory, Victoria University, Australia.
[1] School of Physics, University of Melbourne, Australia.

Abstract: Non-destructive images of type I in-fibre Bragg gratings reveal a complex refractive index distribution within the core of an optical fibre due to the use of a phase mask in the fabrication process. Analysis of features in these images led to the prediction of an additional peak in the reflection spectra of the grating at two-thirds of the operating wavelength. Measurement of the transmission spectra of a type I and IIA fibre Bragg grating confirm the presence of an additional peak, but with differences observed between the grating types. Work is continuing to identify key elements of fibre Bragg grating design and fabrication that affect their lifetime and performance for sensing applications.

1. Introduction

Fibre Bragg gratings (FBG) have become a popular sensor for a variety of applications [1-3]. However a number of questions remain concerning the prolonged use of these devices. For example, it is known that with time the gratings self erase [1, 4]. Currently there exists a wide variety of FBGs having different features which allow application specific gratings to be fabricated. This variety may be characterised in a number of ways: the type of fibre used, the way the fibre is prepared, the means by which the interference pattern is formed within the fibre core and the laser used in manufacture. During the fibre fabrication process a number of dopants may be added into the core region to increase the photosensitivity of the fibre, the most common being germanium, boron, tin and antimony. The photosensitivity may also be increased by holding the fibre in a high pressure hydrogen rich environment, known as hydrogen loading. Continuous wave low power or short high pulse-power lasers may also be used to produce different FBG types. Commonly FBGs are classified as type I, II or IIA [1]. Type I FBGs are characterised by a monotonic increase in the amplitude of the refractive index modulation (up to a maximum of $\sim 10^{-3}$) with writing time and have been related to local electronic defects in the glass matrix. Type IIA gratings evolve from particular type I gratings with ongoing exposure, in which the FBG is first partially erased and then rewritten into the fibre core [5]. Such gratings have been associated with compaction of the glass. Type II gratings are often formed by a single pulse from a high power laser that causes damage or fusion of the glass matrix resulting in large changes in the core refractive index (typically 10^{-3} up to 10^{-2}).

Presented here is a comparison between a type I and a type IIA FBG, in terms of their evolution during writing and their spectral transmission properties. These are complemented by images of the refractive index profile of a FBG.

2. Fibre Bragg Grating Fabrication

2.1 Type I Grating

The FBGs used in this work were manufactured using a standard phase mask technique. Ultra-violet illumination was provided by a continuous-wave frequency-doubled argon ion laser operating at an output power of 106 mW with a wavelength of 244 nm; the fluence was approximately 3.4 W/cm^2 at the fibre core. The phase mask was designed to suppress the zeroth order (manufactured by Lasiris™, the relative zero order contribution was approximately 3%) and had a measured period of 1.059 µm. Visual inspection of the far field diffraction pattern produced by the phase mask during the fabrication process revealed a weak zeroth order, strong first order, moderate strength in the second order and little evidence of the third order. Standard 9 µm core diameter telecommunications fibre (containing germanium in the core) was photosensitised by hydrogen loading for a period of 90 hours under a pressure of 82 atmospheres at a temperature of 80 °C. The transmission characteristics of the fibre were monitored during the 18 minute FBG fabrication process using an erbium fibre amplified spontaneous emission light source in conjunction with an OSA having a resolution of 0.1 nm. The writing history of the FBG is shown in Fig. 1 where the monotonic increase in the amplitude of the refractive index modulation with writing time clearly identifies the grating as type I. The reflectance of the FBG has been inferred from the depth of the transmittance spectrum which, as expected, shifted slightly to longer wavelengths during the writing process. The amplitude of the refractive index modulation, assuming a uniform sinusoidal variation along the axis of the fibre, sharply truncated at each end, and a uniform profile across the fibre core, was calculated using coupled mode theory [1] and found to be approximately 1.7×10^{-4}.

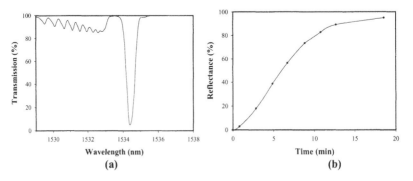

Figure 1. (a) Transmission spectra of a type I FBG, the fabrication method is described in the text. (b) The peak reflectance of the same FBG shown against the writing times through the fabrication process, clearly identifying the FBG as type I.

2.2 Type IIA Grating

The type IIA FBG fabricated in the present work was written into Fibrecore™ fibre (PS1250/1500), without hydrogen loading. This fibre contains both boron and germanium in the core to increase photosensitivity. The total writing time was approximately 80 minutes. The writing power density was ~ 3.2 W/cm², with a grating length of ~ 12 mm.

Figure 2 shows the evolution of the type IIA FBG. In particular, the characteristic increase, decrease and then further increase in the reflectance as a function of the grating writing time is clearly demonstrated in Fig. 2(b).

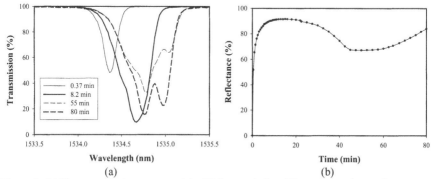

Figure 2. (a) The transmission spectra of the FBG recorded at different times during the writing of the grating. (b) The peak reflectance of the same FBG shown against the writing time through the fabrication process, clearly identifying the FBG as type IIA.

3. Fibre Bragg Grating Analysis

A type I FBG was imaged using a modified form of the differential interference contrast (DIC) technique as described in reference 6. This technique is sensitive to differences in phase between two coherent beams co-propagating through a specimen but separated by 0.5 μm; for this work intensity variation in the image is representative of refractive index change. A series of DIC images of the FBG were taken as the fibre was rotated in 2° increments. This process permitted the identification of the principal measurement planes, parallel and perpendicular to the original FBG writing beam. Fig. 3(a) shows a DIC image of the fibre taken from a direction orthogonal to that of the original UV writing beam, while Fig. 3(b) shows an image taken after the fibre had been rotated by 90° about its axis. Both images in Fig. 3 feature characteristics typical of a DIC image of a fibre. The bright vertical line to the left of centre identifies one edge of the fibre core and the black vertical line the other edge. Evidence of the FBG is seen in the variation of the measured intensity within the core region. The impression given in Fig. 3(a) is of an interleaved variation which, since the image was recorded in a direction perpendicular to the writing beam, may be associated with beating between diffraction orders of the grating.

For the fabrication conditions of this work, a period of 14.3 μm would be realised for beating between the zeroth and first orders, and 4.65 μm for beating between the first and second orders. The observed refractive index variation is therefore consistent with beating between orders one and two [7]. This is also consistent with the qualitative observation of the relative strength of the different orders seen during the fabrication process. From Fig. 3(a) the grating period was found to be 0.535 ± 0.005 μm, taking into account the interleaving of the fringes, consistent with the value inferred from the measured transmission spectrum [Fig. 1(a)]. The variations in the refractive index within the core, as seen in Fig. 3(b), appear as uniform horizontal rows having a periodicity equal to that of the phase mask (2Λ). The DIC image represents a well-defined section through the peak of a single Talbot plane [8] located near the centre of the fibre as seen in Fig. 3(a), and therefore happens to miss every second FBG index ripple.

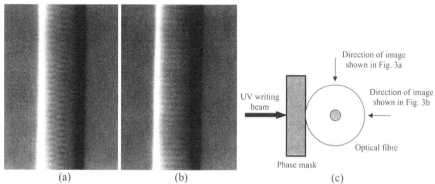

(a)   (b)   (c)

Figure 3: Unprocessed DIC images of the core region of a fibre Bragg grating. (a) The fibre rotation has been selected so that the image is taken from a direction orthogonal to that of the writing beam. (b) The same fibre Bragg grating as is shown in (a) after the fibre has been rotated about its axis by 90°. (The dimensions of the images are approximately 500 × 900 pixels covering 23 × 32 μm.) (c) Schematic diagram identifying the direction of the microscope beam used for the images shown in Fig. 3 (a) & (b).

The complex nature of refractive index variation observed in the core due to the FBG fabrication process has features representative of a periodicity of both 0.535 ± 0.005 μm and 1.07 ± 0.01 μm. However, it has been argued that a wave propagating along the fibre will experience an average of the refractive index variation across the core region [6]. The following paragraphs present a procedure to test this hypothesis.

The fundamental Bragg transmission dip has been designed to be approximately 1534 nm resulting from a grating period of 535 nm. However, we would also expect other reflections to occur at other harmonics of the refractive index periodicity, i.e.:

$$\lambda_m = \frac{2}{m} n_{eff} \Lambda,\qquad(1)$$

where λ_m is the reflected wavelength of order $m = 1, 2, 3, \ldots.$, n_{eff} is the effective refractive index and Λ is the period of the grating. This would result in a second order reflection at approximately 767 nm, with some allowance made for a change in n_{eff} at the lower wavelength. In contrast, if the grating can be described to have a periodicity of 1.07 μm then the fundamental wavelength would be 3.07 μm with the operational wavelength being the second order reflection. Measurement at 3.07 μm was not possible so the third harmonic at approximately 1020 nm was investigated. Reflection at this wavelength would not be apparent if the periodicity of the grating was purely 535 nm.

Measurements of the transmission spectra of the gratings were taken using either an Er^{3+}- or Yb^{3+}-doped fibre amplified spontaneous emission source for illumination of the 1530 and 1020 nm wavelength regions respectively. The fibre samples containing the FBGs were fusion spliced to the output of the fluorescence sources and the transmission spectra over the appropriate wavelength range were recorded on an OSA using a resolution setting of 0.1 nm. To remove the wavelength dependence of the light source and detection system a second reading with the FBG removed was taken as a reference.

3.1 Type I FBG:

The transmission spectra of the FBG shown in Fig. 1 was re-measured at the designed wavelength to see if the spectra had changed in the 6 months since the grating was written. The Bragg transmission dip was found to have shifted to 1535.85 nm ($\Delta\lambda \sim$ -0.55 nm) while the reflection and full width half maximum were found to be very close to the values taken just after writing the FBG. These changes are consistent with that found by other investigators [1]. The transmission spectra of the FBG in the 1020 nm wavelength region shows two dips at 1025.6 nm (reflectance of ~ 2.2%) and at 1026.5 nm (reflectance of ~ 2.8%). Further to this two dips were also observed at half the designed wavelength, one at 772.69 nm (reflectance of ~ 6%) and one at 771.9 nm (reflectance of ~ 4%).

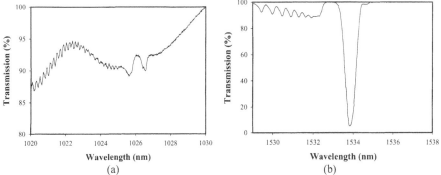

Figure 4: (a) Transmission spectra of a type I grating at two thirds of the designed Bragg wavelength.
(b) The transmission spectra of the same grating at the designed Bragg wavelength.

3.2 Type IIA FBG:

The transmission of the type IIA FBG as a function of wavelength in the 1020 nm region is shown in Fig. 5. Three distinct dips are seen at 1025.9 nm (reflectance of ~ 14%), 1027.3 nm (reflectance of ~ 15%) and at 1028.5 nm (reflectance of ~ 10%).

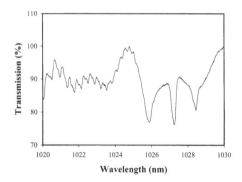

Figure 5: Transmission spectra of type IIA grating at two thirds of the designed Bragg wavelength.

4. Conclusion

A number of interesting observations can be drawn from the data reported here. First, there is clear evidence of reflection at two thirds of the designed wavelength meaning that the contributions of higher refraction orders in the FBG writing beam are having a measurable effect on the spectra. This has been observed previously for a type II grating written using a phase mask where reflections were observed at 1535, 1030, 770, and 620 nm [9]. The reflections were attributed to refractive index variations with a periodicity equal to the phase mask as the 1030 nm and 620 nm wavelengths are not harmonics of the normally expected Λ periodicity. Second, multiple reflections have been observed at wavelengths other than the fundamental. This also has been reported previously and was explained as being due to the presence of higher order modes that are present as the fibre may not be single moded at lower wavelengths [10]. Third, the strength of the reflection near 1020 nm is greater for the type IIA FBG than for the type I FBG. It may be expected that the magnitude of the reflectance due to other harmonics of the fundamental Bragg wavelength will depend upon the level of the saturation of the refractive index variation [11]. It would follow that the transmission dips at the wavelength equal to two-thirds of the design value would be much stronger for the type IIA grating as has been observed in this work.

The data reported here demonstrates that there exists a complex relationship between the fabrication technique used to fabricate a FBG and its spectral characteristics. The development of the DIC technique for imaging FBGs has provided a valuable tool for analysing gratings with small refractive index changes. This technique has enabled us to predict particular features in the FBG spectral characteristics. Work is continuing with these techniques to develop a better understanding of a number of important parameters in the application of FBGs as sensors. These include: ageing, annealing, dopant variety, and laser power. It is expected that this study will result in significant improvement in the fabrication of FBGs for sensor applications. In particular, this work may have direct application for the use of the first and second order diffraction wavelengths for simultaneous measurement of strain and temperature [12].

References

[1]   Othonos A and Kalli K *Fibre Bragg Gratings: Fundamentals and Applications in Telecommunications and Sensing* 1999 (Boston: Artech House)

[2]   Kersey A D, Davis M A, Patrick H J, LeBlanc M, Koo K P, Askins C G, Putnam M A and Friebele E J 1997 *J. Lightwave Tech.* **15** 1442-63

[3]   Rao Y J 1999 *Opt. Lasers Eng.* **31** 297-324

[4]   Erdogan T, Mizrahi V, Lemaire P J and Monroe D 1994 *J. Appl. Phys.* **76** 73-80

[5]   Xie W X, Niay P, Bernage P, Douay M, Bayon J F, Georges T, Monerie M and Poumellec B 1993 *Opt. Commun.* **104** 185-95

[6]   Dragomir N M, Rollinson C, Wade S A, Stevenson A J, Collins S F, Baxter G W, Farrell P M and Roberts A 2003 *Opt. Lett.* **28** (May)

[7]   Dyer P D, Farely R J and Giedl R 1995 *Opt. Commun.* **115**, 327-34

[8]   Mansuripur M 2002 *Classical Optics and its Applications* (Cambridge: Cambridge University Press) pp 251-62

[9]   Malo B, Johnson D C, Bilodeau F, Albert J and Hill K O 1993 *Opt. Lett.* **18** 1277-9

[10]   Hill K O and Meltz G 1997 *J. Lightwave Tech.* **15** 1263-76

[11]   Xie W X, Douay M, Bernage P, Niay P, Bayon J F and Georges T 1993 *Opt. Commun.* **101** 85-91

[12]   Brady G P, Kalli K, Webb D J, Jackson D A, Reekie L and Archambault J L 1997 *IEE Proc. - Optoelectron.* **144** 156-61

A Charge Transport in Sensor Materials Based on Conjugated Polyaminoarenes

O Aksimentyeva, O Konopelnik, M Grytsiv , P Stakhira[1], V Cherpak[1], A Fechan[1]

Chemical Department of Ivan Franko Lviv National University, Ukraine
[1]Electronic Devices Department of Lviv Polytechnic National University, Ukraine

Abstract: The physical and technological properties (temperature dependence of conductivity, thermal stability and structure) of the new sensor materials based on conjugated polyaminoarenes such as polyaniline and its derivatives have been studied in temperature interval 273-773 K.

It's found that temperature dependence of polyaminoarenes specific conductivity is followed by activation law with activation energy of conductivity 1,5-1,6 eV (polyaminohenols) and 0,37-0,59 eV (polyaniline). Change of inclination in temperature dependence is compared with main extrema of thermogravimetrical curves possible entailed by transformation of molecular structure and beginning of polymer thermal decomposition.

By the X-ray powder diffraction is found that polyaminoarenes are amorphously-crystalline materials with content of crystalline phase from 20-22 (undoped polyaniline, poly-*meta*-aminophenol) to 40-44 % (poly-*ortho*-aminophenol, doped polyaniline). Dimensions of crystallites are 15-25 Å. It is manifested that doping of polyaminoarenes by inorganic acid (H_2SO_4, $HClO_4$, HCl) causes to rising of crystalline phase. Characteristics of temperature dependence of conductivity and thermal stability are determined by molecular structure of polymer chain and type of acid used for polymer doping.

1. Introduction

Conjugated polyaminoarenes such as polyaniline and its derivatives have a potential technological application in chemical power sources, electrochromic displays and recently attract a great attention as sensor materials [1-4]. These polymers change the resistance under gases (ammonia, water and alcohol vapor); their electrochemical response is applied to indicate the metallic ions (Ni^{2+}, Co^{2+}, Fe^{2+} and Ag^+), polyaminophenols are used to define the concentration of organic compounds (ascorbic acid, hydroquinone) and complex $K_3Fe(CN)_6$ [3,4]. The rate and mechanism of charge-transport in sensor materials based on polyaniline and its derivatives are connected with their functions, particularly, sensitivity to certain type of physical and chemical influence.

It is known that for conjugated polymers with hydrocarbon backbone (polyparaphenylene, polyacethylene) both donor (n-type) and acceptor (p-type) doping may be used, but the polyaniline and its derivatives achieve the high conductivity mainly in the case of strong proton acids using as doping agents [5]. In these polymers the positive polarons (cation-radicals) are considered as possible charge carriers, under higher level of doping polarons may couple to bipolarons [1,5]. It is found that at the low temperatures (T< 77 K) the temperature dependence of polyaminoarene conductivity may exhibit metallic behavior (positive temperature coefficient of resistance, negative thermo-power and Pauli spin magnetism independent from temperature [1,5]). In temperature range 77-273 K character of polyaminoarene conductivity may be described in the frame of hopping model [1,5,8,]. However, at temperatures T > 273 K such model often does not adequate [1,5-8]. Since the most of polymer sensors operates at the room and higher temperatures, very impotent is an information about changes in conductivity of polyaminoarenes during their application under these temperatures.

Temperature dependence of specific conductivity of some molecular semiconductors at T > 273 K has been described by exponential equation $\sigma = \sigma_o\, exp(-\Delta E/2kT)$, in which ΔE-activation energy of charge transport, σ_o – constant [7]. For the specific resistance of organic semiconductors this equation has a shape: $\rho = \rho_o\, exp(\varepsilon_\sigma/2kT)$, where $\Delta E =\varepsilon_\sigma$, $\rho_o = 1/\sigma_o$ [7]. To examine the possibility of using this equation for new sensor materials we have studied the temperature dependence of conductivity; thermal stability and structure of polianiline and its derivatives – poly-*ortho*-aminophenol and poly-*meta*-aminophenol in temperature range 273-773 K.

2. Experimental Set Up

The powder samples of poly-*ortho*-aminophenol (PoAP), poly-*meta*-aminophenol (PmAP) and polyaniline (PAN) were prepared by the oxidizing polymerization of aminoarene monomers in the presence of equimolar ammonium persulphate in 0,5 M sulfuric acid solution. Obtained product was neutralized by 5% ammonium solution, washed by distilled water, acetone and dried in dynamic vacuum at T=353 K during 8 hours. Acid doping was carried out by exposition of undoped PAN samples in 1N aqueous acid solutions (H_2SO_4, $HClO_4$, HCl) during of 24 hours. Obtained samples were filtered, washed by distilled water, acetone and dried in the same conditions..

Measurements of the specific volume resistance were carried out in dynamic temperature change conditions, commensurable with the heating rate in thermogravimetric (TG) experiments. The sample in pressed pellet ($d=2$ mm, $h =2$ mm) was placed to quartz cylinder between two nickel disk contacts with installed chromel-copal thermo-couple. During the measurements the sample was under pressure 100 kg/cm^2, at which the resistance of organic semiconductors is approached to intrinsic value of specific volume resistance [7]. Studies of polyaminoarenes thermal stability were carried out on derivatograph Q-1500 D in the temperature range 273-1273 K in air and argon atmosphere, with Al_2O_3 standard at heating rate of 10 K/min.

IR-spectroscopy of polymer materials (pressed in KBr pellets) was carried out on spectrophotometer "Specord M-80" in the frequency range 400-4000 cm^{-1}. For structure investigations the method of X-ray powder diffraction at T= 295 K has been used (diffractometer DRON-3.0, FeK_α-radiation). The degree of crystallinity and crystallite size estimation were carried out in according to [8]. The dopant and moisture content have defined from the TG- and elemental analysis data as in [9].

3. Results and Analysis

A polymer formation in the process of polymerization of aminoarene monomers proceeds according to known scheme of oxidizing coupling reaction of aromatic amines and includes the steps of monomer oxidation over aminogen with cation-radicals formation and its coupling accompanied by deprotonation. The IR-spectroscopy of powder polymer material showed for undoped polyaniline the absorption bands at 3330, 3288, 1590, 1497, 1307, 1166 cm^{-1}, which is characteristics for emeraldine base of polyaniline [1-3]. In the case of oxidative coupling of *m*-aminophenol the absorption bands of *para*-substituted aromatic ring (3080, 1520, 760 cm^{-1}), amino-group (3350, 1574 cm^{-1}) and OH-group at 3600, 1410,1200 cm^{-1} were found. This confirms that only the amino-group undergoes oxidation while the hydroxyl group does not intervene. In the case of *o*-aminophenol the subsequent oxidation by two functional groups leading to formation of heterocycle [4]. This is confirmed by absorption bands at 1270-1200 cm^{-1} (ether oxygen) and 3400-3200 cm^{-1} (bound hydroxyl). The molecular structure of polyaminophenols in comparison with polyaniline is presented in the Figure 1.

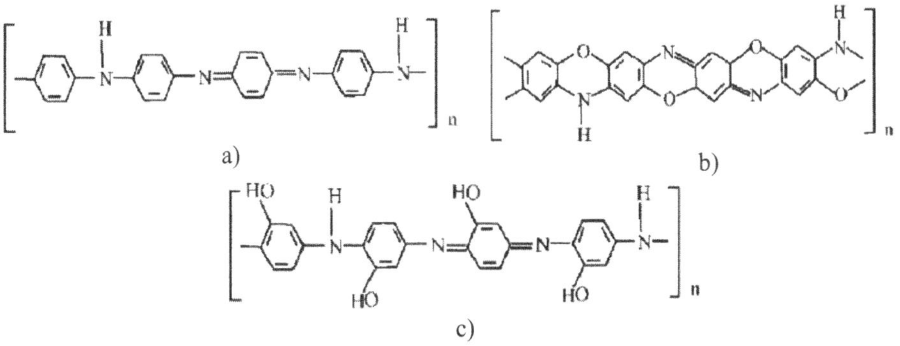

Figure 1: Molecular structure of polyaminoarenes: (a) polyaniline; (b) poly-o-aminophenol, (c) poly-m-aminophenol.

The temperature dependence of polyaminophenols and polyaniline specific volume resistance demonstrate the behavior, which is typical for organic semiconductors [5-7], however the molecular structure of polymer chain and nature of doping acid has a strong effect on the shape of this dependence. As shown in Figure 2, the change of specific resistance, normalized to resistance defined at room temperature, in the range 273-773 K includes some sections: the region of decreasing resistance, transition field and the region of resistance rising

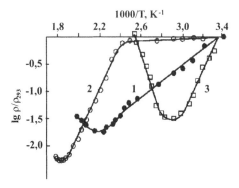

Figure 2: Temperature dependence of specific volume resistanse of undoped polyaniline (1), poly-ortho-aminophenol (2) and poly-meta-aminophenol (3)

At the first section the resistance of polyaminoarenes changes by known exponent low : $\rho = \rho_0 \exp (\varepsilon/2kT)$, where ε - activation energy of charge transport, ρ_0 - constant. Performance of resistance temperature dependence in Arrenius coordinates $\lg\rho$ - $1/T$ for the first section permits to calculate the value of activation energy depended on the molecular structure of polymer and type of doping acid (Table 1)

TABLE 1: Parameters of charge transport in polyaminoarenes

Polymer	Doping level, mol/unit	σ_{293}, S cm^{-1}	ε, eV	Temperature interval, K
Polyaniline base	-	$2\cdot10^{-8}$	$0,59\pm0.02$	293-423
Poly-o-aminophenol	-	$5\cdot10^{-10}$	$1,59\pm0,03$	403-533
Poly-m-aminophenol	-	$7\cdot10^{-9}$	$1,61\pm0,03$	293-333
Polyaniline, doped by H$_2$SO$_4$	0,43	$3\cdot10^{-3}$	$0,35\pm0,03$	293-403
Polyaniline, doped by HClO$_4$	0,45	$7\cdot10^{-3}$	$0,49\pm0,02$	303-383

From the studies of thermal behavior of polyaniline it is found that for all samples the endothermic maximum (T = 383-403 K), associated with loss of chemosorbed moisture is observed (Figure 3). Following, the exothermal peak in the range 433-453 K, typical for aminoarene polymers, is connected with processes of crosslinking of polymer chains [9]. However, considering circumstance that in argon atmosphere the intensity of this maximum is very small, one may suggests that the exothermal oxidation of aminobenzene fragments with formation of iminoquinoide structures take place. This temperature region coincides with going out the temperature dependence of resistance on transition field (Figure 2). The thermal decomposition of doped polymers starts at temperatures 453-473 K depended on the type of doping agent. Attached to temperatures of this maximum the sign change of temperature coefficient of resistance can be observed, that have been associated with a loss of doping admixtures and starting the macrochains destruction processes.

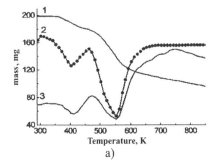 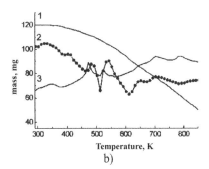

a)   b)

Figure3: The derivatograms of polyaniline (a) and poly-ortho-amonophenol (b):
1-TG; 2-DTG; 3-DTA-curves.

In the case of polyaminophenols the complex character of DTA and DTG curves indicates the running of the structure phase transitions in polymer, which cause the sharp change in specific resistance. At temperature 413-433 K the extraction of chemosorbed moisture proceeds, at T > 513 K the significant mass loss is evidence to destruction of polyaminophenol macrochains (Figure 3,b).

The comparison of the data presented in Figures 2 and 3 permits a suggestion that change in slope on temperature dependence of resistance is observed at temperatures corresponding to start of the polymer thermal decomposition. For poly-*meta*-aminophenol the change in slope of conductivity already at T = 338 K probably is caused by transformation of its molecular structure to ladder polymer (similar to PoAP) with low conductivity.

According to structure study, obtained polyaminoarenes are paracrystalline materials with the different degree of the structural ordering. Diffraction spectra of obtained polymers are characterized by a broad amorphous hallo and crystalline diffraction peaks in the range of $2\Theta = 10-60°$.

The parameters of structure, indeces of crystallinity, calculated on the base of obtained X-ray diffractograms (Table 2) indicate that in result of acid doping a change of diffraction maxima position and accordingly the interplane distances take a place. This is accompanied by a rising of the structural ordering level, probably caused by salt forms of polyaminoarenes formation. However the share of amorphous phase remains reasonably high (60-75 %).

TABLE 2: Parameters of X-ray diffraction, size of crystallites and degree of the crystallinity of polyamonoarenes

Polymer	Interplane distance, d, Å	Crystallite dimension, $1 \pm 10\%$, Å	Index of crystallinity, $I \pm 2\%$
Polyaniline (base)	4,47; 4,23; 3,58	15-20	25
Polyaniline doped by H_2SO_4	5,96; 5,00; 4,37; 3,49	25-30	41
Polyaniline doped by HCl	4,66; 4,31; 3,57	20-30	43
Poly-*meta*-aminophenol	5,19;3,46; 3,26; 2,38;2,38	15-20	23
Poly-*ortho*-aminophenol	7,18 ; 6,66; 5,69; 4,31; 3,56; 3,20; 2,82	10-15	41

Obtained results permit "granular" model of the charge transport [5,10] to applied for doped polyaminoarenes the "domain". According to these performances the ordering areas (domain or crystallites) with high conductivity existence in polymer. Charge transport between these domains occurred across the low-conductive amorphous layers, which create the energetic barrier to conductivity. At the high doping level in the amorphous phase the percolation bridges with high conductivity are forming. When the percolation threshold achived the polymer demonstrates the behaviour similar to conductor. As showed in structural researches, «granules» dimensions for powdery PAN sample compared with film samples (100-200 Å [10]), are small and achived to 20-30 Å (Table 2). Attached to such small domains of crystallinity the charge transport will be determined by resistance of disordered polymer phase.

4. Conclusion

Obtained results on conductivity and thermal stability suggest the significant effect of polymers thermal destruction on charge transport processes. In the case of studied polymers for temperature interval corresponding to linear section of resistance decreasing, the activation equation and some aspects of band theory may be used. The value of activation energy of charge transport is associated with effective band gap, constant ρ_o - with function of carriers free run [7]. The high specific resistance observed for polyaminiphenols in comparison to PAN is caused probably by existence of hydroxyl substitute in molecule structure of polyaminoarenes, which disturbances the system of conjugation along the polymer chain creating the significant energy barrier for charge transport.

References

[1]   Heeger A. Synth.Metals, Vol.125, pp 23-42, (2002)
[2]   Alexe-Ionescu A.-L., Ionescu A.Th., Scaramuzza N., et al, Phys.Rev.E.-Vol:64, pp17081-17088, (2001)
[3]   Boyle A., Genies E.M., Lapkowski M., Synth.Metals, Vol.28 ,pp.C769-C774, (1989).
[4]   Barbero C., Silber J.J., Sereno L., J.Electrochem.Soc., Vol. 291, pp. 81-101 (1990).
[5]   Kohlman R.S., Joo J., Epstein A.J. *"Conducting polymers: electrical conductivity/ Physical properties of polymers handbook"*, ed. J.E.Mark.-Amer. Inst.Phys. Woodbury, New-Yourk, p.453,(1996).
[6]   Raghunathan A., Kahol P.K., McCornic B.J., Solid State Commun., Vol.108 ,pp.817,(1998).
[7]   Inokuty H., Amadu H. 'Conductivity of organic semiconductors' .- Pergamon Press, New-York, pp, 52-56 (1963)
[8]   Wolter A., Rannou P., Travers J.P., Phys.Rev.B : Vol.58, N12 pp.7637-7647 (1998).
[9]   Kulkarni V.G., Campbell L.D., Mathew W.R. , Synth. Metals, Vol. 30, pp.321-327,(1989).
[10]   Savadi H.H.S., Angelopoulos M., Macdiarmid A.G., Epstein A.J., Synth.Metals., Vol. 26, pp.1-9, (1988).

Poster Programme
Paper presented at Sensors and their Applications XII, September 2003
©2003 IOP Publishing Ltd

A Fibre Optic Based Fluorescent Temperature Probe Using A Non-Pulsed White Light Source and Tristimulus Chromatic Detection

I Rallis and J W Spencer

Centre for Intelligent Monitoring Systems, Department of Electrical Engineering and Electronics, University of Liverpool, Brownlow Hill, Liverpool L69 3GJ. UK.

Abstract: A temperature sensor system based on a non pulsed white light source exciting fluorescence in a ruby rod taken from a ruby laser is presented. The outputs from three photo detectors with spectrally overlapping responsivities are used with a chromatic algorithm to quantify changes in the spectral signature. The derived Hue value is uniquely related to temperature from 20°C to 110°C. A simple model is also used to predict changes in the Hue value with temperature to assess the impact of using different types of optical fibres.

1. Introduction

Many optical fibre based temperature sensors, using the principle of measuring fluorescent decay, have been reported in the literature. The technique is well known and involves the use of a pulsed monochromatic laser source directed on to a fluorescent piece of material. The fluorescent emission from the material is detected and the decay time measured. The decay time is related to the temperature of the material decreasing as the temperature increases. As the system is pulsed the detection electronics are electronically locked in to the pulsed sequence of the laser.

This contribution highlights an alternative method for using fluorescence to measure temperature without the use of a pulsed source and time decay measurement. The technique uses a white light source to excite fluorescence in a material and spectrally overlapping photodiodes to detect the spectral emission. Here three photodiodes are used in what is known as a tristimulus system. The three photodiode signals are processed using chromatic algorithms producing three chromatic values known as Hue (H), Light (L) and Saturation (S). The fluorescent material of interest is a ruby rod from a ruby laser. Variations in temperature of the fluorescent material change the fluorescent emission spectra and thereby the chromatic values. Hue values are presented for temperature changes over a 90°C range.

This fibre optic based temperature sensor has a number of applications including measuring the temperature in backfill material intended for use in landfills.

2. Ruby Fluorescent Emission.

The simple representation of emission from ruby as a single wavelength does not accurately present the real situation. Ruby like most fluorescent materials has a far more complex fluorescent emission structure that produces a large number of emission lines. The two major peaks of emission in the laser quality ruby of interest for this sensor are called the R, or resonance, lines. These dominate the fluorescent spectra containing much of the energy of the emitted light. These R lines (692 and 693 nm) are surrounded by a number of other emission lines referred to as near satellites of the R lines.

Changes in the emission spectra with temperature are very complex. Figure 1 show the relative intensity changes between 400nm and 1100nm over approximately a 140°C temperature range. Whilst there is a reduction in intensity below 440nm and between 680nm and 720nm, there is an increase in intensity between 620nm and 680nm and above 840nm. There are two major temperature effects that produce these observed changes in figure 1. The first is a variation in the population of the different energy levels in ruby and the second is line broadening with temperature.

The decay time of the fluorescent emission decreases as temperature increases. This is due to a decrease in quantum efficiency resulting from an increase in collisional de-excitation of electrons and from other non-radiative processes.

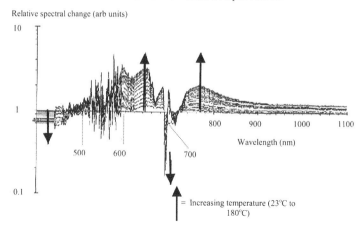

Figure 1. Normalised Spectral changes in fluorescent emission
(Relative changes in normalised spectra referenced against 23.0°C emission spectrum)

3. Chromatic Detection

Chromatic detection has been successfully used over many years for detecting changes in optical spectra [1]. The technique uses either two or three photo detectors with overlapping responsivities. A two-detector distimulus system provides a single value called dominant wavelength [1]. A three-detector tristimulus system provides three values that describe the Hue, Lightness and Saturation of the detected light signal [2]. The tristimulus system has been used for the work described here. The three chromatic values HLS are calculated from the photo detector currents. Given the output from each of the photo detectors is R, G and B respectively. The minimum output of the three detectors is subtracted from each of the other outputs leaving R', G', B'. The Hue value can then be calculated based on which of these is zero. Table 1 shows the formula for calculating Hue [1].

If B' = 0	Hue = 240{ 1 - R'/(R' + G')}
If R' = 0	Hue = 120{ 1 - G'/(G' + B')}
If G' = 0	Hue = 360{ 1 - B'/(B' + R')}

Table 1. Calculation for Hue angle.

Lightness and Saturation values are calculated from:

L = { [max (R, G, B) + min (R, G, B)] / 2 } and

S = { [max (R, G, B) - max (R, G, B)] / [max (R, G, B) + max (R, G, B)] }

These three chromatic values can be plotted on HS and HL polar diagrams. In both cases Hue is represented as an angle and Saturation and Lightness are radii [2]. The sensitivity of the detection depends upon the responsivity of the photodiodes used [1]. The photodiodes chosen were not optimised for this particular application.

4. Temperature Probe

Figure 2 show the arrangement of the ruby rod and optical fibres. The fibres have been arranged orthogonal to each other so that the detection fibre collects a proportion both of the source and fluorescent emission spectra. It is important that the returned signal to the photodiodes is not overly dominated by either. An optical spectrum analyser (OSA) was used to ensure that this condition was met.
A powerful 150W tungsten halogen light source and a 1mm diameter optical fibre were used to deliver the light to the ruby. This arrangement permits the use of long lengths of optical fibre, up to several hundred of metres, for this particular sensor. A second 1mm diameter optical fibre was used to return the modulated optical signal to the detection system where a short length of optical fibre bundle was used to split the light equally to the three photo detectors. The three detectors used, G1735, S1087-03 and G1961 are manufactured by Hamamatsu. They have peak responsivities around 710nm, 560nm and 420nm respectively. The detection unit averages the outputs from the photo detectors over 3 seconds to reduce the noise level.

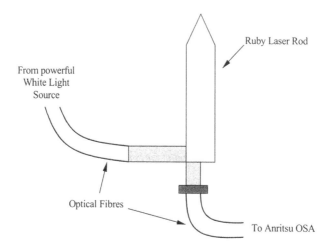

Figure. 2. Schematic of fibre position relative to the ruby rod.

5. Results

The ruby rod was heated slowly from 23°C to 110°C and then allowed to cool to approximately 33°C. A thermal couple attached to the ruby rod was used to measure the temperature. Figure 3 shows the Hue values versus temperature for one complete heating and cooling cycle. The Hue value changes by approximately 4 degrees over the 87°C temperature range. There is a unique relationship between the Hue value and temperature which shows some non linearity. The hysterises between the heating and cooling measurements which appears to be larger at higher temperatures is due to the rapid cooling that takes place of the ruby rod at higher temperatures and the time taken to average the outputs from the photo detectors. No attempt was made to stabilise the temperature during the cooling part of the experiment. It can be seen from figure 3 that the amount of hysterisis reduces as the rate of cooling decreases at lower temperatures. The maximum error in temperature due to this is 3°C.

6. Model

A simple model of the optical sensor system was constructed. This model considers the spectral output of the source, attenuation in the optical fibres and ruby rod, fluorescent efficiency and spectral emission characteristics, responsivity of the photo detectors, other losses due to connectors and noise (detector, amplification and digitisation).

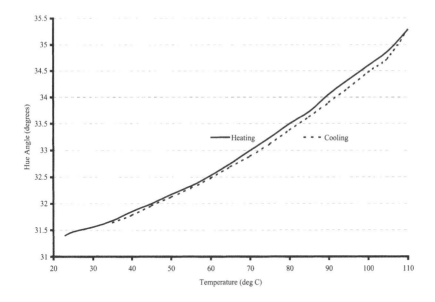

Figure 3. Hue angle changes with temperature.

Figure 4 shows the expected Hue change with temperature over a similar range used in the experimental part of the work. Over the same 90°C range the change in Hue is approximately 5 degrees. The Hue value increases with temperature and there is slight non linearity.

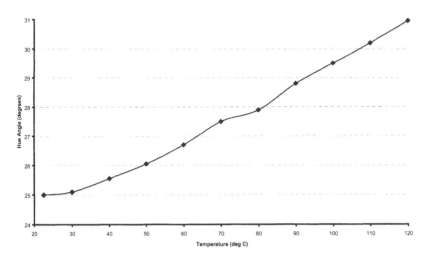

Figure.4. Simulated Hue angle for detectors:
R=G1735, G=S1087-03, B=G1961

7. Discussion

A temperature sensor systems using a continuous white light source, fluorescent emission from a Ruby and tristimulus chromatic unit has been experimentally demonstrated. The simulated results show broad agreement with the experimental results. The differences between the model and the experimental results are small and arise from the simplicity of the model used and the accuracy of the input data (eg fibre specification, responsivity of the photo detectors). The model can be used to optimise the choice of components (eg type of fibre, photo detectors etc) for the temperature sensor system. As an example figure 5 shows the temperature resolution factor over a range of fibre lengths for two different fibre types (HCR – hard clad radiation resistant polymer and HCP – hard clad polymer) for a set of detectors (S1223, S1226-18BK and G1115 with peak responsivities 950nm, 700nm and 610nm respectively). This temperature resolution factor indicates the smallest temperature that can be resolved. In general, figure 5 shows that as the fibre length is increased the temperature resolution worsens and that using HCP fibre rather than HCR fibre results in poorer performance when temperature is calibrated against Hue angle. Furthermore, for short lengths of HCP optical fibre (~300m) temperature resolution is significantly worse than 1°C because of the absorption peak at about 620nm for HCP fibre that is not present for HCR fibre.

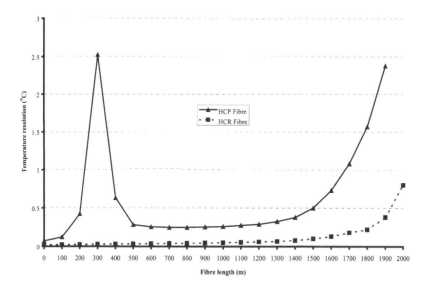

Figure 5. Predicted temperature resolution versus fibre length

8. Conclusion

A non pulsed white light source is used to excite fluorescence in a ruby rod taken from a ruby laser. The spectrum received by the three photo detectors is a mix of source and emission spectra. Processing the output from the three spectrally overlapping detectors with a chromatic algorithm shows that the Hue angle can be calibrated against temperature. A simple model confirms the experimental results and this model is used to show the impact that different fibre types have on temperature resolution in the sensor system.

References

[1] G. R. Jones and P. C. Russell. "Chromatic modulation based metrology", 1993, Appl. Opt., Vol 2, pp87-110.

[2] R. J. Yu, P. J. G. Lisboa, P. C. Russell and G. R. Jones. "Resolution capabilities of chromatic sensing in the monitoring of semiconductor plasma processing systems", 1997, Nondestr. Test. Eval., Vol 13, pp347-360.

Poster Programme
Paper presented at Sensors and their Applications XII, September 2003
©*2003 IOP Publishing Ltd*

A System for Improving the Testability of Luminescent Clad Fibre Optic Sensors

Joseph Walsh, Ian Grout, Colin Fitzpatrick, Padraig Quinn and Elfed Lewis

Department of Electronic and Computer Engineering, University of Limerick, Ireland.

Abstract: This paper describes an automated system that tests the outputs of luminescent clad fibre optic sensors that have been developed by the University of Limericks Optical Fibre Sensor Research Group. These tests involve stimulating the fibre at different points along its length and monitoring the output. The outputs are stored in a SQL database and can provide information on the reproducibility of the sensor fabrication technique and the long term stability of the sensors.

1. Introduction

Work is presented on the development of a diagnostic system to test the overall and spatial output of distributed luminescent clad optical fibre sensors for ultra violet detection. These sensors have been developed to monitor the germicidal ultra violet output (254nm) from a novel microwave plasma ultra violet lamp [1]. Optical fibre sensors have been identified as a very suitable technology for this application as they offer electromagnetic immunity and electrical isolation which is necessary in the dense electromagnetic environments around these lamps. They are also suitable for use underwater and are particularly suited to conducting distributed measurements.

This test rig is designed as a diagnostic tool to indicate the fabrication reproducibility of these sensors. In a controlled environment it irradiates the fibres at different points along its length and monitors and records the outputs for inspection. This allows the user to compare the responses of batches of sensors and thus evaluate the fabrication process. Also, by conducting repeat tests it can be used to test the long-term stability of the sensors, which may undergo degrading after long periods of irradiation.

2. The Sensor

The luminescent clad optical fibre sensor [2] is fabricated from a 1mm core diameter polymer optical fibre. The cladding is stripped from the fibre using solvents and re-clad using a phosphor doped epoxy. When the fibre is irradiated with ultra violet radiation photoluminescence occurs in the new cladding, part of which is coupled to the fibres core. They can be fabricated in positively guiding and negatively guiding configuration depending on the refractive index of the epoxy. A diagram showing the sensor is shown in Figure 1.

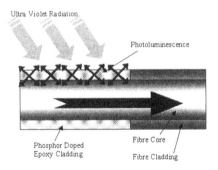

Figure 1. Luminescent Clad Optical Fibre Sensor

The coupling of the emission from the cladding to the core occurs due to the surface roughness of the fibre causing diffuse scattering at the interface between the core and the cladding. Part of the emission can thus become guided mode within the core. Coupling also occurs between the evanescent wave of these guided modes and the cladding photoluminescence [3].

For the reproducible construction of these sensors a number of considerations must be taken into account. The thickness and uniformity of the coating and the phosphor doping density must be identical for each sensor in order to generate identical amounts of photoluminescence. Two coating techniques have been developed to ensure this [4][5]. The condition of the surface of the fibre after stripping is however much more difficult to control and thus the cladding to core coupling efficiency can vary between sensors thus affecting the sensor output. A process is under development to treat the surfaces in a reproducible manner and thus help to generate more consistent results from sensor to sensor.

3. Hardware Design

The system hardware is based around interfacing the UV LED array and sensor circuitry to a PC via the RS-232 serial port. The basic arrangement for the prototype hardware is shown in figure 2.

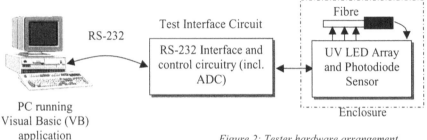

Figure 2: Tester hardware arrangement

The purpose of the hardware is to allow for the 6 UV LEDs within the LED Array to be switched ON/OFF under the control of a software application running on the PC.

Each UV LED is independently controllable, allowing for a range of combinations of light emission onto the fibre sensor. For example, all LEDs ON/OFF or a step sequence is possible. This would have the potential to identify PASS/FAIL scenarios as well as the ability to provide diagnostic capabilities (identifying the position of the failure along the fibre length). For protection reasons, the UV LED array, fibre and photodiode sensor are housed in an enclosure, which is screened from the user and ambient light.

Figure 3 shows a photograph of the UV LED array and sensor within the protective housing. The lid is to be closed when active for protection from the UV LED source.

Figure 3: Hardware Prototype (UV LED array and sensor)

3.1. Test Interface Circuit

The Test Interface Circuit (TIC) provides the interface between the PC and the UV LED array. The TIC contains a UART, an analogue to digital converter (ADC) and signal conditioning circuitry as shown in figure 4. The UART provides a two-way communications link between the *LED Control Circuit* and the (ADC). When the UV light stimulates the phosphor-coated fibre, and produces red (visible light), this propagates through the fibre core and is detected by the photodiode. The amount of leakage current produced by the photodiode varies with the input incident light intensity. The current output from the photodiode is converted to a voltage within a known range, with this voltage acting as an input to the (ADC). The resulting digital code is then sent to the PC via the RS-232 serial port connection for processing by a Visual Basic (VB) application. However due to speed limitations of the serial link the set-up is not suitable for high-speed sampling.

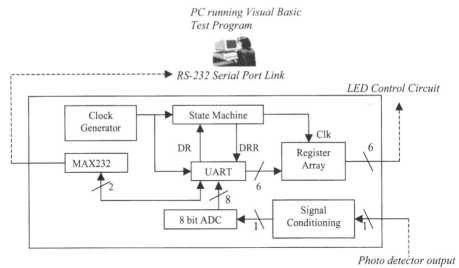

Figure 4. Test Interface Circuit (TIC).

4. Software Design

The software design is based around an application written in Visual Basic (VB). The Graphical User Interface is shown in figure 5. The rationale for using VB is

- Ease of program development
- Access to serial port
- Interfacing to external programs (for example, Microsoft Access for database work)

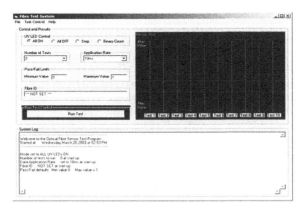

Figure 5: Prototype Application GUI

The GUI enables the user to set the test mode (UV LED control), PASS/FAIL limits and access to the results (database). In addition, in specific modes of operation, a graphical representation of the results is provided.

5. Results

The LEDs were turned ON/OFF in a sequence using the VB application and the output returned a full scale range of 0 to 6 volts. The test was conducted on a negatively guiding sensor and the results were recorded and are shown in figure 6.

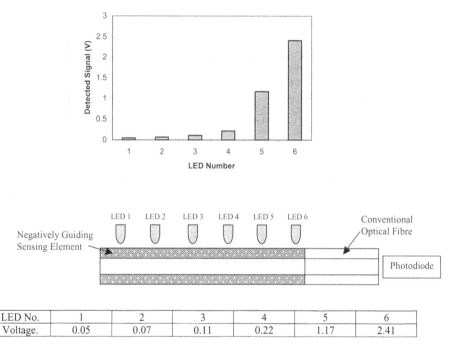

LED No.	1	2	3	4	5	6
Voltage.	0.05	0.07	0.11	0.22	1.17	2.41

Figure 6. Sensor Test Results & Schematic

These results are consistent with those expected from a negatively guiding sensing element. The core coupled emission from the cladding at the test point closest the distal end of the fibre (LED1) undergoes high losses as it travels through the negatively guiding sensor element. This is also true for the second test point (LED2) as it also must propagate a significant distance through this very lossy region. This trend continues with each successive test point producing a greater output than the previous one due to the shorter propagation length in this lossy section. The final test point (LED6) contributes the largest amount of power to the sensor output as it is only transmitted in the positively guiding waveguide which is brought to the photodiode.

6. Conclusions and Further Work

The authors have demonstrated a diagnostic tool for the evaluation of the reproducibility of their luminescent clad optical fibre ultra violet sensors. This system will now be used to provide feedback on the fabrication techniques employed and materials used in the fabrication of these sensors. It will allow the user to evaluate thoroughly potential strategies for the manufacture of commercial standard sensors.

References

[1]   Pandithas I. Al-Shamma'a A. Lucas J., "Low-pressure microwave plasma ultraviolet lamp for water purification and ozone applications", Journal of Physics D: Applied Physics, Vol 34, Sept 2001, pp2775-2781

[2]   Fitzpatrick C., E. Lewis, A. Al-Shamma'a, I. Pandithas, J. Cullen, J. Lucas, "An Optical Fibre Sensor Based on Cladding Photoluminescence for High Power Microwave Plasma UV Lamps Used in Water Treatment.", Optical Review, Vol.8, No.6, Nov/Dec 2001, pg 459-463.

[3]   Marcuse D., "Launching light into fiber cores from sources located in the cladding" Journal of Lightwave Technology, Vol 6 No 8, August 1988, pp 1273-1279

[4]   Fitzpatrick C., E. Lewis, A. Al-Shamma'a, J. Lucas, "A Process for Constructing Novel Optical Fibre Sensors with a High Level of Reproducibility", ICPR-16, 30th July – 3rd August 2001, Prague.

[5]   O'Donoghue C., Fitzpatrick C., Lewis E., Grout I., "A Fabrication Process For Luminescent Clad Fibre Optic Sensors" ODF2002, Tokyo (3rd International Conference on Optics-photonics Design & Fabrication) October 30- November 1, 2002

Improving the Reliability of Eddy Current Distance Measurements with Finite Element Modelling

M R Wilkinson, S Johnstone

School of Engineering, University of Durham, South Road, Durham, DH1 3LE, United Kingdom

Abstract: Eddy current distance measuring devices are used widely in a number of harsh environment applications. However they have limitations and this paper looks at the effect of non-centred targets on the reliability of such devices. It was found that when a target was offset by 1.3 times the sensor coil diameter there was an error of 6.5% in the distance measured.

Finite element electromagnetic field simulations have been employed to model the system and have shown how the field outside the sensor coil changes with target offset. By monitoring this field with small test coils it is possible to correct the distance measurement error. This work has potential to lead to an improvement to the range of applications in which eddy current distance measuring devices may be employed.

1. Introduction

Eddy current sensors (ECSs) [1] are widely employed to measure the distance to conducting targets in harsh environments [2]. Such true position measuring devices comprise a sensor coil and associated electronics. An alternating current is passed through the coil to generate an electromagnetic field. When a conducting target material is placed in this field, eddy currents are generated which produce an opposing field thus reducing the original intensity. This causes an impedance variation in the sensor coil which is detected by the monitoring electronics. The distance to the target is given as a voltage that is directly proportional to the target displacement. Non conducting materials between the ECS and target will not affect the field and so they can operate in environments where dust, oil or humidity, etc are present. This is a major advantage over other distance measuring devices.

The distance range over which an ECS will provide linear results is directly proportional to the diameter of the ECS coil. The magnetic field around ECSs with a shield is less extended than for unshielded designs. This has the effect of reducing the measuring range, but means that smaller targets may be used. For shielded ECSs the target size should be at least 1.5 to 2 times the ECS coil diameter and for unshielded ECSs the target should be at least 2.5 to 3 times the ECS coil diameter [3]. Using a target size smaller than these limits reduces the linearity and long term stability. A similar effect is observed when the target is offset from centre; that is the target extends a greater distance on one size of the ECS coil than on the other. Previous authors have looked at various aspects of ECS reliability [4, 5], but this paper summarises research into a solution to the offset problem and is a subset of wider work on the limitations of ECSs.

Experimental work was carried out using a commercially available eddy current distance measuring device from Kaman Instrumentation. This enabled the researchers to avoid the complication of constructing and fine-tuning such a device [6] and meant that the limitations of a well calibrated system were investigated. The system was modelled using Ansoft's *Maxwell3D*^{TM} finite element electromagnetic field solver using an eddy current module; this is detailed in section two. The results of these simulations have been confirmed experimentally (section three) and section four shows how they were interpreted and exploited.

2. Simulation of Target Offset

There are a number of modelling methods through which more can be learnt about electromagnetic sensor designs. Early analytical work on shielding and eddy current problems [7, 8] is relevant but limited to simple geometries. Probable flux methods [9] are suitable for complex geometries, but often require large assumptions. However with increasing computing power rigorous numerical finite element methods [10, 11] have proven to be useful in three dimensional and high frequency cases [12].

The ECS coil was modelled as a 1 mm wide rectangle swept 360° around a central axis with an alternating current source running at 1 MHz. The magnetic field strength was measured below the ECS coil using a test coil. By treating it as a solenoid, the amplitude of the current in the coil was estimated. This value was then adjusted so that the simulated field strength matched the measured field strength. The ECS coil was located above a square aluminium target that was positioned with various offsets relative to the ECS coil as illustrated in figures 1(a) and 1(b). The target material had a width of 3 times the ECS coil diameter and a thickness of 5 mm. The distance between the ECS coil and the target was fixed at 20 mm.

Finite element field solutions were computed on a fine mesh and the magnitude of the magnetic field in the direction along the ECS coil axis (z-axis in figure 1), $|B_z|$, was measured on a line perpendicular to the ECS coil axis (y-axis in figure 1) in the region between the ECS coil and target.

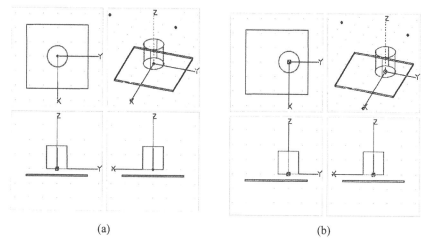

(a)   (b)

Figure 1. Wire frame perspectives of the problem geometry for (a) a target offset of zero and (b) a target offset of 0.6 times the sensor coil diameter.

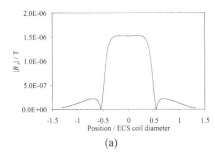 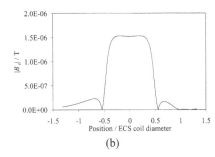

(a) (b)

Figure 2. The field profiles for (a) a target offset of zero and (b) a target offset of 0.6 times the sensor coil diameter.

The field was found to consist of a higher central portion immediately below the ECS coil and two 'lobes' to either side. As the offset was increased (i.e. the ECS coil was displaced laterally relative to the target centre) the lobe on the side closest to the advancing target edge decreased in magnitude and was curtailed in its lateral extension. This is illustrated in figure 2 which shows the field patterns for an offset of zero and an offset of 0.6 times the ECS coil diameter.

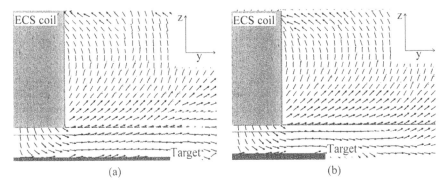

(a) (b)

Figure 3. Direction of the magnetic field for a given phase angle on the y-z plane cross section through the ECS coil, target and profile line for (a) a target offset of zero and (b) a target offset of 0.6 times the sensor coil diameter. Note that the magnitude of the field is not shown by this diagram.

The form of the field profiles may be understood in terms of the direction of the magnetic field vectors as illustrated by figures 3(a) and 3(b). Below the centre of the ECS coil the vectors are vertical (large z-component) as they are within the coil. Moving along the profile line away from the ECS coil centre the flux lines start to curve around and the vector takes on an increasing horizontal component. Immediately below the ECS coil edge the vector is horizontal (large y-component) and the there is a dip on the curves of figures 2(a) and 2(b). For the zero-offset geometry the z-component of the field increases as the flux lines sweep back up to join the upper end of the ECS coil. $|B_z|$ then drops off as the field decays with increasing displacement from the source. For the geometry with an offset of 0.6 times the ECS coil diameter there is also an initial increase as the flux sweeps back up, however in this case the target edge comes into play. Some of the flux lines are swept down and around the target so the field lines are spilt. This has the effect of creating more horizontal than vertical components of the field and thus $|B_z|$ decreases.

3. Experimental Confirmation

The modelling results were confirmed by measuring the magnetic field outside of the ECS coil with small tightly-wound test coils. These test coils were constructed from fine copper wire and an air core was used so as to minimise the disruption to the field. The stranded nature of the devices meant that eddy current interactions were small. An Agilent Technologies impedance analyser was used to match the natural frequency of the test coils closely to the 1 MHz of the ECS coil; this enabled maximum response and sensitivity. The test coils were connected to a digital oscilloscope and the amplitude of the signal recorded.

Two test coils were employed and were positioned either side of the ECS coil on the y-axis in figure 1 so as so track changes in the lobes of figure 2. The offset of the target material was increased and the output from the test coils was recorded. This has been expressed in normalised terms to highlight the variation from the zero-offset case. The ECS output was also recorded and was observed to change as the offset increased, even thought the actual target distance (d_t) remained constant. This is expressed in terms of a percentage distance error, which is simply:

$$\% \text{ distance error} = \frac{d_{t(ecs)} - d_t}{d_t} \times 100$$

where $d_{t(ecs)}$ is the distance given by the eddy current ECS.

Figure 4 shows the normalised output of the test coil on the side of the approaching target edge (i.e. positive y-axis position on figure 1) and the % distance error. For low values of the normalised ECS coil offset (≤ 0.6), the small distance errors were caused by experimental limitations and were not actual device errors.

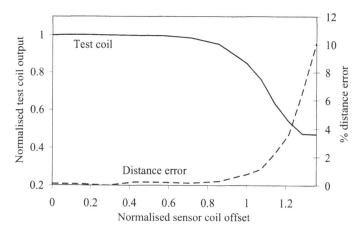

Figure 4. Graph showing percentage distance error resulting from target lateral displacement and the output from the test coil.

When the ECS coil and target edges were aligned (an offset of 1 times the ECS diameter) there was a 6.5% error in the target distance as given by the ECS. The output of a test coil positioned at 1.3 times the ECS coil radius was found to be 0.47 of the value when the target was centred. Thus by monitoring the test coils in this way, it was possible to correct for the errors caused by target offset.

The direction of the target offset was also determined by comparing the output from the two test coils. As can be seen in figure 2 only the lobe closest to the approaching target edge was affected; the opposite lobe remained static.

4. Conclusions

This paper has demonstrated by using finite element field simulations that there is more information available from the ECS system than is extracted by only measuring the impedance of the ECS coil. Outside the ECS coil lobes were observed in the z-component of the magnitude of the magnetic field. As the offset of the target material was increased the lobe on the side of the advancing target edge was observed to decrease in magnitude and its lateral extension was curtailed. By monitoring small test coils positioned to observe these lobes it was possible to counteract the effect on the ECS coil.

With further work the system of test coils and the ECS coil could be combined into a package that was resistant the harsh environment in which ECSs are often employed. The signal from the test coil could be monitored by an electronic system that could directly modify the output from the ECS device. This would make an eddy current distance measuring device that could be used in situations where the target materials were not always centred. Such applications include environments targets move through the path of the ECS (e.g. production lines) and where the target may be knocked out of alignment (e.g. rolling mills).

Acknowledgements

This work is supported by funding from the UK EPSRC reference number GR/R58475/01.

References

[1]   Welsby, S.D. and Hitz, T., *True Position Measurement with Eddy Current Technology.* Sensors Magazine, 1997. 14(11): p. 30-40.

[2]   Aknin, P., Placko, D., and Ayasse, J.B., *Eddy current sensor for the measurement of lateral displacement. Applications in the railway domain.* Sensors and Actuators A, 1992. 31: p. 17-23.

[3]   Kaman Instrumentation, *KD-2300 Instruction Manual Rev. E. (part no. 860021).* 2000, Kaman Corporation, P.O. Box 1, Bloomfield, CT 06002-0001, USA. p. 21-22.

[4]   Tian, G.Y., Zhao, Z.X., and Baines, R.W., *The research of Inhomogeneity in Eddy Current.* Sensors and Actuators A, 1998(69): p. 148-151.

[5]   Vasseur, P. and Billat, A., *Contribution to the development of a smart sensor using eddy currents for measurements of displacement.* Meas. Sci. Technol., 1994. 5(8): p. 889-895.

[6]   Roach, S.D., *Designing and building an eddy current position sensor.* Sensors Magazine, 1998. 15(9).

[7]   Levey, S., *Electromagnetic shielding effect of an infinite plane conducting sheet placed between circular coaxial coils.* P. IRE., 1936. 24(6): p. 923-941.

[8]   Dodd, C.V. and Deeds, W.E., *Analytical solutions to eddy-current probe-coil problems.* J. App. Phys., 1968. 39(6): p. 2829-2838.

[9]   Hugill, A.L., *Probable flux path modelling of an inductive displacement sensor.* J. Phys. E: Sci. Instrum, 1981. 14: p. 860-864.

[10]   Chari, M.V.K. and Silvester, P.P., eds. *Finite Elements in Electrical and Magnetic Field Problems.* Wiley Series in Numerical Methods in Engineering, ed. R.H. Gallagher and O.C. Zienkiewicz. 1980, Wiley-Interscience. 219.

[11]   Silvester, P.P. and Ferrari, R.L., *Finite Elements For Electrical Engineers.* 2 ed. 1990: Cambridge University Press. 344.

[12]   Johnstone, S. and Peyton, A.J., *The application of parametric 3D finite element modelling techniques to evaluate the performance of a magnetic sensor system.* Sensors and Actuators A, 2001. 93: p. 109-116.

Poster Programme
Paper presented at Sensors and their Applications XII, September 2003
©2003 IOP Publishing Ltd

Design for Testability Considerations For High Performance Analogue to Digital Converters

T O'Shea and I A Grout

Department of Electronic and Computer Engineering, University of Limerick, Ireland.

Abstract: In many instrumentation systems, an analogue signal is to be acquired and processed, with the signal processing commonly performed in the digital domain. The conversion process between the analogue and digital relies on high performance Analogue to Digital Converters (ADCs). In order to assess whether the ADC is operating as to the required design specification, suitable test methods are required. Testing an ADC is non-trivial, and with ever increasing performance requirements, the test needs are increasing in complexity. This paper will discuss requirements for the testing of ADCs, in particular relating to production test, and will review a range of test techniques and Built-In Self-Test (BIST) approaches for Integrated Converters.

1. Introduction

In many instrumentation and sensor-based systems, an analogue signal is to be acquired and processed, with the signal processing undertaken in the digital domain. Analogue to Digital Converters (ADCs) provide the link between the analogue physical environment and the discrete time environment consisting of logical 1's and 0's. The ADC is the key building blocks of many sensor systems and the accuracy of the system is in part dependent on the accuracy of the ADC. Understanding the ADC's specifications is essential in selecting the right ADC architecture for a particular application. In order to assess the end-user requirements, these devices can be grouped into a number of basic architectures, providing solutions for a range of sampling speeds and resolution. However, there is a trade off between the speed and resolution of an ADC as Waldron [1] stated:

"*Approximately one bit of resolution is lost for every doubling of the sampling rate*"

The performance of the various ADC architectures can be shown graphically. *Figure 1* shows a plot of resolution versus sampling rate. Waldron also identified a Moore's- type law for ADCs that measured the technological progress "*by the product of the ADC resolution (bits) Vs the sampling rate (time)*". He discovered that the average improvement is approximately "*1.5 bits for any given sampling frequency over the last six – eight years*". This is very slow progress and is a major stumbling block in the development of faster and more sensitive sensor systems.

The costs involved in testing these devices produces a significant impact on the overall costs associated with high volume production of ICs (design, fabrication and test costs). The demand for shorter design and manufacturing cycles increases the need for economical, fast and accurate production test methods. A Built-In Self-Test (BIST) approach, where tester functionality is embedded within the design, would reduce the need of expensive Automatic Test Equipment (ATE), and may allow for a larger number of simple test stations to operate in parallel. This can be a highly desirable situation. This paper aims to discuss ADC architectures and the requirements for testing (predominantly functional) of these devices. The purpose for this is to highlight and investigate the test requirements for these devices, fusing design and test into a Design for Testability (DfT) approach to product development.

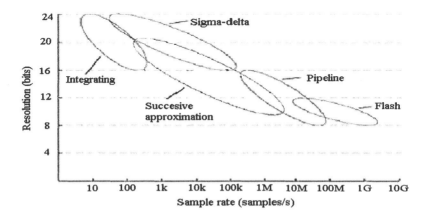

Figure 1: Current ADC Performance Characteristics

The paper is structured as follows. This section introduced the need for accurate ADCs and compares the suitability of different ADCs for particular instrumentation applications. In *Section 2*, a more detailed look is given to the different ADC architectures currently available on the market. *Section 3* outlines the functional testing of ADC and the problems encountered. *Section 4* describes the approach in testing ADCs that incorporates a Self-Test capability, using hardware Built in Self Test (BIST) circuits as an integral part of the converter.

2. Summary of ADC Designs

2.1 Introduction
This section will discuss the following ADC designs currently available and utilised: *Successive Approximation, Delta-Sigma, Flash, Integrating and Pipeline.*

2.2 Successive approximation Register (SAR)
These ADCs [2][3] are used in applications that require a resolution range between 8 bits and 20 bits. They have a conversion speed from 100 sps (Samples Per Second) to 5 Msps (Mega-sps), and are used in low to medium signal frequency domain. The SAR ADC is probably the most popular of all ADC's, partly due to it's good power efficiency. The SAR ADC is centred on a comparator that compares an unknown input voltage (Vx) against a sequence of known voltages that are outputted form a DAC. The

DAC uses a binary search algorithm to determine the best approximation to unknown signal *Vx* that only requires *n* clock periods to complete an *n-bit* conversion.

2.3 Delta sigma converters

This type of ADC [4]-[6] is becoming more popular as its performance is not as sensitive to the matching of components within the DAC. This is due to its ability to trade between speed and resolution by oversampling and digital filtering. A sample and hold circuit holds the continuous signal for the modulator where a low resolution noise shaped digital signal is produced. This signal is linearly related to the input signal but also contains a large amount of high frequency quantization noise. A low-pass filter is then used to reduce the quantization noise. The signal is then undersampled keeping submultiples of the oversampled signal and discards the rest.

2.4 Flash ADCs

Flash ADCs [2][3][7] are the fastest method of converting an analogue signal to a digital signal. Sampling frequencies range from 10 Msps to 2 Gsps. However, they have very low resolution. They are ideal for applications that require large bandwidth such as Video and Radar. Their major disadvantage is that they are large in size and have a high power consumption. A flash ADC is made up of a series of comparators that compares the analogue signal simultaneously against all possible decision of a resistor network that contains 2^n resistors.

2.5 Integrating ADC's

Integrating ADCs [2][3] are frequently used in high precision data acquisition and instrumentation systems. It is much slower than any of the other available ADC's but offers excellent Differential and Integral Linearity.

2.6 Pipeline ADC

These [8] provide a reasonably good balance of size, speed, resolution and power consumption. Pipeline ADCs consists of a number of repeated stages that are used to divide up the conversion process. Each stage except for the last contains a sample and hold circuit, low resolution ADC, DAC and a summing circuit. A coarse M-bit A/D conversion is executed first then a DAC with higher resolution converts the result back into analogue form, which is subtracted from the input. The difference is converted with a fine K-bit flash ADC and the two output stages are combined.

3. Testing of ADC

3.1 Introduction

The testing of the ADC is difficult due to the probalistic nature of the ADC operation. An ADC transfer curve is a "*many to one mapping function*" i.e. many input voltages provide the same output voltage. In an ADC output code, the exact input voltage is unknown, but only the range of the input voltage is known [3]. The range is dependant on the resolution of the ADC. There are a number of tests that can be used to measure and compare the performance of ADCs. These tests can be grouped in 2 sections: (i) Static Tests and, (ii) Dynamic Tests. These tests are functional in nature.

3.2 Static Tests

To perform static tests [3][9], the most important issue is to locate the code edge between each output code for the ADC transfer curve. The most popular method of code edge detection is the sinusoidal method. It tests the ADCs performance in response to rapidly varying input voltages. The waveform applied is slightly higher than the full-scale range of the ADC to ensure that the sine wave passes through all the codes so as to get a histogram with all the code widths. For sinusoids that are clipped, there is a mathematical procedure to find the offset and gain.

3.2.1 Differential Non-Linearity (DNL)

In an ideal ADC, each step size is uniform corresponding to the same voltage rise throughout the ADC conversion range. Differential Non-Linearity (DNL) is a measure that describes the uniformity between ADC codes. DNL is the difference between the measured and ideal Least Significant Bit (LSB) change between the two adjacent codes. When using the sinusoid histogram method to test the ADC, the DNL calculation is made very easy as all code widths are in units of LSBs. The DNL for each code is calculated by subtracting each code width from one.

3.2.2 Integral Non-Linearity (INL)

The deviation of a transition point from its corresponding position on the ideal ADC transfer curve is called Integral Non-Linearity (INL). The ideal ADC transfer curve is obtained by either the end-point line or the best-fit line. The maximum difference is reported as the INL of the converter. It is important to note that the gain and offset is taken into account before measuring the INL.

3.2.3 Monotonicity

As the input code to an ADC is increased the output code should increase. If this does not happen the ADC is said to be non-monotonic. This is very seldom a problem for ADCs as one or more of its codes widths would have to be negative.

3.2.4 DC gain and offset

There are a few ways in which to define the transfer curve of an ideal ADC and may vary from one device data-sheet to the next. Some data-sheets define the ideal location of the first code transition 000 to 0001 as AGnd + 0.5 LSB (AD7476) where as other data sheets define the first code edge as AGnd + 1 LSB (AD7478) to be ideal. It is very important to note which method is being used to define the ideal transfer curve otherwise errors of $\pm$ ½ LSB would be introduced. The offset error is defined as the deviation of the actual first code transmission from the ideal location of the first code transition. There are also a few ways to define ideal gain. In the AD7476, the ideal gain is defined as Vref – 1.5LSB where as in the AD7478, the ideal gain is defined as Vref – 1LSB. The gain error is the deviation of the last code transition 111…110 to 111…111 from the ideal after the offset error has been adjusted.

3.3 Dynamic Tests

The dynamic tests [1][3][9] are very important especially in high-speed applications and determined from fast Fourier transforms (FFT) analysis of sampled ADC output codes.

3.3.1 Signal to Noise Ratio (SNR)

The Signal to Noise Ratio is the ratio of the signal power to the noise power. The only noise power in an ideal ADC is due to the quantization error that is taken in to account by the white noise approximation model. However, in real life applications, there are other sources of error present such as circuit noise, aperture uncertainty and comparator ambiguity. The signal is the root mean squared (RMS) of the fundamental. The noise is calculated by adding all the harmonics together up to half the sampling frequency(fs/2), excluding DC.

3.3.2 Signal to noise plus distortion (SINAD)
The Signal to Noise plus Distortion rate (SINAD) is similar to the SNR except that the SNR does not include the harmonics content, hence the SNR is always better than the SINAD. Both SNR and SINAD are expressed in decibels (dB). Anti-aliasing filters should be placed in front of the ADC to improve the SNR and the SINAD. The SINAD decreases as the input frequency increases towards the Nyquist rate. If the SINAD is tested at low frequencies, on the data sheet there is a high probability that performance of the ADC is worse nearer the Nyquist frequency

3.3.3 Effective Number of Bits (ENOB)
The effective number of bits is a similar way to express the signal to noise and distortion ratio in bits rather than decibels. The ENOB will show up any nonlinearity's in an ADC for AC signals.

3.3.4 Spurious Free Dynamic Range (SFDF)
SFDR is defined as the ratio of the RMS value of the input signal to the RMS value of the largest spur in the frequency domain. Normally, the value of the SFDR can be determined by the largest harmonic in the spectrum but for ADC the harmonics are buried in the noise floor so the largest noise peak will be used. The accuracy of the SFDR calculation can be improved by increasing the number of FFT points or by averaging the results as this will reduce the noise floor level while the amplitude of the spurs shall remain constant.

4. An overview of Built-In Self-Test methods for ADCs

The cost of testing mixed signal circuitry is one of the highest costs associated with high volume production of ICs. The demand for shorter design and manufacturing cycles increases the need for economical, fast and accurate test methods. A Built In Self Test method [10]-[14] would reduce the need of expensive automatic test equipment by removing functionality traditionally required within the ATE and placing it within the Device Under Test (DUT) itself. This would also allow for a larger number of simple test stations to operate in parallel, hence reducing the initial outlay of capital invested when manufacturing an IC. This is of particular importance as the IC may only have a short market lifespan. The current trend in testing ICs is to reduce the workload of the Automatic Test Equipment (ATE) and place additional circuitry on the IC, so it can test itself. The main difficulties associated with designing a Mixed Analog-Digital Built-In Self-Test (MADBIST) is to generate a source on chip which can provide high precision test signals which is insensitive to process variations. This additional circuitry must occupy as little silicon area as possible. As an example, the following BIST scheme, as seen in *figure 2*, was proposed by AT&T in order to verify the monticity of an ADC. Here, a ramp voltage was generated on chip and applied as an input to the ADC during test. The output codes are tested for monotonicity by comparing the present output code to the previous output code. If the present output is greater a counter is incremented. If not a fail flag is raised and the test terminated. The final count for each output code of the ADC can be stored in N-bit registers. A histogram plot of the output code can be obtained to allow for Integral Non-Linearity (INL) and Differential Non-Linearity (DNL) tests to be completed.

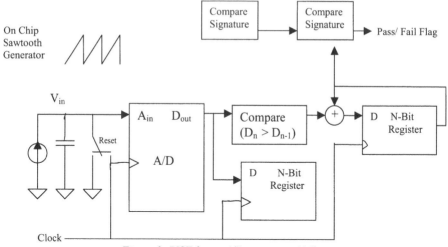

Figure 2: BIST for an A/D converter [14]

A major disadvantage of this method is that the correct functionality of the analogue ramp generator is not tested. This is of importance since the test circuitry itself must be testable. Advances in Digital to Analogue (D-A) and Analogue to Digital (A-D) technologies have however made analogue signalling possible through digital signalling processing techniques. The same concepts are being applied to analogue signal generation through Direct Digital Frequency Synthesis (DDFS). DDFS generates an analogue sinusoidal waveform using a combination of digital logic and digital to analogue conversion. DDFS requires a large silicon area hence has been ruled out as a efficient method of signal generation for BIST. Delta-Sigma modulation techniques can be used very successively in creating a mixed analogue-digital Built-In Self-Test (BIST). A delta-sigma modulation oscillator circuit generates a one bit digital sequence as an analogue test stimulus. This binary sequence contains a sinusoid signal. The signal is applied to the ADC and a Fast Fourier Transform (FFT) analysis is completed on the output. This method is able to generate high quality signals, but it is not very area efficient. There is also another method of generating the one bit pattern through existing digital structures such as boundary scan chain or through a RAMBIST controller. The basic concept is to generate the one bit output from the delta sigma oscillator, capture it and place it in a scan chain with feedback. This method does not generate signals to the same quality as the afore mentioned but it is more area efficient.

5. Conclusions

This paper has described aspects relating to the design and test of high performance Analogue to Digital Converters (ADCs). Such circuits are used in instrumentation systems providing a link between the digital domain in digital signalling process based electronic systems. With the increased performance requirements along with the reduced product costs, ADC test procedures are increasingly "under the spotlight" as a major stumbling block in the product development. In order to improve device testability there is an increasing interest in the device design and test. Both design and test considerations have been covered in this paper.

References

[1]   Walden, R.H. Analog to Digital Converter Survey and Analysis IEEE Journal on Selected Areas in Communications, Vol.17, No.4, April 1999

[2]   Jaeger, R.C. 1997 Microelectronic Circuit Design, International Edition : McGraw Hill

[3]   Burns,M. and Roberts, G.M. 2001 An Introduction to Mixed-Signal I.C. Test and Measurement, New York : Oxford University Press

[4]   Franca, J.E. and Tsividis, Y.1994 Design of Analog-digital VLSI Circuits for Telecommunications and Signal Processing 2nd Edition, New Jersey : Prentice Hall, Englewood Cliff,

[5]   Johns, D. A. and Martin, K. 1997 Analog Integrated Circuit Design NewYork, Chichester, Brisbane, Toronto, Singapore, Weinheim: Wiley,

[6]   Norsworthy, S.R., Schreier, R. and Temes, G.C. 1997 DELTA-SIGMA DATA CONVERTERS: Theory,Design and Simulation IEEE Circuits& Systems Society.

[7]   Maxim: Oct.2001 Understanding Flash ADCs, http://www.maxim-ic.com/appnotes.cfm/appnote_number/810/ln/en

[8]   Maxim:2001 Pipeline ADCs Come of Age, http://www.maxim-ic.com/appnotes.cfm/appnote_number/634

[9]   Mahoney,M.1987 Tutorial DSP Based Testing of Analog and Mixed-Signal Circuits The Computer Society of the IEEE

[10]   Roberts, G.W., 1995. "Re Examining The Needs Of The Mixed-Signal Test Community" In the International Test Conference Proceedings, Oct 21-25, page(s) 298.

[11]   Toner, M.F.; Roberts, G.W.;1995. "On the Practical Implementation of Mixed Analog-Digital BIST." Custom Integrated Circuits Conference, 1995., Proceedings of the IEEE 1995 , 1-4 May Page(s): 525 –528

[12]   Toner, M.F.; Roberts, G.W.; A BIST Scheme for a SNR Test of a Sigma-Delta ADC .Test Conference, 1993. Proceedings., International , 17-21 Oct 1993 Page(s): 805 –814

[13]   Roberts, G.W.; 1997 "Improving the Testability of Mixed Signal Integrated Circuits.", Proceedings of the Custom Integrated Circuits Conference, May

[14]   Roberts, G.W., 1996. "Metrics, Techniques and Recent Developments in Mixed-Signal Testing." Proceedings of the IEEE/ACM International Conference on Computer Aided Design San Jose, California, Page(s): 514-521.

Poster Programme
Paper presented at Sensors and their Applications XII, September 2003
©*2003 IOP Publishing Ltd*

Technology, Principles, and Applications of Inertia Micro-Electro-Mechanical System Technology

Achi Ayomipe Celestine

Kairus Consulting Limited, Lagos, Nigeria

Abstract: This tutorial covers three major topic related to technology, physics, and applications of micro-scale inertial sensors (accelerometers and gyroscopes). The first part provides a basic introduction to technology of MEMS, covering primary questions needed to evaluate micro machining as a technology of choice for inertial micro-sensors. Three technologies, surface micromachining, bulk micromachining, and mixed fabrication processes will be introduced and compared. The second part of the tutorial will be devoted to (i) overview of research and development efforts worldwide; (ii) introduction to the principles of operation and detection; (iii) review of advantages and challenges; and (iv) speculation about possible future design trends. The tutorial will be concluded by highlighting a wide range of applications enabled by inertial micro-sensors and discuss future opportunities for this sensor technology in medicine, robotics and defense.

1. Introduction and overview of MEMS Inertial Sensors

Micro electromechanical systems (MEMS) technology, sometimes also called microsystems technology (MST), is a rapidly emerging field which combines many disciplines, including physics, electrical engineering, electronics, informatics, materials and biochemistry. All industry watchers are united in the view that the technology will have a tremendous impact on the 21st century industrial landscape. Researchers and manufacturers alike are finding new uses for these microscopic devices at a rapid pace and numerous companies from the semiconductor giants to fledgling start-ups are now scrambling for a piece of the action at the microscale. It is therefore not surprising that the European countries, the United States and Japan have invested several billion US dollars in the field since the early nineties.

MEMS devices are typically defined as integrated microscale devices or systems which merge sensing, processing and/or actuating functions to change the way we perceive and control the physical world around us. They are usually fabricated using integrated circuit (IC) compatible batch-processing techniques and vary in size from sub-micrometers to millimeters. Typically they will combine two or more electrical, mechanical, optical, chemical, biological, magnetic or other properties on a single chip or on a hybrid module which incorporates several components.

MEMS technology can be divided into surface micromachining bulk micromachining, LIGA, wafer bonding, material engineering, circuitry, microassembly, micropackaging, and systems integration. Several types of vehicle, such as microsensors, microactuators, microstructures, and microsystems are under development with those core technologies.

The ideas on the potential of MEMS range from devices which are nowadays found in nearly all automobiles (e.g. pressure sensors and accelerometers) to sensors for flow, temperature, force, position, magnetic fields, chemicals, light, IR-radiation, etc., either inexpensive, robust and/or with previously unmet performance. The current and next generations of devices use micropumps, flow sensors, micromixers, microsieves, microreactors, etc. for dosing in medicine, biology, chemical and biochemical analysis systems with numerous applications. Looking further into the future, one can imagine the use of microrobots in a vast number of applications, ranging from microsurgery to fabrication, inspection and repair.

Inertial sensors are widely used in advanced mechanical systems like guidance, robotics, and automobiles. In MEMS, research on inertial sensors has focused primarily on accelerometers and gyroscopes. Of the two, the accelerometers were developed first. Today, MEMS accelerometers enjoy a large commercial market and are considered to be one of the most successful micro sensors ever developed. MEMS gyroscopes, on the other hand, are a relatively new technology. Commercialization of low-grade devices has recently begun while intensive research is still being carried out in laboratories on high-grade devices.

MEMS accelerometers can be fabricated by using either surface or bulk micromachining. Extensive surface micromachining research in the last decade has resulted in the successful development of batch-fabricated accelerometers, which are widely used in the automobile industry as airbag sensors. These low-cost (~us$10 per unit) accelerometers, like the analog device adxl series, are typically comb-drive based and equipped with sophisticated on-chip signal conditioning electronics. The acceleration sensing in the comb-drive devices is carried out through the displacement of the polysilicon comb fingers. Due to the fabrication limitation of surface micromachining, the thickness and therefore the mass of the comb fingers is small, which imposes a severe limitation on the performance of the accelerometers. Typically, the accuracy of the commercially available surface-micromachined accelerometers is in the milli-g range, which is sufficient for impact sensing but does not satisfy the micro-g requirement of inertial navigation.

Bulk-micromachined accelerometers, on the other hand, can provide large proof mass and therefore higher accuracy. A typical bulk-micromachined accelerometer consists of a proof mass suspended on thin beams. Because of fabrication difficulties, the proof mass is usually not monolithically integrated with supporting electronics. As a result, bulk-micromachined accelerometers are typically more expensive than their surface-micromachined counterparts because of the additional manufacturing cost involved in packaging a separate ic chip along with the accelerometer. However, separating the accelerometer and the electronics allows the manufacturer to choose the best driving circuitry for the accelerometer, which is a great advantage in designing high-end accelerometers. Commercial companies like Litton have been actively developing navigation-grade bulk-micromachined accelerometers for the military market. When packaged with gps electronics, these high-end accelerometers are expected to offer a complete solution to advanced navigation in the relatively low price range of us$3000 per unit.

Gyroscopes can be divided into three basic groups according to their performance: rate grade, tactical grade, and inertial grade. Conventional gyroscope technology offers all three-gyroscope types, from the low-end spinning wheel gyro to the high-end ring laser gyro. So the obvious question is "why build another gyroscope?" The answer to this question is that conventional gyroscopes are too bulky and power hungry to meet the requirements of future advanced systems like hand-held navigation and micro satellites. MEMS -based gyroscopes, on the other hand, offer a near-perfect solution because of their compact size, low-cost, and low power consumption. The majority of MEMS gyroscopes currently under development operates in a vibratory mode and measures the angular rate instead of the absolute angle. Their operational principle is based on the coupling of mechanical energy between a vibrating motor element and a sensor element through Coriolis acceleration. Like MEMS accelerometers, MEMS gyroscopes can also be fabricated by using either surface or bulk micromachining. A popular design for surface micromachined gyroscopes uses a comb-drive as both the motor and the sensor. This design leverages the fabrication technology developed for surface micromachined accelerometers and achieves a noise level of about 10/s/Hz1/2, which is accurate enough for automotive applications. However, just like the surface micromachined accelerometers, it also suffers from the 'low-mass' problem and is unlikely reach the level of 10/hr/Hz1/2 required for high-end military market.

The difficulty in overcoming the 'mass' factor in surface micromachined gyroscopes has led to the recent renewed interest in bulk micromachining. Among the main fabrication issues associated with bulk-micromachined gyroscopes, the most important ones are high aspect ratio etching, wafer bonding, and vacuum packaging. In the area of high aspect ratio etching, tremendous progress was made in recent years by companies like STS and Plasmatherm, who successfully developed ICP-based DRIE process chambers for the fast etching of deep and narrow trenches in single-crystal bulk silicon. This technology greatly simplifies the design of high-end gyroscopes by making the fabrication of high aspect ratio beams and proof mass possible. A good demonstration of this technology is the tactical-grade gyroscope developed by Litton. Another trend in bulk micromachined gyroscope that is gaining popularity is the utilization of SOI wafers. With their thick single-crystal device layers, SOI wafers provide the necessary substrates for large-mass mechanical structures, which are critical to gyroscope performance. Additionally, they also allow the possibility of integrating the mechanical structures of the gyroscope with the electronics, as in the case of surface micromachined gyroscope.

Fabrication of bulk-micromachined gyroscopes usually includes the bonding of several silicon wafers. Low temperature processes like anodic and eutectic bonding typically do not provide the necessary strong and thermally stable bonds required for high-end gyroscopes. Currently, the only suitable choice is fusion bonding, which is a high-temperature process (~10000 Celsius). Since this process is incompatible with IC fabrication, another bonding option must be identified before integration of gyroscope and electronics can be realized.

2. Overview of research and development efforts world wide

Micromechanics, Micromachining and Micro-Electro-Mechanical-Systems have been the focus of worldwide research and development efforts for the past ten years. These emerging fields involve fundamental research and the unique design and engineering of sophisticated integrated mechanical, electronic and optical systems on the micro-scale. The scientific field of MEMS is multidisciplinary and involves the collaborative efforts of researchers from diverse fields such as physics, chemistry, mechanics, electronics, materials science and computer science. Extensive research in these disciplines is currently being directed towards obtaining a better understanding of the micro-scale phenomena encountered in MEMS devices.

The economical and practical benefits of MEMS applications are enormous. The technology is an offshoot of standard design techniques and batch fabrication methods used in the microelectronics chip industry. It enables the realization of very small, highly sophisticated, accurate and reliable systems at a very low cost. At present, MEMS sensors can be found in a variety of applications. In the automotive industry, extensive use is made of micro-accelerometers for the deployment of vehicle airbags. Micro-rate sensors are used in Anti-lock Breaking Systems (ABS) and active vehicle stability control. Micro-pressure sensors and micro-flow meters are incorporated in engine control systems. In biomedical applications, MEMS can be found as tiny blood pressure sensors, flow-meter sensors and subcutaneous drug delivery systems. Ink-Jet printers incorporate micro-channels and actuators within printer heads. Future applications of MEMS, which are currently being considered or are already being developed range from micro-spacecraft to applications in the entertainment industry. In addition, military and defense applications form a broad category in their own right.

3. Principles of operation and detection

Like MEMS pressure sensors and accelerometers, silicon ultrasonic sensors use a suspended membrane. From a conceptual standpoint, the sensors resemble thousands of tiny drums on the surface of a silicon chip, where each is only as large as the width of a human hair. The drum structure consists of a thin nitride membrane and aluminum electrode suspended over a cavity. Nitride sidewalls support the membrane above the silicon substrate that forms the bottom electrode.

The drums are capacitive structures that operate under an applied electrostatic field. A DC bias voltage applied across the top and bottom electrodes establishes an electric field that creates tension in the nitride membrane. An AC signal voltage applied across the membrane varies the tension and causes it to vibrate and emit ultrasonic waves. Conversely, during reception of ultrasound, an acoustic wave impinges on the top electrode membrane and causes the membrane to move. The movement alters the capacitance of the sensor and creates an output current. For air or gas applications, the sensor operates in a resonant mode to maximize its sensitivity. However, in immersion or water applications, the sensors are designed to be nonresonant for optimal pulse-echo signal quality and broadband frequency response.

A passive infrared (PIR) sensor as used by the security industry is a special purpose radiometer used to detect the body heat of an intruder. It is an almost ideal sensor because it is passive. Its presence cannot be detected as is the case for active sensors such as ultrasonic or microwave. The design goal is quite simple, To create a radiometer with a high probability of detecting an intruder within a defined area while not responding to anything else. The area of detection is defined by the optics and pyroelectric detector element geometry. The probability of detection is enhanced by the efficient discrimination of the intruders body heat by the optics and electronic signal processing while not responding to anything else is the art and/or science of rendering the detection area background invisible and hardening the sensor against all other stimuli.

For detection, an intrusion sensor is sensitive to changes in infrared energy rather than absolute levels. It *accommodates* itself to the background conditions in the room and perceives the intruder as a change in this state of equilibrium. This change principle is fundamental to the detection process and PIR sensors are designed to maximize this by a process known as *chopping*. Conventional radiometers, those designed to measure temperature, sometimes have a built-in mechanical chopper wheel which alternately cuts the external field-of-view on and off. This renders everything external to the radiometer visible by reference to the chopper wheel, which in effect becomes its reference temperature or background. Another way is use an electronic chopper, which alternatively switches the signal processor from the external source to an internal reference at a suitable frequency modulating the reference.

Intrusion sensors cleverly use the real background as a reference completely avoiding the use of a chopper. By optically dividing the area to be protected into a number of separate and separated fields-of -view, when an intruder moves through the area he appears and disappears from view and by doing so modulates the reference condition. The signal produced is proportional to the difference in temperature between the intruder and the background.

4. MEMS Advantages and challenges

First, MEMS is an extremely diverse technology that could significantly affect every category of commercial and military product. MEMS are already used for tasks ranging from in-dwelling blood pressure monitoring to active suspension systems for automobiles. The nature of MEMS technology and its diversity of useful applications make it potentially a far more pervasive technology than even integrated circuit microchips.

Second, MEMS blurs the distinction between complex mechanical systems and integrated circuit electronics. Historically, sensors and actuators are the most costly and unreliable part of a macroscale sensor-actuator-electronics system. MEMS technology allows these complex electromechanical systems to be manufactured using batch fabrication techniques, increasing the cost and reliability of the sensors and actuators to equal those of integrated circuits. Yet, even though the performance of MEMS devices and systems is expected to be superior to macroscale components and systems, the price is predicted to be much lower.

Many industries, from the automotive to the medical sector are beginning to specify MEMS devices in their products. The technological benefits and market needs driving this rapid growth can be summarized as follows:

> MST and MEMS technology follow the current trend in many sectors towards miniaturization (especially in the Medical / Biotech sectors).
> The technology is based on batch fabrication, derived from integrated circuit technology; hence the systems are easy to mass-produce.
> The nature of the manufacturing process means that production costs are lower than those of comparable large-scale machines. In addition, the cost of ownership is also expected to be lower.
> The materials used are compatible with those used in semiconductor and microelectronic devices. Hence these MEMS-devices are relatively easy to interface with electronics, enabling simple on-board connectivity of sensors and actuators with analogue and digital electronics. This interconnectivity allows more design and functionality freedom.
> Multiple functions can be combined in the same device. The addition of integrated sensors to microelectronics allows information gathering to occur in highly integrated systems in addition to the more traditional roles of information processing and communication.
> The emergence of microactuators promises the ability to exert significant measures of control over non-electronic events at very small scale.
> The ability to do sensing and actuation at low cost in distributed systems promises to significantly extend microelectronic applications.
> MEMS systems promise high reliability compared to large-scale machines.
> The devices are of small volume and low weight. They therefore offer the product designer greater flexibility to include the same functionality in a product.
> There are several property advantages, e.g. increased sensitivity of sensors.
> In the biological and chemical sectors microsystems and microfluidic devices enable new ways of analyzing chemical reactions. These result from:
> Higher reaction speed (as reaction volumes are smaller).
> Lower volumes of (often expensive) reactants required, hence significant reaction in cost.
> Parallelism and multiplicity (many reactions are possible at one time; use of sensor arrays).

5. Industry Outlook

Sensor technology is one of the technologies that is going to play a major role in the future. It can be used in all sectors of industry and can give a product the added value that makes it competitive. In knowledge based economy only companies that are very innovative, grow fast and create most value will do best in tomorrow's competition.

During the past two decades, there has been an unprecedented growth in the number of products and services, which utilize information gained by monitoring and measuring using different types of sensors. The development of sensors to meet the need is referred to as sensor technology and is applicable in a very broad domain including the environment, medicine, commerce and industry. Governments and policy makers throughout the world are realizing the potential benefits of encouraging the growth in sensor technology not only as a result of new technological trends, and hence new products, for the indigenous industry to effect improved product quality and efficiency by broadening the level of control over their processes, but also in support of the implementation and enforcement of government legislation on environmental and safety issues.

Biosensors, studied since the 1960's, will have an increasingly important role in the 21st century biotechnology and medical revolution. Modern biosensors developed with advanced microfabrication and signal processing techniques are becoming inexpensive, accurate, and reliable. Increasing miniaturization of biosensors has led to the realization of complex analytical systems such as DNA micro-arrays and chemical/biological laboratories-on-a-chip.
This rapid progress in miniature devices and instrumentation development will significantly impact the practice of medical care as well as future advances in national defense and the chemical, environmental and pharmaceutical industries, particularly in genomics and proteomics

6. Applications of Sensor Technology

There are numerous possible applications for MEMS. As a breakthrough technology, allowing unparalleled synergy between previously unrelated fields such as biology and microelectronics, many new MEMS applications will emerge, expanding beyond that which is currently identified or known. Many of the fields of technology have been already touched by the MEMS technology. Whilst the most promising applications are at the stage of research, many others are at a point where, technically, they could already be used on a commercial scale. In many cases this is probably the time of transition from development to practical application, after a period of massive research and development effort. In the following list some MEMS and their field of application are presented.

- Textile Industry
- Non-vision sensors to detect defects in fabrics
- Environmental sensors
- Auto Industry
- Exhaust sensors
- Leak location and detection
- Temperature and pressure monitoring
- Oil Industry
- NMR techniques for on-line characterization of residuum in heavy oils
- Biomedics
- Sensors for pulping process
- Information Technology
- Steam generator mockup for round robin and probe evaluation
- Defense

- Aeronautics and Aerospace

7. Conclusions

The predominant technology at present state is surface micromachining, and current developments show that this trend will continue in the future. However the LIGA process will grow in importance, as it is the only method for producing true three-dimensional objects. The practical applications are increasing in number for niche markets as well as for mass-production employment. In the near future strongly innovative products can be expected in the biomedical field. The other industries such as space, aeronautical, and automotive will continue to substitute the conventional sensors with the MEMS equivalents. However some innovative solutions will take form due to the availability of the new sensors, with their advantages in cost, dimensions and reliability. The designer of electromechanical systems should pay attention to the availability of sensors and devices on the market. When possible, the choice have to fall on MEMS devices, as these are commonly cheaper, more accurate and reliable, and less cumbersome. For large number of production, the option of developing a new MEMS component should be considered.

References

[1]   Markus, K W, et al., (1995). 'MEMS infrastructure: The multi-user MEMS Processes (MUMPS),' Karen W. Markus (ed.) *Micromachining and Microfabrication Process Technology*, Vol. **2639**

[2]   G. Stix, *Micron Machinations*, Scientific American, Vol. 267, November 1992, pp. 73-80

[3]   K. J. Gabriel, *Engineering Microscopic Machines*, Scientific American, Vol. 273, September 1995, pp. 118-121

[4]   M. J. Madou, *Fundamentals of Microfabrication*, http://www.er6.eng.ohio-state.edu

[5]   A. M. de Aragon, *Future Applications of Micro/Nano-Technologies in Space Systems*, bulletin/bullet85/mart85

[6]   W. Keny, *Focus on MEMS*, The Engineer, Vol.12, n.3, June 1996, pp. 16-22

[7]   Auto-tech product guide, *Motion Sensing Systems*, Automotive Engineer, Vol. 22, n. 8, October 1997, p. 84

[8]   Product news, *Radar chips get dips*, Automotive Engineer, Vol.21, n. 1, January 1985, p. 14

[9]   R. S. Seley, *Mighty Micromachines*, Compressed Air Magazine, June 1996

Electric phase intelligent sensor and selector

F Udegbue, *Anietie F Nsien

Mobil Producing Nigeria Unlimited, P.M.B. 1001 Eket Akwa-Ibom State
Nigeria.
E-Mail:- fortunatus.udegbue@exxonmobil.com, fudegbue@yahoo.com.
*Dept. of E/ Electronics & Computer Engineering, Fed. University of. Tech.
Owerri.

Abstract: An electric phase intelligent sensor has been designed. This system is designed in modules: the power supply unit; the logic and the final control unit. The power supply unit was designed as a 12vdc full wave rectified source. Voltage regulator integrated circuit (IC) 7805 was used to maintain a constant 5VDC to the logic circuit. The unit is also used to drive the final control unit. The logic unit consists of a network of electronic EXCLUSIVE-OR, OR, AND and NOT gates. This network was developed from the synthesis of a truth table built from all possible results of either presence or absence of power in either or all of the three phases of a domestic electrical power supply system with one phase supplying 240vac. The synthesis of the truth table to form the network was carried out using SUM-OF-PRODUCT TERMS of the elements of the truth table. The final control unit is a power transistor control relay direct current 12vdc relay whose contacts are rated for 30A. During testing of the power isolator, results show that when there was power in either of the three phases only, the phases loop up the other two phases thus supplying power to them. When there was power in the three phases simultaneously, the phases are isolated from each other thus avoiding a line-to-line loop, which would introduce dangerous voltage to the load.

1.0 Introduction

In most under-developed nations such as Nigeria the problem of stable electric power source is still very prevalent. Most middle class homes procure three-phase power source from Nigeria Electric Power PLC. Company when indeed do only need a single-phase source. This gives them the privilege of having electric power as long as there is power in any of their three phases. Figure 1 shows a typical connection usually made to achieve this. The fuse is manually used to change the phase, hence only one fuse is used as against the normal three fuses which ought to be used. The logic behind this is drawn from the fact that a three phase power source is essentially the same as three ordinary single phase sources with each of the sources out of phase with each other by one-third of a cycle (120 degree), (Parker, 1988).

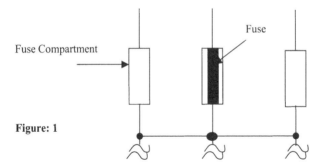

Figure: 1

The phasor relationship of a three-phase power source is shown below:

Figure: 2

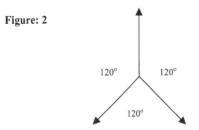

If two fuses are left in phases one and two while there is power in them simultaneously, the resultant power which will be passed to the load for a connection of the type shown in figure 1 is analyzed thus: Using cosine rule as given by (Smyrl, 1980):

$$\left| \, _{b\text{-}a} \, \right|^2 = \left| a \, \right|^2 + \left| b \, \right|^2 - 2 \left| a \, \right| \left| b \right| \cos\theta \text{ ------------------------ (1) where } \theta = 120^\circ.$$

where a =240vac, and b = 240vac.

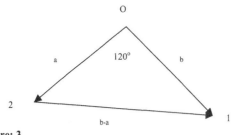

Figure: 3

In Nigeria most domestic or single phase appliances are designed for 240VAC, so this analysis is going to be based on it.

$$| \, _{b-a} \, |^2 = | \, 240 \, |^2 + |240|^2 - 2 \, | \, 240 \, | \, | \, 240 \, | \, (-\cos 60°)$$

$$| \, _{b-a} \, |^2 = | \, 57600| + |57600| - |115200| \, (-\cos 60°)$$

$$| \, _{b-a} \, |^2 = 172800$$

$$| \, _{b-a} \, | = 415 \text{ VAC.}$$

From the analysis above we infer that the logic behind the implementation of the connection shown in figure 1 is to avoid passing a voltage of over 400vac to the home appliances designed mostly for 240vac.

2.0   Problem Statement

Though the electric power connection shown in figure 1 has helped to reduce the rate of electric power down time in middle class homes, it has the following demerits:

(i)   The frequent removal and replacement of the fuses from the fuse compartment will cause the compartment to wear out due to abrasion.

(ii)   Frequent arcing due to installation of fuse when there is power in any of the Phases will gradually destroy the conductivity of both the face of the fuse and its compartment.

(iii)   When no adult is at home or adequate attention paid to the connection showed in Figure 1, the three-phase facility can be immensely under utilized, for even if there is power in the unconnected phases you need somebody to change the fuse to the phases before they can supply power.

3.0   Materials And Method

A one-in-three-phase power isolator has been designed to automatically isolate the phases from each other whenever there is power in two or more of the phases. See table 1 for the logic control philosophy. From Figure 5, it can be seen that the presence of power in any of the phases gives signal to the logic circuitry of the solid state electronic gates, which in turn controls the action of RLS_a and RLS_b via the darlington pair (power transistors). RLS functions to make sure that when there is no power in the single phases, the input to the logic gates does not pick up noise but is proper grounded as recommended by (Udegbue, 2001).

Materials: The following materials were used for the construction of the Isolator.

(a).   12-0-12, 120W step down transformer.

(b).   Bridge rectifiers, voltage regulators, 1000μF and 1μF capacitors.

(d).   Light emitting diodes (red) (yellow) (green) and a normal diode (IN4001).

(e).   Resistors: 1kΩ -1watt, 860 kΩ - 6watt, 920 kΩ - 3watt, 45 kΩ - 1watt

      1.5 kΩ - 1watt.

(f).   Power transistor (TIP31A).

(g).   Logic Chips: XOR (74LS136), OR (74LS132), AND (54LS108),

      INVERTER (74LS104).

(h).   IC PIN holder, Veroboards (25 x 10cm) and (14.5 x 7cm).

(i).   1 amp fuse and heat sinks.

(c).   12VDC and 6VDC relay with contacts rated for 30A. During the construction we realized that NEPA sometimes supply as much as 75A, though this is rare! realizing this will simply damage the contacts of our relays only rated for 30A, we searched for 12VDc relays with contacts up to 70A, but could not find. The relay with the highest contact rating that we could get in the market was that whose contacts are rated for 30A. To solve this problem we redesigned the loop thus:

From Kirchoffs current law the 75A, will split to 25A at C_1, C_2, C_3 respectively, this scenario will also repeat itself at d_1, d_2 and d_3 etc.

TABLE 1: ELECTRIC PHASE INTELLIGENT SENSOR
LOGIC CONTROL TRUTH TABLE.

	I_1	I_2	I_3	a	b	c	d	e
1	0	0	0	0	0	0	0	0
2	0	0	1	1	1	0	0	1
3	0	1	0	1	1	0	1	0
4	0	1	1	1	0	0	1	1
5	1	0	0	1	1	1	0	0
6	1	0	1	1	0	1	0	1
7	1	1	0	0	1	1	1	0
8	1	1	1	0	0	1	1	1

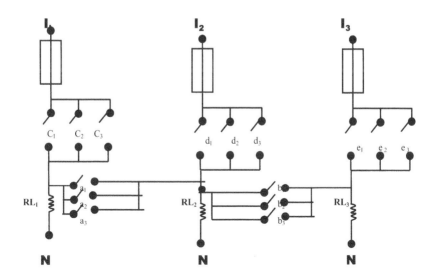

Figure 4: Design For A Lower Rated Relay.

Figure: 5

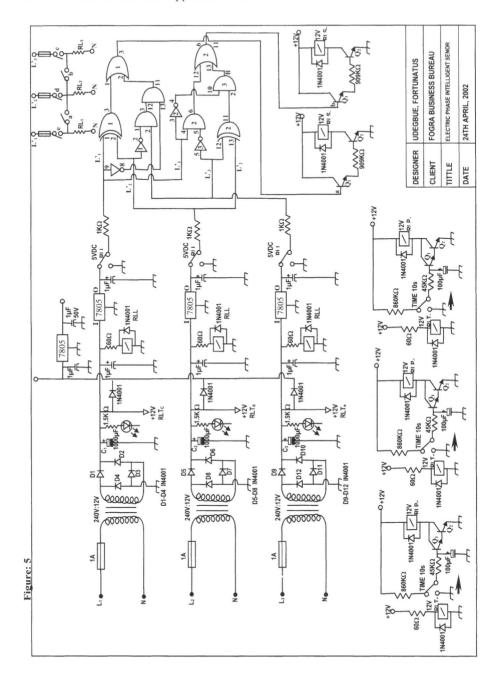

Design Calculations for the one – In – Three Phase Power Isolator.

(i.)   Power transformer: step-down transformer 240VAC 50HZ

     12Vac 500mA

(ii.)   Control Logic input control relay RLL. Voltage: 6VDC. Coil resistance: 60Ω contacts

     current rating: 1A.

(iii.)   LED used:

     Current : 15mA. Voltage :2vdc.

(iv.) RLT_C; RLT_d; RLT_e; All equal

     Voltage = 6VDC   Coil resistance = 60Ω

(v.) RLP_C; RLP_d; RLP_e; are the same as RLS_a & RLS_b.

     Voltage =12VDC   Coil resistance = 100Ω   Contacts = 30A.

(vi.)   Output current of the 74LS32 is 8.0mA.

(vii.)   TIP31A. (medium linear power switching transistor). See data attached.

From the truth table of the one – in three-phase power isolator logic control the table below
is formed. The expression for the sum of product terms (equation 2) for logic whose output is
denoted a is derived from the product terms in each row where a=1 as given by (Bartee, 1991).

Table 2: Sum Terms and Product Terms

I_1	I_2	I_3	a	b	Sum terms	Product terms
0	0	0	0	0	$I_1 + I_2 + I_3$	$\bar{I}_1 \bar{I}_2 \bar{I}_3$
0	0	1	1	1	$I_1 + I_2 + \bar{I}_3$	$\bar{I}_1 \bar{I}_2 I_3$
0	1	0	1	1	$I_1 + \bar{I}_2 + I_3$	$\bar{I}_1 I_2 \bar{I}_3$
0	1	1	1	0	$I_1 + \bar{I}_2 + \bar{I}_3$	$\bar{I}_1 I_2 I_3$
1	0	0	1	1	$\bar{I}_1 + I_2 + I_3$	$I_1 \bar{I}_2 \bar{I}_3$
1	0	1	1	0	$\bar{I}_1 + I_2 + \bar{I}_3$	$I_1 \bar{I}_2 I_3$
1	1	0	0	1	$\bar{I}_1 + \bar{I}_2 + I_3$	$I_1 I_2 \bar{I}_3$
1	1	1	0	0	$\bar{I}_1 + \bar{I}_2 + I_3$	$I_1 I_2 I_3$

From the table above, the sum of products expression for the output a is:

$(\bar{I}_1 \bar{I}_2 I_3) + (I_1 \bar{I}_2 I_3) + (\bar{I}_1 I_2 \bar{I}_3) + (\bar{I}_1 I_2 I_3) + (I_1 \bar{I}_2 \bar{I}_3) = a$ -------------- (2)

$(\bar{I}_1 \bar{I}_2)(\bar{I}_3 + I_3) + (I_1 \bar{I}_2)(\bar{I}_3 + I_3) + (\bar{I}_1 I_2 I_3) = a$ ----------------------- (3)

From the boolean postulate $a + a = 1$ as given by (Whitaker, 1996), $I_3 + \bar{I}_3 = 1$

∴ $\{(\bar{I}_1 I_2) + (I_1 \bar{I}_2)\} + (\bar{I}_1)(\bar{I}_2 I_3) = a$ -------------------------------- (4)

The equation above is what we used to design the control for power relay with contact identified as "a" in the drawing, i.e. RLS_a.

From the table above the sum of products expression for the output "b" is below:

$$(\bar{I}_1 \bar{I}_2 I_3) + (\bar{I}_1 I_2 \bar{I}_3) + (I_1 \bar{I}_2 \bar{I}_3) + (I_1 I_2 \bar{I}_3) = b \text{ --------------------------- (5)}$$

$$(I_1 \bar{I}_3)(\bar{I}_2 + I_2) + \bar{I}_1 \{(\bar{I}_2 I_3) + (I_2 \bar{I}_3)\} = b \text{ ------------------------------- (6)}$$

$$(I_1 \bar{I}_3) + \bar{I}_1 \{(\bar{I}_2 I_3) + (I_2 \bar{I}_3)\} = b \text{ --- (7)}$$

The equation above is what we used to design the control for power relay with contact identified as "b" in the drawing, i.e. RLS_b.

Series resistor to the LED have its value calculated below:

$$12V = I_{LED} R_{SLED} + V_{DLED} \text{ --- (8)}$$

Where $I_{LED} = 15mA$ $V_{DLED} = 2VDC$ (Manufacturer's rating)

     $12V = 0.015A \times R_{SLED} + 2V$

     $10V / 0.015A = R_{SLED.}$

     $R_{SLED} = 10,000 / 15 = 666.57\Omega$.

The control logic input control relay (RLL) draw the magnitude of current calculated below:

coil resistance $\implies$ ($R_{RLL} = 60\Omega$),   Voltage = 6VDC

     $V_{RLL} = I_{RLL} R_{RLL} \text{ ------------------------------------ (9)}$

     $I_{RLL} = 6 / 60 = 0.1A = 100mA$.

The resistor in series to RLL is R_{SRLL}.

     $12V = I_{RLL} R_{RLL} + I_{RLL} R_{SRLL} \text{ --- (10)}$

     $12V = 6V + 0.1 R_{SRLL}$

     $R_{SRLL} = 6V / 0.1A = 60\Omega$. --- (11)

The time constant of an RC circuit is defined generally as the time it takes the voltage across a capacitor to rise up to 67% of the charging voltage. The capacitor voltage V_c during charging can be generally expressed mathematically thus:

$$V_c = V_{cc} (1 - e^{-t/RC}) \text{ --(12)}$$ (Horenstein, 1996).

Where V_{cc} is supply or charging voltage

     V_c is voltage across the capacitor.

When $V_c = 0.67 V_{cc}$, t in equation 5 becomes the time constant T of the capacitor.

$0.67 = (1 - e^{-t/RC})$, $0.33 = (e^{-T/RC})$, $\ln 0.33 = -T/RC$, $-1.1 = -T/RC$

T (time constant in seconds) = $1.1RC$ $\text{ ------------------------ (13)}$.

The maximum base current for TIP31A is 1A. (Fairchild, 2000)

     $I_B R_B + 0.6V = 12V \text{ ---------------------------------- (14)}$

$I_B R_B = 11.4V - 0.6V$

$R_B = \{(11.4)/1\} = 11.4\Omega$

R_B is the minimum base resistance for the transistor to operate within safe limits

Let the delay to make time for RLP_C be 10secs; then $1.1\ R_B\ C = 10sec$.

Let $C = 10\mu F$.

Then $R_B = 10 / 1.1 \times 10 \times 10^{-6} = 909.090K\Omega$.

We desire the discharge of capacitor should take between $0.5 - 1secs$.

$$R_{DS} = 0.5 / 1.1 \times 10 \times 10^{-6} = 45.455K\Omega.. \text{------------------ (15)}$$

Base current (I_B) $\Rightarrow$

$$I_B = 11.4V / 909.090K\Omega = 1.254\mu A. \text{--------------------------- (16)}$$

hfe is given as 25 at $I_C = 1A$ and $V_{CE} = 4V$. (Fairchild, 2000)

Since I_B of $Q_1 = 1.254 \times 10^{-6}A$.

$I_C = 25 \times 1.254 \times 10^{-6}A$

Total collector current (It_c) of Q_1 equals I_C + (current from the coils of RLP_C)

$(It_c) = 10.032 \times 10^{-6}A + 12V / 100\ \Omega = 0.12003135A. \text{--------------------------- (17)}$

The base Current of Q_2 =Emitter current of Q_1.

Emitter current of Q_1:

$\Rightarrow I_{CQ1} + I_{BQ1} = I_{EQ1} = 0.120032604A. \text{--- (18)}$

Therefore: $I_{BQ2} = 120.03260mA$.

But h_{FE} is given as 25 at

$I_C = 1A$ and $V_{CE} = 4V$.

$\Rightarrow$ That at 1.2V V_{CE}, I_C will be equal to $1.2V / 4V = 300mA$.

V_{CE} is a max of 1.2V at saturation. h_{FE} at saturation is given as 8. (Fairchild, 2000)

$\Rightarrow I_c = 300mA / 8 = 0.3 / 8 = 0.0375A$.

$\Rightarrow$ The minimum base current required for saturation is 37.5mA.

A base current of 120.0362mA is a conservative case and is ideal for our kind of operation here, because we needed excess of base current, since our load is inductive.

Direct Current Voltage to be produced by the power supply unit in each Phase.

$\Rightarrow V_{RMS}$ from transformer is 12VAC

$V_{RMS} = V_P / (\sqrt{2})$ $V_P = 2\ V_{RMS} \times 1.41$

$V_P = 1.41 \times 12V = 16.92VAC. \text{----------------------------------- (19)}$

For full wave rectification with capacity filtering

$V_{DC} \cong V_P - 0.5V_r$ (Whitaker, 1996) $\text{----------------------------------- (20)}$

Where V_r is the ripple voltage expressed mathematically as

$V_r = V_P / 2FCR$. (Whitaker, 1996) ------------------------------------- (21)

F is the input AC frequency.

R is the total load resistance.

C is the filter capacitor.

   Load resistance of Phase 1 includes

R_{RLL} ; R_{SRLL} ; R_{GRLTc} ; R_{GRLPc} ; R_{GRLPc} ; $R_{GRLSa.}$

While neglecting the resistance of the IN40001 during forward conducting state and

the resistance of 7805 (IC).

Resistance of the IN40001 during the forward conduction of the Darlington Pair.

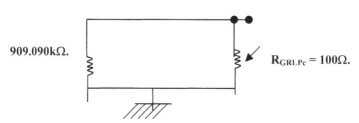

909.090kΩ.

$R_{GRLPc} = 100Ω.$

Fig 6: Equivalent circuit of the resistance of the darlington pair, while neglecting forward
   Resistances at the junction.

$1 / R_{GRLPc} = \{1 / 909090Ω \} + 1 / 100 = 0.0000011 + 0.01.$ ------------------------------------- (22)

$1 / R_{GRLPc} = 0.0100011$

$R_{GRLPc} = 99.98Ω$

$1 / R_T \cong 1 / (R_{RLL} + R_{SRLL}) + 1 / R_{GRLTc} + 1 / R_{GRLPc} .$ ------------------------------------- (23)

$1 / R_T \cong 1 / 120 + 1/120 + 1 /99.98$

$1 / R_T = 0.0083 + 0.010002 + 0.0083 = 0.0266$

$R_T = 37.6Ω.$

   $V_r = V_p / 2FRC$

   F =50HZ,   R = 37.6Ω $V_p = 16.92V.$

   $V_r = 16.02 / (2 \times 50 \times C \times 37.6) = 16.92 / 3760C.$

Recall that $V_D \cong V_P - 1/2 V_r$.

But we require V_D to be 12VDC

$V_r/2 = 16.92V - 12V = 4.92$ V -- (24)

$\qquad V_r = 4.92V$.

From the equation above

$C = 16.92/3760 \times 4.92V = 914.6$ µF. ------------- (25)

4.0 Result

Result shows that when there is power in all of the phases, all the series relay contacts open while all the parallel relay contacts close. When there is power in phase one and no power in any of the other phases, the parallel contacts and the series contacts close. When there is power in phase one and two but no power in phase three, the series contact (a) opens while the series contact (b) closes, also the parallel contacts (e) and (d) closes while the parallel contact (c) opens.

5.0 Discussion And Conclusion

During the testing of the initial design we discovered that after there was no power to the any of the three phases, then power suddenly comes to all the phases, the series relay contacts closes transiently for about 3 – 5 seconds, then opens. The danger here is that within this interval, high voltage of over 400VAC would have passed into the load designed for a voltage of 240VAC maximum. This necessitated a change in our original design to include a time delay circuit (delay to energize) of about 10seconds to the phase relays; this 10second delay eliminated the transient period within which the series relays were energized. The observed phenomenon is as a result of the characteristics of the logic circuit used to control the series relays. The phenomenon observed is common in microprocessors, it is however a surprise to notice it in an ordinary logic gate circuitry!

Reference:

[1] Bartee, Thomas, C. 1991. Computer Architecture and Logic Design. McGraw-Hill International Editions. McGraw-Hill Inc. NJ. P. 67.

[2] Fairchild Semiconductor International, 2000. TIP31 Series(31/31A/31B/31C) Medium Power Linear Switching Transistor Datasheet. Rev. A. February 2000.

[3] Horenstein, Mark, N. 1996. Microelectronics Circuit and devices, 2nd edition. Prentice Hall Inc. Ohio.

[4]   Parker, Sybil, P. 1988. McGraw-Hill Encyclopedia of Electronics and Computers. 2nd Edition. McGraw-Hill Book Company, New York, St. Louis San Francisco. P.7.

[5]   Smyrl, J.L. 1980. An Introduction to University Mathematics. Hodder and Stroughton Educational publishers, London. P.132.

[6]   Udegbue, Fortunatus. 2001. Design of an Intelligent Hydrocarbon Well Watcher using pressure, level, vibration gas leakage and Intruder detection as parameters. Unpublished M.Sc thesis. Physics Department University of Calabar, Nigeria. P.24.

[7]   Whitaker, Jerry C. 1996. The Electronics Handbook. CRC Press and IEEE Press. Technical Press Inc. Oregon. P. 46 and 942.

Intelligent Process Monitoring and Management – Time for a Return to Sensor-Based Methods?

R I Grosvenor and P W Prickett

Intelligent Process Monitoring and Management (IPMM) Group, Cardiff School of Engineering, Cardiff University, Wales

Abstract: Condition monitoring research at Cardiff University evolved from sensor-based methods to a Petri-Net-based method that primarily uses embedded machine signals. Most recently distributed monitoring systems have been developed that are applicable to sequentially controlled machines and processes. Microcontroller systems with CAN bus inter-chip communications and internet connectivity have been used and provide remote display of monitored and processed information and connection to a server-based database. The authors feel that it is timely to consider the integration of traditional and smart sensors into such a distributed system. This paper will present some of the considerations and possible applications.

1. Introduction

Condition monitoring research at Cardiff University, as applied in the Intelligent Process Monitoring and Management (IPMM) group, is well established and has evolved over the past 20 years. In the earlier part of this period the research was almost exclusively applied to machine tools and used heavily sensor-based methods. This was exemplified, and will be illustrated in this paper, by the MIRAM project ('Machine Management for Increasing Reliability Availability and Maintainability')[1]. The principal role in this project was the design and deployment of a Data Acquisition System (DAS). DAS systems were actively used in 3 European industrial sites and collected monitoring information over a significant timescale. The information was then used by partners concerned with both developing fault-based mathematical models and diagnostic / expert systems. Table 1 provides evidence as to the extent of the number and range of analog sensor signals that were interfaced to the DAS. The associated programming, technology and logistical problems are discussed more fully in the following section.

The bulk of the development of a hybrid approach was then achieved during the FMS-Maint project ('Integrated Condition and Machine Process Monitoring Systems for Flexible Manufacturing Systems and Stand-Alone CNC Machines')[2]. Within this project, industrial, sequentially controlled machines and processes were monitored. Petri-Net models were adopted as the vehicle for producing structure and logic, and for providing a graphical diagnostic user display. The vital need for intelligent data manipulation and management was compounded during this work. The monitoring system was designed to use embedded machine signals, the majority of which were digital (controller I/O signals, switch inputs, operator inputs etc.). These were supplemented, as appropriate, by a small number of key analog sensor signals (vibration, motor current/power, cutting forces for example).

MIRAM Data Acquisition System (DAS) Summary [1] 1991 -1994
PC: 486SX, 20 MHz, 2Mbyte RAM, 40 MByte HDD + 40 MByte tape drive,
Commercial & In-house interface cards

to monitor the following machine tools signals in remote locations:

Existing Machine Signals:	3 X axis signals (demand, current, tachogenerator)
	3 Y axis signals (ditto)
	3 Z axis signals (ditto)
	3 spindle drive signals (ditto)
	2 motor drive signals
	46 digital signals
Transducer Signals:	6 Pressure
	5 Flow
	5 Power
	2 Vibration
	3 Temperature

Table 1. Sensor-Based Machine Tool Condition Monitoring

A previous paper [3] also reflected on the condition monitoring approaches and discussed the impact of a set of rapidly emerging technologies. Several of these technologies have been investigated and are central to the latest distributed, low-cost, microcontroller-based systems being researched within the IPMM group.

2. Sensor-Based Approaches and Their Disadvantages

During the MIRAM project the, mainly logistical, difficulties of employing a heavily sensor-based approach became apparent. Industrial partners were often unable or unwilling to release their key production machines for the time required to install and commission monitoring systems with significant numbers of sensors, other than in annual shutdown periods for example. The hybrid approach was thus, in part, developed in response to this, with the aim of reducing the amount of extra monitoring hardware (and its associated wiring and isolated interfaces etc.) to be installed. Thus, where possible, the current state of the machine or process was inferred by tracking the logical sequential during cycles of operation. The examination of the logical inputs and the setting of time-out alarms for example, then allowed for efficient fault diagnosis/location (and user interfaces) for 'hard' faults. By adding date and time stamping to the sequence transitions and database storage and manipulation, progressive or 'soft' fault conditions could also be monitored. This regime has typically been supported by the retained use of a limited number of key transducer signals. Examples of these, for a variety of applications, have been reported elsewhere [4].

3. The Hybrid Petri-Net-Based Monitoring Approach

As previously indicated, this approach evolved due to a number of factors and was established to use as many existing, or relatively easily interfaced, machine or process signals. Thus, the use of existing digital controller signals, along with any sensor signal already available on the machine or process, plus the fitting of a very limited number of additional sensors eased the logistic problems. The task still depends upon the age of the existing machine/process controller(s) and on the extent of communications infrastructure in the industrial location. With older generation PLCs, for example, the monitoring system has to 'tap into' the existing signals via a hardwired arrangement. With more recent PLCs, with communications modules fitted, the interfacing to the monitoring PC is a greatly simplified task.

The terminologies, definitions and symbols used with the Petri-Nets are outside the scope of this paper and have been previously reported [3][5] and an overview may be obtained by reference to the IPMM research group web pages [6].

The migration to this approach allowed for more diverse applications. The method is applicable to sequentially controlled machines and processes and initially entailed a PC based deployment for each monitoring application. More recent project work has focussed around the application of the monitoring techniques within industries classified as Small-to-Medium Enterprises (SMEs). Typically such SMEs may have, relatively, limited computing resources and may not wish to dedicate a PC to each monitoring exercise. In response to this, distributed, microcontroller-based, systems have been developed.

4. Distributed IPMM Systems and Their Applications

This approach is centred on the deployment and implementation of a Petri-Net based monitoring structure employing the Microchip family of PIC microcontrollers rather than the previously utilised PCs. Figure 1 shows a typical system architecture for such distributed monitoring systems. Figure 2 shows more detail of the typical capabilities of one of the networked monitoring modules [7]. For a given application, an appropriate number of monitoring modules will be deployed and the inter-chip communications achieved via the Controller Area Network (CAN) capabilities of the microcontrollers. Other networked PIC modules will be included if required to provide, for example, specific data processing or library functions. The network includes a connectivity module to provide the internet connection to a server database and to allow dynamic web pages to be implemented. The protocol employed within the connectivity module is the User Datagram Protocol (UDP). The Petri-Net techniques are retained, but are applied more to provide structure to the programming of the distributed system rather than the direct graphical user interface. The monitoring module shown in Figure 2 has a 2 chip CAN-bus implementation. A single chip solution is now available.

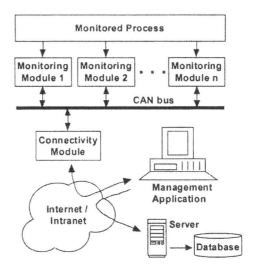

Figure 1. Distributed Monitoring System Architecture [7]

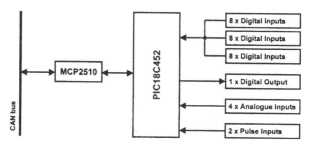

Figure 2. Monitoring Module Hardware Block Diagram [7]

The distributed monitoring system has been deployed and tested on 3 diverse systems to date. These are an industrial control trainer (mimics an industrial assembly process with a flexible sequence depending upon the loading sequence of the two constituent components), an IPMM model of a hydraulic press (mimics control sequence of real industrial press but allows system testing under deliberately introduced fault conditions) and the tool changer sequence of a machine tool.

5. A Return to Sensor-Based Monitoring?

In carrying out research into the distributed monitoring systems refinements were required to the established Petri-Net models. These were primarily in response to the need to refer to context-based sensor inputs. In the simplest case a logical input to the Petri-Net model is required by comparing a current sensor input level (at appropriate time and conditions) to a pre-determined threshold level. The practicalities of implementing analog sensor inputs in such a PIC microcontroller regime, in terms of sample speed and memory availability restrictions have been investigated through MSc projects [8][9]. The potential use of relatively simple, low-cost sensors has similarly been investigated. One possible application investigated [10] was the use of a microphone input. The likely usage would be restricted to a 'crude' and quick overall check that no unusually audio signals were present or developing on the monitored process. If other than the normal sound levels and constituent frequencies were present these would identify the need for the system to instigate more detailed monitoring for the ensuing period.

The authors have also reflected on the relative merits of the distributed monitoring approach, particularly for SME industrial applications. These coupled with other developments in the general condition monitoring field, indicate that it is timely to incorporate more sensor inputs into the distributed monitoring system. It is probably not possible to indicate more than a number of possible development areas at this stage. A summary of some of the potential areas is as follows:

➤ **Closed Loop Performance Monitoring (CLPA)-** there is a need to monitor the performance of typical control loops, possibly via demand and feedback signals. These signals could be used with the distributed Petri-Net system to provide more context-based monitoring.
➤ **Process Instrumentation-** there appears to be synergy between the IPMM work and the modern generation of process instrumentation, with these now typically allowing internet-based remote monitoring.
➤ **Lower Level Distributed Systems-** if the trend towards lower cost chips with sufficient functionality and interconnectivity continues, then a monitoring system using a relatively large number of distributed chips either embedded or interfaced to industrial machines or processes will become feasible and viable.
➤ **Integration with Modern/Intelligent Sensors-** in a similar fashion the widespread availability of small-scale sensors will result in their likely incorporation in a range of monitoring applications.

6. Remote Monitoring and User Interfaces

These remain as important considerations in the design and deployment of monitoring systems. The IPMM group, as stated, are considering the use of a variety of technologies [3] in providing timely an appropriate alarms to users of different levels and requirements. The outputs from the distributed monitoring system have been formatted and incorporated in dynamic web pages to allow the remote 'browsing' of live monitored process performance. Of particularly promise, in this respect, is the use of parameters such as overall equipment effectiveness (OEE).

Acknowledgements

The authors acknowledge the particular contributions of A.D. Jennings (IPMM project manager) and M.R. Frankowiak. The latter is supported by CAPES, a Brazilian Federal

Agency for post-graduate education. The contributions of Q. Ahsan and W. Amer, initially via their MSc projects and now as PhD students are also acknowledged.

References

[1]   Drake P R, Grosvenor R I and Jennings A D 1995 *Proc. Sensors and their Applications VII, Dublin* 433-437
[2]   Prickett P W and Grosvenor R I 1995 *J. Qual.Maint.Eng.* **1** 47-57
[3]   Frankowiak M R, Grosvenor R I, Jennings A D, Prickett P W and Turner J R 2001 *Proc. Sensors and their Applications XI, London* 39-44
[4]   Grosvenor R I, Prickett P W, Jennings A D and Frankowiak M R 2002 *Proc. Control Loop Performance Assessment Seminar, IEE, London* 9/1-9/5
[5]   Jennings A D, Notwatschek D, Prickett P W, Kennedy V R, Turner J R and Grosvenor R I 2000 *Proc Comadem 2000* 643-650
[6]   www.processwatch.cf.ac.uk *accessed April 2003*
[7]   Frankowiak M R, Grosvenor R I, Prickett P W and Jennings A D, *2003 submitted to* Comadem2003, August 2003, Växjö University, Sweden
[8]   Ashan Q, 2002, *MSc thesis, Cardiff University*
[9]   Amer W, 2002, *MSc thesis, Cardiff University*
[10]   Taylor J P, 2001, *MSc thesis, Cardiff University*

Poster Programme
Paper presented at Sensors and their Applications XII, September 2003
©*2003 IOP Publishing Ltd*

Optical Sensing Systems for Primary Level Science Education

Adam Markey[*1], Bernard Tyers[1], Emma O'Brien[1], Aisling K McEvoy[1], Roderick Shepherd[1], Brian D MacCraith[1], Bakhtiar Mikhak[2,3], Carol Strohecker[2]

[1] School of Physical Sciences, National Centre for Sensor Research, Dublin City University, Glasnevin, Dublin 9, Ireland;
[2] Media Lab Europe, Sugar house Lane, Off Crane St., Bellevue, Dublin 8, Ireland;
[3] MIT Media Laboratory, 20 Ames Street, Room E15-020, Cambridge, MA 02139, United States

Abstract: We report the development of a range of miniaturized optical sensors, for use with the LEGO Mindstorm platforms and related technology. This work was carried out as part of a collaborative initiative to stimulate interest in Science at Primary School level through a project-based learning approach.
Sensors have been developed and fabricated for use with the LEGO platform. A pilot "Sensors for Science Education" programme has been established in three Irish primary schools. In particular a working oxygen sensor has been designed and fabricated. The principal design features were compatibility with the programmable LEGO platforms and robustness for classroom use. This sensor uses the method of intensity quenching to determine oxygen concentration.

1. Introduction

The "Sensors for Science Education" programme is a collaborative project between Media Lab Europe (MLE), MIT Media Lab (MITML) and the National Centre for Sensor Research (NCSR) at Dublin City University. It started in early 2000 with the objective of extending the LEGO Mindstorms/Cricket technology by incorporating new sensor modules (e.g. physical and chemical) for implementation in the Irish science education environment. In particular, we aimed

- To produce simple, robust, and low cost sensors.
- To build user-friendly software interfaces to accompany the sensors.
- To encourage children to develop a keener interest in science through activities which make use of the sensors and software.

This programme has been established in three Irish Primary Schools to date:

- St. Brigid's National School, Castleknock, Dublin
- Clontubrid National School, Co. Kilkenny

- Holy Spirit National School, Ballymun, Dublin

Each school has welcomed this initiative enthusiastically and these schools have already received simple sensors (e.g for temperature and pH) and software under this programme. The next stage in this programme is to introduce the teachers and students involved to a range of more advanced sensors, among which are the oxygen sensors described in this paper.

In 1998, LEGO released a robotic construction kit, known as LEGO Mindstorms. This package is the result of technology and ideas developed by Seymour Papert through research at Massachusetts Institute of Technology (MIT). The Logo brick was the result of early research in this field. This was essentially a micro-computer. Sensors and motors could be attached to it in order to allow it to monitor and control motion in its immediate surroundings.

A programming language (Logo) was specially designed for the purpose of communicating with and programming the Logo brick. Using this programming language, children are able to control the movement of various mechanical projects by connecting the "programmable brick" to an interface box and using Logo to programme them. The first "programmable bricks" had drawbacks in that they needed to be hardwired to a desktop to facilitate programming and data retrieval. This greatly limited the potential for truly autonomous machines.

The new improved version of the Logo brick was then developed, and is called the LEGO RCX brick. This has the advantage that it does not need to be physically connected to a desktop computer. Programming is carried out remotely using an infra-red communication link. In addition, information gathered by the brick through the use of sensors can be relayed to the desktop via the IR link.

2. LEGO RCX Brick Overview

LEGO have now produced many Mindstorms kits [1], which are readily available on the retail market. Although these kits differ in size, content and project capabilities, they all consist of the following components:

- RCX brick
- Motors and various sensors
- LEGO bricks, including axles, gear wheels and other mechanical components

The RCX brick is at the centre of any project built from these kits. Figure 1 shows the RCX brick. It is essentially a miniature computer with an LCD screen, and contains the circuitry required to drive motors and collect data using a collection of miniaturized sensors. Any project built around the brick can be almost completely autonomous. The brick converses with the desktop by means of an IR "tower". This is directional, however, and the IR transceivers of both the brick and the tower must be lined up to allow communication.

Figure 1: LEGO RCX brick

Figure 2: RCX programming environment

The programming environment [2] used with the brick is shown in Figure 2. Programmes are "written" by piecing together the various shape icons, each of which has a specific command associated with it. This simplifies the task of writing programmes and also makes it more appealing to children, since the software is graphic-based and programmes do not have to be manually typed.

3. Cricket Architecture

3.1 Cricket overview

One of the predecessors to the RCX brick is the cricket [3] (Figure 3). This is a tiny programmable computer which is the size of a nine volt battery that can directly control motors and receive information from a collection of specialized sensors. The cricket is based around a PIC microcontroller.

Figure 3: Cricket with battery

The cricket may be programmed, and can communicate with a desktop computer by means of a bi-directional infra-red link. Like the RCX brick, the cricket is suitable for standalone applications since it does not need to be hardwired to the desktop.

A software environment, known as Cricket Logo, is used to programme the cricket. This environment is shown in Figure 4(a). Procedures may be written in the right-hand window and downloaded onto the microcontroller situated on the cricket, so that an immediate response from the cricket may be initiated. This window may also be used in order to record and log sensor readings from the sensors plugged into the cricket.

For remote applications, if the cricket is to be used away from the desktop, procedures may be downloaded to the cricket. To initiate the programme, a button on the cricket must be pressed, thus the programming environment is not essential in order to initiate procedures at a distance from the desktop computer.

A newer programming environment has been developed for use with the cricket platform. This is known as Logo Blocks. It is an iconic system much like the environment used to programme the RCX brick. This is more appealing to children, since having to remember commands is not necessary. The blocks will only fit into each other in a logical order, much like a jigsaw puzzle. This programming environment is shown in Figure 4(b).

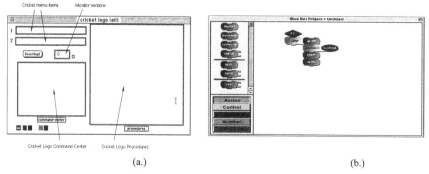

(a.)   (b.)

Figure 4: *(a.) 'Cricket logo' programming environment   (b.) 'Logo blocks' programming environment*

3.2 Bus port

Many of the early sensors used with the brick were based on changes in resistance using, for example, light dependent resistors   (LDRs) as detection element. Sometimes the resistance of a sensor cannot be made change when measuring an entity. Voltage and current are also indicators of a changing signal. The measured signal needs to be changed into a more useful format. An expansion port, or bus port, is included on the cricket platform. This allows more complex sensors (i.e. non-resistance-based) and actuators to be integrated into the cricket system. The addition of bus devices greatly enhances the crickets' use as a learning and development tool.

3.3 Communication

At the heart of every bus device is a microprocessor, one pin of which is dedicated to bus communication. Each bus device has its own identification, since many bus devices can be daisy-chained to the cricket. A master-slave method is implemented when using such bus devices. The cricket acts as the master, and the sensors as slaves, waiting until a command is sent from the cricket before doing anything.

Figure 5 shows a graphic representation of the bus protocol. The bus line is held in the high state, with the pins dedicated to communication from each device connected together. A transmission consists of a 100 microsecond pre-start pulse, where the bus line is driven low, followed by a start bit, 8 data bits and a 9th bit to indicate whether the bitstream was data or a command. A 'one' signifies a command, while a 'zero' indicates data.

The pre-start synchronization pulse allows all connected bus devices to prepare to capture and process the proceeding bus data. Every device should be able to dedicate its full attention to the bus data before the synchronization pulse ends. Since every bus device has its own identification address, only the device with which the cricket wants to communicate will dedicate its time to the cricket after processing the bus data. The rest of the bus devices connected will return to their previous state, either collecting data, or waiting for their address to be called.

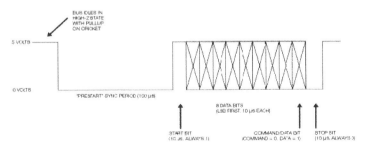

Figure 5: *Bus protocol*

4. Oxygen sensor [4]

As part of our pilot programme in Primary Schools, we have been developing a range of chemical sensors. We report here the development of an oxygen sensor based on fluorescence quenching of an oxygen-sensitive fluorophore. Sensors such as these had not previously been developed for use with either the RCX brick, or cricket platform. Many applications demand the precise determination of oxygen concentration, e.g. wastewater monitoring or blood gas analysis. For educational purposes, exact concentration measurement is not essential, and the operational range of the sensor is limited to oxygen concentrations around ambient. The requirement specifications are quite different to those for industrial applications:

- A high degree of accuracy is not essential – but should be able to distinguish changes of 1 – 2%
- The sensing range should be optimised for oxygen concentrations between 0 and 21%
- Robustness is a must if the sensor is to be implemented in a learning environment
- The sensor must be sufficiently small as to fit into a wide range of educational projects

The sensing element of the oxygen sensor comprises an oxygen indicator entrapped in a porous hydrophobic sol-gel. The oxygen indicator is a ruthenium dye complex, $[Ru^{II}-tris(4,7-diphenyl-1,10-phenanthroline)]^{2+}$, which when illuminated with blue light, fluoresces in the orange region of the spectrum. The sensing mechanism involves porous hydrophobic sol-gel. The oxygen indicator is a ruthenium dye complex, $[Ru^{II}-tris(4,7-diphenyl-1,10-phenanthroline)]^{2+}$, which when illuminated with blue light, fluoresces in the orange region of the spectrum. The sensing mechanism involves quenching of the luminescence of the excited state dye complex by oxygen. As the concentration of oxygen increases, the fluorescence decreases. Figure 6 shows the sensor output for concentrations of 0, 5, 10, 15 and 20% oxygen. The signal is greatest for 0% and lowest for 20% oxygen. Because the sensor deduces the oxygen concentration by measuring the light intensity output of the ruthenium complex, and converts it into a voltage, the sensor needed to be realized in the form of a bus device.

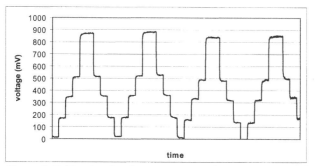

Figure 6: *Calibration data for oxygen sensor for concentrations of 0, 5, 10, 15 and 20% oxygen, where the signal is greatest for 0% and lowest for 20% oxygen.*

The following points indicate the program flow that is necessary to take a reading and send it back to the cricket on command:

1. The sensor waits until its address has been called by the cricket
2. The sensing circuitry is powered up
3. The intensity of the emitted light is measured and converted to a number between 0 and 255
4. The sensor drives the bus line low for 100 microseconds (pre-start synchronization pulse)
5. The sensor reading is sent to the cricket via the bus line
6. The cricket sends the reading to the desktop pc via the infra-red data link where it is displayed in the monitor window of the Cricket Logo programming environment

The microprocessor used with the oxygen sensor is a *PIC16F872*. The reason for using this particular microprocessor is the fact that it has analogue to digital capability. This allows the raw sensor reading to be fed directly into one of the input pins on the microprocessor and then it is converted to a digital signal, which is transmitted to the cricket. Alternatively, the microprocessor could process the reading and use a look-up table to evaluate the oxygen concentration. The percentage of oxygen present could then be sent to the cricket, rather than an arbitrary number.

Since the sensor works on the basis of measuring light intensity, it is adversely affected by ambient light. To overcome this problem, the detection circuitry includes a lock-in chip (AD630) which enables lock-in detection and discriminates against ambient light.

To reduce the size of the sensor circuit, the microprocessor is used to modulate the excitation light source, instead of a dedicated frequency generating circuit. This results in very stable frequency modulation of the LED used to illuminate the ruthenium complex, due to the fact that a crystal oscillator is used for the timing of the microprocessor. The modulated signal ranges from 0 to 5 volts. This is channelled directly to the reference port on the lock-in amplifier chip. It is also driven through a current-limiting resistor in order to modulate the LED. This results in a very clean and stable excitation light output.

A photodiode with integral amplifier is used as the means of detection of the emitted light. The signal produced is first filtered, and then fed into the AD630. After the modulated signal is extracted from the noise, it is demodulated and turned into a dc signal. This is then amplified once more before being fed into one of the analogue ports on the microprocessor.

The above arrangement can also be used with the RCX brick. The analogue signal can be plugged directly into one of the sensor ports on the brick, since the RCX has its own internal analogue to digital converter. The only modification needed is the arrangement of the microprocessor. A bus line is not needed, although an interrupt signal could be used in order to initiate the sensor. The *PIC16F84* microprocessor would be an ideal component in this application, because the analogue to digital conversion capability is not necessary, and with respect to size consideration, the *PIC16F84* is much smaller than the *PIC16F872*.

The goal of the development of the above sensor is to introduce primary school children to a wider range of scientific tools and the concept of measurement. The application focus addresses specific learning objectives of the new Primary School Science Curriculum (e.g. Testing the Environment). Applications of the oxygen sensor will include experiments such as the monitoring of the respiration process of plants and humans as well as combustion.

5. Applications & future work

Further development of this sensor is envisaged. Oxygen sensors based on measurement of the fluorescence intensity are susceptible to light source, detector drift and changes in the optical path [5]. The lifetime is an intrinsic property of the fluorophore, which is virtually independent of these external changes. Phase fluorometry is a low-cost time-based technique that could be used where the oxygen-sensitive phase difference is measured between the modulated fluorescence signal and the modulated reference signal. The use of compatible carbon dioxide sensors [6] in conjunction with the reported oxygen sensor will further expand the possible applications and experiments. A range of other sensors is under development and will be introduced to schools involved in our pilot programme.

References

[1] F. Martin, B. Mikhak, M. Resnick, B. Silverman, R. Berg, To Mindstorms and Beyond, evolution of a construction kit for magical machines, Morgan Kaufman / Academic Press, San Francisco, March, 2000

[2] F. Martin, D. Butler, W.M. Gleason, Design, story-telling, and robotics in Irish primary education, IEEE Systems, Man and Cybernetics conference, Oct. 2000

[3] F. Martin, B. Mikhak, B. Silverman, Metacricket: a designers kit for making computational devices, IBM Systems Journal, Volume 39, Numbers 3 & 4, 2000

[4] G. O'Keefe, B.D. MacCraith, A.K. McEvoy, C.M. McDonagh, J.F. McGilp, Development of a LED-based fluorimetric oxygen sensor using evanescent wave excitation of a sol-gel immobilized dye, Sensors and Actuators B 29 (1995) 226-230

[5] C. McDonagh, C. Kolle, A.K. McEvoy, D.L. Dowling, A.A Cafolla, S.J. Cullen, B.D. MacCraith, Phase fluorometric dissolved oxygen sensor, Sensors and Actuators B 74 (2001) 124-130

[6] C. von Bültzingslöwen, A.K. McEvoy, C. McDonagh, B.D. MacCraith, I. Klimant, C. Krause, O.S. Wolfbeis, Sol-gel based optical carbon dioxide sensor employing dual luminophore referencing for application in food packaging technology, *Analyst*, 2002, *127*, 1478

An Optical Fibre Sensor for Underwater Proximity Measurement

G Dooly, D Toal, E Lewis, W B Lyons

Department of Electronics and Computer Engineering, University of Limerick, Ireland.

Abstract: The development of low cost short range and tactile optical fibre sensors for marine applications is described. An example application would be on unmanned underwater vehicles (UUVs) operating in confined spaces. The fibre sensors augments the sensory abilities derived from ultrasonic and other sensors employed for marine proximity measurement / detection. Of particular interest is proximity detection in the "near" (less than 1m) and tactile areas. The paper describes the basic principle of operation and alternative sensor configurations. Preliminary results are given based on testing in the laboratory and deployed on a mini autonomous submersible in a test pool.

1. Introduction

In the realm of submersible technology the Autonomous Underwater Vehicle (AUV) is coming to commercial maturity, [1]. Up till now the workhorse of sub sea operations for off shore oil and gas and marine survey has been the remotely operated vehicle (ROV). In the case of both ROVs and manned submersibles, a human pilot controls the vehicle based on visual information (video relayed to surface for ROV) and sonar sensor data. In the instance of AUVs, an onboard controller is responsible for autonomous detection of the vehicle's surrounding environment and must be capable of reacting to that environment appropriately based on sensor input. The vehicle requires sensors that are capable of detecting objects and obstacles / potential hazards, both at close range and remotely. Vision systems might be thought to be an obvious choice. However, due to the variability in water turbidity, colour and light levels, solutions employing artificial vision are fraught with difficulty [2].

Ultrasonic sensor systems have become the system of choice for object detection in the marine environment for UUVs [3] and allow detection of objects far beyond the range of video, even in clear water conditions. Most applications for which AUVs are being employed involve remote sensing using multibeam, side-scan or sub bottom profiling sonars and in such cases the vehicle is flown remote from the bottom and other obstacles. Ultrasonic systems, however, do not lend themselves to all round coverage of autonomous craft in very close proximity to potential undersea hazards. If the AUV is deployed for close up video inspection [4] the controller must be able to detect objects in all directions from the vehicle to allow vehicle control/manoeuvring without becoming entangled or snagged. The aim of this work is the development of sensors for the detection of objects within a short (<0.1m) range in the aquatic environment. A principal objective is to construct a sensor that will allow an autonomous submersible to detect objects close to it and react accordingly, e.g. by stopping or turning to avoid collision.

2. Background

Ultra-sonic sensors are commonly used as marine sensors for submersibles and boats for the detection of the seabed and other obstacles and targets such as shoals of fish. There are various classes or types of ultrasonic sensors used in marine robotics including side scan sonar, multi-beam, simple narrow beam echolocation (e.g. altimiter sonars), etc. With the basic echolocation type sensors distance is calculated based on the time delay to reception of reflected ultrasonic pulse. These sensors are however not useful for detection / resolution of close up objects as typical minimum detection range for altimeter sonars for example is of the order of 1m (receiver must be blanked or switched off during pulse transmission). Separate echolocation sensors could be integrated for proximal detection for vehicle operation in confined spaces but to guarantee all round coverage a significant number or array of these devices would be required. Cross talk between sensors can also be a problem giving spurious multi-path distances to objects and this is especially so in close up operation or operation in confined spaces. The cost of marine hardened ultrasonic sensor systems is also high. The requirement of many sensors would make for expensive all round obstacle detection.

For autonomous craft operating close to hazards (e.g. drilling platforms, risers, sea cages) or in tight confined spaces (caves or wrecks) we aim to develop "near-touch" sensors, which are capable of instantaneous detection of proximal objects within 1m of the craft. The craft can then be given all round coverage with an array of these sensors. With the aim of giving autonomous submersible craft 4π steradians of solid angle coverage, many sensors will be required, even on small craft, so the sensor cost must be low. Optical fibre sensors may provide a cost-effective solution, as many may be used on a single craft without a serious weight burden. The sensors must be immune to the relatively hostile marine environment and again optical fibre sensors readily fulfil this requirement.

3. Sensor Design

Various configurations of both fibre optic extrinsic (where the fibre is used only as a conduit to guide the light), and intrinsic (where the light propagating through the optical fibre is modulated indirectly by optical path-length changes within the fibre), sensors are being developed and investigated. Some of these configurations transmit light pulses from the end of the fibre/wave-guide and rely on reflection of light energy by the target (extrinsic type), see figure 1. In other configurations the pulsed light is not transmitted into the water but is contained within the fibre. In these configurations deformation of the wave-guide has a detectable effect on the light pulses within the fibre, see figure 2 (intrinsic type).

3.1 Simple Open End Whisker

As illustrated in figure 1, pulsed light reflected off a target and coupled back into the whisker can be detected.   Various configurations of such a "whisker" were experimented with; (a) whisker with separate transmit and receive fibres, (b) three fibre whisker with third fibre used for sensor correction/normalization for ambient light levels, (c) single transmit/receive fibre with integral coupler/splitter at emitter/ receiver end. The results of this experimental work has been reported in [5].

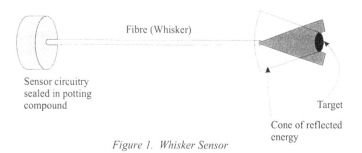

Figure 1. Whisker Sensor

The primary interest is in detecting proximal objects with relative motion with respect to the submersible. Consider relative motion between the whisker sensor and the target such that they approach each other at constant velocity. As the obstacle approaches the whisker, sensor output varies/increases with falling distance until the whisker bumps. Returned light intensity is inversely proportional to the fourth power of the sensor-obstacle distance (i.e., $Range^4$). The intensity of returned light coupled back into the fibre depends on many parameters, most of which cannot be controlled a priori. Examples of parameters affecting the intensity include water turbidity, target colour, target surface smoothness and geometry, etc.   The expected returned light intensity for a given target distance cannot therefore be predetermined. Nevertheless variations in these various parameters will give a family of similar curves of sensor output versus time/distance.   Each of the curves will show a similar discontinuity at the point of contact.

The rate of change of the intensity signal could be used for detection of object proximity as against using intensity alone for detection. Use of rate of change of intensity would overcome one potential problem with the sensor system. Detection would be made independent of the "colour" (i.e., absorption and reflection characteristics) of the target. If the sensor system relied on intensity alone for range measurement it would give different readings as the submersible approached two different targets made of different material. However, if we analyse the rate of change of intensity signal we find that it is independent of the absorption/reflection characteristics of the target (a family of similar curves).

By differentiating the intensity signal and comparing the differentiator output with some threshold level the sensor system could be set up to give a switched output for a given target distance. By sampling and storing the intensity signal, further more complex signal processing possibilities could yield extra information about the target, e.g. target velocity relative to the submersible (useful in the instance of moving targets).

At the point where the whisker hits the target, the sensor will exhibit a detectable discontinuity in the intensity versus time signal. This discontinuity gives us the opportunity to fabricate a very simple micro-switch type device.   This "micro-switch" type device could give a fail-safe sensor output in the event that the more complex signal processed output is not wholly reliable. In this scenario of bump detection, the otherwise smooth intensity curve has a very distinct discontinuity.   For example, in a two-fibre whisker (separate transmit and receive fibers), if the sensor fibre ends are cleanly blocked off by the target the received intensity will instantly fall to near zero. If the surface characteristics of the target are such as not to cleanly block the fibre ends there will nevertheless be a distinct discontinuity in the received intensity signal. These abrupt changes may be used to derive a switched output indicating whisker collision.

Various electronic detector circuits have been built and tested to investigate the most reliable characteristics of the output versus distance trace for distance/proximity measurement. The simplest detector circuit detects the discontinuity at the point of whisker contact.

Whisker length and stiffness can be varied for a given UUV application. A given craft will have a minimum stopping distance and thus whisker length and stiffness will need to be varied to suit a particular craft. If the fibre whisker is flexible and of sufficient length it should allow the sensor end to bump off obstacles while still allowing the submersible craft time to slow down and come to a halt safely. Sensors to the side of the craft, where lateral velocities are much less than forward and reverse motion velocity, would typically be much shorter than sensors forward and aft.

3.2 Loop Sensor Configuration

In the loop sensor configuration the pulsed light is contained within the fibre and not propagated into the surrounding aquatic environment. This has the distinct advantage of making a loop sensor immune to the parameters of the aquatic environment that affect the whisker sensor described in the previous section.

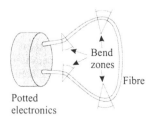

Figure 2. Loop Sensor

The loop sensor operates in the following manner. Light pulses from a light-emitting-diode are transmitted from the launch end of the fibre/wave-guide loop to a photo-detector located at the receiving end. It is well established that optical fibres suffer radiation losses due to bends or curves on their paths due to energy in the evanescent field at the bend exceeding the velocity of light in the cladding [6]. Such radiation losses are experienced in the fibre loop sensor due to the gross deformation or bending of the fibre when it hits an obstacle in the aquatic environment. The photo-detector is capable of detecting the resulting changes in transmitted light intensity within the core.

By experimentation it has been found that gross bending (>70°) over short ~5mm sections of the fibre are necessary for reliable detection. By stiffening or reinforcing the fibre along most of its length, by pre-stressing the fibre to certain bend angles, and by controlling the length of fibre bend zones the sensor can be designed to give the required detectable response over short contact distances. The sensor can also be fabricated by coiling the fibre thus giving many short bend zones.

4. Results & Conclusion

Various configurations of whisker sensors with 1, 2 and 3 fibre (third fibre for ambient light compensation) have been fabricated and tested in the laboratory to date [5]. Figure 3 shows testing output traces versus target distance test results for one of these configurations   Notice the distinct trace discontinuity At the whisker touch point.

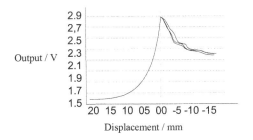

Figure 3. Sample Output versus Target Distance Trace

Various configurations of loop type intrinsic sensors have also been tested. Figure 4 shows one prototype sensor configuration with associated test results. The light source used in this sensor is an Infenion SFH450 infrared surface emitting LED pulsed with 1A, while the detector is an Infineon SFH250 PIN photodiode. The fibre loop is reinforced to make it rigid along length F in the figure while B represents the length of the sensor, which bends or deforms after contact with target/obstacle.

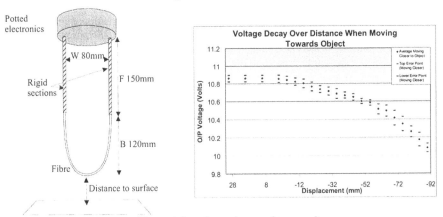

Figure 4. Sample test loop and test results

Various sensor configurations offering good results on bench tests are being reconstructed to reduce size and weight for testing on the mini submersible illustrated in figure 5. Sensor electronics are being sealed in potting compound. Preliminary tests of these sensors on the submersible are ongoing. The sensors are interfaced to the onboard controller and incorporated within the obstacle avoidance behaviours implemented on the autonomous craft. The mini autonomous submersible has undergone testing with sonar sensors and mechanical micro-switch bump sensors in experimentation into reactive and behaviour based control for UUVs [7,8]. Testing of such behaviours employing simple fibre sensors is ongoing. More complex sensor configurations employing advanced signal processing techniques are yet to be tested.

One potential hazard in employing the intrinsic loop sensor is that the loop may snag in the underwater environment. Intrinsic sensors have an advantage over the extrinsic whisker form as the light is contained within the sensor. In the extrinsic whisker light is transmitted from the sensor end, reflected and received back in the fibre. Sensor end fouling, chipping, wear and tear and other problems relating to the sub aqua environment are avoided with the intrinsic sensor which can be protected with covering material. It is planned to test intrinsic whisker sensors rather than loop sensors to avoid the snagging and other problems. To make the whisker intrinsic will require a mirror surface at the whisker end.

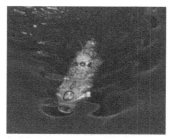

Figure 5. Mini Autonomous Underwater Vehicle
Used for Sensor Experimentation

The successful development of optical fibre proximity sensors should give some distinct advantages over mechanical micro-switch sensors and sonar sensors. Sealing and corrosion problems of switch elements are obviated (fibre sensors have no mechanical moving parts). The sensors are not limited in terms of minimum range or the number, which can be employed in an array as are sonar sensors. Sensor cost will allow many to be deployed on a single craft and sensor cross-talk problems associated with using multiple ultrasonic sensors are much less a problem.

Acknowledgments:

We wish to thank the "Irish Research Council for Science, Engineering and Technology" for the Government of Ireland Post-Doctoral Fellowship funding.

References

[1] Jones D., "The AUV Marketplace" UnderWater Magazine Article: July/August 2002
[2] Arrabales R., Flanagan C. and Toal D., "An Adaptive Video Event Mining System For An Autonomous Underwater Vehicle", Proc., ANNIE 2002, St.Louis.
[3] Everett H.,R., "Sensors for Mobile robots, theory and Application", A K Peters Ltd., 1995, ISBN 1-56881-048-2.
[4] Toal D., Flanagan C., Molnar L., Hanrahan G. "Autonomous Submersible Development for Underwater Filming Employing Adaptive Artificial Intelligence." Sensors and Their Applications XI, September 2001, London. pp 163-168. ISBN 0-7503-0821-4.
[5] Toal D.J.F., Lewis E.,Flanagan C., Nolan S.,Gawley A.N.., Lyons W., "Approaches to opticla fibre based proximity sensing for autonomous underwater vehicles", Proc. 3 Wismarer Automatisierungssymposium, pp 2.2-3, Sept 2002, Wismar.
[6] Wolf, H. F., Handbook of Fibre Optics: Theory and Applications, Gartland STPM Press, New York, 1979.
[7] Nolan S., Toal D., Flanagan C., "Evolving Miniature Robots For The Underwater Environment", Proc. Mechatronics 2002, Twente.
[8] Toal D., Flanagan C., Molnar L., Nolan S., Love T., "Reactive Control In The Design Of An Autonomous Underwater Vehicle" Proc. ANNIE 2002, St Louis.

Poster Programme
Paper presented at Sensors and their Applications XII, September 2003
©*2003 IOP Publishing Ltd*

The use of Microelectrode Sensors in the Investigation and Modelling of Sensing Electrodes in Industrial Electrical Impedance Tomographic Applications

M Gallagher, A McNaughtan, R O Ansell

School of Engineering, Science & Design, Glasgow Caledonian University, Cowcaddens Road, Glasgow G4 OBA.

Abstract: Oxidation, reduction and electrode corrosion processes have been investigated at the surface of a Hastelloy C276 sensing electrode currently being used in an electrical impedance industrial process tomography measurement system. Although degradation and corrosion of the sensor surface was not found to be significant electrode corrosion potentials as large as 330mV were measured. Linear ramp cyclic voltammetry with microelectrode sensors was used to investigate these processes and a novel equivalent circuit model has been developed to aid measurement system optimisation and image reconstruction.

1. Introduction

Over the last decade tomographic imaging has emerged as a valuable technique for interrogating industrial processes to reveal the distribution of materials[1]. Electrical impedance tomography based on resistance (ERT), capacitance (ECT) and inductance (EMT) techniques have been successfully demonstrated and both ERT and ECT instruments are now commercially available. Historically the focus of impedance process tomography has been towards industrial applications involving relatively passive materials. However, as this technology progresses into the chemical production arena, and in particular the agrochemical and pharmaceutical industries, issues associated with electrochemical reaction mechanisms present in chemically active environments must also be considered[2]. Failure to address these issues generally leads to sub-optimal imaging, incorrect signal interpretation, accelerated electrode corrosion and corruption of the chemical matrix. Preliminary investigations by Cilliers et. al[3] identified offset voltages and transient signals associated with electrochemical reactions, which degraded the tomographic signal.

The application of this work is to support the optimisation of an ERT tomographic system currently being used within a pressure filtration vessel as illustrated in figure 1. The filtration system is used in an agrochemical process and incorporates Hastelloy C276 sensing electrodes. Hastelloy C276 is a high temperature corrosion resistant stainless steel alloy which has been found to be very stable in extremely hostile environments. The bulk impedance is measured by injecting a current at one electrode and the potential difference is measured between pairs of electrodes in sequence using an LCR bridge. This work is concerned with the investigation and modelling of the electrochemical processes associated with the sensor electrode interface when used in chemically active media.

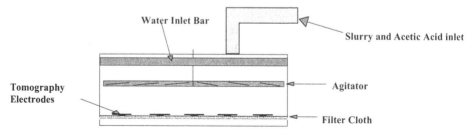

Figure 1. Pressure filtration vessel.

2. Voltammetric investigations

2.1 Experimental Set-up

Linear ramp cyclic voltammetry was used to investigate electrochemical processes associated with corrosion reactions at the electrode surface. A block diagram representation of the voltammetric measurement system is illustrated in figure 2. The voltammetric response, normally represented by a graph of the applied voltage versus electrode current, can provide a valuable insight into a wide range of electrochemical processes occurring at the electrode surface.

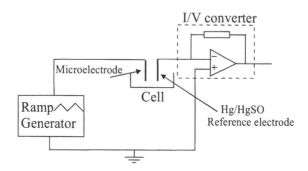

Figure 2. Voltammetric Measurement System.

A unique feature of this work is the use of a 100μm diameter Hastelloy microelectrode to investigate these corrosion processes. Hastelloy C276 is known for its corrosion resistant properties which makes this type of investigation very time consuming when using large electrodes. However, the higher current densities associated with a microelectrode[4] allow these processes to be investigated in a very short time scale. The microelectrode was constructed by sealing a 100μm diameter Hastelloy wire in a glass capillary tube using a micro-flame butane burner. The end of the capillary was then polishing to expose the cross-section of the wire. A custom built current to voltage converter[5] was used to measure the small currents associated with the microelectrode. A mercury/mercury sulphate reference electrode was used to avoid contamination by the conventional saturated calomel reference electrode since chloride can have a significant impact on the corrosion processes associated with stainless steels[6]. The analyte used in this investigation was acetic acid which was a major constituent in the filtration process.

2.2 Results

The cyclic voltammogram obtained using a 100μm diameter Hastelloy microelectrode in 1M acetic acid solution is shown in figure 3. The key feature on the voltammogram is the plateau region between –300mV and +350mV which is known as the passive region where corrosion of the electrode is at a minimum. At applied potentials outside this range corrosion of the electrode will be greatly accelerated resulting in a reduction in the signal to noise ratio and the possibility of product contamination. There is no evidence of redox processes associated with breakdown of the analyte in this application.

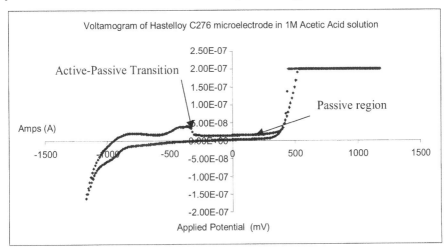

Figure 3. Voltammetric Response. Solution 1M acetic acid, Scan range –1.2V to +1.2V, scan rate 10mVs⁻¹, working electrode 100μm diameter Hastelloy, reference electrode mercury/mercury sulphate.

3. Investigation of Background dc Potentials

3.1 Experimental Set-up

Background dc potentials previously reported by Cilliers et. al.[3] , which can degrade the tomographic signal, were investigated using a concentration gradient cell and measurement system as illustrated in figure 4.   These voltages, normally associated with dissimilar electrode materials, can arise in this application due to the non-homogeneous chemical matrix surrounding the electrode during filling, washing and mixing phases of the filtration process.   The measurement system uses a high impedance buffer amplifier to ensure no current flows during the measurement process.

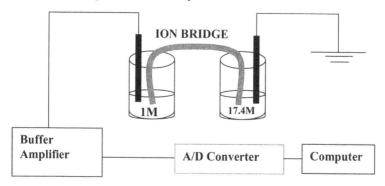

Figure 4. Concentration gradient cell and measurement system.

3.2 Results

Figure 5 shows the potential generated between two Hastelloy electrodes exposed to different concentrations of acetic acid and is of the order of 300mV. Potentials associated with the ion bridge were not investigated in this work. The magnitude of these potentials are generally in the mV range and were considered not to be very significant due to the magnitude of the potentials observed. Independent measurements on the filtration vessel without an ion bridge also showed potentials within the range of 300mV to 500mV and further supports the view. These dc potential sources can occur between any of the 24 electrodes within the vessel and can subsequently result in additional currents at the ERT sensing electrodes.

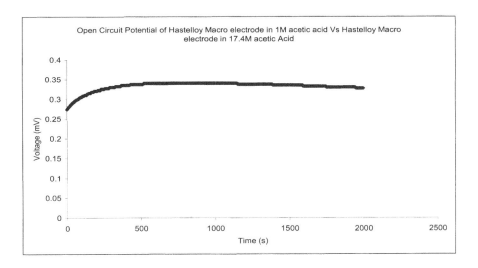

Figure 5. Open circuit potential between two Hastelloy electrodes.

4. Equivalent circuit model

An electrochemical cell consisting of two electrodes immersed in an electrolyte has traditionally been represented by the Randles equivalent circuit model shown in figure 6. The bulk impedance which is of interest in the ERT measurement is represented by Z_{bulk}. The electrode solution interface is represented by Z_{ct} and Z_{diff} in parallel with the double layer capacitance C_{dl}. The double layer capacitance can provide a low impedance path across the electrode/solution interface for the ac tomographic signal.

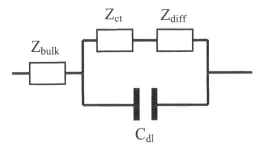

Figure 6. Randles Equivalent circuit.

A novel equivalent circuit model has been developed as shown in figure7. The new model includes an additional current source which arises from the corrosion processes within the vessel. The diffusion impedance Z_{diff} is represented by a Warburg impedance at a macroelectrode. Future work will involve the integration of this model within the Tomographic reconstruction algorithms.

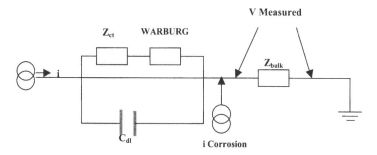

Figure 7. The modified equivalent circuit model for a macroelectrode.

5. Conclusions

The novel use of microelectrodes in the investigation of corrosion processes at highly corrosion resistance stainless steel electrodes has been demonstrated. The information gained from technique can been used to identify the optimum excitation potentials for an electrical resistance tomographic measurement system in order to minimise corrosion of the sensing electrodes, minimise breakdown or contamination of the product and otimise the signal to noise ratio of the tomographic signal. The source of dc background potentials previously reported by other workers has been identified and a new equivalent circuit model has been developed. This model will be used in future work to optimise the tomographic signals and image reconstruction algorithms.

References

[1]   Proceedings of the 1st World Congress on Industrial Process Tomography, April 1999, Buxton, UK.
[2]   A. McNaughtan, K. Meney, B. Grieve, Chemical Engineering Journal, vol. 77, (2002), 27-30.
[3]   J J Cilliers, W Xie, S J Neethling, E W Randall, A J Wilkinson, Meas. Sci. Technol. 12 (2001) 997–1001
[4]   A. M. Bond, Analyst, 119, (1994), R1-R21
[5]   A. McNaughtan, R. O. Ansell, J. R. Pugh, Meas. Sci. Technol., 5 (1994), 789-792.
[6]   Ph. Botella, C. Frayet, Th. Jaszay, M. H. Delville, J. of Supercritical Fluids, 00 (2002), 1-10.

Poster Programme
Paper presented at Sensors and their Applications XII, September 2003
©2003 IOP Publishing Ltd

A simple process for the rapid prototyping of elastomer MEM structures.

Thomas J. Russell, Vincent Casey

Physics Department, University of Limerick, Limerick, Ireland

E-mail: vincent.casey@ul.ie

Abstract. Photolithography is expensive, slow and not very suitable for the production of 3D MEM structures. Conversely, it is claimed that softlithography is relatively inexpensive, accessible and ideally suited to the formation of 3D structures on both planar and non-planar supports at both micro- and nanoscale. In this paper we describe a simple process which may be used to generate the elastomeric stamps or moulds which are key to the softlithography technique. Details are presented on artwork design, photomask printing, preparation of photosensitive PDMS, exposure and development. Tests structures with reliable minimum feature size of the order of $120\mu m$ were developed.

1. Introduction

One of the major limitations of micromachining technology from the point of view of sensor technology, particularly in the chemical and biological/biomedical fields, is the long prototyping time taken to implement new designs or to modify existing designs. Other limitations include cost (capital and operating costs are high), limited 3D ability, restricted materials base and confinement to planar surfaces. In the mid nineties the Whitesides group [1] at Harvard University developed a new approach to micro and indeed nanofabrication of 3D structures known as softlithography since it involves the use of a soft patterned elastomeric stamp. The stamp has the required relief features patterned on its surface. It is brought into conformal contact with the substrate to be patterned and transfer of the features from the stamp to the substrate is achieved using techniques such as microcontact printing[2, 3], replica moulding[4] and embossing[5]. Each of these techniques is procedurally simple and inexpensive. In addition they are applicable to a wide range of materials, can be used with non-planar surfaces, are applicable to manufacturing/low cost mass production, can be used to pattern large areas and can be carried out in an ambient lab environment.

Rapid Prototyping.

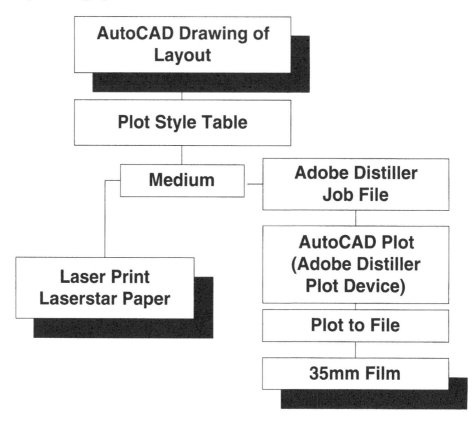

Figure 1. Artwork Generation

2. Experimental

2.1. Artwork Generation

The artwork generation process involves creating accurate 2D features that can be transferred to a physical medium suitable for use as a photomask. A wide range of drawing packages is available which will accomplish this task. AutoCAD was used in this work as it allows complete control of linewidths and also facilitates a wide range of file output formats. Accurate control over linewidth is achieved in AutoCAD by choice of parameters set in a `plot-style-table` which is used in conjunction with the `Plot` option. The minimum linewidth value achievable in this way is 1nm. However, the ultimate plottable linewidth is set by the minimum printable width of the plot/print device, figure(1). Unfortunately, AutoCAD produces files in vector format and although it has the facility to convert the vector file to bitmap and postscript format the parameters which may be controlled are limited and resolution may be impaired. Adobe Distiller is a high-resolution file plotting software package which may be incorporated

Rapid Prototyping.

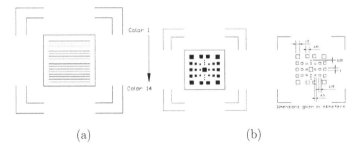

(a)   (b)

Figure 2. Test patterns, (a) Linewidths ranging from $1\mu m$ to $280\mu m$, (b) Solid squares.

into AutoCAD as a plot device. Similar to a postscript printing device the Adobe Distiller interprets postscript code and creates a PDF file according to the parameters that are setup in a control file called `job-option`. By editing this `job-option`-file control over the resolution (up to 4000dpi), page setup, compression and conversion processes can be achieved. A customized work file can then be created which provides the optimum output for high-resolution artwork generation/printing. Alternatively, bitmap files suitable for 35mm slide film printing may be generated.

2.2. Mask Production

Direct laser printing of artwork onto transparent sheet (acetate or special paper) and photographic printing are the main low cost alternatives to quartz photomask production. For laser printed masks the resolution of the laser printer is the critical parameter although print medium is also important. Features as small as $20\mu m$ can be achieved using a 3200dpi laser printer. The common office laser printer with a resolution of 1200dpi can resolve feature sizes of the order of 100μ m. However, laser-printed artwork on transparent acetate sheet is prone to pinhole defects and lack of definition in the feature formed. Better definition and contrast are claimed for masks generated by direct printing from file to photographic film. A comparative evaluation of both approaches relative to quartz photomasks is given in ref[6] and ref[7]. In this study, a 1200dpi laser printer was used to print a test layout pattern onto A4 Laserstar film, which is used in high resolution pcb production. PDF files of the layout were printed onto Liptip 35mm film (Repro35 Ltd.). Slide mounts were used in both cases to support the masks produced.

2.3. Photolithography

An improvised resist spinner was made by replacing the drive on a household blender with a disc shaped chuck machined from PTFE (Teflon). A variac was used to adjust the mains drive voltage in order to control the revolution speed. The spinner was calibrated using an optical tachometer. The spinner speed could be controlled continuously in the

Rapid Prototyping.

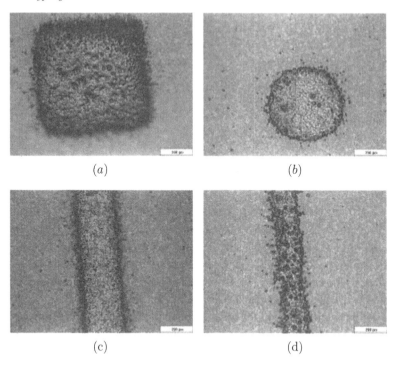

(a) (b)

(c) (d)

Figure 3. Test pattern detail obtained using 1200dpi laser printing onto Laserstar film, (a) 1mm square, (b) 0.5mm diameter disc, (c) 200μm line, (d) 120μm line. The scale bar corresponds to 100μm

range $1000 - 3000rpm$. The stamp polymer, polydimethylsiloxone (PDMS)[8] available commercially as Sylgard 184 from Dow Corning, was rendered photosensitive by the addition of the photo-initiator dimethoxyphenylacetophenone (DMAP). DMAP was first mixed with xylene and then added to PDMS and left to stand overnight after thorough mixing. A minimum of 0.4 g of DMAP in 10g PDMS, i.e. 4% by weight, was required in order to get satisfactory photoinitiation. A primer solution consisting of $2cm^3$ TMSM in $18cm^3$ of toluene to which one drop of de-ionized water had been added was used to prime silicon substrates cut from $100mm$ wafers which had been wet oxidized by placing in concentrated H_2SO_4 prior to resist spinning. A drop of the PDMS/DMAP mixture was placed on the silicon substrate which was spun at 1200 rpm for 20 seconds. The mask was placed in direct contact with the PDMS coated substrate in order to prevent air contact with the PDMS. UV exposure of the photoresist was carried out using a mercury UV light source. The PDMS was developed by placing the substrate in xylene for 30s, rinsing with IPA followed by drying in nitrogen.

Rapid Prototyping.

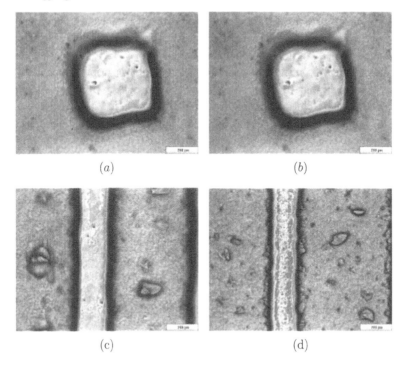

(a) (b)

(c) (d)

Figure 4. PDMS elastomer stamp elements obtained using Laserstar mask, (a) 0.5mm square, (b) 0.5mm diameter disc, (c) 140μm line, (d) 120μm line. The scale bar corresponds to 100μm

3. Results and Discussion

It was found that extremely long exposure times (24-48 hours) were necessary in order to get satisfactory patterning in the developed PDMS film using the laser printed Laserstar masks. There was no evidence of exposure for the 35mm film masks even after 48 hours. UV-visible transmission spectra of the various mask media were obtained using a Cary IE spectrometer in the spectral region of interest, i.e. around 365nm. The Laserstar paper was found to be only 7.2% transmitting at this wavelength while the 35mm film was as low as 2%. Polyester film (Melinex) was found, on the other hand, to be 82% transmitting. The particularly poor transparency of the Laserstar and 35mm film to UV light at 365nm accounts for the difficulties experienced with exposure using these mask media.

It can be seen, figure(3) that the features formed have good definition and contrast (printed regions are well defined) and are good representations of the features drawn in AutoCAD. Both the 200μm and 120μm lines, figure(3c,d), show good edge definition and contrast between the printed region and acetate. However, pinholes and poor edge definition occurred for the 100μm lines.

Rapid Prototyping.

The stamp features formed from the line width test pattern are shown in figure(4). These structures represent ribs that are formed in the polymer. Shown in figure(4d) is the minimum usable rib width produced using this approach.

Structures formed on the Laserstar by the 1200dpi laser printer appear to correlate well with the dimensions of the artwork drawn in AutoCAD. Structures showed good contrast and edge resolution for feature sizes greater than $120\mu m$. Structures with dimensions below this value deteriorated due to an increase in pinhole density and a decrease in edge definition largely as a result of the limited resolution of the print device.

4. Conclusions

Generation of the artwork using AutoCAD2000 proved to be successful. Control over feature sizes in the micron region was easily achieved using the plot styles function. The generation of a high-resolution output file containing the required artwork was also possible by setting up Adobe Distiller as a plot device in AutoCAD2000. The major restriction on minimum feature size is imposed by the print device/process used. Liptip slide film absorbs heavily in the UV region critical to DMAP photoinitiation and so this film type is unsuitable for photomasks. Patterns formed using a 1200dpi laser printer on Laserstar paper gave satisfactory stamp features sizes down to $120\mu m$. However, because of the low UV transmission of the Laserstar paper, 7%, long exposure times are required. It should be possible to reduce exposure times significantly using polyester (Melinex) film masks and a more intense UV source. PDMS prepolymer photosensitized with DMAP can be directly patterned as a softlithography stamp.

References

[1] Y Xia and G M Whitesides, Soft Lithography, Annual Reviews Materials Research, 28 (1998) 153–184.

[2] R Hull, T Chraska. Y Liu and D Longo, Microcontact printing: new mastering and transfer techniques for high throughput, resolution and depth of focus, Materials Science and Engineering, C19 (2002) 383-392.

[3] Xiao-Mei Zhao, S P Smith, M Prentiss and G M Whitesides, Fabrication of waveguide couplers using microtransfer molding, CME6, CLEO'97.

[4] L Malaquin, F Carcenac, C Vieu and M Mauzac, Using polydimethylsiloxane as a thermocurable resist for a soft imprint lithography process, Microelectronic Engineering, 61–62 (2002) 379-384.

[5] F Arias, S R J Oliver, B Xu, R E holmlin and G M Whitesides, Fabrication of metallic heat exchangers using sacrificial polymer mandrils, J Microelectromechanical Systems, 10 (2001) 107–112.

[6] P S Gwozdz, IEEE Trans Edu 39 (1996) 211–

[7] A M Christenson, B H Augustine, Rapid prototyping of masks from various 35mm film types for use in photolithography, Proc Nat Conf Undergrad Res (NCUR) 2000, University of Montana, April 26-29, 2000.

[8] J C Lotters, W Olthuis, P H Veltink and P Bergveld, The rubberelastic polymer polydimethylsiloxane applied as spring material in micromechanical sensors, Proceedings of MST'96, Potsdam. Germany, September 1996, pp. 73–78.

Electrostatic Monitoring of AEROSPACE BEARINGS

I Care, R J K Wood[1], T J Harvey[1], S Morris[1], L Wang[1], H E G Powrie[2]

Rolls-Royce plc, ML-77, PO BOX 31,Derby, DE24 8BJ, UK
[1] School of Engineering Sciences, University of Southampton, Southampton, SO17 1BJ, UK,
[2] Smiths Aerospace Electronic Systems - Southampton, School Lane, Chandlers Ford, Hampshire SO53 4YG, UK

Abstract: Previous work has shown that electrostatic charge signals can be used to detect the onset of wear in lubricated tribocontacts. Preliminary investigations have shown the viability of this system when tested on laboratory-based equipment [1-4]. These preliminary experiments have investigated steel to steel contacts and ceramic (silicon nitride) to steel contacts with a variety of aerospace lubricants [5-8]. Several charging mechanisms could be involved, namely tribocharging, surface charge variations, debris generation and exo-emissions, which have been monitored using an electrostatic sensor.

This paper details the studies looking at using the signals from the electrostatic sensor to detect the onset of bearing deterioration (wear). Charge levels under fully lubricated condition will be detailed and discussed in relation to wear rates. Interpretation of the results will be with reference to contact potential differences induced by the presence of incomplete additive films, phase transformations, oxide formation/removal and other wear processes associated with boundary/mixed lubrication such as ceramic spallation and debris formation. We conclude with the potential use of the electrostatic sensor as a monitor for bearings with non-ferrous components that cannot be detected by the usual aerospace magnetic particle detectors.

Keywords: Charge, wear, electrostatic, ceramic, bearing steel, lubrication.

1. Introduction

The use of lightweight ceramic rolling elements in gas turbine engine bearings gives the advantages of high hardness, high temperature operation, small centrifugal loads and smaller gyroscopic moments. This enables bearings to support a heavier load with a higher speed of 2 to 4 million DN (bore diameter in mm × shaft rotational speed in rpm) [9-10]. In engine terms this helps to keep the bearing compartment smaller and gas path efficiency higher by utilising higher shaft speeds. The use of ceramic elements is not always favourable, their main benefit arises in low or boundary lubrication, high rotational speeds, and areas where loads vary such that there are times of high slip. Under normal running the ceramic elements give lower heat generation compared with the equivalent steel on steel [11]. This type of bearing is often referred to as a hybrid bearing.

Changing an existing bearing design to one with ceramic rolling elements requires minimal changes to the surrounding architecture to achieve a maximum of benefits. On Gas Turbine Engines the typical way to detect the onset of wear is to use a magnetic particle collector or analyser. This will not work for ceramic elements, so another method has to be found. Ideally a system is needed that discriminates between non-metallic debris and debris ingested by the engine that finds it's way into the secondary air system (such as sand). Thus, new condition monitoring techniques have to be developed for hybrid bearings that can be incorporated into engine health management systems. One such technique uses electrostatics to detect ceramic wear [10-12].

The gas turbine lubricant is required to cool the oil lubricated components and in doing so transport any debris away that may enter the lubrication system (e.g. through air seals). The lubricant chemistry has been designed over the years to work with steel, magnesium, aluminium and titanium metal components. The chemistry of the lubricant has to be compatible with the new bearing materials [7,8,12], and also work with any distress or wear sensing methods that are used.

Electrostatic Sensing The electrostatic sensing technology is based on the assumption that a change of the static charge levels in the system reflects certain changes in conditions of the dynamic system. The technology and charge theory has been overviewed in [11].

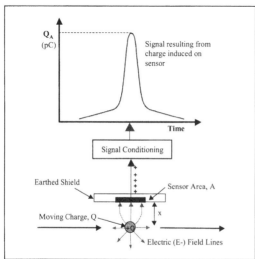

Figure 1 includes a schematic of the electrostatic charge sensing system to explain its operation. The electrostatic monitoring system comprises a passive sensor connected to signal conditioning (a charge amplifier) from which voltage signals may be recorded and processed. If electric (E-) field lines due to charge Q passing under the sensor terminate on the sensor face, the electrons in the sensor redistribute to balance the additional charge in the vicinity of the sensor and hence the presence of the charge is measurable. The charge Q is shown as positive in Figure 1. The signal conditioning converts the detected charge into a proportional voltage signal, which is collected and processed. When a surface charge or charged particle moves across a sensor face, an equal and opposite electrostatic charge is induced on the sensor face. The magnitude of induced charge is proportional to the flux of field lines terminating on sensor face.

Research aims. We have looked at the electrostatic response for steel on steel contacts and for ceramic on steel [6,7]. As well as bench tests under pure sliding conditions it has also looked at bearing rig tests and full engine tests [5]. The purpose was to determine whether the electrostatic sensor was suitable for all types of bearing and whether it could detect signs of wear at the same time or before conventional instrumentation. Detecting bearing overload distress was also an aim.

2. Experimental

Pin-on-disc tribometer tests. A lubricated test ball was loaded against a bearing steel disc under pure sliding conditions on a Pin-on-Disc (PoD) tribometer.

Figure 2 shows the schematic of the rig used. To enable an extensive investigation of the wear as well as the charge generation, additional instrumentation was installed on the pin-on-disc including:

a. A tachometer to monitor the disc rotation speed;

b. An electrostatic sensor (ESP) to monitor the charge level on the wear track (see Figure 9);
c. A strain gauge to monitor the contact friction;
d. An LVDT (Linear-Variable-Differential-Transformer) to monitor the vertical displacement of the pin towards the disc from which the linear pin and disc wear is calculated;
e. An infrared thermometer to monitor the disc surface temperature (~10mm from contact point).

The sliding speed was kept at 7m/s. Experiments start with no load. After the disc surface is fully lubricated by an oil spray, an initial running-in load of 10N is applied for 5 minutes. The load is increased in regular increments, with 5 minutes between each increase to allow additional running-in at the new load. Maximum load applied was 190N, which produced a contact pressure of 4.15 GPa. The initial μ value was 2.8, i.e. within the EHL region. Tests were run until either scuffing occurred or for 2.5 hours. Scuffing is recognised when a rapid rise in contact friction occurs, together with severe damage to the rubbing surfaces.

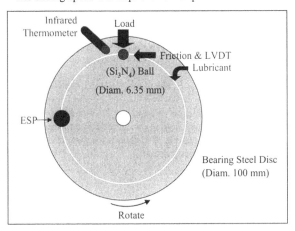

Figure 2. Schematic of Pin-on-Disc rig with associated instrumentation

Post-test wear scars and wear tracks were measured by 3-D laser profilometry. The specific wear rates were then calculated and compared. The primary correlation between the electrostatic sensor responses and the material wear rate were observed by on-line monitoring. Wear mechanisms were investigated through FEG-SEM, EDX, and optical microscopy.

Figure 3. Bearing rig system schematic   *Figure 4. Instrumentation schematic*

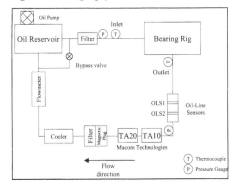
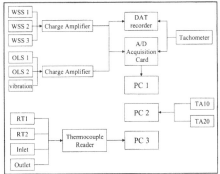

Bearing Rig Tests

The bearing test rig consists of two test chambers (one of which has been commissioned) a drive motor, hydraulic loading system and oil recirculation system. The test chamber employs four taper-roller bearings, two support and two test bearings. The test chamber has been modified to accommodate three electrostatic wear-site sensors (WSS) in close proximity to the bearings (see Figure 8). Two sensors monitor the two support bearings (bearings #1 and #4); and the third (WSS-2) monitors both of the test bearings (#2 and #3). An electrostatic oil-line sensor (OLS) pair (see Figure 10) has been installed in the lubrication recirculation to monitor any debris produced during testing.

In addition to the electrostatic monitoring of both the wear-site and oil-line, vibration monitoring has been employed to corroborate detection of bearing deterioration. In the lubrication scavenge line there are two on-line Macom Technology debris monitoring devices, which use eddy current technology and ferromagnetism to sense debris entrained in the lubricant.
The vibration sensor is mounted on the housing containing the #1 (support) bearing.

Test conditions and methods. Bearings are not designed to fail, but in order to have material for the sensors to detect, an accelerated test was required. After initial test running, commissioning and calibration of the various systems, a defect was indented onto the number 2 test bearing and the bearings loaded to 20 kN (200% dynamic loading). Lubricant flow was maintained at 4 litre min^{-1}.
The bearing components were weighed before and after testing. Previous testing at 15kN for 48 hours showed that accelerated failure initiated from the artificial defect.

3. Results and Discussion
Pin-on-disc rig. The electrostatic responses together with the linear wear of both pin and disc versus time from a typical test is shown in Figure 5. Only the results between 140 minutes and 150 minutes are shown here to have a closer look at the area when scuffing occurred. Before 143.7 minutes, steady conditions are seen. At 143.7 minutes, the wear rate increased, and at the same time, the electrostatic sensor gives intermittent high levels of charge response. Under steady state conditions, the rms charge level from the electrostatic sensor was 0.07pC while it gives bursts over 4 pC after scuffing occurred.

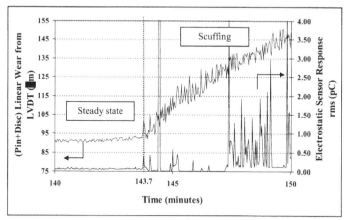

*Figure 5. Electrostatic sensor detects the wear of lubricated hybrid contact:
Si$_3$N$_4$ on M50NiL, used oil sample UO1*

The relatively low friction coefficients (μ~0.06-0.08) indicates that an adequate load-bearing film has been formed. This film helps to redistribute the contact stresses and aids the lubrication process. Within this time period, the surface temperature increase steadily from 80°C to about 105°C before the contact breaks down and a rapid increase in temperature is seen. This also confirms that an anti-wear film is formed and effective which is correlated by the low wear rate and controlled friction over this period. Scuffing occurred (see Figure 5) at 143.7 minutes of this test and the friction and surface temperature increased dramatically.

Bearing Test. In the test data shown in Figure 6 there are two periods of interest: about test time 50 hours and prior to failure at 68 hours (expanded data 60-72 hours is shown in Figure 7).

At test time 48hr, a period of bearing wear occurs indicated by the TA20 and electrostatic WSS-2 data. The period of activity occurs over a 2-3 hour period, but appears not to be detrimental at this stage, to the bearing life.

The failure at 68 hours leads to seizure of the bearing rig and stalling of the motor. The experimental data indicates bearing distress up to four hours (test time 64 hours) prior to failure. The oil-line sensors (TA10, TA20 and electrostatic OLS) detect debris production increasing from 64 hours, with the TA20 reaching its' upper limit 40 minutes before seizure. From the wear-site sensor and vibration sensor there is some indication of a change around test time 62 hours, however this is not appreciably above background levels.

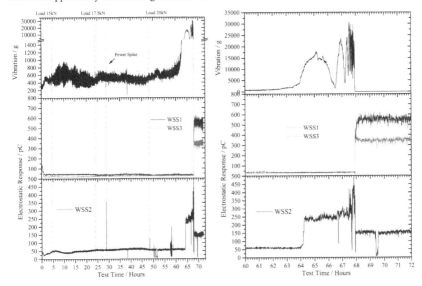

Figures 6 & 7 Vibration and Electrostatic wear site sensor data from bearing test rig analysed by inner race defect frequency normalisation

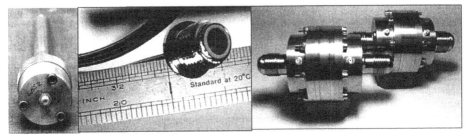

Figures 8 & 9. Wear Site Sensors (WSS)   *Figure10. Pair of oil-line sensors (OLS).*

Electrostatic sensing. The possible charging mechanisms responsible for this dramatic increase in charge are not well defined, but are linked to the sudden increase in the linear wear rate. Previous work [7, 13] has shown the likely dominant mechanisms for hybrid contacts is associated with the generation of large quantities of charged material. Secondary mechanisms such as tribo-emissions associated with crack formation and growth within the steel wear track and contact potential differences formed at phase transformed areas (white layers) of the wear track may also contribute to charge levels. The on-line electrostatic sensor shows it can detect particles in the lubricant, and using conventional software can trend and alarm the information gained from the sensor.

The disadvantage of the electrostatic sensors is that they need to be monitoring in real time, whereas a system that uses a magnet to 'catch' data can do periodic inspections / measurements on the 'catch'.

4. Conclusion

For pure sliding wear:
- A correlation between the charge level and the specific wear rate can been seen
- A correlation between the wear of silicon nitride and bearing steel under oil lubricated contacts and electrostatic charge levels has been found
- A correlation between the wear of bearing steel on steel under oil lubricated contacts and electrostatic charge levels has been found
- Electrostatic charge levels increased with the onset of scuffing and measurable electrostatic charge levels were recorded even in the presence of highly conductive used oils.
- Incomplete additive films in used oil give rise to higher levels of wear and increased electrostatic charge levels

From the bearing tests:
- All monitoring techniques employed detected bearing deterioration prior to seizure.
- Oil-line electrostatic sensing detected the increased production of debris 4 hours prior to seizure.
- Wear-site sensing detected bearing deterioration 4 hours prior to seizure, from electrostatic wear-site sensor 2 and the vibration sensor.

Electrostatic sensing is a viable method of detecting wear and distress in a continuously lubricated rolling element bearing system.

Electrostatic sensors can be made that work in the temperature and pressure environments of a gas turbine bearing chamber and can withstand aviation lubricants.

5. Further work

A full-scale hybrid bearing test is underway with one of the partners in Germany and another with a partner in Italy. A full gas turbine engine test with hybrid bearings is being prepared.

Electrostatic sensors will probably work for grease packed components (such as railway bearings) but have not yet been tested on such arrangements.

Acknowledgements

This research was supported by the European Commission within the GROWTH Programme, Research Project 'Advanced Transmission and Oil System Concepts (ATOS)', contract G4RD-CT-2000-00391, and by the U.S. Office of Naval Research (ONR), grant numbers: N00014-00-1-0437 (UoS) and N00014-00-C-0248 (Smiths). This financial support is gratefully acknowledged. Acknowledgements also go to the project partners: Rolls-Royce Derby, FAG and Hispano-Suiza for supplying the test materials, and FAG for their help with SEM scan information on silicon nitride. Smiths Aerospace Electronic Systems - Southampton for supplying the electrostatic sensors, parts for the bearings test rig, and for their help with analysing and interpreting the electrostatic sensor data. Timken Company for the supply of test bearings and valuable support on the bearings testing.

References

[1] O.D.Tasbaz, H.E.G.Powrie and R.J.K.Wood, Electrostatic Monitoring of Oil lubricated contacts for early detection of wear, *International Conference on Condition Monitoring, 12-15th April 1999.*

[2] R.J.K. Wood, M. Browne and M.T. Thew, Electrostatic charging precursor to scuffing in lubricated contacts, *IMechE World Tribology Congress, 8-12th Sept 1997, London.*

[3] O.D. Tasbaz, R.J.K. Wood, M. Browne, H.E.G. Powrie and G. Denuault, "Electrostatic monitoring of oil lubricated sliding point contacts for early detection of scuffing", *Wear 230/1 (1999) 86-97.*

[4] H.E.G.Powrie, C.E.Fisher, O.D.Tasbaz and R.J.K.Wood, Performance of an Electrostatic Oil Monitoring system during an FZG gear scuffing test, *International Conference on Condition Monitoring, University of Wales Swansea, 12-15th April 1999.*

[5] H.E.G. Powrie, R.J.K. Wood, T.J. Harvey, L. Wang & S. Morris, Electrostatic Charge Generation Associated with Machinery Component Deterioration, *2002 IEEE Aerospace Conference Proceedings, ISBN 0-7803-7232-8*

[6] L. Wang, R.J.K. Wood, T.J. Harvey, S. Morris, H.E.G. Powrie and I. Care: Feasibility of using electrostatic monitoring for oil lubricated ceramic to steel sliding contacts, *29th Leeds-Lyon Symposium on Tribology, Leeds, Sept 2002*

[7] L. Wang, R.J.K. Wood, T.J. Harvey, S. Morris, I. Care and H.E.G. Powrie: Wear performance of oil lubricated silicon nitride sliding against various bearing steels; *The 14th International Conference on Wear of Materials, Washington DC, USA, March 30th – April 3rd 2003*

[8] L. Wang, R.J.K. Wood, T.J. Harvey, S. Morris, I. Care and H.E.G. Powrie: Performance evaluation of advanced aircraft engine oils for ceramic on steel contacts; *STLE/ASME International Tribology Conference, Florida, USA, October 26-29, 2003*

[9] K. Fujii, Y. Fujii, M. Mori and S. Yamamoto: Cylindrical Roller bearing under 4M DN and 300°C using advanced materials; *2nd world tribology congress, Vienna, Sept 2 - 7, 2001*

[10] H. Yui, S. Aihara, S. Yamamoto and M. Yamazoe: The performance of Hybrid ball bearings under high temperature and high speed conditions; *Proceedings of the international Gas Turbine Congress 1999, Kobe, 14-19 November 1999*

[11] J.M. Reddecliffe and R. Valori: The performance of a high speed ball thrust bearing using Silicon Nitride balls; *ASME journal of lubrication Tech. No. 98 (1976) pp553-563*

[12] E.E. Klaus, J.L. Duda, W.T. Wu, Lubricated wear of silicon nitride, Journal of the Society of Tribologists and Lubrication Engineers, 47(8), August, 1991, pp679-684

[13] S. Morris, R.J.K. Wood, T.J. Harvey, & H.E.G. Powrie, Use of electrostatic charge monitoring for early detection of adhesive wear in oil lubricated contacts, Transactions of the ASME: Journal of Tribology, 124 (2002), 288-296

[14] D.M. Taylor and P.E. Specker, Industrial Electrostatics: Fundamentals and Measurements, 1994, Research Studies Press, Taunton, UK, ISBN 0 86380158 7

[15] P. Hammond, Electromagnetism for Engineers - An introductory Course, 1978, Pergamon Press Ltd., Headington Hill Hall, Oxford, UK, ISBN 0 08022104 1

Poster Programme
Paper presented at Sensors and their Applications XII, September 2003
©*2003 IOP Publishing Ltd*

Measurement of oxygen concentrations at ppm levels using solid-state sensors

W C Maskell[1], J A Page[2] and D Tzempelikos

Middlesex University, Bounds Green Road, London N11 2NQ, UK

Abstract: Amperometric zirconia oxygen sensors are normally designed for use in the percentage oxygen range. There are many applications requiring a sensor to operate in the parts per million (ppm) concentration range. In this work zirconia sensors were constructed with diffusion barriers of 0.5-2 mm diameter and tested in nitrogen dosed with small amounts of oxygen. The sensors worked well and showed good performance down to 20 ppm oxygen.

1. Introduction

Zirconia oxygen sensors can be constructed to measure concentrations in 1-10% oxygen and higher [1, 2]. Currents that can be drawn from the pump electrodes are limited mainly by the electrode kinetics but also by the conductivity of the solid electrolyte. In order to operate in the percentage oxygen range the diffusion barrier must be sufficiently restrictive of oxygen diffusion because the current output of the sensor is directly proportional to the rate of transport of oxygen through the barrier. This can be achieved by using a diffusion hole of typically 30-100 μm diameter. However, for applications requiring measurements in the ppm range such diffusion barriers result in extremely low currents which may be similar or smaller than background values. In order to investigate the application of these sensors at ppm oxygen concentrations, sensors were constructed with diffusion barriers with diameters in the range 0.5-2 mm.

2. Experimental

Sensors were constructed similarly to those described previously using zirconia-plastic materials [3]. The diffusion barrier was engineered by pre-drilling the zirconia disc onto which electrodes were not applied, prior to assembly of the sensor. Drilling was accomplished using conventional steel twist drills of 0.5-2 mm diameter. The sensors were fired at 1450°C to burn off the plastic and sinter the ceramic.

---

[1] Now of Sensox Ltd at the address shown above, w.maskell@mdx.ac.uk
[2] Present address: Flight Refuelling Ltd, Brook Road, Wimborne, Dorset BH21 2BJ, PageJA@flight-ref.com

Sensors were tested by inserting the silica tube shown in Fig. 1 into a furnace, with the water-cooled end plate protruding. Gas was introduced continuously into the silica tube through the end plate and it flowed out via a bubbler. This allowed the atmosphere surrounding the sensor to be adjusted and the sensor temperature to be controlled.

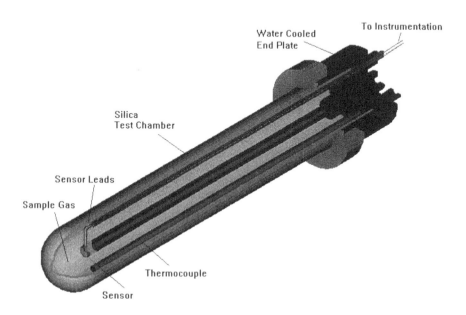

Fig. 1 Sensor test rig

Nitrogen gas (oxygen-free) was dosed with oxygen in two ways: (i) Mass flow valves were used (maximum 200 and 5 ml per minute for nitrogen and air respectively) to control the flow rates and set the oxygen concentration of the mixture. This method was suitable for generating mixtures down to 250 ppm. (ii) Oxygen was introduced electrochemically into a nitrogen flow using a zirconia tube with platinum electrodes inside an out and pumping in oxygen. This is a very precise method; the rate of pumping of oxygen is proportional to the current flowing and is given by Faraday's law. The resulting oxygen concentration can be calculated from the applied current and the nitrogen flow rate. This enabled concentrations down to 10 ppm or lower to be generated.

3. Results and Discussion

Current-voltage curves are shown in Fig. 2 for various hole sizes and oxygen concentrations. Curves are typical of those for amperometric sensors; as the voltage is increased from zero, the current rises and then comes to a limiting value as the concentration of oxygen in the internal volume approaches zero. At these concentrations the limiting current is expected to be directly proportional to the oxygen concentration and to the ratio S/L where S is the cross-sectional area of the hole and L is its length. In the case of these sensors L was the thickness of the sintered zirconia disc.

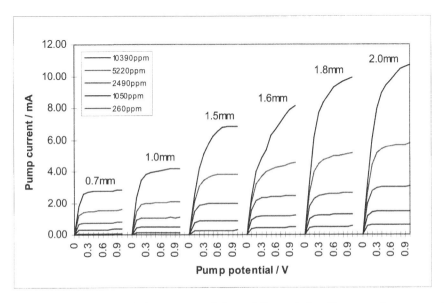

Fig. 2 Characteristics of the sensors with large diffusion holes (diameters shown as cut in unfired material). Each curve was recorded at the oxygen concentration shown in the box, decreasing continuously from top to bottom. Sensors were operated at 700 °C.

Limiting currents are plotted versus oxygen concentration in Fig. 3. Linearity, predicted from theory, is good. Clearly the sensor worked well in the target oxygen concentration range.

It is readily shown by invoking Fick's law, and assuming that diffusion is controlled solely by the diffusion hole without end effects, that the slope of the lines in Fig. 3 should be proportional to the ratio S/L. This is tested in Fig. 4. The deviation from linearity indicated on the points may be a consequence of the high aspect ratio of the holes (diameter greater than or comparable with the length) causing some deviation from the ideality assumed.

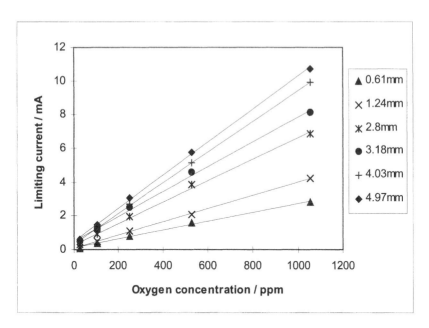

Fig. 3 Limiting current of the sensors versus oxygen concentration. The S/ L values of the diffusion holes are shown in the box. Sensors were operated at 700 °C.

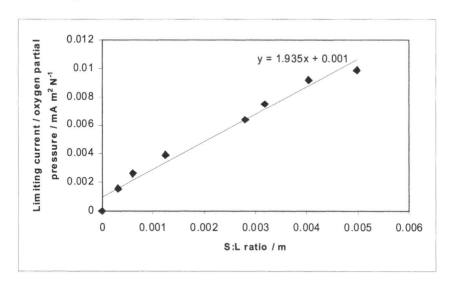

Fig. 4 Slopes of lines in Fig. 3 plotted versus the S/ L ratio

4. Conclusions

Amperometric zirconia oxygen sensors can measure oxygen concentrations at ppm levels provided the diffusion hole is engineered appropriately. This can be achieved by incorporating a diffusion hole of diameter around 0.5-2 mm. Behaviour was shown to be close to the theoretical based upon ideal diffusion along a cylindrical hole and ignoring end effects.

References

[1]   Miniature amperometric oxygen pump-gauge, WC Maskell and BCH Steele, *Solid State Ionics* **28-30** (1988) 1677-1681

[2]   Progress in the development of zirconia gas sensors, WC Maskell, *Solid State Ionics*, **134** (2000) 43-50

[3]   Zirconia oxygen sensors constructed from a ceramic-plastic sheet material, WC Maskell and JA Page, in "*Sensors and their Applications VIII*", AT Augousti and NM White (Ed), Institute of Physics Publishing, Bristol, UK, 1997, pp 177-182

Continuous monitoring chirp encoded ultrasound sensor for flood detection of oil rig leg cross beams

Rito Mijarez [1] Patrick Gaydecki [1] and Michael Burdekin [2]

[1] Department of Instrumentation and Analytical Science
[2] Department of Civil and Construction Engineering
UMIST
PO Box 88
Manchester M60 1QD
United Kingdom
Tel: [UK-44] (0) 161 200 4906
Fax: [UK-44] (0) 161 200 4911
E-mail: patrick.gaydecki@umist.ac.uk

Abstract. Structural flooded member detection of offshore oil installations implicates the detection of seawater in their normally air-filled steel bracing members. Both Radiographic and ultrasonic testing methods have been used to inspect for the presence of water in these applications, often in conjunction with remote operating vehicles. As an alternative, an ultrasonic sensor system is now being developed which can be permanently attached to a sub-sea installation and that can be powered by the action of the seawater. Upon activation the transducer is able to transmit encoded information to monitoring systems at deck level. This paper presents encouraging results that have been attained using chirp signals and suitable correlation algorithms.

1. Introduction

Crossbeams use to strengthen and support the legs in offshore oil rigs are hollow steel beams [1]. A main concern is the penetration of water into these beams, leading to impaired condition of the structure and the onset of corrosion. Recently, there has been considerable interest in flooded beam detection as a method of inspection to detect whether any brace beams have developed inward thickness cracks. The advantage of this method is the capability of identification of any inward thickness cracks regardless its location in the length of the member [2]. Therefore, underwater nondestructive testing (NDT) methods such as x-rays imaging systems and ultrasonic arrays [3] are often used for flood detection. Reasonably good results are produced; however, these techniques are expensive to employ and demand special enclosures and strict security precautions, requiring the deployment of a diver or a remotely operated vehicle.

To cope with this need an automatic and continuous monitoring ultrasonic sensor system is now being developed which is permanently attached to the inner wall of the lower attachment point of a cross beam, and whose long-life battery, normally inert, is activated by the ionic action of the seawater. Upon activation, the sensor transmits an encoded signal to a hydrophone, mounted in a permanent location close to the water's surface. The encoded signal provides information to a base receiving station on the location of the sensor and hence the flooded member.

In this paper the use of low-power, narrow-beam, encoded-chirp signal and signal processing detection technique is presented. Experiments have been carried out, using 45kHz PZT ultrasound transducers, based on two possible communication media: steel or water. Trials in large water tanks have shown that utilising simple hardware detection, it is possible to detect chirp-encoded signals with a drive voltage of 300mV, over distances greater than 10m; and applying correlation algorithms, detection of the chirp signal embedded in noise can be achieved reducing significantly the driving voltage.

2. Theory

Ultrasonic energy is generated and received by devices call transducers or probes; these convert acoustic energy to or from such types of energy as electrical, mechanical and thermal. Reversible transducers have the capability of convert energy in either direction. The piezoelectric effect is based in the relationship:

$$dt_a = p_m \times Vt \qquad (1)$$

Where dt_a is the resultant change in thickness, Vt is the applied voltage and p_m is the piezoelectric modulus appropriated for the crystal in question. For a given crystal, the applied voltage determines the amplitude of the generated ultrasound wave. Very small voltages of 300mV and 5 mV were employed in the assessment of the system. Mechanical quality Q is a measure of the increase in amplitude if a transducer is vibrating at its resonant point, fr, as part of an oscillating circuit. It is higher when the oscillation losses of energy in the material are lower. The quality Q is also linked with the bandwidth B of the resonance curve. Measuring B at 70% of its maximum value, a good approximation can be defined by [4]:

$$B = \frac{fr}{Q} \qquad (2)$$

For the purposes of this application a high Q and narrow bandwidth was employed. Acoustic impedance, Za, defined as the product of density δ and sound velocity c, determines the transmission and reflection of sound energy between two different materials. This is a source of signal attenuation and it is covered in the next topic. All ultrasound transducers produce sound fields that diverge in the far field. In this application, the divergence was of the order of 42° at –3dB.

3. Signal detection restrictions

The ultrasound transducers, steel and seawater comprise the transmission channel for the remote system. Various and significant restrictions imposed in this signal channel exist, the most serious of which are noise and signal attenuation, leading to poor signal to noise ratios.Mismatched acoustic impedance between steel and water leads to reflection of waves. The reflection and transmission coefficients of pressure for steel 4340 and water (20°C) are –0.938 (-0.556 dB) and 0.062 (-24.62dB) respectively. The sound pressure of the transmitted ultrasonic wave reduces to 6.2% of the incident wave [4]. Limitation due to the medium is an important consideration, since this largely controls the signal-to-noise ratio. It is reported that for signals below 100 kHz this may be noise due to the sea state, however above 100 kHz it is essentially thermal noise resulting from molecular agitation of water [5]. Absorption of sound energy in the sea increases with frequency and distance and is dependent on the local temperature and salinity. However, the lower the chosen signal frequency, the smaller the possible bandwidth that can be attained with the ultrasound transducers. Taking into consideration these restrictions, the selected frequency for driving the ultrasound transducers was 45 kHz. Summarising, a signal channel that has low power output, limited bandwidth, and absorption and noise difficulties resembles the problems encountered in radar [6]. To overcome this problem, an excitation signal widely used in radar was selected, namely the pulse compressed signal, also known as the chirp signal.

4. Pulse Chirp signal generations

Long pulses of constant carrier frequency possess narrow bandwidths and thereby poor resolution properties. Nevertheless, introducing frequency modulation in these long signals, the bandwidth can be broadened considerably. When the modulated pulse uses a linear frequency sweep, it is known as the chirp pulse [6]. The Chirp pulse is defined as:

$$F(t) = \cos(\omega_o t + \mu t^2/2) = \cos\{\psi(t)\}; 0 \le t \le T \tag{3}$$

Where $\omega_0 = 2\pi f_0$ is the initial frequency in rad/s; $\mu = 2\pi B/T$, is the scan rate in rad/s, B is the bandwidth of the signal and T is the duration of the pulse. $\Psi(t)$ is the instantaneous signal phase as a function of time. The instantaneous angular frequency is given by differentiation of its argument, which yields $d/dt\,[\Psi(t)] = f_0 + Bt/T$. The bandwidth-duration product defined as the dispersion factor, D, is given by $D=BT$. The percentage of the total signal energy contained within the chirp depends on the dispersion factor. This retains a high value (~93%) even for low values of the dispersion factor ($D=10$). Therefore an increase of this factor results in an increase of the total energy content of the chirp [7].

5. Apparatus and experimentation

An experimental arrangement was set up for the transmission and reception of chirp signals, as shown in Figure. 1. This included a TTI TGA1230 30 MHZ Synthesised arbitrary waveform generator and its WaveCad windows software, two ultrasound transducers, a receiver and a Tektronix digital oscilloscope. The ultrasound transducers were immersed in a large water tank and separated by 10m. The elements comprising the receiver are shown in Figure 2. The chirp signal, as depicted in Figure 3, was designed with 1024 samples and had a bandwidth B of 20 KHz, a starting frequency of 35 KHz, pulse duration T of 1 ms and a dispersion factor D of 20.

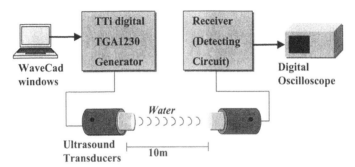

Figure 1. Experiment set up for transmission and reception of chirp signals in water.

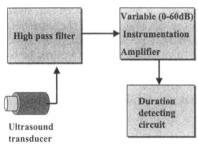

Figure 2. The Receiver system.

Table 1. Frequency readings according to time separation of transmitted chirps

Mark-space of transmitted chirps (ms)	Received frequency readings (Hz)
4	244
5	194
6	160
7	136
8	119

The minimum amplitude employed in the transmitter to detect clear frequency readings was 300 mV with a 52 dB gain in the receiver. The longitudinal waves emitted by the generator were propagated through the water and the digital oscilloscope displayed the received signals. In order to simulate signals from unique transmitters, the mark-space ratio of the chirp bursts was altered; a typical sequence is shown in Figure 4. The oscilloscope displayed a set of different frequencies, according to the transmitted signals. Some of these frequencies are shown in Table 1.

6. Correlation and encoded chirp signal detection

Cross correlation of two signals may be defined as:

$$r_{XY}(l) = \sum_{n=-\infty}^{\infty} x(n)y(n-l) \qquad (4)$$

Where x(n) can represents the sampled version of the transmitted signal, y(n) the sampled version of the received signal at the output of the analog-to-digital converter (ADC), l is the (time) shift (or lag) parameter and $r_{xy}(l)$ indicates the sequence being correlated. However, dealing with finite-duration sequences, it may be expressed in terms of finite limits on the summation

$$r_{XY}[l] = \sum_{n=i}^{N-\lceil k \rceil - 1} x[n]y[n-l] \qquad (5)$$

Where x[n] and y[n] are casual sequences of length N; and i=l, k=0 for $l \geq 0$ and i=0, k=l for l< 0.

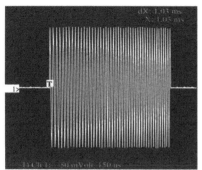

Figure 3. Chirp signal, bandwidth 20 kHz, pulse duration 1ms and central frequency 45KHz

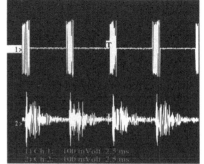

Figure 4. Transmitted (upper) and received chirp signals waveforms 52 dB gain (lower)

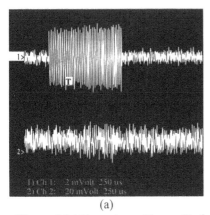 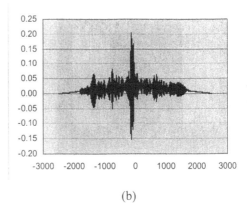

(a) (b)

Figure 5 (a) Upper trace: Transmitted chirp signal 5 mVpp, lower trace: Received signal gain 40 dB, (b) Correlation function in Matlab, maximum point 0.21

In order to test that chirp signals can be encoded digitally in the remote system depicted in figure 1, an experiment was carried out including the same elements, except that the distance of separation between ultrasound transducers was 1m, the amplitude of the chirp signal was 5mV and the receiver circuit was an amplifier followed by a digital oscilloscope using 2500 samples and a sampling period of 1 MS/sec. The received signal can be represented as:

$$y[n] = \alpha x[n - D] + w[n] \qquad (6)$$

Where α is some attenuation factor representing the signal loss involved in the trip transmission of the chirp signal x[n], D is the trip delay, which is assumed to be an integer multiple of the sampling interval, an w[n] represents the additive noise that is picked up by the transducers and any noise generated by the electronics components contained in the front end of the receiver. Applying a correlation algorithm in the computer using a Matlab program, was possible to extract important information from y[n], even though the received signal was embedded in noise as is shown in figures 5. Digital encoding the chirp signals was carried out with two different transmission signals, $x_0[n]$ and $x_1[n]$ representing the logic 0 and 1 respectively. In this case $x_0[n]$ was defined as one chirp signal (used formerly) and $x_1[n]$ was selected to be two chirp signals separated by 2msec as is shown if figure 6. The received signal may be represented as:

$$y[n] = x_i[n] + w[n]; i = 0,1; 0 \le n \le N - 1 \qquad (7)$$

Where N is the number of samples in each of the two sequences (in this case was 2500). The uncertainty is whether $x_0[n]$ or $x_1[n]$ is the signal component in y[n]. The comparison process is performed off line by means of correlation as was described previously. The discrimination between these signals and noise is shown in table 2. The logical values of these readings were obtained considering normalised values and a threshold of 5.5, where Z stands for a not defined logical state.

7. Discussion

The evaluation of chirp encoded ultrasound signals, used with ultrasound transducers for flood detection of oil rig leg cross beams suggests this method has much to recommend it for practical circumstances. The use of very low-voltage signals (300 mV), simple hardware detection and moderate initial range (10m) is encouraging.

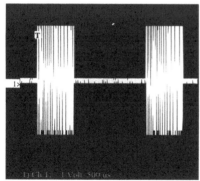

Figure 6 Transmitted signal $x_1[n]$

Table 2. Logical values of encoded chirps signals using correlation

Casual sequences $x_i[n]y_i[n]$	Correlation value $r_{xy}[n]$	Logical value
$x_0[n]$ and noise	0.17	Z
$x_0[n]$ and 1 Received chirp	1.00	0
$x_0[n]$ and 2 received chirps	0.18	Z
$x_1[n]$ and noise	0.50	Z
$x_1[n]$ and 2 Received chirps	0.85	1

Moreover, implementing correlation algorithms for chirp signal detection embedded in noise can dramatically reduce driving voltages. However, difficulties associated with real structures include longer distances, interference of steel structures and transducers coupling factors, which have to be considered for further work.

8. Conclusion

An experimental system employing ultrasound transducers has shown that it is possible to transmit coherent chirp signals through water, with excitation voltages as low as 300mV, over distances of 10m. In addition, by encoding chirp signals, it is possible to encode binary information using correlation algorithms. This represents an important step along the road to developing autonomous sensors for flood detection in oil rig cross beams.

Acknowledgement

The authors wish to express their thanks to the Electrical Research Institute (IIE) and CONACYT in Mexico for financially supporting this work.

References

[1]   www.offshore-technology.com
[2]   Burdekin FM et al. Experimental validation of the ultimate strength of brace members with circumferential cracks. Offshore technology reports 2001/081. UMIST and University College London for the Health and Safety Executive.
[3]   Stirling DG et al. A novel ultrasonic inspection system for flooded member detection offshore. NDT&E International Volume 30, October 1997
[4]   Krautkramer J Ultrasonic testing of materials, 4th edition Springer-Verlag.
[5]   Mitson RB, Review of high-speed sector-scanning sonar and its application to fisheries research. IEE proceedings, Vol. 131, Part F, No. 3, june 1984
[6]   Rahman MS, An investigation into spectral analysis using a chirp signal matched filter. PhD thesis University of Manchester, October 1992.
[7]   Pollakowski M and Ermert H, Chirp signal Matching and Signal Power ptimization in Pulse-Echo Mode Ultrasonic Nondestructive testing. IEEE trans. on ultrasonics, ferroelectrics and frequency control. Vol. 41, No 5, Sep. 1994.

Poster Programme
Paper presented at Sensors and their Applications XII, September 2003
©2003 IOP Publishing Ltd

Moisture transport studies in building materials

D Bailly[1], M Campbell[1], N Poffa[1], J Sun[1], G H Galbraith[1], R C McLean[2], C Sanders[3] and G G Nielsen[4]

[1] School of Engineering, Science and Design, Glasgow Caledonian University, Glasgow, UK
[2] Department of Mechanical Engineering, University of Strathclyde, Glasgow, UK
[3] Building Research Establishment, Scottish Laboratory, East Kilbride, Glasgow, UK
[4] GNI Ltd, Denmark

Abstract. Moisture transport through building envelopes is a subject which, to date, has not been adequately addressed, despite being of national, and indeed international, importance. The most likely reason for this situation is probably due to the absence, until now, of suitable investigative tools which are capable of measuring moisture transport within relatively large volumes of material in realistic climatic conditions.

A new experimental X-ray system, built by GNI of Denmark, has been developed at Glasgow Caledonian University, which is capable of investigating materials of dimensions up to 420 mm thick under a wide range of temperature and humidity regimes. Both water content and moisture diffusion coefficients have been measured for Danish yellow brick and these are in good agreement with previous X-ray studies which were carried out in Denmark at the end 2002.

1. Introduction

The subject of moisture transport in building materials is of great importance to the construction industry since it impacts on both the acceptability of new materials and the maintenance and viability of older materials. Recent changes in weather patterns to wetter climatic conditions has led, for example, to the deterioration of the external walls of historic buildings. In particular, the de-lamination of sandstone is proving to be a difficult process to halt and it is very costly to rectify. Moisture transport through building materials is a subject which has not been adequately addressed despite being of national, and indeed international, importance. The absence, at least until recently, of suitable investigative tools which are capable of analysing realistic volumes of building material in 'real' environmental conditions, is probably the main reason for this situation. A new experimental X-ray system, built by GNI of Denmark, which is capable of investigating materials of dimensions up to 420 mm thick under temperature and humidity regimes, has been further developed at Glasgow Caledonian University. This system is not only capable of determining moisture content and diffusion coefficients but it can also image moisture fronts. This opens up the possibilities of studying the movement of moisture fronts at material interfaces such as mortar layers.

2. Theory

The attenuation of X-rays in a medium is governed by Beer's law, $I = I_o \cdot e^{-\mu \rho x}$, where I is the transmitted X-ray intensity, I_o is the incident X-ray intensity, μ is the mass attenuation coefficient (m^2/kg), ρ is the density of the medium (kg/m^3) and x is the thickness of the sample in metres. It follows that for a composite material: $\ln(I_o/I) = \sum_i \mu_i \rho_i x_i$. In this work, the material chosen was Danish yellow bricks due to the consistency and homogeneity of the dry material.

2.1 Moisture content

In order to determine the moisture content it is necessary to examine the different experimental configurations in order to isolate the relevant factors which contribute to the overall absorption process. Four configurations are shown below together with the appropriate components of Beer's Law.

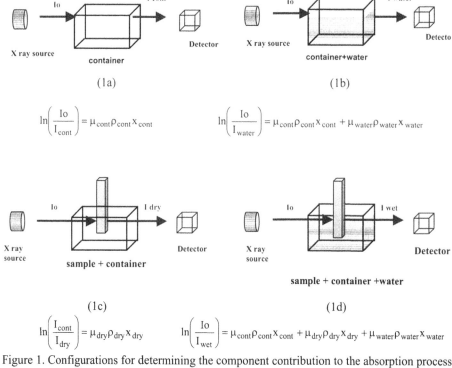

$$\ln\left(\frac{I_o}{I_{cont}}\right) = \mu_{cont}\rho_{cont}x_{cont}$$

$$\ln\left(\frac{I_o}{I_{water}}\right) = \mu_{cont}\rho_{cont}x_{cont} + \mu_{water}\rho_{water}x_{water}$$

$$\ln\left(\frac{I_{cont}}{I_{dry}}\right) = \mu_{dry}\rho_{dry}x_{dry}$$

$$\ln\left(\frac{I_o}{I_{wet}}\right) = \mu_{cont}\rho_{cont}x_{cont} + \mu_{dry}\rho_{dry}x_{dry} + \mu_{water}\rho_{water}x_{water}$$

Figure 1. Configurations for determining the component contribution to the absorption process

From figures 1a and 1b it follows that $\ln\left(\frac{I_{cont}}{I_{water}}\right) = \mu_{water}\rho_{water}x_{water}$ and, in conjunction with

figures 1c and 1d, it follows that $\ln\left(\frac{I_{dry}}{I_{wet}}\right) = \mu_{water}\rho_{water}x_{water}$.

By definition, the moisture content is given by $u = \dfrac{m_{water}}{m_{dry}} = \dfrac{\rho_{water} x_{water}}{\rho_{dry} x_{dry}}$ thus, by mass

$$u\left(kg \cdot kg^{-1}\right) = \frac{\ln\left(\dfrac{I_{dry}}{I_{wet}}\right)}{\rho_{dry} x_{dry} \mu_{water}}$$ and by volume: $\theta\left(m^3 \cdot m^{-3}\right) = u \dfrac{\rho_{dry}}{\rho_{wet}}$

2.2 Diffusion coefficients

To a good approximation, moisture transport in a porous media under isothermal conditions can be described by a non-linear diffusion equation $\dfrac{\partial \theta}{\partial t} = \dfrac{\partial}{\partial x}\left(D_\theta \dfrac{\partial \theta}{\partial x}\right)$. Moisture diffusion coefficients may be calculated from the plots $\theta = f(x)$ (x being the distance from the water along the sample) using a Boltzman transformation [1,2] In the Boltzmann transformation, wavelength values were calculated from the corresponding times and distances of the measurement points using the relation: $\lambda = \dfrac{x}{\sqrt{t}}$. When θ is plotted against λ, each series of data fall onto a single curve, from which the diffusion coefficient D_θ at a particular moisture content can be calculated using a moving average:

$$D_\theta = -\frac{1}{2} \cdot \frac{1}{\dfrac{(d\theta)}{(d\lambda)_\theta}} \cdot \int_{\theta_o}^{\theta} \lambda d\theta$$

3. Experimental arrangements

Some results had already been obtained, towards the end of 2001,using the DTU system in Denmark, i.e. prior to the installation of the GCU system. The GCU system can produce much higher currents but cannot quite achieve the voltage of the DTU installation. Nevertheless, both systems had approximately same X-ray tube settings i.e. a voltage of 70 kV, and a current 15 μA so that meaningful comparisons could be made. Each single measurement produced a 256 channel pulse height spectrum from a CZT detector. For data analysis purposes, an option of summing the scaler contents of (i) all the channels or (ii) groups of channels (minimum one) was available for calculating X-ray intensity ratios.

4. Results

The results for X-ray attenuation and moisture diffusion coefficients are as follows.

4.1 Mass attenuation coefficients of aluminium and water.

The values of mass attenuation coefficients for aluminium and water were determined by calculating the ratios of intensities as explained above. The results were obtained using single channel values are shown in table 1. An additional complication was that measurements for water were taken using different containers.

mass attenuation coefficients	GCU	DTU	accepted value [3] @60 keV
aluminium (cm2/g)	0.247	0.239	0.255
water (cm2/g)	0.240	0.201	0.197

Table 1. Single channel results

The results obtained using the summation of 160 channels in each spectrum gave poorer results for the aluminium samples as shown in table 2.

mass attenuation coefficients	GCU	DTU	accepted value [3] @60 keV
aluminium (cm2/g)	0.310	0.465	0.255
water (cm2/g)	0.262	0.221	0.197

Table 2. Sum of 160 channel results

4.2 Diffusion coefficient of sample of bricks:

Prior to X-rays measurements, the samples were dried overnight in a oven at 70°C, and their dimensions and weights were measured. After precisely locating the sample within the x-ray chamber by finding the coordinates of each edge, the samples were scanned using a 5*20 mm matrix of measurement points starting at 5 mm from the wetting point. The measurements were then repeated 5 times at regular time intervals. Moisture contents were calculated using the ratio of the intensities, which had been summed for all channels. The moisture profiles, obtained at specific distances along the bricks during vertical wetting in the GCU and DTU systems, are shown in figures 2(a) and 2(b) respectively. A time stamp is added for each curve to enable a meaningful interpretation of the measured profiles.

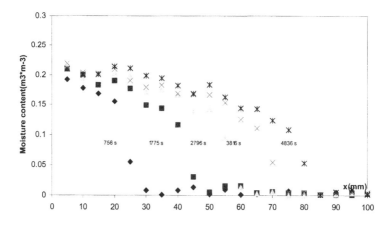

Figure 2(a) Moisture profile during vertical wetting in the GCU system

b) DTU sample2 vertical wetting

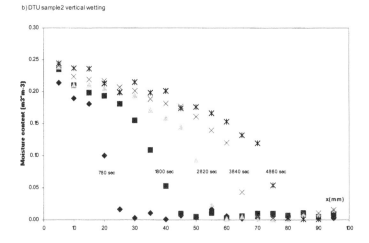

Figure 2(b) Moisture profile during vertical wetting in the DTU system

Not only were both sets of data very similar but comparable curves were also obtained for moisture profiles in porous building materials using other techniques such as NMR [1-2,4-5].

The Boltzmann transformation of each set of moisture profiles are shown in figures 3(a) and 3(b) respectively and it may be seen that they are clearly collapsing onto one distinct curve.

a) GCU sample2 vertical wetting

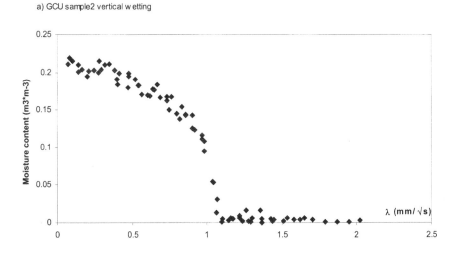

Figure 3(a) Boltzmann transformation for the GCU vertical wetting data

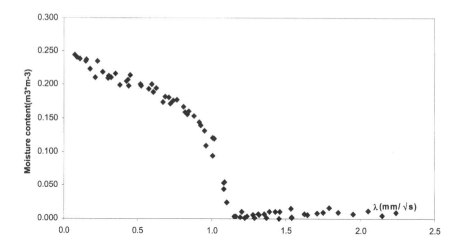

Figure 3(b) Boltzmann transformation for the DTU vertical wetting data

Once again, the two curves obtained at both GCU and DTU are very similar with a slightly steeper slope in the former case.

Moisture diffusion coefficients were calculated by integrating the data from the highest moisture contents to the intersection of the set of data with the λ-axis, and then by dividing the result with the series of gradients that formed the closest fit to the curve. Log_{10} (diffusion coefficient) is plotted against the moisture content for both sets of results as shown in figure 4.

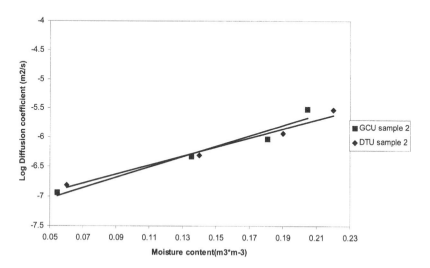

Figure 4. Log_{10} (diffusion coefficient) as a function of volume moisture content

The small number of points available is due to the limited number of gradients that can be fitted to the curves. However, the diffusion coefficients show a good linearity over the measured range of moisture contents. This is consistent with reports [1] showing that the overall behaviour of D_θ can be approximated with an exponential function.

5. Conclusions

Preliminary results on moisture transport in a porous material show reasonable consistency when using similar X-ray systems at DTU and GCU. The diffusion coefficients were also found to be reasonably close to values in the literature for similar kinds of building material using NMR methods. However, further work is needed to get a better approximation of the curve $\theta=f(\lambda)$ in order to improve the estimation of $(d\theta/d\lambda)$ at any particular moisture content.

References

[1]   Pel L, Moisture transport in porous building materials, 1995, PhD Thesis, Eindhoven University of Technology, The Netherlands.
[2]   Galbraith G.H. et al , Materials and Structures, 2001, 34, 389-395
[3]   Olszycka B, 1979,Gamma-ray determination of surface water storage and stem water content for coniferous forest, PhD thesis, University of Strathclyde.
[4]   Krus M, PhD thesis, 1995, University of Stuttgart, Germany.
[5]   Carpenter et al, Materials and structures, 1993, 26, 286-292.

High-temperature sensor potential and sustainability of fibre Bragg gratings written in highly photosensitive Sb/ Er/Ge codoped silica fiber

Yonghang Shen[1,2], Tong Sun[2], Kenneth T V Grattan[2] and Mingwu Sun[3]

[1] Department of Physics, Zhejiang University, Hangzhou 310027, China.

[2] School of Engineering and Mathematical Science, City University, Northampton Square, London EC1V 0HB, UK

[3] China Building Materials Academy, Beijing 100024, China

Abstract: The photosensitivity properties and the high temperature sustainability of fibre Bragg gratings (FBGs) created in a specialized Sb/Er/Ge codoped slica fibre were examined and compared, showing a much higher photosensitivity than those produced in a Sn doped silica fibre. The very high temperature sustainability, of 850 ^{0}C, of the fibre is of interest for a number of FBG-based sensor applications.

1. Introduction

Many kinds of photosensitive fibres have been developed and used in fibre Bragg grating (FBG) fabrication in recent years, in addition to the use of hydrogen loading [1]. This has included Boron/Germanium (B/Ge) codoped silica fibre [2], the rare-earth ions (such as Ce^{3+} and/or Er^{3+}) into doped silica fibre [3][4] and high Germanium (Ge) doped silica fibre. These fibres are quite photosensitive and are able to show a large refractive index change after exposure to the light from a uv excimer laser. The FBGs fabricated in these fibres, however, are often not the most suitable for the many potential sensing application at high temperatures as they would be easily "washed out" at these elevated temperatures, of over ~350 ^{0}C [5]. Compared to these fibres, the tin (Sn) doped fibre developed by Dong et al [6] has been found to possess a quite significant level of photosensitivity, and the FBGs created in this fibre were also shown to be able to sustain a temperature as high as 800 ^{0}C, in work reported in recent years [7]. These properties of high photosensitivity and strong high temperature sustainability make the FBGs created in this tin doped fibre strong candidates for sensing probes used for high temperature monitoring or used at high temperatures e.g. for strain determination.

The photosensitivity of the tin doped fibre is, however, still not as high as that of B/Ge codoped fibre. To fabricate a FBG with a reflectivity greater than 95% in tin doped fibre normally takes around half an hour, while those FBGs created in B/Ge fibre with a reflectivity greater than 99% take less than one minute to create under the same experimental conditions (with 12 mJ pulses from a KrF excimer laser at 100Hz), thus making it difficult to write a significant number of FBGs in this tin doped fibre.

In this work, a novel highly photosensitive fibre using Antimony/Erbium/Germanium (Sb/Er/Ge) is considered and the details of the composition, fabrication technique and the performance, at high temperatures, of the FBGs made using this fibre are presented in this paper. Antimony has mainly been considered on the basis of the electron charge effect of the outer shell electrons in the atom, in a similar way to the situation in P-doped fibre, with the large Sb^{3+} ion size (76 pm, greater than the 70 pm of Sn^{4+}) a feature of its use. It is believed that the larger size of the doping ion may effectively stabilize the performance of the material concerned, as in the situation of the fluorescence performance of Tm-doped Yttrium Aluminum Garnet (YAG), where the doping of the Tm ion created a garnet with a very stable fluorescence intensity and decay lifetime when experiencing annealing at temperatures as high as 1200 ^{0}C [8]. The choice of Er in this work as the major dopant is mainly due to its well-known fluorescence properties, as well as its photosensitivity. The doping of the third element, Ge, is included mainly to enhance the photosensitivity and increase the refractive index of the fibre core. Such fibre may also be used for sensing purposes through the utilization of its fluorescence characteristics [9, 10] and involving FBG-based schemes [11].

2. Fibre fabrication

The method of modified chemical vapor deposition (MCVD) was used for the fabrication of the fibre preform. Initially, the highly purified Er_2O_3 and Sb_2O_3 powders used were dissolved in hydrochloric acid and then diluted with de-ionized water. The two liquids were then mixed in a known ratio to form an Er^{3+}/Sb^{3+} solution. A silica tube was used for the deposition of a SiO_2 inner cladding layer, at a temperature of 1750 ^{0}C. A loose opaque layer of SiO_2—GeO_2 was deposited on the SiO_2 layer at a comparatively low temperature of 1200 ^{0}C, following the deposition of the SiO_2 layer. The silica tube was then soaked in the Er^{3+}/Sb^{3+} solvent for one hour so that the Er and Sb ions were absorbed in the loose layer of SiO_2—GeO_2. The silica tube was placed again into the MCVD lathe and heated to 600 ^{0}C for dehydration. After that, it was sintered at 2000 ^{0}C to transform the loose SiO_2—GeO_2 layer into a transparent glassy core layer. To deter the evaporation of GeO_2, a small amount of $GeCl_4$ and O_2 gases were allowed into the tube during the process. The silica tube was finally heated to 2100 ^{0}C and became a solid fibre perform after the collapse of the silica tube, from which a fibre with a diameter of 125 μm was then drawn. The composition of the fibre thus created was: GeO_2 ~15 wt%, Er^{3+}: 500 ppm, Sb^{3+}:5000 ppm.

3. Fibre Bragg Grating (FBG) fabrication

Two samples of the fibre were used in the experiments to fabricate the FBGs, by exposing them to the UV emission from a KrF excimer laser (Braggstar-500 by Tuilaser AG) at 248 nm through a phase-mask (pitch period:1060 nm, supplied by OE-Land Inc., Canada). A cylindrical plano-convex lens (focal length 20 cm) was used to converge the laser beam onto the photosensitive fibre through the phase-mask. For comparison, another fibre sample was also used, which was a Sn co-doped high germanium fibre (made in China, with a composition of GeO_2 ~15 wt% and SnO_2 of ~ 0.2 mole%). The writing times of the FBGs were controlled individually for each sample until the FBG achieved a high level of reflectivity, this having been monitored using an Optical Spectrum Analyzer (OSA-Agilent HP86140A).

The reflectivity of the FBGs written into the Sb/Er/Ge fibres increased quickly during the initial stage. By the time the UV exposure exceeded 5 minutes, a reflectivity of greater than 96% had been reached. From the FBG length (of 6.5 mm), it could be deduced that the refractive index modulation was 1.9×10^{-4}. Beyond that time, the reflectivity of the FBGs continued to increase at a slower rate and finally, after a UV exposure of 12 minutes for the 1st sample and 14 minutes for the 2nd sample, the gratings reached their highest reflectivity of 99.6% and 99.3% respectively (which corresponded to refractive index modulations of 2.75×10^{-4} and 2.46×10^{-4}). Using the same experimental configuration, the FBG written in the Sn fibre achieved a highest value of reflectivity of 97.8%, in 30 minutes, which corresponds to a refractive change of 2.0×10^{-4}. The increase of the reflectivity of the FBGs, as well as the modulation of the refractive index with time, for one of the Sb/Er/Ge fibre samples and the Sn doped silica fibre during the UV exposure, is illustrated in Figure 1 from which it can be seen that the Sb/Er/Ge co-doped fibre has a much higher photosensitivity than the Sn doped fibre.

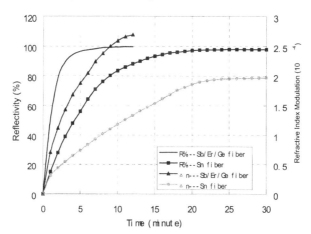

Figure 1 Reflectivity and refractive index modulation increase for an FBG written in Sb/Er/Ge fiber, with time, compared to the performance of an FBG written into an Sn doped silica fiber under the same experimental conditions

4. FBG characterization at high temperatures

The FBGs fabricated in this way were then placed loosely in a silica tube and put into a Carbolite tube oven to observe their thermal decay characteristics over a period of time, with an annealing time of 24 hours at each temperature increment (shown as a dot on Figure 2). A fast decay of the grating characteristic was seen initially, followed by a more substantial slow decay being observed. At lower temperatures, this decay was so slow that there was almost no observable change in the reflectivity, and thus of the modulation of the refractive index of the FBGs after annealing for 24 hours. This is illustrated in Figure 2.

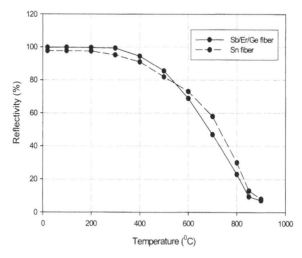

Figure 2 Annealing results for FBGs written into the Sb/Er/Ge fiber with temperature, compared to a FBG written into an Sn doped silica fiber under the same experimental conditions

The experimental results on the annealing tests on the FBGs shown above clearly indicated that the gratings fabricated in the Sb/Er/Ge fibre had a similar high temperature sustainability to those in the Sn doped fibre. They could still retain a particular value of reflectivity after annealing at 850 ^{0}C for 24 hours and 900 ^{0}C for 4 hours (no further annealing test was performed at 900 ^{0}C), as shown in Figure 3. The FBG samples, after annealing at 900 ^{0}C for 4 hours, were then cooled to room temperature in the oven and then heated step by step, until a temperature of 850 ^{0}C was reached. The peak wavelengths of the reflective spectra were recorded after the temperature was stabilized at each one hundred degrees Celsius point. This process was repeated several times to test the reproducibility of the peak wavelength determined. The results show an excellent repeatability of the peak wavelength in the reflection spectra with a deviation of less than 0.02 nm, which was equivalent to less than 2 ^{0}C of temperature change and is satisfactory for many measurement purposes. The data for the peak wavelength shift with temperature are presented in Figure 4. Obviously, the graph is not fully linear over the whole temperature range, as is often the situation for other fibres. The temperature sensitivity of the peak wavelength for the FBGs in Sb/Er/Ge fibre is about 12 pm/^{0}C at room temperature and 18 pm/^{0}C at around 800 ^{0}C.

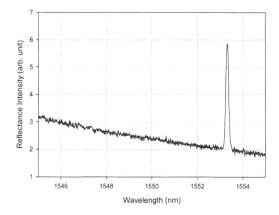

Figure 3 The reflectance spectrum of an FBG written into the Sb/Er/Ge fiber at 900°C, after long term annealing from room temperature to 850°C

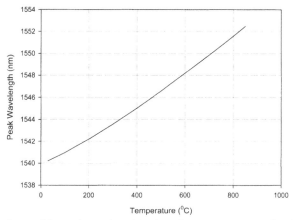

Figure 4 Dependence of the peak wavelength in the reflectance spectrum of an FBG written into the Sb/Er/Ge fiber with temperature

The fluorescence properties of the Sb/Er/Ge fibre were also examined. By using a laser diode (LD) working at 980 nm, emitting into a pigtailed fibre a power output of 20 mW, the strong fluorescence on the 1550 nm band could be readily detected by using an OSA, as illustrated in Figure 5, where a small dip in the fluorescence spectrum due to the reflectivity of the FBG could be clearly seen. Even though there was no aluminium involved in the composition of the Sb/Er/Ge fibre, the fluorescence performance still seemed quite satisfactory. Further experimental work on the fluorescence properties of the fibre will continue to be performed and reported in more detail in the future.

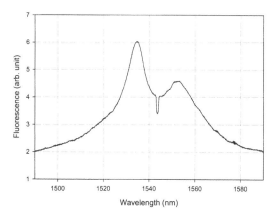

Figure 5 Fluorescence spectrum of the Sb/Er/Ge fiber with a FBG written into the fiber under 400°C, when excited by a LD working at 980 nm

There are a number of relevant sensor applications, including the monitoring of strain and temperature at elevated temperatures, for example in furnace linings or in fire detection and alarm systems. Fibre gratings have not been used extensively at these very high temperatures and their versatility and simplicity as in-line, intrinsic sensor system elements makes them very appealing for a range of applications. The evidence of this paper is that a range of new high temperature applications can be expected in the future.

5. Conclusions

In summary, a highly photosensitive Sb/Er/Ge co-doped silica fibre in which FBGs were written, showing a strong high-temperature sustainability, was presented. The properties of high photosensitivity and high temperature sustainability of the FBGs written in this fibre were examined and compared with those in a Sn doped silica fibre. The results showed that the Sb/Er/Ge fibre had a much better photosensitivity than the Sn doped silica fibre and a similar temperature sustainability (850 ^{0}C) for the FBGs created. It is believed that the high photosensitivity of the fibre is mainly due to the doping of Sb^{3+} and partly due to the co-doping of the Er^{3+} ions, while the high temperature sustainability of the FBG written into this fibre is mainly due to the doping from the Sb^{3+} ion. It is of particular value for the volume production of this kind of FBG, for sensing applications in the high temperature range. The work presented in this paper is also valuable in the design and development of photosensitive fibres with strong temperature sustainability for use in high temperature measurement situations.

6. Acknowledgements

This work was supported through several schemes and was partly funded by the Science & Technology Project of Zhejiang Province, China (Project No.011106205). Funding from the Engineering and Physical Science Research Council (EPSRC) through a number of schemes and for the collaborative research work between City University and China is also appreciated.

7. References

[1] P.J. Lemaire, R.M. Atkins, V. Mizrahi, et al. "High-pressure H_2 loading as a technique for achieving ultrahigh UV photosensitivity and thermal sensitivity in GeO_2 doped optical fibers," Electronics Letters, 29, 1993, pp1191-1193

[2] D.L. Williams, B.J. Ainslie, J.R Armitage, et al. "Enhanced UV photosensitivity in boron codoped germanosilicate fibers," Electronics Letters, 29, 1993, pp45-47

[3] M.M. Broer, R.L. Cone, and J.R. Simpson. "Ultraviolet-induced distributed-feedback gratings in Ce^{3+}-doped silica optical fibers," Optics Letters, 16, 1991, pp1391-1393

[4] F. Bilodeau, D.C. Johnson, B. Malo, et al, "Ultraviolet-light photosensitivity in Er^{3+}-Ge-doped optical fiber," Optics Letters, 15, 1990, pp1138-1140

[5] S. R. Baker, H. N. Rourke, V. Baker and D. Goodchild, "Thermal decay of fibre Bragg gratings written in boron and germanium codoped silica fiber," IEEE J. Lightwave Technol., 15, 1997, pp1470-1477

[6] L. Dong, J.L. Cruz, J.A. Tucknott, et al, "Strong photosensitivity gratings in tin-doped phosphosilicate optical fibers," Optics Letters, 20, 1995, pp1982-1984

[7] G. Brambilla, V. Pruneri and L. Reekie, "Photorefractive index gratings in SnO_2:SiO_2 optical fibers", Appl. Phys. Lett., 76, 2000, pp807-809

[8] Y. Shen, W. Zhao, J. He, et al. "Fluorescence decay characteristic of Tm doped YAG crystal fiber for sensor applications, investigated from room temperature to 1400 ^{0}C," IEEE Sensors, accepted for publication, 2003

[9] K.T.V. Grattan, Z. Y. Zhang, Fiber Optic Fluorescence Thermometry, Chapman & Hall (1995)

[10] D.I. Forsyth, S. A. Wade, T. Sun, et al. "Dual temperature and strain measurement with the combined fluorescence lifetime and Bragg wavelength shift approach in doped optical fiber," Applied Optics, 41, 2002, pp6585-6592

[11] A Othonos "Fiber Bragg gratings for sensor applications" in Optical Fiber Sensor Technology: Advanced Applications, Eds K T V Grattan and B T Meggitt, Pub Kluwer Academic Press, Dordrecht, The Netherlands, 2000

Poster Programme
Paper presented at Sensors and their Applications XII, September 2003
©2003 IOP Publishing Ltd

Application of new graphical tools and measurands for tensiography surface science studies in BSA protein adsorption on a stainless steel substrate

G Dunne, D Morrin, M O' Neill, N D McMillan, B O' Rourke, C I Mitchell

Institute of Technology, Carlow, Kilkenny Road, Carlow
Tel: 0503 70400 E-mail: Dunnega@itcarlow.ie

Abstract. Major environmental problems exist with bio-contamination of water, beverage, biotechnology and other industries. Measurement problems for instrumentation and health problems for humans arise from adsorptions of proteins/enzymes. Characterisation of the adsorption of proteins and enzymes on surfaces cannot be done in a sensitive real-time way. Four empirical straight-line relationships recently discovered are used here in BSA protein surface studies. Two equations are of particular interest and results on various albumins are reported: A straight-line equation applying only to pure liquids and a second to surfactants. These equations allow a new classification of the variable surface activity of proteins and open up a new surface science methodology that should have importance in the classification of the surface activity of proteins and elsewhere in surface science. A new protein classification is suggested based on these empirical equations and are compared with traditional methods of classifying surface activity. The bulk measurement potential of tensiography is well established and this is explained with examples from measurements on various liquids. The potential of these bulk measurements in protein science are also discussed. The practical aim is to deliver a rapid and thus early warning of bio-contaminants such as slurry, silage and sewage into watercourses.

1. Introduction

The drop-volume technique has always required experimentally determined correction factors to give an accurate value of the surface tension of a liquid. The work presented here, describes how established stalagmometric techniques can be combined with the new measurement science of optical tensiography to provide the surface tension of a liquid of known refractive index directly.

The tensiograph has been shown to be able to determine surface tension without the necessity of correction factors. Historically the drop volume method has been developed on the basis of tables of these correction factors. These values are notoriously unreliable and despite the very considerable efforts of Miller [1] and his group in recent times to improve these correction factors, they still are THE limiting factor on the accuracy of the drop volume measurement. It is true to say therefore, that if this contention is shown to be true, that tensiography avoids the use of correction factors, then **this development in surface science is one of great significance**. This paper will give the leading edge results on surface tension measurements using the drop volume method without correction factors.

The remnant drop, which is observed on nearly every drophead design, is a complicating factor for the drop volume method, which in other respects is a simple measurement procedure. The size of the drop just before it separates from the supporting drophead is determined by simple physics, namely the weight of the drop on a cylindrical drophead is equal to:

$$2\pi r^* \gamma \tag{1}$$

The division of this drop after separation from the drophead complicates this and the correction factor gives what is only an estimate of the size of the remnant drop, which then allows the measurement of the surface tension.

The basic approaches to advanced instrumental methods in surface science require complex measurement strategies such as in the drop volume technique (tensiometry), varying the delivery rates of the sample to measure variations in the volume (or weight) of the drop brought about by changes in the conditions of surfaces. The diffusion rates of surfactant molecules leads to variations in the surfaces and through such 'dynamic surface tension' measurements the thermodynamics of the molecular processes is studied. This methodology essentially relies on only one active measurand, namely the drop volume or weight.

2.   Apparatus

The multianalyser tensiograph [2] is a new instrument that has been developed on the principles of stalagmometric instruments. It is a fiber optic instrument based on some simple principles of physics. An infra red or visible beam reflects light through a drop while it is forming. The instrument monitors the optical coupling between source and collector fibers placed in the drop-head and the optoelectronic signal produced is known as a tensiotrace. Every liquid has a unique drop shape and hence a unique tensiotrace, which leads to the fingerprinting capability of the instrument. Therefore, the instrument is fundamentally a development of surface and interfacial science because it is based on the analysis of an optical signal determined fundamentally by the shape of the liquid on the drop-head.

The present study is concerned only with the analysis of surface activity, but the technique has been used in a number of other application areas, namely, sugar analysis [3], brewing [4], distilling [5] and pollution monitoring [6]. From current work it is clear that there are a wide range of other potential application areas for the multianalyser such as adhesive manufacture, in food analysis where it can be used to analyse oils and other liquid products and in pharmaceutical fingerprinting for the forensic identification of drugs.

The drop-head used in this work is a flat design. The design employs standard 1mm polymethylmethacrylate (PMMA) fibers. The fibres are polished using down to 0.3 micron lapping film before gluing them into the drop-head. The drop-head is made from stainless steel. The diameter of the head is 9mm with the fibers separated by 6mm. A HPLC capillary glued into the centre of the head is used to deliver the liquid. The head is designed such that it wets (i.e. the suspended liquid covers the entire lower surface of the drop-head) when liquid is delivered. Light from an LED source is injected into the drop-head through the source fiber and the signal is picked up by the collector fiber and goes to the photodiode, which then produces the trace.

The principal features of the tensiotrace obtained for a water sample are illustrated in Figure 1. Water is taken as the reference liquid for most applications. The principal features of the trace seen here are the rainbow peak (RPP), the tensiograph peak (TPP) and the drop period (DP). The drop period is the length of time taken for a drop to form on the drop-head and then fall off. It is directly related to the surface tension and density of the analyte.

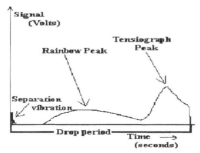

Figure 1. Typical water tensiotrace showing the characteristic features.

3. Experimental

Surface activity was studied using: (i) Methanol (BDH Chemicals), (ii) Sodium Lauryl Sulphate (BDH Chemicals), (iii) Bovine Albumin (fraction V, A-7906, Sigma-Aldrich Co.). All experiments were carried out at 30°C and 950nm.

3.1 Calibration using methanol and Sodium Lauryl Sulphate (SLS)

A series of methanol/water solutions were made, 5, 10, 25, 40, 60, 80 and 100% v/v. 6 traces of each concentration were obtained with the pump speed set to deliver 250μl/200 seconds. Water was used after each dilution to clean the drop-head. The results of drop period versus surface tension and (TPP-RPP) versus surface tension were plotted.

A series of SLS solutions were prepared by dissolving different amounts in 100cm^3 of distilled H$_2$O. The concentrations were 0.005, 0.007, 0.01, 0.02 and 0.05% w/v. 6 traces of each concentration were obtained at two different pump speeds 250μl/200 & 600 seconds. Water was used after each dilution to clean the drop-head. The results of drop period versus surface tension were plotted for the two different pump speeds.

3.2 Bovine Serum Albumin (BSA) analysis

A series of BSA solutions were prepared by dissolving different amounts in 100cm^3 of distilled H$_2$O. The concentrations prepared were 50, 100, 200, 300, 400 and 500 ppm. 6 traces of each concentration were obtained at two different pump speeds 250μl/200 seconds and 250μl/600 seconds. Water was used after each dilution to clean the drop-head. The results of (TPP-RPP) versus surface tension at the 200 second pump speed and drop period versus surface tension at the 600 second pump speed were plotted. The surface tension of each compound was determined using a Du nouy ring Torsion balance.

4.   Results and Discussion

4.1   Pure liquid analysis

The drop period, DP, measured with the tensiograph, explained above, allows a calibration measurement of surface tension. As can be seen in Figure 2(a) the DP versus surface tension relationship for methanol is not linear but more a parabolic relationship. This is due to the influence of density on the forming drop. At low concentrations (high surface tension) the drops are lighter and remain on the drop-head longer thus giving a curved feature to the relationship.

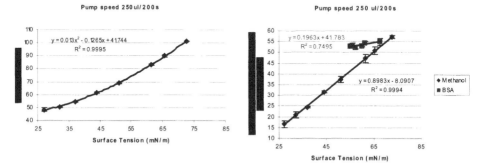

Figure 2 (a). A series of methanol – water dilutions in a plot drop period versus surface tension (mN/M) at a pump speed of 200s. (b) Tensiograph Peak Period – Rainbow Peak Period for methanol-water solutions and a series of Bovine Serum Albumin (BSA) solutions versus surface tension (mN/m) run at a pump speed 250μl/200s.

To remove the influence of density from the relationship an equation was derived to eliminate both the initial and concluding stages of drop development. The equation is based on measurements of tensiopeak period (TPP) and rainbow peak period (RPP). These measurement positions in the drop formation are used as markers for surface tension and remove the influence of density that mainly influences the concluding part of drop development. This new approach is shown in Figure 2(b), TPP-RPP versus surface tension for methanol-water solutions. The error bars in determining these drop period differences belies this excellent correlation coefficient. The error bars shown for Figures 2-3 are the range values (3σ). The syringe volume used in all the experiments was 250μl.

4.2   Surfactant analysis

Studies with surfactants using this relationship have been undertaken for drop period versus surface tension and the results are shown in Figure 3(a) for a slow drop delivery. Experimentally there is an excellent relationship shown by the very small error bars and the high correlation coefficient in the surface tension range 37-72mN/m. It was however found that for fast drop deliveries this simple relationship did not hold due to the fact that a diffusion process was complicating the situation. Each molecule has of course a different diffusion coefficient and this effect has been thoroughly investigated in a series of studies using pharmaceuticals and various protein molecules, but a 600s drop delivery rate has proved adequate to achieve a tight linear relationship for all surfactant molecules. The regression value however markedly deteriorates with increasing speed for this measurement with surfactant molecules as can be seen from Figure 3(b). A second order relationship is then obtained that asymptotically tends to a straight line for low concentrations of the surfactant, that is here for high values of surface tension. The curve increases away from this asymptote towards low surface tensions and high concentrations of surfactant. The curves for Sodium Lauryl Sulphate (SLS) show the quadratic form revealing it as being a surface active molecule. For solvents, no diffusion can occur, and variations with pump speed are not observed, as these are obviously homogenous liquids.

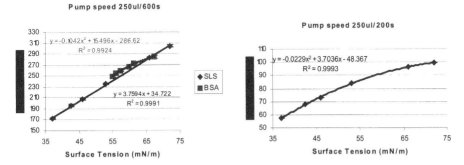

Figure 3(a) a linear relationship shown for drop period(s) for SLS solutions plotted against surface tension (mN/m). Second order relationship for BSA. Measurements taken at a pump speed 250μl/600s.
(b) Drop period (s) for a series of SLS dilutions versus surface tension (mN/m) run at a pump speed 250μl/200s.

4.3 Protein analysis

Bovine serum albumin (BSA) was used in this study to establish if the surface activity of the protein could be determined depending on its position relating to the two graphs (Figures 2(b) & 3(a)). As can be seen from Figure 2(b), BSA cannot be classified using this graph as it does not have the same relationship as the methanol solutions. The surface tension of the BSA solutions varies as can be seen from the dispersion across the x-axis. There is only a small variation in the y-axis direction. This is due to the fact that all the traces for the concentrations examined have the same characteristics in the central part of drop formation (from rainbow peak to tensiopeak). The main part of drop formation where BSA has an influence is the concluding part (from tensiopeak to drop separation).

The protein solutions were also analysed using a lower pump speed (250μl delivered in 600 seconds). The results of drop period versus surface tension were plotted and compared to the results for the surfactant SLS. These results are show in Figure 3(a). The results show a linear relationship between BSA and SLS. The results would indicate that BSA has a high surface activity because of its closeness to the calibration line.

5.   Conclusions

The important conclusion to draw from this analysis is that molecules can be effectively characterized in their surfactant behaviour based on these two graphs, namely (i) the graph of drop period versus surface tension for surfactant solutions and (ii) the graph of (TPP-RPP) versus surface tension for non-surface active molecules.

      The protein studies that we have undertaken have at the first level been able to demonstrate the utility of this approach. The molecular diffusion effect can also be studied by stopping the pump at a specific time (i.e. using the tensiopeak period as a marker) and thus determining the length of time it takes for the drop to fall. Experimentally, it seems that this time is determined by the amount of molecules that are left to arrive at the surface (liquid/air interface). This implies that with higher pump speeds there are more molecules available to arrive at the interface. Thus these molecules after the pump has stopped change the surface and influence the time the drop stays suspended. At lower speeds where the molecules have had time to arrive at the surface, the effect is less marked. This work will be described in a later report.

      In most cases the tensiotrace has at least two well-defined peaks. There is an advantage in tensiography over tensiometry. The tensiotrace has more measurement capability than tensiometry as this depends on the single drop period (or drop weight) measurement. The conclusion that might be drawn from the work begun in this project is that tensiography may have a very significant potential in classifying the surface activity of liquids in a way that is perhaps more convenient than traditional drop volume methods.

6.   References

[1] Möbius, D. and Miller, R., Drops and Bubbles in Interfacial Research, Elsevier, 139-186, 1998.

[2] Möbius, D. and Miller, R., Drops and Bubbles in Interfacial Research, Elsevier, 593-705, 1997.

[3] McMillan, N.D., Finlayson, O., Fortune, F., Fingelton, M., Daly, D., Townsend, D., McMillan, D.D.G. and Dalton, M.J., A fiber drop analyser; a new analytical instrument for the individual, sequential, or collective measurement of the physical and chemical properties of a liquid, Rev. Sci. Instrum., 63(6), 216-227, 1992.

[4] McMillan, N.D., Reddin, M., O'Neill, M., Jordan, R., Phillips, D., Goff, D., Nolan, J., Harnedy, R., Mitchell, W., Harkin, J., Lawlor V. and McMillan, L.R.L., The tensiograph – A novel instrument for the fingerprinting and analysis of multiple physical attributes of beer, Journal of the Institute of Brewing, **106:3**, 147-156, 2000.

[5] McMillan, N.D., Lawlor, V., Nolan, J., Lo, W.Y., Harnedy, R. and O'Neill, M., The application of the tensiograph D-functions to quality control in whiskey manufacture, Colloids and Surfaces A: Physiochem. and engineering aspects 143, 421-427, 1998.

[6] McMillan, N.D. Tensiograph studies of pollution in rivers and water courses, Private consultancy report, Carlow, September 1997.

Poster Programme
Paper presented at Sensors and their Applications XII, September 2003
©2003 IOP Publishing Ltd

Geometrical Analysis and Experimental Characterisation of a Figure-of-Eight Coil for Fibre Optic Respiratory Plethysmography

A T Augousti, F-X Maletras, and J Mason

Faculty of Science, Kingston University, Penrhyn Rd, Kingston, Surrey, KT1 2EE UK

ABSTRACT: An optical fibre sensor based on the macro-bending loss effect has been redeveloped for thoracic and abdominal circumference measurements in non-invasive plethysmographic respiratory monitoring. The primary novelty of this paper is the use of a figure-of-eight loop, which displays increased linearity and less mechanical resistance, as well as other benefits. A comparison is made with earlier implementations of this technique, and the newer sensor compares favourably, indicating a higher resolution and a simpler mechanical construction.

1. Introduction

The use of optical fibers for plethysmography, with particular application to respiratory monitoring, has been reported earlier [1-5]. This earlier work has demonstrated the feasibility of using the macrobending loss effect (MBLE) produced in fibre coils for sensitive measurements of chest circumference. As the fibre experiences increasing curvature, the optical power transmitted by the fibre is altered by the eradication of high order modes. This effectively results in a lowering of the total mode power emerging from the fibre end. However, one of the drawbacks of this earlier system was the requirement of using several coils of fibre in series (see Figure 1), so that one could achieve sufficiently high sensitivity, whilst at the same time remaining within a relatively linear region of the response characteristic of a single coil. One problem in using a series of coils in this way is partly mechanical, in that it is difficult to ensure the even distribution among individual coils of an extension that is applied across the coil series as a whole. As a result, individual coils will suffer excursions into non-linear regions of their response characteristic. Additionally, as the extension is released the coils may retract in a different sequence other than the reverse of that which occurred when they expanded. The consequence of these two effects is to reduce the linearity of the overall response and to increase its hysterisis.

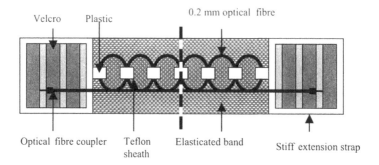

0.2 mm optical fibre

Optical fibre coupler   Teflon   Elasticated band
                         sheath   Stiff extension strap

Figure 1: Previous FORP Sensor

Consequently, one potential design improvement would be to utilise a coil construction that provided a linear response over a wider extension range (while largely preserving the lateral dimensions, in other words, without using a much wider coil), thereby reducing the overall number of coils required to achieve the same loss. It would also be desirable to make the coil movement smoother, thereby reducing microscopic "stictional" effects (where the coil expands or contracts in a series of jumps, rather than by smooth movement). Both of these advantages are embodied through the use of a figure-of-eight design (see Figure 2a) that is described in more detail below. The remainder of this paper provides an analysis of such a coil, allowing its performance to be characterised in terms of this analysis.

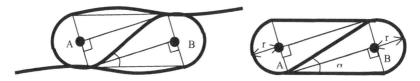

Fig. 2a: Geometrical decomposition of the figure-of-eight coil

Fig. 2b: Model of the figure-of-eight coil

2. Sensor Development

2.1 Simple coil configuration

The earlier FORP sensor was based on a multiple simple coil design, as can be seen in Figure 1. The details of construction of this device have been published elsewhere [1-3]. The response of this sensor to elongation was essentially non-monotonic due to the presence of points of inflexion in the response curve. A linear approximation over a given range was acceptable because the residual deviation was small. The non-monotonic response was attributable to the complex mechanical assembly of the fibre, which generates non-linear behaviour as the radius of each coil can vary out of phase with one another in a random manner.

The fibres were held in place using plastic collars, which forced them to take up a two-dimensional coil configuration. This configuration did not involve any tying of the fibres and was therefore mathematically equivalent to an unknot. The tendency of the fibre would be to come back to its naturally untwisted state if the plastic collars did not restrain it, thereby leading to the production of forces transverse to the plane of the coil. These forces lead to increased friction between the coils and the collars, which in turn limits sensor resolution. This configuration is mechanically complex in terms of mounting the fibre and the collars on the band because it contains multiple coils that are in a condition of forced equilibrium.

The friction was minimised in this configuration by reducing the fibre stiffness by using thinner fibres, typically with a diameter of 0.2 mm. This raises two issues: firstly, coupling light into small diameter fibre is delicate because of the possible alignment errors between the fibre ends and the light source and detector; secondly, a smaller core diameter restricts the maximum possible optical power being carried by the fibre. Both issues contribute to the same effect, namely the output optical signal is diminished. This is easily compensated by appropriate amplification of the optical signal at the photo-detection stage but this also inevitably produces a higher noise level. Ultimately, the most appropriate solutions are firstly to use a larger optical fibre and secondly a more powerful light source at the fibre entrance.

2.2 Figure-of-eight configuration
The solutions suggested above were put into practice with the redevelopment of the sensor. The light source is now pulsed in synchrony with the acquisition circuit to produce a higher instantaneous input optical power than is available in continuous mode and the sensor itself uses a 0.6 mm core diameter fibre. The transducer itself also has a simpler mechanical construction, as it is comprised of a single knot only, designated as the figure-of-eight configuration.

The figure-of-eight has a central point of symmetry and two-dimensional geometry. Since it has only two linked coils, this configuration has numerous mechanical advantages. Lower friction at the points of contact allows the use of a larger fibre core diameter, which leads to a better output signal level. The sensor response is now truly monotonic and its deviation from a linear approximation over a limited elongation range is significantly reduced.

This mechanical simplicity also confers robustness and reliability, as it is less prone to fibre breakage. Even more importantly, when stretched the figure-of-eight behaves rather like a spring by providing a recoil force on each of its extremities. This recoil force arises from the torque produced by the two circular arcs of the fibre. This recoil force offers the useful advantage of actively minimising the sensor hysterisis by reducing the effect of friction at the fibre crossing points. This provides excellent repeatability and reproducibility. Also, symmetric forces interacting at both fibre crossing points generate planar stability, eliminating a further source of friction. Once stretched, the transducer remains self-sustained in a plane and a guiding structure is no longer required. This makes it relatively easy to mount on a band, since the fibre needs only to be tethered at both ends. Two plastic cubes are used for this purpose. Soft and robust skin compliant polymer rubber is used as a mounting band. It has the same elastic properties as Tubigrip band but with the advantage of greater longevity, through the absence of any cutting and sewing operations.

Following polishing, each end of the optical fibre is glued into an drilled brass cylinder that serves as a coupling element. It has an inner diameter that matches the fibre outer diameter, and an outside diameter corresponding to the inside diameter of the source and detector plastic housing.

3. Geometrical Model

A simple geometrical model of the figure-of-eight coil has been derived which has met with considerable success in explaining the observed results. The model assists in understanding the behaviour of the radius of curvature as a function of the elongation applied to the coil. It is based on a simplification obtained by decomposition of the fibre path inside the figure-of-eight into a collection of segments that are either circular arcs or straight lines. Figure 2a represents the superimposition of the actual fibre path and the decomposition of this path, whereas figure 2b shows the model only, based on the decomposition seen in figure 2a.

Two identical isosceles triangles and two semicircles of diameter equal to the base of the triangles are used. Note that the diameters of the two semicircles are parallel. The figure-of-eight has a central point of symmetry and therefore it can be divided into two identical substructures, shaped like pendant drops. Each drop is constituted of a semicircle of diameter equal to the base of one of the isosceles triangles.

The length of fibre that forms the shape of the transducer is termed L. With reference to the description above, we can describe L as a sum of straight paths (AB) and semi-circular paths (πr)

$$L = AB + \pi r + AB + \pi r + AB = (2 \pi r) + (3 AB) \tag{1}$$

The triangle side AB can be expressed as a function of r via the angle α of the isosceles triangle. We may say

$$AB(r) = r / \sin(\alpha/2) \tag{2}$$

By visual inspection one may observe that the figure-of-eight coil appears to be scale invariant. In other words it suffers only minimal qualitative alteration in its shape while undergoing stretching, and for this reason the ratio of width to height may be assumed to be constant. According to the model, the width and height are given by

$$W = 2 r + AB \tag{2a}$$

$$H = AB \sin(\alpha) \tag{2b}$$

By employing this assumption of scale invariance to the model, we may obtain the following expression

$$W / H = (1/\sin(\alpha)) * (1 + 2r/AB) = constant \tag{2c}$$

One consequence of scale invariance is that the angle α remains constant. This is confirmed by visual inspection. We may then derive another constant expressing the scale invariance in the following way

$$K = 3 AB(r) / 2 \pi r \tag{3}$$

The constant K here represents the ratio of the sum of linear paths to the sum of curved paths in the shape. According to equation 2, the constant K can be expressed as

$$K = 3 / 2 \pi \sin(\alpha/2) \tag{4}$$

Using the constant K, we can now simplify the expression for L by combining equations 1 and 3

$$L = 2 \pi r (K + 1) \tag{5}$$

Equation 5 represents a direct relationship between the radius of curvature of each semicircle and the overall length of the optical fibre transducer. The aim of this model is to express the radius r in terms of the quantity Δx, the elongation applied to the sensor since Δx, unlike L, is directly measurable. One may therefore define the total length of fibre in the figure-of-eight sensor as L_{max}, where L_{max} is defined as the sum of L and Δx, with L the length of fibre in the transducer itself and Δx being the elongation applied across both ends of the fibre.

$$L_{max} = L + \Delta x \tag{6}$$

Using equation 5, L_{max} is defined in the same manner as L but for a fixed radius r_{max} and we may thus write

$$L_{max} = 2 \pi r_{max} (K + 1) \tag{7}$$

An expression for Δx may be obtained by combining equations 5, 6 and 7.

$$\Delta x = L_{max} - L = 2 \pi (r_{max}-r) (K + 1) \tag{8}$$

The quantity $r - r_{max}$ is termed Δr and we now have an expression for the change in the radius of curvature change as a function of elongation

$$\Delta x = -\,2\,\pi\,\Delta r\,(K+1) \tag{9}$$

The form of K is a function of the angle α and to determine K, α must be measured, since it it not uniquely defined. It was found in this case to be approximately equal to $\pi/4$ radians. Equation 9 can now be expressed as

$$\Delta r = -0.07\,\Delta x \tag{10}$$

Equation 10 expresses the fact that the radii of both semicircular lobes diminish in a linear manner when an extension is applied to the sensor. This equation is experimentally verified by applying an extension Δx to a 0.6 mm core diameter glass optical fibre in the figure-of-eight configuration and simultaneously measuring Δr with electronic callipers. The results were recorded and a linear regression line of Δr as a function of Δx was obtained (Figure 3). Over a variation of 150 mm for Δx, the regression and the actual measurements have a Pearson correlation coefficient of 0.997 which indicates a high degree of correlation and therefore strong linear behaviour within this region.

The regression equation provides an experimental value of the sensitivity coefficient $\Delta r/\Delta x$ of -0.0815, indicating reasonably good agreement with the model. The theoretical and experimental values of the sensitivity coefficient differ by 14% of the larger value. This result is encouraging given the relative simplicity of the model.

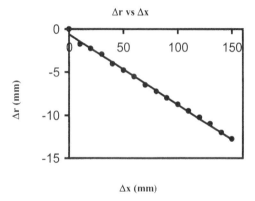

Fig. 3 Curvature radius variation in figure-of-eight as a function of elongation variation

In the present state of the model, the most probable source of error is the value of K, which has a strong dependence on α. A plot of the relative error between the theoretical and experimental values of $\Delta r/\Delta x$ for different values of α is given below (Figure 4). This plot reveals that a 0% percent error between the model and the measured value of $\Delta r/\Delta x$ is obtained when α reaches 60°, which is unrealistic: this value of α was not observed in the figure-of-eight coil. In other words, the relative error is not minimised by more careful measurement of α, which indicates a more fundamental discrepancy between the model and the reality. This discrepancy is likely to lie in the geometrical simplification that the model imposes on the real system.

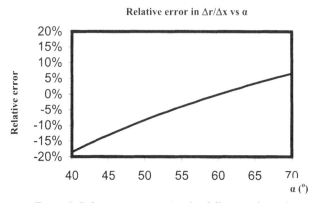

Figure 3. Relative error in Δr/Δx for different values of α.

4. Optical Model

The figure-of-eight sensor is based on the MBLE. D. Gloge [7,8] has derived a mixed wave and geometrical optics model of the MBLE. The model is based on the use of a coefficient, α, that represents the power loss attenuation per unit length of circular path. This coefficient is a function of the propagation mode angle and the fibre radius of curvature. It reflects the attenuation of the electric field in the cladding during bending loss. As can be seen below in equation 11, this coefficient α (not to be confused with the angle α in the figure-of-eight model described above) has an exponential form that is a strong function of the propagation angle θ.

$$\alpha = 2nk(\theta_c^2 - \theta^2)\exp(- 2/3 \; nkR(\theta_c^2 - \theta^2 - 2a/R)^{3/2}) \tag{11}$$

where k is the angular wave number $2\pi/\lambda$, n is the core index, θ_c is the critical angle, a is the core radius and R is the fibre radius of curvature. By inspection, it is easily shown that all modes above an apparent critical angle θ_f are eliminated by the coefficient. This apparent critical angle θ_f may be expressed as:

$$\theta_f = \theta_c * sqrt(1 - 2a/R\theta_c^2) \tag{12}$$

Equation 12 is very important as it expresses the modification of the angular condition for total internal reflection at the core-cladding interface due to fibre curvature. In essence, the MBLE generates an apparent increase of the real critical angle due to the bend. Figure 6 below shows the example of a guided mode propagating with angle θc in a straight fibre that suddenly becomes extinguished when the condition for total internal reflection cannot be satisfied at the onset of the bend.

Boechat and co-workers [9] have derived an approximation that is independent of α for the relative power attenuation by MBLE based on the Gloge model. The approximation is adapted for short lengths (circa 5m) of large core (>0.2 mm) multimode fibre. The Boechat approximation uses the MDP of the fibre at the optical launch point. For such short fibre, it is assumed that natural mode coupling (i.e. coupling that occurs without fibre bending) does not occur sufficiently for the mode distribution function to reach an equilibrium profile. The distribution therefore depends on local conditions such as the nature of the launching optics. The MDP cannot be derived theoretically because it depends only on these experimental parameters which are particular to this circumstance; it must necessarily be measured and eventually modelled for simplicity of use. After measurement, modelling work by Boechat indicates that a best fit approximation of the MDP for large core multimode fibre can be approximated by a trapezium. This approximation is very close to the real MDP except for small propagation angles where the rapidly oscillating distribution is simplified as a flat distribution.

To calculate the relative power attenuation, the local attenuation must be multiplied by the MDP and then integrated with respect to the circular optical fibre path length. The resulting expression is then summed over all modes to provide a global mode power. The relative power attenuation Pr is defined as the difference in the global mode power before and after a local curvature, divided by the global mode power before the curvature.

Since α varies extremely rapidly with θ and the overwhelming majority of modes beyond θ_f are completely extinguished, α is more appropriately considered to represent a simple low-pass function (and can therefore be simply represented as a step function) rather than as a continuously-varying modulating factor. Using this characteristic, Boechat proposes a simplified relative power attenuation P_r that does not contain the attenuation coefficient α. An exact expression for P_r and an approximate expression are given below:

$$\mathrm{Pr} = \frac{\int_0^{\theta_c} P_0(\theta)d\theta - \int_0^{\theta_r} P_0(\theta)\exp(-\alpha l)d\theta}{\int_0^{\theta_c} P_0(\theta)d\theta} \approx \frac{\int_{\theta_f}^{\theta_c} P_0(\theta)d\theta}{\int_0^{\theta_c} P_0(\theta)d\theta} \tag{13}$$

In equation 13, l represents the curved length of optical fibre and $P_0(\theta)$ is the MDP. Despite the absence of α, the simplified relative power attenuation differs from the exact form by a minimal amount, between 0% to 0.5 %. This justifies the approximation. Boechat's work also indicates that the length of the curved path has a negligible effect on bending loss compared to the radius of curvature itself, since the approximation does not include the fibre curvature length l specifically. This suggests that most of the MBLE takes place at the onset of a circular bend. It therefore follows that a proliferation of coils, as in the previous version of the sensor, is not necessarily required to produce a stronger MBLE, depending on what occurs to the MDP in the straight sections between coils.

The behaviour of P_r may be calculated from the MDP and the apparent critical angle θ_f. The expression for θ_f was originally associated with α and therefore is the only remaining contribution to the original model by Gloge in the approximation for P_r. By inspection, we can say that θ_f decreases extremely rapidly for small radii of curvature, therefore quickly expanding the integration domain towards small propagation angles as R diminishes. This sudden rapid expansion in θ_f creates a similarly rapid and sudden increase in P_r for small radii of curvature. This emphasises the importance of the accuracy of the MDP when calculating P_r for small radii of curvature.

In the case of Boechat's experiment, the optical source is a laser and therefore has a non-uniform distribution profile for high order modes. The MDP can be approximated by a trapezium, meaning that a large proportion of the mode population is constituted of axial modes. P_r was calculated using this trapezium approximation and then compared to actual measurements. The comparison revealed that the calculation was in good agreement with the measurement but that the model overestimates the MBLE for small radii of curvature. One must remember that the Gloge model was developed with the idea of calculating high order bending losses and therefore extrapolation to low order modes does not guarantee accuracy.

The MDP in the first few meters of the fibre is uniquely dependent on experimental conditions and essentially on the quality of the optical components that are used. For high quality optical fibre systems with good metallic optical connectors, one can assume that the measurement of the MDP will be fairly consistent. This is not necessarily the case for lower quality plastic connectors such as the ones that were used in this work [Siemens SFH 450V,

250V]. Such connectors, because of their high mechanical tolerance, do not provide consistency in the fibre connections and as a consequence alter the MDP in ways that are difficult to reproduce. For this reason, the Boechat approximation is strongly system dependent and therefore not easily generalised. In other words, it can be used to assess the MBLE of an existing optical fibre system but not to forecast the loss of a system to be built, prior to MDP measurements. Whatever method is used, the MBLE calculation problem will remain strongly dependent on experimental parameters and no purely theoretical model can deliver a definitive quantitative answer as to how much power is lost during curvature.

5. Experimental Conditions

To obtain a large dynamic response in the figure-of-eight sensor, we can contribute actively to maximising the MBLE. This is achieved in practice by overfilling the fibre at the optical launch point in order to exaggerate the number of cladding modes, thereby encouraging the fibre to lose these higher order modes during bending. The source wavelength also contributes to the MBLE. From a geometrical optics viewpoint, selecting a short wavelength source may increase the maximum possible number of modes. The chosen wavelength must, however, conform to the optical fibre transmission spectrum. In our case we have used an infra red LED (SIEMENS SFH450) of wavelength 950 nm as a compromise between ease of transmission and number of modes.

In order to test the sensor for MBLE response to elongation, an automated test bench was developed. The bench consists of a computer controlled tensiometer that applies very accurate elongation variation to the sensor. The sensor was illuminated by a pulsed IR LED source and the optical power received at the sensor output during elongation was converted in synchrony with the source to an electric signal via analogue signal conditioning electronics and then sent to the control computer for digitisation and acquisition. Code for the control computer was developed using Visual Basic 5 (for the tensiometer control) and C (for the optical source pulse and acquisition control). The tensiometer, the source pulse generator and the acquisition system were synchronised by means of CPU interruptions. Matlab^{TM} was used for signal processing.

A silica optical fibre of core diameter 0.4mm and NA of 0.37 (Newport sensor grade optical fibre F-MBC) was used in the construction of the new FORP sensor. The fibre diameter was chosen as a compromise between the sensitivity of the attenuation with curvature and mechanical stiffness. Stiffness is required to provide the recoil force that reduces the hysteresis. The fibre was connected via 2 brass cylinders to the source and detector, both housed in plastic, for which external and internal diameters were tailored to match the inside diameters of the plastic housings and the outside diameter of the fibre. Once fitted with these diameter converters, both fibre ends were polished to maximise the planarity of the optical junctions. The fibre was then tied into a figure-of-eight knot and mounted on the tensiometer.

6. Sensor Stretch Results

The test bench permitted us to investigate the behaviour of different types of optical fibre undergoing bending loss and to compare them with the MBLE model that was discussed above. Primarily, we were interested in assessing the new sensor. Figure 7 shows a comparison of the responses of the earlier and the present sensor configuration, using the same optical source and receiver, the same fibre input optical power and the same offset and magnification settings.

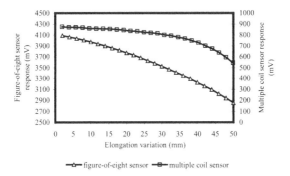

Fig. 5 Comparison between earlier and present FORP sensor responses.

One may observe that the earlier sensor has a relatively large static attenuation (with a small dispersion from the average and a small offset) and a limited range. We attribute these two characteristics to the large number of coils in the sensor configuration: the sensor response is the product of the responses from each individual coil. Since each coil receives only a fraction of the elongation applied to the sensor, the power attenuation per coil is small and the product of these modulations remains small too. This explains the limited dynamic range of the earlier sensor. Also, having the coils aligned in series will necessarily produce a non-linear sensor response, even if the individual coil response is relatively linear in the first place. This is because the overall response characteristic is the product of the responses of the individual coils, which will have the effect of maximising any deviations from linearity. This effect is noticeable in Figure 7: the earlier sensor response curve exhibits significant nonlinearity whereas the newer sensor, because it has a single coil, exhibits a more linear response curve. Over a small elongation range one may reasonably adopt a linear approximation for the sensor response.

More importantly perhaps, one may also observe that the new sensor has a response amplitude approximately 2.5 times greater than the earlier one. Since the elongation applied to the figure-of-eight coil is distributed over a single coil only, the variation in the radius of the coil is maximised and consequently so is the MBLE. Such an improvement to the resolution thereby increases the signal to noise ratio (SNR).

In addition, Figure 6 demonstrates the greatly improved performance of the system as far as hysterisis is concerned.

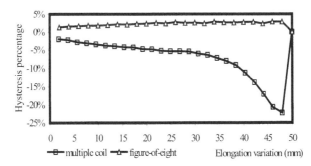

Figure 6 Comparison of multiple coil and figure-of-eight coil sensors relative hysteresis

The test bench also offered the opportunity to compare the predictions of the MBLE model and the figure-of-eight MBLE response obtained under given conditions. The graph in Figure

7 shows the result of this investigation. Close agreement over most of the measurement range is observed. However, we note here that, unlike the case of the Boechat experiment, the calculated MBLE always underestimates the measured MBLE for small radii of curvature. Since we have calculated P_r in the same manner as Boechat, it is unlikely that this calculation could be the source of the discrepancy. This difference therefore is likely to arise from variability in the experimental conditions as alluded to in the discussion of the model above.

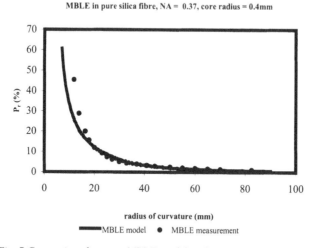

MBLE in pure silica fibre, NA = 0.37, core radius = 0.4mm

radius of curvature (mm)

━━━━MBLE model   ● MBLE measurement

Fig. 7 Comparison between MBLE model and figure-of-eight transmission

In conclusion, then, the newer FORP sensor shows an immediate improvement in the response, indicating that the sensor re-development has been successful.

7. Discussion and Conclusion

The FORP sensor has been successfully redeveloped. The new sensor, based on a simpler coil configuration resembling a figure-of-eight knot, has proven very satisfactory in terms of MBLE response amplitude, achieving a response amplitude that is 2.5 times larger than the response amplitude of the old sensor, for the same elongation range. The new sensor response curve exhibits a markedly more linear behaviour than the previous one. The optical fibre in the figure-of-eight configuration behaves like a spring, and therefore provides its own recoil force. This mechanism helps to reduce substantially the sensor hysteresis and increase the precision of the sensor. For the purpose of conducting these elongation tests, an automated test bench was developed to apply a continuous elongation variation to the sensor while measuring and recording its response.

In order to understand the linear relation between the elongation applied to the figure-of-eight coil and the resulting variation in curvature radius, a geometrical model of the coil has been developed. Model and experimental proportionality coefficients differ by 14%, which is acceptable given the simplicity of this geometrical model.

The experimental MBLE response of the coil under elongation has been compared to a theoretical model of the MBLE given by Boechat for short length large core optical fibre. It has been noted that the model is in good agreement with experimental data, except in the region of small radii of curvature where discrepancies appear due to the nature of the model itself. It was pointed out that the calculation of MBLE response is largely dependent on experimental conditions and therefore no purely theoretical model could make an accurate prediction, or rather that several of the terms involved in any prediction required measurement by experiment. Despite these discrepancies, both model and experimental results can be approximated as linear responses for small elongation variation, as occurs in normal quiet breathing.

References

[1]   The development of a fibre-optic respiratory plethysmograph (FORP) **A.T.Augousti and A. Raza** p401-6 *Sensors VI: Technology, Systems and Applications* **K.T.V. Grattan and A.T. Augousti (Eds)** IOP Publishing 1993

[2]   Design and characterisation of a Fibre Optic Respiratory Plethysmograph **A.T. Augousti, A. Raza M. Graves** SPIE p250-7 **2676** Proceedings of *BiOS 96 - Biomedical Sensing, Imaging and Tracking Technologies* 27 January - 1 February 1996 San Jose, California

[3]   Calibration protocols for a fibre optic respiration monitor **A. Raza and A.T. Augousti** *Sensors and Their Applications VIII* **A.T. Augousti and NM White (Eds)** p33-38 IOP Publishing 1997

[4]   A new sensor for monitoring chest wall motion during high frequency oscillatory ventilation. **C. Davis, A. Mazzolini, J. Mills, and P. Dargaville** *Med. Eng. & Phys.* **21** 619-623 2000

[5]   A new fibre optic sensor for respiratory monitoring. **C. Davis, A. Mazzolini and Murphy D**. *Autralasian Physical & Engineering Sciences in Medicine.* **20** 214-219 1997

[6]   Signal processing considerations in the use of the fibre optic respiratory plethysmograph (FORP) for cardiac monitoring **F-X. Maletras, A.T. Augousti and J. Mason** *Sensors and Their Applications XI* **K.T.V Grattan and S. H. Khan (Eds)** p371-376 IOP Publishing 2001

[7]   Weakly Guiding Fibers **D. Gloge** *Applied Optics* **10** 2252-2258. 1971

[8] Bending Loss in Multimode Fibers with Graded and Ungraded Core Index **D. Gloge** *Applied Optics* **11** 2506-2513 1972

[9]   Bend Loss in large core multimode optical fiber beam delivery system **A. A. P. Boechat, D. Su, D. R. Hall and J. D. C. Jones** *Applied Optics* **30** 321-327 1991

Drop shape sensor modelling and imaging: new measurement approaches for refractive index monitoring

D Carbery[1]*, N D McMillan[1], M O'Neill[1], S R P Smith[2], S Riedel[3], B O'Rourke[1], A Augousti[4] and J Mason[4]

[1] Institute of Technology Carlow, Kilkenny Road, Carlow, Ireland.
[2] University of Essex, Wivenhoe Park, Colchester CO4 3SQ, United Kingdom.
[3] Kelman Ltd., Kelman Ltd, Lissue Industrial Estate East, Lissue Road, Lisburn, Co. Antrim, N. Ireland BT28 2RB
[4] School of Life Sciences, Kingston University, Kingston upon Thames, Surrey KT1 2EE, UK

Email: carberyd@itcarlow.ie

Abstract: Various mathematical methods to model drop shapes are critically evaluated. The aim of future work is to get the theoretical curves to accurately match the camera images of various known liquids, supported on a 9mm diameter drophead. Many traditional numerical methods provide adequate fits to the drop image using non-linear numerical methods. This work explores the limitations of the traditional methods on these large diameter drops. The work has also led to an explanation of the complex image produced by backlit drops. From the analysis of these backlit images, refractive index measurements for the liquids have been obtained. The paper ends by presenting a broadened set of research objectives for future work based on the experimental and theoretical evaluation of drop imaging.

1. Introduction

For two centuries drop profiles have been calculated from the Laplace-Young [1] equation and experimental technologies were later derived from the subsequent work of Bashforth and Adams [2] who developed alternative formulations of these equations. Fordham [3] and Mills [4] devised tables and data for sessile and pendant drops that are used to define drop shapes. A fuller systematization of this work to produce handy tables came from the work of Padday [5]. Numerical approaches to this problem can be traced to the work of Rotenburg et al [6] who defined an error function as the perpendicular point between the measured points and the calculated profile, which they minimized using conventional Newton-Raphson technique. A different approach using numerical methods was devised based on fourth-order Runge-Kutta numerical methods such as those of Huh and Reed [7] and is also used in this paper to show the variance with real camera images The failure of the various proliferation of approaches to drop shape modelling however to give accurate solutions to the shape of equilibrium drops perhaps has been somewhat obscured by the simplicity of the Laplace-Young equation, or perhaps this fact has been masked by the power of emerging computer technologies to deliver computer perturbation methods to these drop shapes such as those of O'Brien [8]. Various attempts have been made to find solutions to the shape of equilibrium drops by finding local energy minima and other approaches to the minimization procedures such as the Maze-Burnett algorithm [9]. There exists several major review studies of drop shape methods such as that by Hartland and Hartley [10] who dealt fairly comprehensively with the well established methods and the recent publication edited by Möbius and Miller [11] which has dealt with both the historic and more recent work.

Experimental work on a backlit drop is also reported in this paper, showing the path of light through the drop, which is the basis for refractive index measurements.

2.   Theory

Bashforth and Adams equation represents the starting point for most of the drop shape methods developed so far. Figure 1 shows an appropriate diagram to which the equations below refer. The model profile is calculated from the Laplace equation. This equation can also be represented in the form of a set of three first-order differential equations:

$$\frac{dx}{ds} = \cos(\theta) \tag{1}$$

$$\frac{dz}{ds} = \sin(\theta) \tag{2}$$

$$\frac{d\theta}{ds} = -\beta z + \frac{2}{b} - \frac{\sin(\theta)}{x} \tag{3}$$

where z is the vertical distance of point S above the apex. s is the arc length and θ is the normal angle. b is the radius of curvature at the apex. β is known as the shape factor defined as $\beta = \Delta\rho g / \gamma$ where $\Delta\rho$ is the density difference, g the gravitational acceleration constant and γ is the surface tension of the liquid.

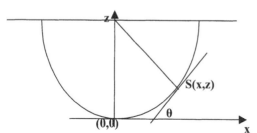

Figure 1. *Sketch of the section of a pendant drop.*   **Figure 2.** *Real image of a water drop*

The differential equations are solved by a fourth-order Runge-Kutta integration algorithm. Runge-Kutta methods [7] refer to a large family of one-step methods for numerical solution of initial value problems. The differential equations are numerically integrated using a step length Δs. Boundary conditions are applied to the three governing equations.

3.   Experimental

3.1   Drop modelling

Two different approaches were adopted to model drop profiles using 1) the Bashforth and Adams equations and the Runge-Kutta procedures for solution and 2) Fordham & Mills equations. All models were compared to a real digital camera image of a water drop formed on a 9mm orifice, Figure 2.

3.1.1   Model 1: Bashforth and Adams

Two different methods were implemented to model the drop profiles. Firstly, the traditional approach was programmed in MatLAB, where you start at the apex of the drop and move upwards using Runge-Kutta integration procedures. The boundary conditions here are x=z=θ=0 and dθ/ds = 1/b. Figure 3 shows the predicted shape for a water drop superimposed on the real camera image. Secondly, the drop is modelled starting at the top of the drop. For the boundary conditions the x and z coordinates are known and the contact angle is determined from Young's equation. The drop shape predicted from this process is shown in Figure 4.

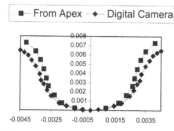 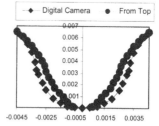

Figure 3. Drop profile starting from apex.   **Figure 4.** Drop profile starting from the top

3.1.2   Model 2: Fordham & Mills

Fordham & Mills tables of data were used to create a model, which was used to test the validity of the programmed models (using the Runge-Kutta algorithm). This data was obtained from Bashforth & Adams equation. The values of θ, x and z were determined by interpolating the data from the Fordham & Mills tables for a β-value (determined from the physical properties of the liquid). Typically the β-value for a water drop does not exactly equal any particular value of β given in the tables.

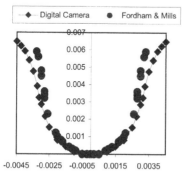

Figure 5. *Drop shape predictions (Fordham & Mills) for a Water drop.*

Therefore, further accuracy was obtained by interpolating values of x/b or z/b between the two relevant columns of the table of data. These tables enable values of x/b and z/b between θ = 0 and 180° for a range of β-values from 0.125 to 100 to be obtained. To plot the drop shape from the x/b and the z/b data one needs to estimate the value of b. The radius, b, is obtained by using the equation:

$$2\gamma/b = \rho gh \qquad (4)$$

where h is the vertical distance between the apex and the drophead (i.e. here the pressure head ρgh). A program was written in MatLAB incorporating a nested loop to determine x and z for different height values (i.e. different drop volumes) at each value of s/b, using the determined equations for θ, x/b and z/b. The predicted drop shape using this method can be seen in Figure 5.

3.2   Refractive Index Studies: - Backlit drop

Figure 6 shows a CCD image of a 90.4 % ethanol-water drop. The experimental arrangement, for the backlit drop study, (see Figure 7) was a simple white diffuser illuminating the drop from the back, with a 220V standard bulb used to provide the illumination and thus does not deliver collimated illumination.

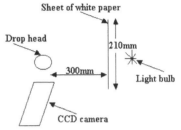

Figure 6. *90.4% ethanol drop image*   **Figure 7.** *The elevation view of the backlit drop.*

As usual in these backlit images, there is a concentration of light, which looks somewhat like the early positive phase of the moon. This image is too complex here to explain, but has been analysed by one of the authors in her Ph.D. thesis [13].

As can be seen from Figure 2 the water drop base has a dark band that follows the contour of the drop. The bright centre region is clearly seen at a certain point in the drop growth. The camera has recorded the entire history of the drop growth from the remnant drop initial stage to the fully developed pendant drop. Across the top is a clear dark edge, which is produced by the dark TIR interface, beyond which all rays are totally reflected.

In addition, a dark rim tracks around the edge of the drop. The experimental situation is not ideal in that a white light source incident on the back of a diffuser will produce 'diffuse scatter' from every point on the diffuser. The back of the drop is effectively illuminated then by a cone of light at every point defined by the size of the diffuser. It is a simple matter therefore giving consideration to the geometry, to determine the two extreme limits of the scattered rays from the diffuser, which in our set-up subtended a cone angle of approximately 36°.

The angle of total internal reflection (TIR) is calculated for water as 48.5°. Figure 8 shows scaled ray constructions (based on an image similar to that seen in Figure 2) for a number of important rays incident on the back of the drop. The image seen in Figure 6 is obviously from the front of the drop and here this represents the RHS of the drop. The dark TIR line can be seen clearly in this image. In Figure 2 various reflection processes occur within the drop but at positions W, S and Q (Figure 8) rays are refracted up from the base of the drop. A TIR dark edge will be produced at the point Q. Simple physics tells us that ray construction here will give us a measure of refractive index. The angle is measured at 48.5°, which not unsurprisingly give a refractive index of 1.333. This graphical construction in our study was done by hand on large-scale photographs. Modern ray tracing packages would of course do this job without any human intervention. A collimator could clearly improve the experimental set-up. The ray reflections from the inside of the hollow drop head produces a second TIR position above Q at S, and thus produce a dark band in the image as seen in Figure 2.

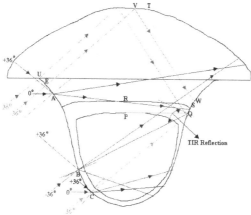

Figure 8 Ray construction using the extreme rays of -36° to +36°.

The dark rim on the base of the drop edge is produced by minimum deviation, which was explained centuries ago by Descartes. The extreme ray that comes from the very edge of the drop is refracted more than a Descartes ray inside this position, which marks the position of minimum deviation. All rays between these rays refract to positions inside the edge marked by the Descartes ray. All rays beyond the Descartes ray refract similarly to positions inside this Descartes ray. As a consequence, a bright centre to the backlit drop is seen during part of the drop growth delineated clearly by a dark rim around the base of the drop. Mathematical investigations of this phenomenon lead to a complex analytical condition for the minimum deviation ray and there is not space here to describe this calculation. It never the less can yield a good measurement of refractive index from the Descartes edge seen in these images, which again we have shown from hand ray tracing is 1.333. Again it is hardly worth pointing out very accurate ray tracing packages exist to do such jobs and secondly collimators should be used to get improved measurement.

4.   Discussion

A start has been made but this modelling still requires some work, as the shapes are quite different from actual shapes recorded with camera studies. The intention of the authors is to integrate these drop models with ray tracing technologies. The following ideas are under discussion and should be implemented into the program: a) relationships between the radii of curvature of the drop profiles and b) the volume/ area ratios. These give important insights into the mechanics of the drop.

5.   Conclusion

The paper has shown that the standard methods for drop shape modelling do not provide adequate fits for pendant drop shapes except by using such artifices as non-linear error computer approximations. The intention of the research will be to obtain accurate fits to equilibrium or quasi-equilibrium drop shapes for tensiography. There have been some important discoveries in drop modelling, which are not described here, but it appears may lead to a simple analytical approach to these problems. In addition, the ray tracing considerations described above of the backlit drop, somewhat by way of a bonus to the main tensiograph study has highlighted the potential for the development of a very accurate drop imaging refractometer. The results obtained with graphical methods were to an accuracy that approached that of the standard laboratory Abbé refractometer, but this technology could perhaps have advantages in on-line monitoring perhaps for brewing industry to measure alcohol content during fermentation, or indeed elsewhere, where small volume monitoring is a particular advantage.

References

[1]   Laplace, P.S., Mechanique Celeste, Supplement 10, 1806.
[2]   Bashforth, F., and Adams, J.C., An attempt to test the theories of capillary action by comparing the theoretical and measured forms of drops on solids, Cambridge university press, Cambridge, 1883.
[3]   Fordham, S., Proc. Roy. Soc. (London), 194A, 1, 1948.
[4]   Mills, O.S., Brit. J. Appl. Phys., 4, 24, 1953.
[5]   Padday, J.F. in: Matijevic (Ed.), Surfaces and Colloids Series, Vol.1,161 (Wiley Interscience, New York, 1969)
[6]   Rotenberg, V.A., Boruvka, L., Neumann, A.W., , J. Colloid Interface Sci. 93 ,1, 169 (1983).
[7]   Huh, C. and Reed, R.L., J. Colloid Interface Sci. 91,2,472 (1983).
[8]   O'Brien, S.B.G.M.., J. Fluid. Mech.,LII, 519-537 (1991).
[9]   Maze, C., Burnett, G., Surface Sci. 24, 335 (1971).
[10]   Harland, S. and Hartley, R., Axisymmetric Fluid-Liquid Interfaces (Elsevier, Amsterdam, 1976)
[11]   Möbius and Miller 1998
[12]   O'Neill M 2001 New tensiograph signal processing and multivariate methods with confirmatory measurement programmes in brewing bio-engineering and pure liquid applications Ph.D. Thesis(NCEA Dublin).

Poster Programme
Paper presented at Sensors and their Applications XII, September 2003
©*2003 IOP Publishing Ltd*

Ecological tensiography: Initial application study on water colour from an Irish bog, unpolluted river and rivers receiving mine effluent

D D G McMillan[1], A C Bertho[2], B O'Rourke[2], D Phillips[2], M Neill[3] and D Morrin[2]

[1] SHE Department, Lisheen Mine, Tipperary, Ireland
[2] Department of Applied Biology and Chemistry, Institute of Technology, Carlow, Kilkenny Road, Carlow
[3] Regional Water Laboratories, EPA, The Butts, Kilkenny, Ireland

Abstract: This paper constitutes the first ecological assessment through field trials of the use of the new multimeasurand tensiograph sensor technology with reports of the first measurements on water colour. These trials have been conducted as part of a European research project to produce early warning surveillance technologies for natural water bodies. The solid reference background knowledge for this study derives from an assessment of the biological and physico-chemical status of four water bodies: a raised bog prepared for peat moss harvesting, an unpolluted river and two rivers receiving effluent from a commercial heavy metal mine. This involves detailed analysis of these waters by traditional methods in water laboratories and some extensive ecological studies of the four habitats. Important lessons have been learned for water monitoring despite the test programme being confined to the single measurand of colour and the paper ends with a discussion of the wider potential of tensiograph monitoring and some of the practical issues involving its deployment.

1. Introduction

The invention of the optical tensiograph as a result of considering what John Tyndall would have done with the modern fiber optic, which was heralded by his 1854 demonstration of the light pipe[1]. Tensiography began in Carlow as a result of experiments based on knowledge of his pioneering work, with the first studies of light injection inside drops using fibre optics to study the rainbow-type reflections produced inside drops distended by gravitational forces [2].

2. Theory

The tensiotrace effectively provide a unique fingerprint of the liquid being studied. Figure 1(a) shows the convention for heights and time of peaks used in the tensiograph analysis. Various characteristic feature of the tensiotrace provides information about the physical properties of the liquid and details of how to derive these measurements from the tensiotrace are described by McMillan, Lawlor et al [3]. For measurements of colour and turbidity, it is the tensiograph and rainbow peaks heights that are used as the measurands. The rainbow peak height is determined in the first instance by the refractive index of the Liquid Under Test (LUT), whilst the tensiograph peak shows a strong dependence on the absorbance of the liquid. Both peaks however depend on the refractive index and absorbance of the LUT.

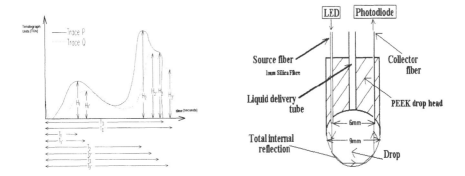

Figure 1(a): Convention for heights and times of peaks used in the tensiograph analysis
Figure 1(b): Diagram of the concave drophead

In order to determine the tensiographic absorbance, it is convenient to assume that the tensiograph peak maximum is in an equivalent position for each trace. The tensiographic absorbance is then given by

$$A_T = \log_{10} \frac{H_m}{H_{m'}} \, or \, \frac{H_{m'}}{H_m}$$

(1)

where the peak height is that of the tensiograph peak.

It has been shown that both the tensiographic absorbance and turbidity obey the Beer-Lambert Law [4] very closely, and hence the following equation applies

$$A = \varepsilon c \int \partial l = \varepsilon c l_{av}$$

(2)

where ε is the tensiograph molar absorptivity (A-units $mol^{-1} \, cm^{-1}$), and c is the concentration in mol l^{-1}.

3. Apparatus

The simplest form of the instrument is a two-fibre drop-head system where light from a LED (Light Emitting Diode) source is injected into the drop-head via the source fibre and the signal is picked up by the collector fibre which passes it to the photodiode that produced the trace.

The multianalyser operates by recording just a single tensiotrace that is scissored from the incoming A/D detector signal that is produced by the light collected from the collector fibre. Data acquisition is started once the optical eyes situated below the drop-head send a control signal triggered by the detection of a falling drop. Data recording continues until the fall of the second, 'measurement' drop. The measurement drop data is converted to a digital form by the A/D card and it is stored in the computer's archive system.

4. Experimental

4.1 Traditional approaches to water monitoring

The health of a river can be monitored by chemical or biological means or preferably by a combination of both. This is because in general, physico-chemical analysis is better at measuring the causes of pollution (i.e., the pollutants) whereas biological monitoring is the only way by which the ecological effects and extent of pollution can be measured.

Biological monitoring: In the presence of pollution well-documented changes are induced in the flora and fauna of rivers and streams, particularly on the macro-invertebrate community, which shows a reduction in community diversity and a replacement of what are termed 'clean-water' species by more tolerant fauna. These changes are due to the varying sensitivities of the different components of the community to pollution stress.

The Environmental Protection Agency's Biological Quality Index (BQI) procedure divides benthic macroinveretebrates into five arbitrary 'Indicator Groups' ranging from the Group A 'sensitive' forms to Group E, the 'most tolerant'. Once the relative proportions of various organisms in the sample are determined the water quality is inferred by a comparison of this data with what would be expected from unpolluted habitats of the type under investigation, to assign a quality (Q) rating, ranging from Q1 which indicates high quality, unpolluted water to Q5 or seriously polluted water [5].

Physico-chemical monitoring: For the assessment of organic pollution the more commonly measured parameters are dissolved oxygen (DO), biological oxygen demand (BOD), ammonia, oxidised nitrogen, (nitrites plus nitrates) and phosphates. Continuous records of concentration and flow form the ideal basis for water quality assessment but in practice reliance is usually on discrete samples. In contrast with biological assessment which evaluates the incidence and intensity of pollution based on the degree to which the chosen community deviates from its expected natural diversity, the physico-chemical assessment is usually based on a comparison of the measurements made with water quality criteria or standards derived from such criteria.

4.2. On-line Monitoring (OLM) and Obstacles to OLM

Various European Directives provide the framework for water quality and these controls are closely related to analytical measurements. A handful of key organic and inorganic substances are the most important pollutants, species such as nitrate, phosphate, ammonium, volatile organic carbons (VOCs) and pesticides and while laboratory analysis of these species are for the most part well developed, many field analysis problems still exist ranging from troubles of fouling to lack of equipment reliability and excessive cost of ownership making them unavailable to OLM [6]. However, OLM is urgently needed as only this can capture the spatial- and time-dependent variability of the water body characteristics. Furthermore, while individual species can be identified and quantified for pollution incidents there are difficulties in rapidly fingerprinting the exact nature of the pollutant involved with present procedures taking up to 2 days making effective targeted preventative action to the spread of the pollution impossible.

4.3 Ecological tensiography: a new technique for environmental monitoring

For the reasons outlined above, there is a growing trend towards stand-alone, low-maintenance OLM systems to substitute for more labour-intensive traditional methods of monitoring rivers. The object of the present work was to assess the capability of the tensiotrace to measure colour and turbidity in natural water bodies. Only the work on colour is sufficiently advanced to report here.

The main organic sources of colour in water systems are complex organic molecules derived from vegetable matter such as humic and fulvic acids while colour also arises from colloidal iron or manganese. The colour of natural water bodies varies widely in space and time with highest levels occurring during flood conditions.

4.4 The Study Sites

Abbeyleix Bog: Random quadrats were used to assess the vegetation on Abbeyleix bog, which has been prepared for peat moss harvesting by the installation of parallel drainage ditches spaced 15 metres apart. The vegetation survey showed that the west side is more substantially degraded than the east side. Consequently, the west side has much in common with dry heath while the east side still contains large areas of typical raised bog habitat with substantial growth of the *Sphagnum* moss crucial to bog development.

Animal fauna was very limited in the highly acidic bog drains but quite diverse in the moderately acidic perimeter drains. Grab samples were taken from the same drainage ditches and perimeter drains on the west and east side of the site. Chemical analysis of these indicated waters typical of bogs, that is to say, very acidic, highly coloured by humic complexes and with only trace levels if any of most ions and metals.

Lisheen Rivers: The Drish and Rossestown rivers, both tributaries of the River Suir, receive peat silt runoff from the peat works through which they flow before they receive effluent from the Lisheen zinc-lead mine discharges with its potential for heavy metal contamination. BQI surveys carried out on in 2002 and 2003 consistently indicated a BQI rating of 3 (moderately polluted). Grab samples for physico-chemical assessment were also taken upstream and downstream of the discharges. Analysis profiles the rivers as (generally) highly coloured with slightly alkaline pH, average DO levels, moderate levels of nutrients and BOD and high background levels of iron, aluminium, barium and ammonium.

Fishogue River: The Fishogue river which is a tributary of the River Barrow receives no major effluent sources and is flanked for most of its course by woodland vegetation. It was assessed for its macroinvertebrate fauna in October 2002 and April and June 2003 at three sites, 2 upstream of the Killeshin Road bridge and one just below. This gave a pristine BQI rating of Q5 for the site. Grab samples for physico-chemical assessment were taken at the same sites and analysis of these indicated a moderately hard water with slightly alkaline pH, high DO levels, low levels of nutrients and BOD and slightly elevated levels of aluminium and iron.

4.5. Analytical results

Preparation of colour standards: Colour standards were prepared by dissolving 1.246g of potassium chloroplatinate and 1.000g of cobalt (II) chloride hexahydrate in 1 litre of 0.1 M HCl to give a 500 Hazen Units (HUs) stock solution. A set of standards was prepared from this stock solution ranging from 25 to 200 rising in increments of 25 HUs and then from 200 to 500 rising in increments of 100 HUs.

Analysis of standard colour solutions: Ten replicate tensiotraces of each standard were taken at 470 nm using a PEEK drophead at a temperature of 20°C. Average values of rainbow and tensiograph peak heights were calculated together with standard deviation and range for each parameter. Each of the ten test tensiotraces were compared to the ten distilled water reference tensiotraces to improve the statistical validity of the results. At the same time, 10 replicate readings were taken of the same standards using a UV-visible spectrophotometer and the average, standard deviation and range were calculated for these.

Tensiograph peak heights from the tensiotraces gave errors generally about twice as large as for the UV-visible but there was a demonstrably improved sensitivity of the tensiograph over the UV-visible spectrophotometer. As errors are generally consistent whatever the colour concentration the technique is less accurate in the colour range 0 to 15 HUs but above this level its improved sensitivity outweighs the disadvantage of greater errors.

Analysis of site water samples: The same grab samples taken for physico-chemical assessment were also measured as for the standard colour solutions using both stainless steel and PEEK dropheads and the UV-visible spectrophotometer. Both dropheads gave effectively the same readings. In terms of colour these represented a complete range from practically zero HUs in the Q5-rated Fishogue to 200 plus in the bogwater samples giving an extensive testing of the instrument's range and sensitivity.

Again, both PEEK and steel dropheads demonstrated an improved sensitivity over the UV/visible spectrophotometer for samples with colour concentrations above 15HUs (see Figure 2, Fishogue river and Abbeyleix bog samples).

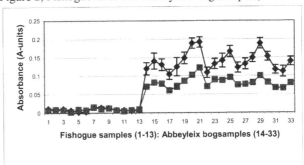

Figure 2. *Typical monitoring results of both tensiographic (Upper) and UV-visible spectrophotometer (Lower) absorbance taken on Fishogue and Abbeyleix bog samples.*

There have been in the present study a full set of measurements taken on both colour and turbidity at the three sites involved in these trials. The data is currently being analysed with a number of data mining tools and will hopefully demonstrate a range of further tensiographic biological and other ecological measurements on these waters.

5. Conclusions – Prospects for Ecological Tensiography

This paper has shown the usefulness of ecological tensiography for monitoring colour on real water samples to an improved level of sensitivity over the standard UV-visible analytical method for samples greater than 15 HUs and upwards. Initial work on turbidity standards has shown acceptable sensitivities and detection levels, but significant and increasing errors for solutions greater than 75 NTUs. Mixed colour and turbidity solutions also indicated that simultaneous measurement of both is possible with a small refinement of the method. The next phase will involve careful studies to overcome these limitations for turbidity and attempt to define assays for mixtures of colour and turbidity. Furthermore, a theoretical assessment of the instrument and its capabilities will be undertaken in an attempt to outline the potential and future lines of research to be undertaken. Part of this will involve a study on the suitability of adapting various spectrophotometric techniques, which fall within the current range of 370-1200 nm of the instrument. The present study has obviously not in any way developed colour through standard water chemistry procedures [7] for tensiographic measurements. However, such techniques are of course already widespread and are currently used in OLM systems. The fact that tensiography has an existing requisite sensitivity to conduct such measurements could enable the technique to be used in such systems.

The great prospect for tensiograph in water monitoring, apart from what looks as some real potential to combine OLM of colour and turbidity, is through its unique ability to deliver multi-measurand profiling. This ability could provide fingerprint information on complex organic water pollutants through the development of the differential tensiographic technique [8]. The technique can be readily adapted to field conditions and an easily-cleaned quartz drophead has been developed to allow for efficient cleaning of the drophead. It is also conceivable that the existing temperature control could be removed for OLM applications if temperature is measured and corrections to tensiotrace measurements and other computer-based calibration procedures are employed to remove the temperature factor. The technology is presently being tested for these possibilities in the research laboratory of IRH, Nancy and the work reported here is a component part of this European-wide feasibility study to develop the instrument for early-warning monitoring and surveillance. Finally, important hardware improvements will see the existing 11-bit resolution of the tensiograph in the immediate future raised above 13-bits, with a corresponding improvement in signal-to-noise, sensitivity and detection limit and this will extend both its monitoring and expand its application capabilities.

References

[1]   Tyndall J On some phenomena connected with the motion of liquids. 1854. Proc. Roy. Inst G.B. 446, Jan. 27, 1854.

[2]   McMillan N D Finlayson O.Fortune F Fingleton M Daly D Townsend D McMillan D D G &. Dalton M 1992. The fiber drop analyser: a new multianalyser analytical instrument with applications in sugar processing and for the analysis of pure liquids. Meas. Sci. Technol, Vol. 3, 766-764, 1992.

[3]   McMillan N.D. Lawlor V. Baker, M. & Smith s 1998. Drops and Bubbles in Interfacial Research. (eds) D. Möbius and R. Miller. Elsevier Press, 593.

[4]   McMillan N D O'Neill M O'Mongain E O'Rourke B Riedel S. Smith S. Wüstneck N & Wüstneck R Quantitative drop spectroscopy: An experimental and theoretical investigation into the instrumental advantage and utility of a fiber optic drop-head spectrophotometer. To go to press July 2003.

[5]   Environmental Protection Agency. 2002. Water Quality in Ireland 1998-2000. EPA, Ireland.

[6]   Colin F & Quevauviller P (Eds). 1998. Monitoring of Water Quality – The contribution of advanced techniques. Elsevier Science Ltd, The Netherlands.

[7]   Standard methods for the examination of water and waste water. 20th Edition 1998 Ed. Clesceri L S Greenberg A E Eaton A D Section 2-1.

[8]   Bertho A C McMillan D D G McMillan N D. O'Rourke B. Doyle G O'Neill M and Neill N A feasibility study into differential tensiography for water pollution studies with some important monitoring proposals. Adjacent article in these proceedings

Paper presented at Sensors and their Applications XII, September 2003

Possible new tensiograph drop volume method for surface tension measurement without correction factors: Further studies on tensiotrace relationships

D Morrin[1], N D McMillan[1], K Beverley[1], M O'Neill[2], B O'Rourke[1], G Dunne[1], S Riedel[3], A Augousti[4], J Mason[4], F Murtagh[5] and M Köküer[6]

[1] Institute of Technology Carlow, Kilkenny Road, Carlow, Ireland
[2] Carl Stuart Ltd., Whitestown Business Park, Tallaght, Dublin 24
[3] S. Riedel, Kelman Ltd., Lissue Industrial Estate East, Lissue Road, Lisburn, BT28 2 RB, N. Ireland
[4] School of Life Sciences, Kingston University, Penryn Road, Kingston upon Thames, KT1 2EE, Surrey, U.K.
[5] School of Computer Science, The Queen's University, Belfast BT7 1NN
[6] School of Information Sciences, University of Coventry, Armstrong Siddley Building, Priory Street, Coventry, W. Midlands CV1 5FB, U.K.

Abstract: The stalagmometric method is based on drop period (or weight) measurements and rests on the use of empirically derived correction factors. That important extended measurement potential of the new optical tensiographic technique over tensiometry in both bulk and surface measurement capabilities are demonstrated. Significantly, it is shown that drop-volume surface measurements can be obtained without the use of correction factors. The new approach also suggests methods of classifying surface activity based on a single drop measurement. The paper concludes with a discussion of the possible applications of this new approach in surface, pharmaceutical, biotechnology and brewing science

KEYWORDS: surface-tension; stalagmometry; tensiometry; tensiography; drop-volume method; correction-factor(s).

1. Introduction

The drop-volume technique has always required experimentally determined correction factors to give an accurate value of the surface tension of a liquid based initially on the work of Harkins and Brown [1]. Significant work has been done subsequently and Lando and Oakley [2] made important improvements to these correction factors. As new volume dispensing apparatus appeared Rowe [3] made some improved estimates on the correction factors, but perhaps the most notable work was that of Jho and Burke [4]. Jho and Carreras [5] subsequently worked on improving the drop volume techniques in the light of subtle viscosity effects. Today the further improvement in these drop volume instruments is due to Miller and co-workers [6] who have pioneered the technique as a leading edge of a range of new drop volume approaches, many of which are now standard laboratory surface science equipment. The development of tensiography by McMillan, Fortune et al [7] and his co-workers began a new chapter in stalagmometry in that from the outset provided improved measurement capabilities over the tensiometer being based on a graphical output. McMillan, Lawlor *et al* [8] reported that these measurements were principally of the bulk liquid properties. In the work presented here, we described some interesting and new developments, which develop new bulk measurement capabilities. For the first time a real new potential in surface measurements is demonstrated in that surface tension measurements are obtained without any correction factor.

2. Apparatus

The optical tensiograph is based on the fiber drophead and for the simplest case of a two-fiber system as shown in Figure 1.

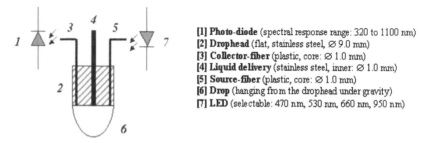

[1] **Photo-diode** (spectral response range: 320 to 1100 nm)
[2] **Drophead** (flat, stainless steel, ⌀ 9.0 mm)
[3] **Collector-fiber** (plastic, core: ⌀ 1.0 mm)
[4] **Liquid delivery** (stainless steel, inner: ⌀ 1.0 mm)
[5] **Source-fiber** (plastic, core: ⌀ 1.0 mm)
[6] **Drop** (hanging from the drophead under gravity)
[7] **LED** (selectable: 470 nm, 530 nm, 660 nm, 950 nm)

Figure 1. *Fiber drophead of a two-fiber system*

The liquid under test (LUT) is pumped at uniform rate via the liquid delivery tube to the drophead using a motor stepper pump. The drophead supports the liquid drop until the gravitational forces overcome the surface forces acting on the drop, at which point drop separation occurs. Throughout the lifetime of the drop, light from a light emitting diode (LED) is injected into the drop through the source-fiber. The reflected signal is picked up by the collector-fiber and transmitted to a photo-diode. Amplification of the low photocurrent into an adequate voltage signal is necessary before the signal is passed to an acquisition board incorporated in a personal computer (PC). The embedded analogue to digital converter (ADC) of the acquisition board converts the analogue voltage into a digital signal, which is subsequently transferred to the memory of the PC. This digital form is stored for later analysis. A data set recorded over the time of a life cycle of the drop is called a tensiotrace. Figure 2 shows a tensiotrace with characteristic features highlighted.

The form of the tensiotrace is dependent on the physical and chemical properties of the LUT, and shows particular sensitivity to the surface tension, absorbance and refractive index. The positioning of the fibers in the drophead is fundamentally important in determining the form of the tensiotrace and has been engineered to optimise the measurement of these properties.

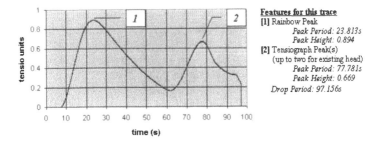

Figure 2. Typical tensiotrace showing the characteristic features

Four tensiotrace measurands are of relevance to the work presented here, namely the drop period (T_1), rainbow peak period (P_R) and rainbow peak height (H_R) and the first tensiograph peak period (P_P). McMillan originally discovered the rainbow peak by using concave dropheads, as it does not show generally in flat drophead optrodes. The magnitude of H_R was found to be extremely sensitive to refractive index (η). Smith [9] has shown by his computer modelling that this peak arises from a second order reflection and not a triple reflection as originally suggested. The name rainbow peak was suggested to highlight the sensitivity of H_R to η and from an analogy taken from rainbow physics where a measurement of the angle of the bow to that of the observer provides a measurement of the refractive index of the liquid.[10] The principal use of the tensiograph is however as a fingerprinting technique. It has been demonstrated frequently that the combination of the physical and chemical properties of the drop provides a very sensitive fingerprint technique of liquids or solid samples dissolved in a solvent [11].

3. Theory

The actual surface tension is given by a modified version of Tate's law thus: -
$$T_1 q = 2\pi\gamma/\Delta\rho g - \varepsilon \qquad (1)$$
here the weight of the drop $W = T_1 q$, where q is the flow rate. R_D the radius of the drop head and $(-\varepsilon)$ is the intercept in the experimental fit to the equation obtained below. The equation enables an experimental value for the Harkins and Brown.[12] correction factor ϕ to be determined. Lohnstein [13] in three papers had earlier calculated correction factors f as a function of $a = (2\gamma/\Delta\rho g)^{1/2}$ and R_D. Lohnstein [14] subsequently further developed this work in two further papers. His work was based on a physical model. Hartland and Srinivasan [15] specified the intervals were Lohnstein's calculations were accurate and analysed how deviations from Lohnstein's calculations of the experimental data obtained by Harkins and Brown. Wilkinson [16] showed that the correction factors can be presented in the form $R_D/a = f(z) = f(R_D/V_D^{1/3}) = z(A + z(B +z(C + zD))) + E$. The constants A, B, C, D and E are respectively 0.50832, 1.5257, -1.2462, 0.60642 and –0.0115. The physics of drop detachment is of course very complex and an error analysis of the whole problem has been made in recent times by Earnshaw et al.[17] The tensiograph drophead radius (R_D) in the present study is 4.5mm and in all cases delivers drops with a $R_D/(V_D)^{1/3}$ in range 0.9 to 1.6, which it should be

noted is outside of both the recommended Harkins and Brown range of 0.6 to 1.2 and also above the Earnshaw upper limit of 0.85.

4. Experiment and discussion

4.1 Bulk properties studies

Tensiotraces for pure water were obtained on a 9 mm drophead at a wavelength of 950 nm over a range of temperatures from 15-40°C. The drop period and rainbow peak period were recorded. The surface tension of the water was measured using the du Nuoy ring method and compared with literature values to ensure purity. Ten tensiotraces were obtained for each temperature, to ensure repeatability, and the average values were used for analysis. The pump was set to deliver 250 μl in 200 s, which gives a liquid delivery rate of $q=1.25 \times 10^{-9}$ m^3 s^{-1}.

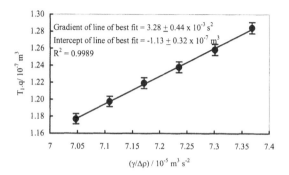

Figure 3. *Drop period versus $\gamma/\Delta\rho$ for water from 15 – 40°C*

Ten tensiotraces for the following liquids were obtained at a wavelength of 470 nm and a temperature of 20°C:- water, methanol, ethanol, 1-propanol, 2-propanol, 1-butanol, 2-methyl 1-propanol (iso-butanol), 2-methyl 2-propanol (tert-butanol) and 1-pentanol. The drop period, rainbow peak period and rainbow peak height were recorded for each tensiotrace. The drophead size and liquid delivery rate were as given for the pure water tensiotraces. Purity was estimated by comparison between the surface tension obtained using the du Nuoy ring method and literature values. Averaged values were used for the analysis. It should be noted that the variation in wavelength between the traces taken for pure water and traces taken for the pure liquids was to ensure the minimum absorbance for the different liquids. In all cases, absorbance in the visible region is negligible.

A plot of $(T_1.q)$ versus $(\gamma/\Delta\rho)$ is expected to be a straight-line plot with gradient $(2\Pi r/g)$ and intercept $-\phi$. The linearity of the plot is confirmed by the high R^2 value. The relationship was checked by calculation of the drophead radius from the gradient of the line of best fit, which was shown to agree within experimental error, which here was unfortunately very large at 13.5%, resulting from error propagation of the various terms in this relationship.

Work has been done on quantifying the relationship of the rainbow peak height with refractive index and other properties of the LUT. The best measurements to use for this study are perhaps temperature studies on pure liquids as the values of the LUT are accurately known. The dependence of rainbow peak height on the refractive index of a liquid however was initially believed to give a universal relationship by McMillan, Finlayson *et al* [18]. Later work by McMillan, Reddin [19] *et al* showed that there was a more complex dependence. Straight chained alcohols in the rainbow peak height (H_R) graph versus molecular weight fitted the data with a single second order curve. The branched mono-ol tertiary butanol however lay well removed from this curve for the homologous mono-ols. The improved instrumental performance of the tensiograph since these measurements has shown more clearly this subtle dependence of rainbow peak height on the molecular structure. Clearly there is not a single relationship for all alcohols as reported in the original paper, and there exists in fact a series of lines for each family of alcohols as shown in Figure 4(a). The water measurement point also does not fit onto the main line plotted here for any of the alcohol curves. Further, investigation of this data produced a plot with an improved relationship in the plot of (η/ ρ) versus H_R shown in Figure 4(b). It should be noted that this plot gives a very strong linear   fit   for   the   primary   alcohols.

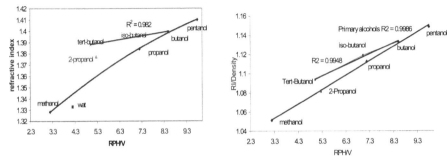

Figure 4(a) *Plot of refractive index versus rainbow peak height (Volts):*
Figure 4(b) *Plot of refractive index/density*10^{-3} ($kg^{-1}m^{3}$) versus rainbow peak height (Volts)*

4.2 Surface studies

It has been found that solvents with higher densities including water solutions, have a very strong relationship between the T_1 and γ. In the case of surfactant solutions in water this graph (T_1 vs γ) can be used to explore the surfactant molecules diffusion properties. The results on sodium lauryl sulphate (SLS) such as shown in Figure 5(a) but delivered with a pump speed of 1.25µl/s, produces a graph which is discernibly curved, that fits with a second order polynomial. With a reduced pump speed of 0.41625µl/s a straight-line graph is obtained obviously as this gives sufficient time for these surface-active molecules to diffuse to the surface. We interpret this result as showing it is possible to obtain an extremely good calibration graphs of surfactant species, but also to investigate surface activity in terms of this graph and the divergence from the straight line. It is estimated that the accuracy of this measurement procedure can give a surface tension measurement of ±?%. It is believed that this discussion of SLS applies to all surfactants.

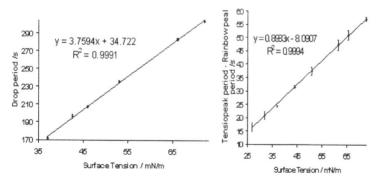

Figure: 5(a) and (b) *Series of SLS dilutions v's surface tension (mN/m), run at two different pump speeds of 200s and 600s.*

For pure liquid, binary mixtures of liquid and solutions of non-surfactant molecules a different approach is required for a surface tension measurement.

5. Conclusions

The drop period of a tensiotrace has been shown to be dependent on the $(\gamma/\Delta\rho)$ ratio of a given liquid and the volume of liquid remaining on the drophead after drop detachment. The volume of the "remnant drop" must be experimentally determined, and the drop period alone does not afford a simple measurement of $(\gamma/\Delta\rho)$. A relationship between the rainbow peak period and drop period for a related series of liquids has been experimentally determined. This relationship is dependent on the surface tension, density and refractive index of the liquid tested. The relationship has been used to determine the $(\gamma/\Delta\rho)$ ratio for liquids of known refractive index belonging to different homologous series to within 10%. Recent improvements to the tensiograph acquisition software are expected to reduce this error significantly. Work is ongoing to determine the relationship between rainbow peak height and refractive index, and it is hoped that a combination of these three parameters will give a self-calibrating method for determining $(\gamma/\Delta\rho)$ from a drop volume technique.

References

[1] Harkin WD and Brown FE 1919 J. Amer. Chem. Soc.: **41** 499
[2] Lando JL and Oakley HT, Tabulated correction factors for drop-weight-volume determinations of surface and interfacial tensions 1967 J. Colloid. Int. Sci. **25**,526-530
[3] Rowe EL May 1972 Automated drop volume apparatus for surface tension measurements J. Pharmaceutical Sciences Vol. 61 No.5 781-782
[4] Jho C and Burke R Sept. 1983 Drop weight technique for the measurement of dynamic surface tension J. Colloid and Interface Sci. **95** No.1 61-70
[5] Jho C and Carreras M June 1984 The effect of viscosity on the drop volume technique for the measurement of dynamic surface tension J. Colloid Interface Sci. **99**, No.2, 542-548

[6] Miller R and Fainerman VB 1998 'The drop volume technique' in 'Drops and Bubbles in Interfacial Research, Eds. D. Mõbius and R. Miller, Elsevier 139-186

[7] McMillan ND Fortune FJM et al , June 1992 A fiber drop analyser: A new analytical instrument for the individual, sequential, or collective measurement of the physical and chemical properties of liquids Rev. Sci. Instrum., 63 (6) 3431-3454

[8] N.D. McMillan, V. Lawlor, M. Baker and S. Smith 1998 From stalagmometry to multianalyser tensiography: The definition of the instrumental, software and analytical requirements for a new departure in drop analysis, in "Drops and Bubbles in Interfacial Research" Vol.6 in Series 'Studies in Interface Science' Edited by D. Mõbius and R. Miller (Amsterdam:Elsevier) 593-705

[9] Smith S 13 December 2002 to McMillan ND Private communication 'On ray model of fiber drop analyzer

[10] McMillan ND Davern P Lawlor V Baker M Thompson K. Hanrahan J Davis M Harkin J 1996 The instrumental engineering of a polymer fibre drop analyser for both quantitative and qualitative analysis with special reference to fingerprinting liquids Colloids and Surfaces (A) **114** 75-97

[11] McMillan ND Riedel SM Augousti AT and Mason J 2001 The development of a multivariate laboratory and process control measurement system with industrial and medical applications, Sensors and their Applications XI, (Bristol:IOPP) 191-196

[12] *op.cit*, Note 1

[13] Lohnstein T 1906 Ann. Physik, 20 237 and 20 606 and (1907) 21 1030

[14] Lohnstein T 1908 Z. phys. Chem. 84 686 and 1913 84 410

[15] Hartland S and Srinivasan PS 1974 J. Colloid Interface Sci. 49 318

[16] Wilkinson MC 1972 J. Colloid Interface Sci. 40 14

[17] Earnshaw JC. Johnston EG Carroll BJ and Doyle PJ 1996 J. Colloid Interface Sci. 177 150

[18] McMillan ND Finlayson O Fortune F Fingelton M Daly D Townsend D McMillan DDG Dalton MJ and Cryan C 1992 A fiber drop analyser: A new analytical instrument for the individual, sequential, or collective measurement of the physical and chemical properties of a liquid, Rev. Sci. Instrum. **63**(6), 216-227

[19] N. McMillan ND Reddin M O'Neill MM 2000 et al, J. Inst. Brewing, Vol. 106, No.3 147-156

Digital Sampling Processor-Based Measurement of Fabry-Perot Interferometer peaks in Large Multiplexed Fibre Optic Bragg Grating (FOBG) sensor schemes

W J O Boyle, Y M Gebremichael, B T Meggitt[1], K T V Grattan

Electrical, Electronic & Information Engineering, School of Engineering & Mathematical Sciences, City University, Northampton Square, London EC1V 0HB
[1] EM Technology, Rouse Gardens, PO Box PO Box 4191, SE21 8YY, London

Abstract: This paper reports the development and evaluation of a Digital Sampling Processor (DSP) scheme for large multiplexed Fibre Optic Bragg Grating (FOBG) sensor schemes. The DSP scheme has been developed using electronically synthesised pulses to simulate those produced by Fibre Bragg Gratings. This scheme is then evaluated with real sensors in laboratory and field trials. The resulting system can measure over 800 peaks per second with a resolution better than 3pm over a 20nm range in field use. Evaluation of the DSP electronics and software shows a potential sensitivity of 110 dB of full scale, corresponding to 0.05pm for a 20nm range.

1. Introduction

In Fabry-Perot interferometer-based Fibre Optic Bragg Grating (FOBG) sensor schemes, the interferometer is typically scanned with a ramp voltage and the positions of the peaks resulting from light reflected back from Bragg gratings in a sensor array are recorded. An accurate measurement of the wavelength of a Bragg grating requires synchronous knowledge of the ramp voltage and the intensity of the detected light. For slowly scanned schemes with only a few gratings, a PC-based data acquisition system will suffice as the ramp voltage can be slowly stepped through its range and the signal intensity of each step can be duly recorded to yield an accurate spectrum containing the Bragg grating wavelength data. PC-based acquisition, however, presents difficulties when a high throughput of measurements is required since PC processors and the Windows[tm] software are inherently asynchronous. The systems currently under development comprise upwards of x8 channels of x8 Wavelength Division Multiplexed (WDM) multiplexed grating sensors sampled at 20 samples per sensor per second at sub-pm sensitivity. This acquisition rate puts considerable demands on the system timing, beyond the reach of PC-based data acquisition.

Before discussing the design and evaluation of the DSP system, the optical arrangement which is to be controlled is presented below. This comprises the multi-channel

Wavelength Division Multiplexed WDM scheme [1] shown in Fig.1. A fluorescent fibre system is used as the source with a bandwidth of some 20-40nm from 1520-1560nm. The tuneable filter is a fibre-coupled, free space, Fabry-Perot filter actuated by piezo-electric transducers and operated over the bandwidth of the source at up to 250 scans per second. Light from the source is filtered by the FP filter and reflected back from the Bragg gratings in the N x M array, through the optical couplers, to eight photodiode detectors. These detect the resulting time domain spectra of the gratings in each of the serially connected arrays. Fig.2 shows a reflected spectrum for one channel from such an arrangement, from an array of gratings with varying reflectance.

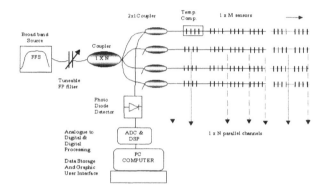

Figure 1: Schematic of the Multiplexed Multi-Channel WDM Architecture [1]

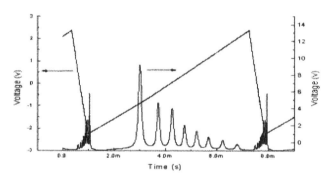

Figure 2: Reflected Spectrum from one Scan Channel, with 8 serial Array Gratings (left hand) and the ramp voltage (right hand). Vertical scales arbitrary

In this work, a DSP processor is used both to generate a temporally accurate ramp voltage to control the interferometer scan and synchronously, to sample the analogue signals from the FOBG sensors and hence determine accurately the spectral positions of the gratings. This scheme is evaluated, firstly with DSP-generated synthesised Gaussian peaks and then later in laboratory and field measurements.

2. Measurement Scheme and Setup

The measurement scheme is depicted in Fig.3, comprising three sections. The first is a PC computer system with a GUI software and a fast serial interface connected to the DSP system. The synchronous DSP system, in turn, controls and takes data from the optoelectronic system. This is typically connected to the FOBG sensor arrays in a multiplexed sensor system with the WDM architecture depicted in fig.1.

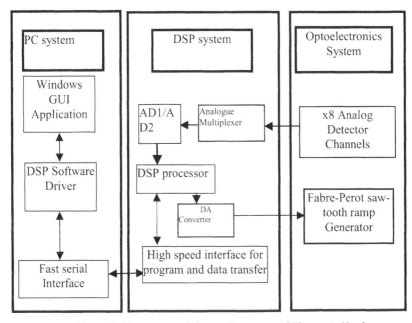

Figure 3: Measurement Scheme: Computer and Electronic Hardware

DSP software generates the saw-tooth control signal for the FP filter. The signals from these photodetectors are then analysed by the DSP system. This determines the position of the peaks relative to the scan of the FP filter using the mean centroid method. This results in approximately 1000 wavelength measurements per second. Such measurements are then transferred to the controlling PC using FIFO buffering on a fast serial data link.

The DSP hardware used is built around an Analogue Devices ADSP-21061 DSP chip. This floating point DSP operates at 40MHz and executes at approx 100Mflops. The DSP unit has two 12 bit analog-to-digital (AD) converters operating at up to 1.2MHz, connected to signal-conditioned analogue input channels by analogue multiplexers which operate at approx 1.0 MHz. Four analogue channels are electronically multiplexed to each AD converter in the DSP controller. The controller also has two digital-to-analogue (DA) converters that operate at up to approximately 100 kHz. One of these is used to provide a software-generated scan voltage signal for the Fabry-Perot cavity spectrometer.

The main function of the DSP software is to execute program code at fixed time intervals under the control of an interrupt timer. All program instructions for the DSP have a fixed duration, and thus with careful programming accurate recording of the time of the measurement can be ensured. The interrupt timer is taken from the rising edge of the 40MHz DSP system clock pulse and has an accuracy of about 10% of the clock period, ~ about 2.5ns. The DSP program can be repeatedly executed in multiples of the clock period and is typically set to 1000 clock periods or 25µs duration. During each of these 25µs intervals the DSP software executes the following tasks:

- a measurement is taken from one of the channels and the channel index is incremented to the next channel
- a reading is taken from the AD converter and the value of this voltage is compared to a threshold to determine if it is from a peak. If it is, then all values until the signal falls below the threshold are added to an accumulator. These are also multiplied by the value of the scan voltage and added to a second accumulator. As the signal level falls below the threshold, the program determines the end of the peak. If the number of samples accumulated is greater than a minimum, the program takes this as an indication of a "good" peak. The ratio of the two accumulators is then taken to find the centroid. This value for the centroid is then normalised to the time to sweep one spectrum. Finally the grating index which indicates to which grating the peak relates, together with the centroid measurement and a time stamp, are placed on the FIFO buffer for later retrieval by the PC
- if the last channel has been reached, the channel index is reset and the scan voltage for the FP interferometer is incremented to its next value. On reaching its maximum, the voltage starts to be decremented, as depicted in the voltage scan shown in Fig.2. Here the difference in the values of the system clock from the interrupt to the time for generating the scan voltage are used to compute the voltage value.

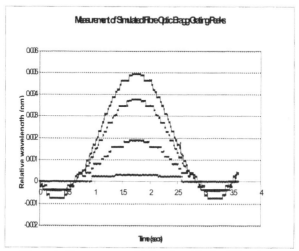

Figure 4: Measurement of 4 Simulated Peaks. FSD=20nm

To evaluate the DSP hardware and software before adding the complexity of the optoelectronics, Gaussian shaped analogue voltage peaks are generated to model those from the Bragg grating sensors. These Gaussian peaks are generated by the DSP using a second interrupt clock-driven program. The characteristics of these peaks are as follows: peak voltage = 6Volts (60% FSD), peak width = 0.1% of total scan (0.2nm for a scan of 20nm), and scan positions of 4%, 30%, 60% and 80% of the total scan. The Gaussian peaks were sinusoidally modulated at 0.1% of their scan position value with a period of approximately 2.9s to test the dynamic response and linearity of the acquisition system. The scan rate is 100 scans per second.

Fig.4 shows the 4 simulated peaks as generated and measured by the DSP system and plotted on a relative scale. The main point to notice from this figure is the potential resolution with the Gaussian peaks resolved to better than 0.1pm over a range of 20nm. The discrete values of the wavelength position in the figure are a consequence of the Gaussian peak data being stored as an array of discrete values.

3. Results and Analysis

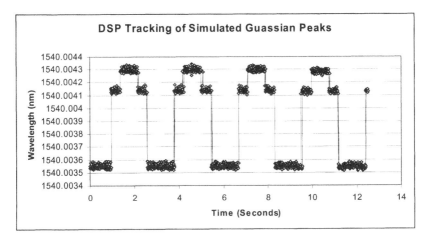

Figure 5: Illustration of the Resolution of the DSP Electronics

Fig.5 focuses in on the potential resolution of the system. This figure shows data from the first Gaussian peak which has a modulation of depth of about 1pm. The sensitivity of the system in these measurements should be noted, where data are resolved to within 0.05pm of the wavelength scale. This represents a sensitivity of better than 3 parts in 10^6 of the full 20nm scan, corresponding to a strain sensitivity of better than $0.1\mu\varepsilon$. In the figure, the standard deviation of the 2nd group of measurements with values between 1540.0035nm and 1530.0036nm is less than 0.006nm.

Finally Figs.6 and 7 put this instrumentation system in the context of real measurements of strain. Fig.6 shows laboratory-based continuous measurements [3] from a grating under stepwise changing strains in $50\mu\varepsilon$ steps. Fig.7 shows the response of one of

28 sensors to a 50 Ton truck moving onto a steel box-section road bridge in long-term trials in Norway[4]. The step change in strain here is approximately 60με. Both measurements show sensitivities of better than 5pm, at over 20 samples per sensor per second.

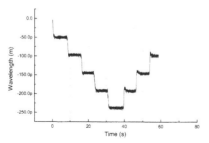

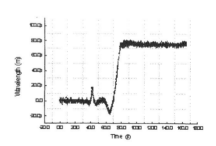

Figure 6: Response of FOBG sensor to a 50με stepwise changes in strain

Figure 7: Response of fibre sensor to a 50 Ton truck on the bridge

4. Summary

This paper discusses the development of a DSP system suitable for data acquisition from large multiplexed arrays of FOBG sensors. An evaluation of the acquisition system shows its potential to resolve the position associated with over 800 measurements from simulated spectral peaks per second with a sensitivity of 5 parts in 10^6, thus potentially allowing high speed measurement to sub-microstrain (με) levels from multiplexed FOBG sensors. Further, results from laboratory and field based measurements show a sensitivity achieved of approximately 3με for a large multiplexed system. Such sensitivity, with many multiplexed measurements, would not be possible without the level of synchronisation between the scanning voltage on the Fabry-Perot spectrometer and the timing of the sampling of the optical signals allowed by the DSP processing used.

Acknowledgement

This work was supported by an EU MILLENNIUM project under the Framework IV Initiative, and EPSRC through several schemes including the INTErSECT Faraday Partnership.

References

[1] Y M Gebremichael, B T Meggitt, W J O Boyle, W Li, K T V Grattan, B McKinley, L Boswell, K A Aarnes & L KVenild & P Y Fonjallaz, *"Field trials of multiplexed fibre Bragg grating sensor system for online integrity monitoring of the Mjosund road bridge, Norway: EU Millennium project"* Optical Techniques for Sensing, The Physics Congress p6, 18-22 March 2000, Brighton, UK

[2] A D Kersey *et al* Fiber Grating Sensors, J. Lightwave Tech. 15, 1442-63, 1997

[3] Y M Gebremichael, B T Meggitt, W J O Boyle, W Li, K T V Grattan, B McKinley, L Boswell, K A Aarnes, L Kvenild & P Y Fonjallaz, "*Comparative field study of fibre Bragg grating strain sensors and foil gauges for structural integrity monitoring*", Proceedings of the 14th International Congress on Condition Monitoring and Diagnostic Engineering Management (COMADEM 2001) p943-951, 4-6 September 2001, Manchester, UK

[4] Y M Gebremichael, W Li, B T Meggitt, W J O Boyle, B McKinley, K T V Grattan, L Boswell, CA D'Mello, KA Aarnes & L Kvenild, "*Bragg grating based multi-sensor system for structural integrity monitoring of a large civil engineering structure: a road bridge in Norway*". Proceedings of SPIE on Photonics and Applications, Advanced Photonics Sensors and Applications II, p343-348, 27–30 November 2001, Singapore

Fibre Bragg Grating Sensor Arrays Structurally Embedded In Europe's First All-Fibre Reinforced Polymer Composite Public Road Bridge

Y M Gebremichael, W Li, B T Meggitt[1], W J O Boyle, K T V Grattan,
B McKinley, G F Fernando[2], G Kister[2], D Winter[2], L Cunning[3], S Luke[3]

School of Engineering and Mathematical Sciences, City University, London
EC1V 0HB, UK
[1] EM Technology, PO Box 4191, London, SE21 8AG, UK
[2] Sensors and Composites Group, ESD, RMCS, Cranfield University, Shrivenham, Swindon, UK
[3] Mouchel, Advanced Engineering Group, West Byfleet, Surrey KT14 6EZ

Abstract: A novel bridge structure has been instrumented during construction using optical fibre sensor diodes. Forty fibre Bragg gratings were integrated during the construction phase of a new fibre reinforced polymer composite road bridge. This was done during performance testing, design verification and the commissioning programme, to provide temperature compensated real-time strain and temperature data.

1. Introduction

Fibre Bragg grating (FBG) technology [1-5] has enabled the implementation of a relatively simple and compact sensing approach for strain or temperature metrology. These sensor systems can be readily retrofitted to existing structures and in this regard work has been previously reported on their attachment for integrity monitoring on a new 350m steel road bridge in Norway [6] for remote and on-line structural health monitoring. Fibre optic sensor technology [1] offers a number of advantages for real time structural health monitoring in a range of engineering structures and materials. For example, a large number of sensors can be multiplexed along a single length of optical fibre, enabling single or multi-parameter measurements to be made. Significant advances have been made over recent years on the design and deployment of optical fibre-based sensor technologies for structural health monitoring. These devices have been integrated into fibre reinforced composite [2] and concrete structures [3]. However, there is comparatively little information on the integration and use of optical fibre sensors in 'real world' civil structures during their construction and commissioning phases.

From an end-user perspective, the availability of an on-line and remote structural health monitoring holds the potential for significant economic and safety benefits resulting from the introduction of informed *condition maintenance*. The unique nature of a FBG measuring system for such applications arises due to the spectral nature of the signal reflected from the Bragg grating sensor allowing absolute wavelength shift measurements, without the need for optical phase measurements.

Furthermore, since gratings of this type can be written at different wavelengths, many individual sensors can be wavelength-division-multiplexed and integrated onto a single fibre optic strand and furthermore the small diameter of the fibres (250μm) means that the sensors can be embedded with minimum intrusion into composite materials, thus without degrading the mechanical performance of the structure.

The recent advances made in implementing multi-FBG array sensor technology for structural integrity monitoring applications are reported for a new type of fibre reinforced engineering structure and on the results obtained from the design verification during commissioning tests on a newly constructed, West Mill Bridge in Oxfordshire, UK. In addition to utilising established FBG sensing technology for this application, the major additional challenges of this particular programme which were met included: (i) the design of the instrumentation system and meeting the on-line software requirements for remote monitoring (ii) sensor integration and protection technology on the extruded section composite structure and (iii) on-site sensor system installation and cabling methodologies.

2. Fibre Bragg grating sensing principles

A fibre Bragg grating is a periodic perturbation of the refractive index of the core of an optical fibre written usually in a single mode fibre with a UV light interference pattern [4][5]. The combined strain and temperature sensor response (given by a single combined Bragg wavelength shift, $\Delta\lambda_B$) can be represented by the linear relationship:

$$\frac{\Delta\lambda_B}{\lambda_B} = (\alpha + \xi)\Delta T + (1 - \rho_e)\Delta\varepsilon \qquad [1]$$

where $\Delta\varepsilon$ is change in strain, ΔT is the change in temperature, α (=0.55 $\times10^{-6}/^{\circ}$C) is the fibre linear thermal coefficient, ξ (= $8.3\times10^{-6}/^{\circ}$C) is the thermo-optic coefficient and ρ_e (= 0.26) the strain-optic coefficient. Once these coefficients are known for the specific type of fibre on to which the FBG sensor is written, the sensing method is self-calibrating and it allows drift-free long-term strain and temperature measurements.

Under the programme of funding for this research, the authors have developed, tested and assembled a high-performance strain monitoring system based on the wavelength division multiplexing (WDM) technique [7] by interrogating fibre Bragg grating sensors, creating a system which meets the desired performance specifications. The system uses a tuneable Fabry-Perot filter, which scans across a 35nm free spectral range of a broadband superluminescent diode source. The relatively wide spectral width of the SLD source allows demultiplexing of an array of x8 FBG sensors with >±2000με operating region per sensor when deploying an 8 parallel channel architecture which has proved to be a versatile configuration for use on a wide variety of structural platforms and industrial processing plant. The interrogation system was specifically designed for use on the West Mill Bridge and links the network of x40 Bragg grating sensor elements at seven different sections on the bridge structure. In addition, x11 resistive strain gauges were attached to the bridge structure at various points, allowing comparative measurements to be made. The optical sensor system was configured with a DSP data acquisition and processing system permitting a sampling rate per sensor of up to 200 Hz and is capable of being accessed and controlled remotely via an Internet-enabled software which calculates and reports the strain data from transient road traffic and thermal loading with online, web-based graphical user interface (GUI) for real-time data visualization. The data collected are currently being used as part of a continuous monitoring programme on its structural performance, design verification and predictive lifetime assessment study.

3. Prior laboratory evaluation

Initially, before the systems were installed on the bridge, laboratory testing was conducted to evaluate the performance of the advanced monitoring system developed. An investigation was then made into developing sensor attachment techniques for optimum stain transfer, sensitivity and sensor protection methods for sensor isolation to ensure survivability during bridge construction and deployment in the harsh engineering environment of the bridge. In this regard, a series of quasi-static and dynamic loading and heat cycling tests were carried out in the laboratory environment utilising composite structural test elements, representative of the bridge structure. In most of this work, the Bragg grating sensors (with small pre-tension) were directly attached to the composite structure providing for maximum strain transfer. Various bonding techniques were investigated with the adopted bonding technique which provided a constant ~100% (<±2% variation) strain transfer efficiency.

4. Composite bridge instrumentation and sensor system installation

The new fibre reinforced polymer composite road-bridge was constructed from a series of extruded, inter-locking sections and provided a direct replacement for an aged steel and concrete decked bridge across the River Cole in Oxfordshire, England. This bridge is the first of its kind in Europe to be constructed from advanced composites and was instrumented with fibre Bragg grating sensor technology for design verification and long term monitoring.

Figure 1. West Mill Bridge (10m long, 7m wide), during loading tests

The bridge designed by Mouchel, UK was constructed on x4 longitudinal polymer beam supports with the composite transverse decking being fabricated in large extruded profile sections (by Fibre Line, Denmark) and delivered to the site for bonding by the contractors (Skanska, Sweden). During the construction stage, prior to bonding of the composite profiles sections, several sections of the composite decking and longitudinal composite support beams were instrumented with an array of single axis and rosette FBG sensors strategically placed at pre-selected sites of maximum strain. The sensors were attached to the structure with a proprietary bonding agent, after applying a small pre-strain to allow for better strain transfer and attachment. Prior to bonding, surface preparation was carried out with a small area of the composite surface being carefully ground and de-greased with a suitable solvent to achieve high integrity bonding. Similarly the sensor section (~10mm) of the fibre is stripped of its acrylate coating for good surface contact and efficient strain transfer. Most of the FBGs used were attached to the structure between the bond lines with appropriate sensor protection (to prevent unwanted transverse stain effects) to measure both the strain and temperature effects. With all the sensors attached and bridge profiles bonded, the bridge deck was then lifted into place on prepared bridge abutments for road surfacing and alignment.

5. On site strain measurement

Mandatory commissioning tests were carried out on the bridge before it was officially opened for public use. These controlled loading tests were carried out with a 30T lorry positioned at various points on the bridge as specified under the bridge design specifications. The tests were conducted in three stages. In the first, the lorry was located centrally on the carriageway and proceeded from one end of the bridge to the other, stopping momentarily at one meter steps along the bridge while continuous measurements were taken by the FBG sensor data logging system. Resistive strain gauge readings were also taken at every stop.

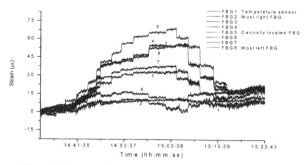

Figure 2. Strain readings from 8 serial FBG sensors on the underside of the deck

Two similar tests were then carried out with the lorry wheels positioned 50mm from the curb at the left lane, followed by the same pattern on the right lane. As the lorry progressed along the bridge, the strain readings were seen to increase monotonically until the lorry reached the mid section, across which all the sensors are located, this being followed by a decrease in the strain as the lorry moved away from the mid section. The difference in strain measured by each sensor, as shown in Figure 2 correlates with the relative position of the sensors and the loading point. Trace 5 in Figure 2 corresponds to the most centrally located FBG, while Traces 2 and Traces 8 are readings from the sensors at either edges of the bridge, close to the curb. It should be noted that during these tests, the vehicle load was centrally positioned on the middle of the carriageway. Trace 1 is from a strain isolated temperature compensation sensor.

Readings from electrical strain gauges (ESG) were also taken, as illustrated in Figure 3. Here three strain gauges corresponding to the FBG sensors, labelled FBG3, FBG7 and FBG5 located on the bottom side of the deck are shown illustrating the close correlation of the two sets of data. The readings from the FBG sensors have stepwise increments in strain reflecting the strain levels measured continuously as the truck progressed and stopped momentarily, while the resistive gauges readings (Figure 3) were only taken at intervals where the truck was stationary.

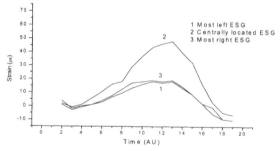

Figure 3. Strain readings from three resistive strain gauges on the underside of the deck

Similarly, in Figure 4, the readings were taken from FBG sensors located on the side of the longitudinal beam, and the different strain levels relate to the position of the sensors relative to the neutral axis with the highest strain reading being for the sensor furthest away from the axis.

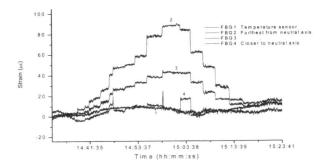

Figure 4 Strain readings from 4 serial FBG sensors on the side of the longitudinal beam

Dynamic loading was also recorded as the truck moved across the bridge at speed. This is depicted in Figure 5, which displays transient strains recorded by the 8 FBG sensors located transversely at the mid section on the underside of the bridge deck. Data was calibrated with the laboratory tests carried out prior to installation and by comparison to data from the resistive strain gauges. The measurement precision is typically ±5µε. The temperature was near constant at 18°C during these tests, with the strain results being compensated for the small temperature fluctuations.

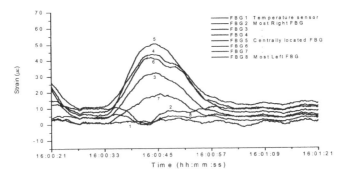

Figure 5 Strain readings from 8 serial FBG sensors on the underside side of the deck profile

6. Summary

Eight channels of sensor arrays encompassing x40 FBG sensors were successfully embedded into the structure of a new fibre polymer composite public highway road bridge during its construction phase and used successfully to monitor traffic loading and environmental temperature changes (for compensation purposes) at various locations across the structure. Controlled loading tests were carried out using a 30T test vehicle during initial commissioning tests prior to the opening of the bridge for public use. Quasi-static as well as dynamic strains were measured to a precision of <10με. Resistive strain gauges were also used in these tests to give comparative readings and the results obtained showed close correlation between the two sets of data from the two types of devices (to within 15με of peak values). The sensor system deployed on the bridge will be monitored online as part of a long-term condition integrity assessment programme of the new composite structure.

Acknowledgements

This work is sponsored by the Engineering and Physical Sciences Research Council (EPSRC), UK under the Faraday INTErSECT partnership and through other funding schemes.

References

[1] KTV Grattan & BT Meggitt (2000) "Optical Fibre Sensor Technology: Fundamentals" Kluwer Academic Press.

[2] RM Measures, AT Alavie, R Maaskant, M Ohn, S Karr & S Huang (1995) "A structurally integrated Bragg grating laser sensing system for carbon fibre pre-stressed concrete highway bridge" Smart Mater. Struct. 4, p.20-30.

[3] J Meissner, W Nowark, V Slowik, T Klink (1997) "Strain Monitoring at a Prestressed Concrete Bridge" Proceedings of 12th International Optical Fibre Sensors Conference, (OFS 12), Williamsburg, USA 408- 411, Pub: IEEE

[4] AD Kersey, MI Davis, HJ Patrick, M LeBlanc, KP Koo, CG Askins, MA Putman and EJ Friebele (1997) Fiber Grating Sensors J. Lightwave Technol. 15 No. 8 1442 -1463.

[5] KTV Grattan & BT Meggitt (Eds) (2000) "Optical Fiber Sensor Technology: Advanced Applications" Kluwer Academic Publishers. Dordrecht, The Netherlands, 79 – 187.

[6] YM Gebremichael, BT Meggitt, WJO Boyle, W Li, KTV Grattan, L Boswell, B McKinley, KA Aarnes & L Kvenild. (2000)" Multiplexed fibre Bragg grating sensor system for structural integrity monitoring in large Civil Engineering applications" Proceedings of the 11th Conference on Sensors and Applications XI, London. p. 341 – 345. Pub: Institute of physics publishing, Bristol, UK, 2000.

[7] AD Kersey, TA Berkoff and WW Morey (1993) "Multiplexed fiber Bragg grating strain-sensor system with fiber Fabry-Perot wavelength filter" Optics Letters 18, No. 16 1370-1372.

©2003 IOP Publishing Ltd

Diver/UUV underwater measurement system

D A Pilgrim[1], A Duke[2], G Symes[2] and D M Parry[1]
QinetiQ Bincleaves, Newtons Road, WEYMOUTH, Dorset, DT4 8UR
[1] Institute of Marine Studies, University of Plymouth, Plymouth PL4 8AA
[2] DERA Bincleaves, Newtons Road, Weymouth, Dorset DT4 8UR

Abstract: Sea mines laid by an aggressor pose a deadly threat to both military operations and commercial shipping. Advances in technology have transformed relatively simple devices into a diverse range of complex and intelligent weapon systems.

An important requirement when investigating a contact is to confirm that it is indeed a mine and to derive features such as overall size, protrusions, graphical descriptors and the like. This acquisition of visual intelligence assists the subsequent process of deciding what action needs to be taken.

Video images obtained by sub-sea cameras have variable perspective and scale but this problem may be solved by the use of structured lighting comprising an array of diode lasers. A 5-spots laser system is being developed jointly between University of Plymouth and DERA Bincleaves for use by divers and unmanned underwater vehicles (UUVs). An important aspect of this research is the ongoing development and optimisation of verifiable measurement and sampling techniques, which may then be employed with confidence in hostile waters.

1. Introduction

The UK Royal Navy uses remotely operated vehicles (ROVs) as the principal means of sea-mines disposal; the need being to remove the diver from this hazardous task wherever possible, and to increase the depth at which mine disposal can be conducted beyond that which can be achieved by non-saturation diving methods. The vehicle is first deployed to conduct a video survey to confirm the target identity. It is then used to place a demolition charge close to the mine before returning to the deploying platform where it is recovered prior to remote detonation of the charge.

The maturity of current vehicle technology, compared to that available to the RN in the 1980s, offers the prospect of remote solutions for a wider variety of military objectives. These objectives encompass every ocean environment from operations at 200-300 m depth to those in shallow and turbulent waters, including the surf zone, with typical water depths of less than 10 m. Central to any operations is the need to survey and identify target features, and a variety of sensors are being considered for this purpose. Image acquisition using video cameras remains a highly attractive option since the technology to produce good quality pictures is readily available. These cameras are robust, inexpensive and easily interfaced with existing systems. A method that would allow dimensional data to be extracted from captured images would therefore be very welcome and of significant benefit to the RN.

The analysis of underwater photographs to study and record seabed phenomena has been a technique in common use since the 1940s (e.g. Ewing *et al*, 1946). Early quantitative methods were limited usually to the use of downward-looking cameras in which a known altitude and fixed angular field of view were used to calculate image dimensions. A major problem associated with the analysis of ROV photographs of the seabed is that camera height is unknown and the use of zoom lenses results in a variable field of view. Also, oblique camera angles introduce perspective whereby the scale changes from the bottom to top of the image. This change in scale can be accommodated by overlaying a perspective grid, sometimes known as a Canadian Grid after the use of this technique in mapping Canada's Laurentian Plateau by oblique aerial photography (Crone, 1963), (Fig.1).

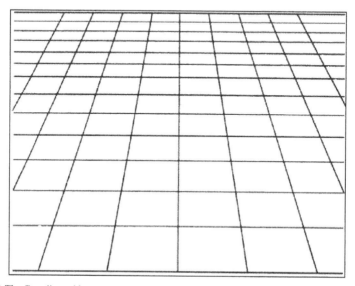

Fig.1 The Canadian grid

2. Canadian grid

The trigonometry of constructing such a perspective grid for use in oblique deep-sea photographs has been presented, in detail, by Wakefield and Genin (1987). *Abiss* (Automated Benthic Image Scaling System) presently under development at the University of Plymouth (Pilgrim *et al*, 2000) includes the option of overlaying the image with a perspective grid but the approach of Wakefield and Genin (1987) was not adopted since their analysis assumes that the camera has a fixed field of view, is situated at a known altitude above the seabed and is directed at a known perspective angle. *Abiss* makes full use of existing camera zoom and tilt facilities to accommodate variable and unknown fields of view and perspective angles. Moreover, the body of the manoeuvring camera platform may lie at a variable angle to the non-horizontal seabed, and at a variable altitude. It is possible to find these unknowns through the use of structured lighting.

The simplest system of structured lighting comprises a single pair of parallel lasers that project two spots onto the seabed in a camera's fixed field of view (i.e. no zoom). Since the separation of the two spots on the image represent a known distance, the image can be scaled approximately. This has been the most widely used laser spot system for scaling photographs but suffers two significant disadvantages: it works for the whole image only if the camera is set at right angles (in both planes) to the seabed or similar target; and it cannot be used with a zoom lens (variable field of view) unless there is a separate input of camera-target distance.

Two improved systems, following different approaches, have been used in work at the Monterey Bay Aquarium Research Institute (MBARI). In the first (Davis and Pilskaln, 1992), the camera provides real-time readouts of focus distance and zoom (field width). In the second approach (Davis, 1999; personal communication), an arrangement of four lasers is used, three parallel to the optical axis to give perspective angle, and one crossing two of the others to give a measure of range and hence scale. The trigonometry involved is solved using the *Laser Measure* program (Davis, 1998) which runs in the Optimus image measurement and processing system.

The approach taken at the University of Plymouth in the development of *Abiss* is to fit the camera with an array of five diode lasers (Figs 2 and 3).

Fig.2 Abiss laser scaling array mounted on the camera of a Phantom XTL

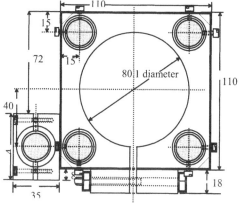

Fig.3 Schematic of the Abiss laser scaling array

Four of these lasers are set parallel and 8·0 cm apart; they project a pattern of four dots onto the seabed. A fifth laser is set at an angle to, but in the same plane as, the bottom pair so that a dot appears between these two at a horizontal position which depends upon the camera-target range (Figs 4 and 5).

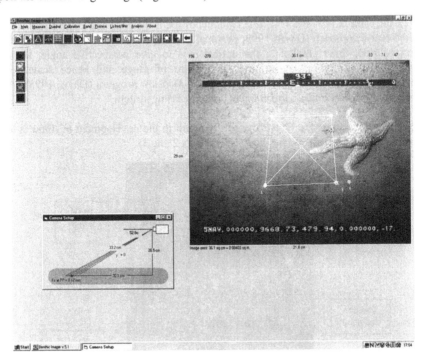

Fig 4 Screen display of Benthic Imager showing 5 laser spots on the seabed

In Figure 5, distance *CB* is calculated in this way during image analysis, and further calculation gives the required distance *CP* between the camera and the principal point.

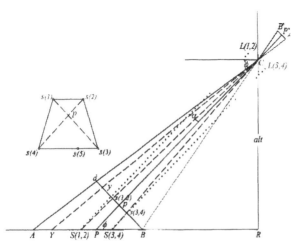

Fig.5 Conversion of virtual to real, and real to virtual distances in the vertical plane

The four parallel lasers form a square only if the camera is normal to the plane of projection (seabed or whatever); usually the dots form a trapezium (Fig.5). The camera's perspective angle (ϕ) measured downwards from a plane parallel to the seabed, is calculated from the ratio of the lengths of the upper (smaller) to lower (longer) sides of the trapezium.

The varying scale of the image may be ascertained by further analysis such as that illustrated in Figure 5, in which the camera is at C, and has field of view ACB in the vertical plane. P is the principal point on the optical axis. $A'P'B'$ is the projection of seabed APB onto the image plane. For trigonometrical calculations, it is convenient to place the image plane at Bd (clearly triangles BCd and $A'C'B'$ are similar). $L(1,2)$ and $L(3,4)$ represent pairs of lasers set equidistantly above and below the optical axis and parallel to it. Projected spots $S(1,2)$ and $S(3,4)$ on the seabed appear on the image at $s(1,2)$ and $s(3,4)$. Any point Y on the seabed appears at point y on the image. The problem is to convert virtual distance py, measured in pixels, into a real distance PY measured in, say, centimetres. The trigonometrical solution to this problem is fairly straightforward. Essentially, the connection between PY and py is the common angle α in triangles PCY and pCy, and our approach has been to use angles at C to convert between virtual (pixel) measurements in plane Bd and real (cm) measurements in equivalent plane BA. For example, it is easy to show that since:

$$\alpha = \tan^{-1}[py / Cp]$$

where Cp, calculated separately, is a virtual (pixel) distance, then:

$$PY = (\tan[\alpha + \pi/2 - \phi] \times alt) - RP$$

where ϕ is the perspective angle of the camera, alt is the altitude of the camera above the bed, and RP is the distance measured along the seabed between the ROV and the principal point, P. To convert from real to virtual distances in the vertical plane the equivalent equations are:

$$\alpha = \tan^{-1}[(RP + PY) / alt] - \pi/2 + \phi$$

and

$$py = \tan[\alpha] \times Cp$$

A similar treatment resolves the real to virtual and virtual to real conversions in the horizontal plane. (The term 'horizontal' and 'vertical' as used here refers only to the up-down and left-right planes apparent to the camera; for example seabed plane AR need not be horizontal in the gravimetric sense.)

3. The University of Plymouth *Abiss* system
3.1 Platform and camera
The platform used by the University of Plymouth is a Deep Ocean Engineering Phantom XTL ROV fitted with an Osprey OE1366 CCD colour TV camera. The colour zoom camera has a minimum field of view of $4.2° \times 3.2°$ (focal length 64·8 mm) and a maximum field of view of $48° \times 37°$ (focal length 5·4 mm). Unfortunately, the field of

view between these limits cannot be ascertained during underwater operations unless the zoom is at maximum in or out.

3.2 Laser system

The laser system (built in house) comprises four parallel lasers in an 8·0 cm square, plus a fifth laser set at an angle to the bottom pair. Different configurations have been reported in the literature; for example, it is quite possible to scale the image using a 3-spot rather than 4-spot 'box' (Tusting and Davis, 1992). However, it has been found that the 4-spot arrangement is much more useful to the ROV pilot and other real-time observers to get a direct sense of scale and perspective angle. The laser assembly was machined from a block of ABS (Figs 2 and 3). The 5 VDC 0·9±10% mW Imatronic LDM115/63/1 miniature laser diodes were sealed in tubes machined from an acrylic rod and fitted with a window cut from an acrylic sheet. Wiring to the ROV's unused sonar power supply is through watertight glands threaded into the other end of the tubes and sealed with a silicon sealant.

These diode lasers are red (633 nm) so ostensibly not ideal for underwater use. Light at this wavelength has a *half distance, ω,* (at which its intensity is attenuated to 50% of its starting value) of less than 2 m in clearest ocean water whereas the half distance for blue light (470 nm) is about 35 m (Pilgrim, 1998). However, the red spots contrast very well in images of an otherwise red-deficient world.

In building the assembly the question arose: how far apart should the lasers be fixed in a rectangular (though not necessarily square) array? The further apart the spots are, then clearly the more accurate will be the system, but if they are too far apart then spots will not appear at all on close-up, zoom-in images (for example, the zoom-in image at a distance of 1m is only about 8 × 6 cm). In fact, the major constraint is the physical structure of the ROV. The laser assembly must be fitted to the camera and there is limited space between the camera and the body of the ROV. With a separation of 8·0 cm the array just fits between the camera and ROV body, and an 8·0 cm square is about right for the turbid, and hence short range, conditions in which we usually operate. For longer range (clearer water) work it would be necessary to mount the lasers on extensions well forward of the camera to remain clear of the ROV body whilst tilting the camera. It is absolutely essential that the laser beams are set parallel to the optical axis of the camera and that they remain so throughout a survey operation; in our assembly (Fig.2) each laser is mounted within an adjustable, lockable universal joint.

3.3 Benthic Imager software

The scaling of the grabbed video image is calculated using a PC program, *Benthic Imager*, written by one of the present authors in Microsoft Visual Basic 6.0 and incorporating Inventions Software Ilib libraries nos. 1 and 2 for optional image enhancement and TeeChart ProActivex 3.0 for graph drawing. The image is displayed on the screen, as in Figure 4, and the five laser dots are located manually using the cursor (we have developed a sub-program to do this automatically, but have found it of little advantage compared to manual selection). The program is then able to calculate the camera's perspective angle, camera-target distance and image scale. These, and other geometric details, are displayed in a diagram beneath the image (Fig.6). Around the image are displayed its *real* dimensions in centimetres.

Measurements may then be made between any two points on the image by clicking with the cursor (Fig.6). In biological surveys, it is required frequently to count the number of organisms in a given sample area. To facilitate this procedure, *Benthic Imager* is able to overlay the image with a selection of variously sized grids (similar to Fig.1), single squares (projected as trapezia) and circles (projected as ellipses).

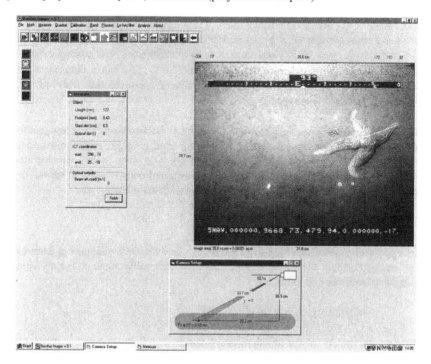

Fig.6 Taking a measurement with Benthic Imager

If required, image colour analysis is provided by displaying the x, y and R, G, B values of the current cursor position above the upper right and left corners of the image, or by selecting the Histogram or Image Statistics displays.

Pre-processing of the image is possible through a selection of image enhancements (mostly from the Ilib libraries) and include band selection (R, G, B, monochrome, etc), high frequency/edge filters, low frequency filters, histogram stretches, binary division, etc. During the early stages of testing the image analysis function of *Benthic Imager* it was discovered that the optical axis of the Osprey OE1366 camera shifts slightly during zoom-in-out operation. It was not possible, therefore, to assume that the principal point (point *p* in Fig.5) was at the geometrical centre of the image. This problem was solved easily within the *Benthic Imager* program which now ascribes the principal point to the geometrical centre of the major four laser spots (Fig.5), and re-assigns all co-ordinates accordingly.

3.4 Image resolution and definition of optical conditions
A crucial aspect of the system assessment work underway at the University of Plymouth is the measurement of image resolution and definition of the optical conditions in which the image is obtained. It is essential that this be done if comparisons are to be made

between the details derived from different images. *Benthic Imager* measures resolution in terms of image footprint. This is a term borrowed from the field of remote sensing and is the size of a single pixel "in the real world". In a satellite image, footprints may be measured in hundreds of metres; in our ROV images the scale is millimetres. Footprint sizes, which vary with position over the oblique image, are displayed in *Benthic Imager* for the principal point, top corner of the image (maximum size) and bottom centre of the image (minimum size). The footprint is also displayed for the mid-point of any measurement (Fig. 6). Footprint sizes may vary a hundred-fold for typical ranges of fields-of-view (FOV) and camera to target distances. For example, for a 560 × 420 pixel image with the camera zoomed in (FOV: 4·2° × 3·2°) at a distance of 10 cm, the footprint at the principal point is 0·013 mm, whereas zoomed out (FOV: 48° × 37°) at a distance of 10 m it is c.1·6 cm.

From time to time during imaging operations, a beam transmissometer is lowered to the depth of operation to measure the beam attenuation coefficient (c) effectively a measure of the optical turbidity of the water. Collimated light (L), such as that which transmits an optical image through the water, is attenuated exponentially at a rate governed by the beam attenuation coefficient. This is expressed by Beer-Lambert's law:

$$L(r) = L(0).\exp[-c.r]$$

where r is the horizontal distance of propagation. Similarly, the decrease in contrast, C, of an image follows the same exponential law, generally known as the Duntley-Preisendorfer contrast reduction law:

$$C(r) = C(0).\exp[-c.r]$$

A detailed discussion of the attenuation of underwater light and the instrumentation deployed to measure it, including the beam transmissometer, is presented by Pilgrim (1998, 1999).

When dealing with underwater visibility, a useful unit of measure is the optical distance ($\gamma = c.r$), that is the geometric distance multiplied by the beam attenuation coefficient. A given optical length/depth of water will result in a particular attenuance. For example, light attenuates (and image contrast is reduced) equally when propagating through 5 m of water of $c = 0.3$ m^{-1} (5 m × 0·3 m^{-1} = 1·5 = γ) and when propagating through 10 m of water of $c = 0.15$ m^{-1} (10 m × 0·15 m^{-1} = 1·5 = γ). Provision has been made in *Benthic Imager* to input values of c and calculate certain values of γ. For example, *Benthic Imager* displays the camera to principal point distance (CP in Fig.5) in both metres and optical distance. Similarly and importantly, the optical distance is given for any objects measured by *Benthic Imager* (Fig.6).

An even more useful and comprehensive approach to the measurement of the optical quality of the water as a transmitter of an image is the *modulation transfer function, MTF(f)*, which is usually presented as a graph of modulation (effectively contrast) against image frequency. A simple approach is to obtain an image of a half-black, half-white target. In pure optical conditions, the variation in brightness across the image would be a step function in signal processing terms. In turbid water, this edge function, (*EF*), is a ramp. The differentiation of *EF* is the point spread function (*PSF*) and the

Fourier transform of *PSF* is the *MTF(f)*. A black and white bar has been fitted to the front of the ROV and occasional images of it are recorded during ROV operation. *Benthic Imager* is able to analyse this image and calculate the *MTF(f)* as illustrated in Figure 7.

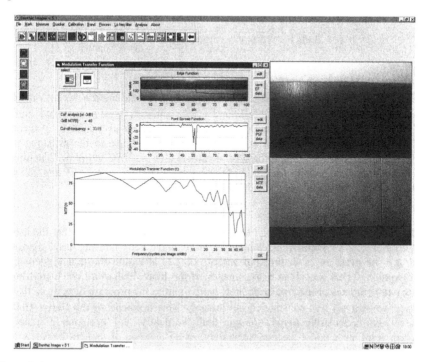

Fig.7 Display of MTF(f) curve in Benthic Imager

4. System calibration

It is essential that all aspects of a measuring system, such as the laser system described here, be thoroughly calibrated before use. We discuss here three different types of calibration. The first, calibration of the laser beams, is critical and must be done before every imaging operation. The second, testing the overall measuring capability of the system, is done initially and occasionally as a check, particularly if any changes are made to the system. The third, calibration of the camera inset distance and image morphing to allow for lens distortion, probably needs to be done only once for any particular camera.

4.1 Laser system

The viability of the whole system depends upon the four main laser beams being parallel to the optical axis of the camera and being at a known, fixed distance from it.

Since there is significant refraction at the air-acrylic-water interface of the diode laser housing, it is essential that the calibration be done under water. There is only a small difference between the refractive indices of seawater and fresh water so our own calibrations were undertaken in fresh water in a 3·6 × 1·5 × 1·2 m laboratory tank. With the ROV resting on bottom of the tank, the camera is made level with a spirit level. At the far end of the tank there is a target with a 8·0 cm square, the corners of which represent the relative positions of the four main lasers. A mark at the centre of the square represents the principal point (centre) of the camera. The position of the target is adjusted to make the centre mark appear at the exact centre of the video screen. The four lasers are then adjusted so that their spots fall on the four corners of the square. The fifth laser is then adjusted so that its beam crosses the plane of the bottom two lasers of the square. The angle set on this laser depends upon the turbidity of the water, and hence the maximum range at which the system is likely to be used. The position of the fifth laser beam is then recorded as three offsets for input into the *RovImager* package.

4.2 Overall accuracy

To test the overall accuracy of the system, a waterproof calibration board, on which were painted 10.0×10.0 cm black and white squares, was constructed and set horizontally on the bottom of our ROV test tank. A number of images were recorded and the camera-to-target distance (CP) and horizontal distance to the principal point (RP) were measured with a tape for several camera altitudes $(RC$ in Fig.5). The images were analysed using *Benthic Imager*, and the calculated values of RC, CP and RP were checked against the actual values.

4.3 Camera backset correction

As is apparent in Figure 5, the trigonometry of *Benthic Imager* implies that the camera is at a point (C). Of course, the camera within its underwater housing has a significant size, and the problem is to find out where point C is within the housing-lens-CCD configuration. This was done using images of the black and white chequered board. The x,y co-ordinates of the five laser spots were input to the program directly so that the perspective analysis was constant for all images. Measurements of the 10.0×10.0 cm squares were taken using *Benthic Imager* both 'vertically' and 'horizontally' since the program resolves x and y vectors separately, using a range of camera backsets from $+12.0$ cm to -12 cm. The % errors in *Benthic Imager* measurements in both 'vertical' and 'horizontal' directions were calculated for each backset correction and plotted in Figure 8. This shows that 'horizontal' measurement errors were no more than about 2%, whilst a backset of 2 cm gives the optimum accuracy in the 'vertical' direction.

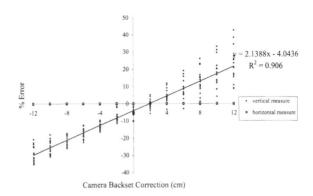

Fig.8 Graph of measurement errors (%) from Benthic Imager as a function of camera backset correction

4.4 Lens aberration and distortion

Since *Benthic Imager* works to pixel accuracy, its performance will be affected adversely by camera lens aberration and any distortion at the water-glass-air interface of the camera housing.   This problem has been investigated (Saikko, 2000) and the findings have been incorporated into an image morphing routine in *Benthic Imager.*

5. Research projects at the University of Plymouth using *Abiss*

According to Robert Ballard, celebrated discoverer of hydrothermal vents and *Titanic*: "Less than one tenth of one percent of the deep sea has been explored" (Earle, 1996). Today, there is acknowledged to be a great and growing need to better understand the structure and nature of the world's seabeds, which comprise some seventy percent of the earth's surface, and underwater photography (or its modern video equivalent) is one of the major survey techniques being used. It is probably true to say that the range and quality of scientific exploration has always been tied to advances in technology; certainly this must be the case in the field of underwater exploration. Of course, access to this level of technology is intrinsically tied to the availability of research funding, and the costly equipment available to Robert Ballard and the major oil exploration companies is out of reach to most ocean scientists. However, there are a number of small, simple and fairly inexpensive ROVs now available on the market, and it is possible to modify these basic machines for scientific work in shallow coastal waters. A programme, *SeeBed*, has been established at the University of Plymouth, with the principal aim of developing such an ROV and, more importantly, to study and develop proper techniques that can be employed in remote survey and experiments to produce results that will be robust and acceptable to the scientific community. To this end the imaging capabilities of *Abiss* is being assessed in two related benthic surveys, one biological and one geological/geotechnical in Plymouth Sound and River Dart (Dartmouth), and the adjacent waters along the South Devon coast (Pilgrim *et al,* 2000). In each case a significant aim of the particular project will be to assess quantitatively the value and viability of the system in comparison with other, traditional approaches.

6. Clear water operations

It is apparent that *Abiss* is in the early stages of development, and that there is scope for improvement and expansion. One crucial area of development currently under way is the adaptation of *Abiss* to the clear water conditions that may be expected in some areas of mine-clearance operations. The system described here incorporates an 8.0 cm laser spot square which has been found to be ideal in short-range work in the turbid waters of the Tamar and Dart estuaries. However, in clear water deployment the four spots would appear in the captured image in a configuration too small for accurate analysis and scaling. Clearly, the lasers must be moved much further apart and to this end a new, wider jig is being built to carry the five diode lasers. The completed unit will be designed to be operable by both a diver as a handheld system, and to be retrofitted to a number of UUV variants.

As mentioned earlier, red is the colour most attenuated by water (half distance, $\omega < 2$ m). However, the red spots contrast well with the red-deficient seabed. For this reason, the red 0·9 mW diodes have worked reasonably well in our short-range operations. It is anticipated that the high attenuation of red will render these diodes less effective in long-range work, so experiments are planned to assess the performance of red diodes of greater power as well as those of different colours.

7. Conclusions

Video imaging techniques such as those incorporated into *Abiss* offer significant benefits in terms of increasing the capability to confidently gather accurate underwater target intelligence. These will both speed and enhance the identification process.

Further, it has been demonstrated that modestly priced equipment and robust, low-complexity hardware designs can be utilised to deliver excellent results, though it is apparent that there is scope for further development and optimisation.

It is also recognised that successful target identification has much to do with well developed, tested and practised techniques. It is in shallow-water work, such as that described here, that remote techniques can be critically assessed and measurements can be verified by more direct methods. In this way, techniques will be perfected and subsequently used with confidence as remotely operated machines operate in hostile waters, or descend to otherwise inaccessible depths.

Acknowledgements

The authors are pleased to acknowledge the support and collaboration of Sonardyne International Ltd; Britannia Royal Naval College, Dartmouth; and Plymouth Marine Laboratory.

References

Crone, D.R. (1963). Elementary photogrammetry, Frederick Ungar, New York. 197pp.

Davis, D.L. (1998). Laser Measure: Users' Guide Version 2.0. Monterey Bay Aquarium Research Institute, Moss Landing, California.

Davis, D.L. and Pilskaln, C.H. (1992). Measurements with underwater video: camera field width calibration and structured lighting. *Marine Technology Society Journal*, **26**(4), 13-19.

Earle, S.A. (1996). Sea change—a message from the oceans. Constable, London, pp.361.

Ewing, M., Vine, A.C. and Worzel, J.L. (1946). Photography of the ocean bottom. *Journal of the Optical Society of America,* **36**(4), 5-12.

Pilgrim, D.A. (1998). The observation of underwater light - part 1. *The Hydrographic Journal*, **90**, 23-27.

Pilgrim, D.A. (1999). The observation of underwater light - part 2. *The Hydrographic Journal*, **91**, 13-18.

Pilgrim, D.A., Parry, D.M., Jones, M.B. and Kendall, M.A. (2000). ROV image scaling with laser spot patterns. Submitted to *Underwater Technology.*

Saikku, R.M. (2000). Calibration of lens distortion of an ROV camera. BSc honours project, Institute of Marine Studies, University of Plymouth.

Tusting, R.F. and Davis, D.L. (1992). Laser systems and structured illumination for quantitative undersea imaging. *Marine Technology Society Journal*, **26**(4), 5-12.

Wakefield, W.W. and Genin, A. (1987). The use of a Canadian (perspective) grid in deep-sea photography. *Deep Sea Research*, **34**, 469-478.

AUTHOR INDEX

TITLE INDEX